Lecture Notes in Artificial Intelli

Edited by J. G. Carbonell and J. Siekmann

Subseries of Lecture Notes in Computer Science

Harrie de Swart   Ewa Orłowska
Gunther Schmidt   Marc Roubens (Eds.)

Theory and Applications of Relational Structures as Knowledge Instruments II

International Workshops
of COST Action 274, TARSKI, 2002-2005
Selected Revised Papers

 Springer

Series Editors

Jaime G. Carbonell, Carnegie Mellon University, Pittsburgh, PA, USA
Jörg Siekmann, University of Saarland, Saarbrücken, Germany

Volume Editors

Harrie de Swart
Tilburg University, Dante Faculty
P.O. Box 90153, 5000 LE Tilburg, The Netherlands
E-mail: h.c.m.deswart@uvt.nl

Ewa Orłowska
National Institute of Telecommunications
ul. Szachowa 1, 04–894 Warsaw, Poland
E-mail: orlowska@itl.waw.pl

Gunther Schmidt
Universität der Bundeswehr München, Institute for Software Technology
Department of Computing Science, 85577 Neubiberg, Germany
E-mail: nataliagunther.schmidt@t-online.de

Marc Roubens
Faculté Polytechnique de Mons
Rue de Houdain 9, 4000 Mons Belgium
E-mail: roubens@belgacom.net

Library of Congress Control Number: 2006938750

CR Subject Classification (1998): I.1, I.2, F.3-4, D.2.4, H.2.8

LNCS Sublibrary: SL 7 – Artificial Intelligence

ISSN   0302-9743
ISBN-10   3-540-69223-1 Springer Berlin Heidelberg New York
ISBN-13   978-3-540-69223-2 Springer Berlin Heidelberg New York

Springer is a part of Springer Science+Business Media

springer.com

© Springer-Verlag Berlin Heidelberg 2006

Typesetting: Camera-ready by author, data conversion by Scientific Publishing Services, Chennai, India
Printed on acid-free paper   SPIN: 11964810   06/3142   5 4 3 2 1 0

Preface

This book is a follow-up of LNCS volume 2929 with the same title, and presents the major results of COST action 274 (2002-2005), TARSKI: Theory and Applications of Relational Structures as Knowledge Instruments.

Relational structures abound in the daily environment: relational databases, data-mining, scaling procedures, preference relations, etc. Reasoning about, and with, relations has a long-standing European tradition, which may be divided into three broad areas:

1. *Algebraic Logic*: algebras of relations, relational semantics, and algebras and logics derived from information systems.
2. *Computational Aspects* of Automated Relational Reasoning: decidability and complexity of algorithms, network satisfaction.
3. *Applications*: social choice, AI, linguistics, psychology, economics, etc.

The main objective of the first TARSKI book (LNCS 2929) was to advance the understanding of relational structures and the use of relational methods in applicable object domains. There were the following sub-objectives:

1. To study the semantical and syntactical aspects of relational structures arising from 'real world' situations
2. To investigate automated inference for relational systems, and, where possible or feasible, develop deductive systems which can be implemented into industrial applications, such as diagnostic systems
3. To develop non-invasive scaling methods for predicting relational data
4. To make software for dealing with relational systems commonly available

We are confident that the present book will further the understanding of interdisciplinary issues involving relational reasoning. This book consists of papers which give a clear and self-contained overview of the results obtained by the TARSKI action, typically obtained by different persons from different work areas. The study and possible integration of different approaches to the same problem, which may have arisen at different locations, will be of practical value to the developers of information systems.

The first three papers concern *applications*. In the first paper a fair procedure for coalition formation is given. The software tool MacBeth for multi-criteria decision making is used to determine the utilities of the different alternatives to parties and the RELVIEW tool is used to compute the stable governments and to visualize the results. If there is no stable government, graph-theoretical results are used to find a government as stable as possible and if there are several stable governments negotiations or consensus reaching may be used to choose one.

In computer science, scenarios with interacting agents are often developed using modal logic. The second paper shows how to interpret modal logic of knowledge in relation algebra. This allows the use of the RELVIEW tool for the purpose of investigating finite models and for visualizing certain properties. This approach is illustrated with the well-known 'muddy children' puzzle using modal logic of knowledge.

The authors of the third paper use a regional health care perspective on maintenance and analysis of data, information and knowledge. Examples are drawn from cardiac diseases. Analysis and development are viewed from the by-pass surgery point of view. Association rules are used for analysis, and they show how these rules take logical forms so as to prepare for development of guidelines.

Computational aspects are treated in the next four papers. The fourth paper gives a generalization of the Hoede–Bakker index, which is a measure for the power of players in a network, taking into account the mutual influences between the players.

The fifth paper gives a relational presentation of nonclassical logics, providing a general scheme for automatic translation. The translation process is supported by a flexible Prolog tool.

The sixth paper provides a translation of the multimodal logic of qualitative order-of-magnitude reasoning into relational logics and presents a sound and complete proof system for the relational version of the language.

Logics of binary relations are presented in the seventh paper, together with the proof systems in the style of dual tableaux. Applications of these logics to reasoning in nonclassical logics are mentioned.

The remaining papers may be classified in the field of *algebraic logic*.

Papers 8 till 11 deal with different aspects of fuzzy preference relations. Fuzzy information relations and operators are studied in paper 8, where an algebraic approach is given based on residuated lattices. The authors of paper 9 give an overview of results on the aggregation of fuzzy relations and the related property of dominance of aggregation operators. The authors of the next paper, paper 10, address the added value that is provided by using distance-based fuzzy relations in flexible query answering. The last paper in this group gives a state-of-the-art overview of general representation results for fuzzy weak orders.

The next four papers deal with *lattices*. Relational representation theorems for lattices endowed with various negation operations are presented in a uniform framework in paper 12. The next paper gives relational representation theorems for classes of algebras which may be viewed as weak relation algebras, where a Boolean part is replaced by a not necessarily distributive lattice. Paper 14 treats aspects of lattice and generalized pre-lattice effect algebras. And the last paper in this group presents a decision procedure for the quantifier-free satisfiability problem of the language BLmf of bounded lattices with monotone unary functions.

Paper 16 addresses the relation of dominance on the class of continuous t-norms with a particular focus on continuous ordinal sum t-norms. Geometrical insight is provided into dominance relationships involving prototypical Archimedean t-norms, the Łukasiewicz t-norm and the product t-norm.

The last paper in this volume addresses the problem of extending aggregation operators typically defined on [0,1] to the symmetric interval [-1,1], where the '0' value plays a particular role (neutral value).

Referees

Radim Belohlavek	Irina Perfilieva	Viorica Sofronie Stokkermans
Ferdinand Chovanec	Alberto Policriti	Harrie de Swart
Stéphane Demri	Ingrid Rewitzky	Dimiter Vakarelov
Ivan Kojadinovic	Marc Roubens	Peter Vojtáš
Beata Konikowska	Wolfgang Sander	Hui Wang
Angelo Montanari	Gunther Schmidt	Michael Winter
Mirko Navara	Andrzej Skowron	
Wojciech Penczek	Roman Slowinski	

Acknowledgements

We owe much to the referees mentioned above and are most grateful to them. Jozef Pijnenburg was instrumental in editing this book because of his highly appreciated expertise in LaTeX. The cooperation of many authors in this book was supported by COST action 274, TARSKI, and is gratefully acknowledged.

Editors

Harrie de Swart, Chair, Tilburg University, Netherlands
Ewa Orłowska, Institute for Telecommunications, Warsaw, Poland
Gunther Schmidt, Universität der Bundeswehr, Munich, Germany
Marc Roubens, Université de Liège and Faculté Polytechnique de Mons, Belgium

Table of Contents

TARSKI: Theory and Applications of Relational Structures as Knowledge Instruments II

Social Software for Coalition Formation[*]

Agnieszka Rusinowska[1,2], Rudolf Berghammer[3], Patrik Eklund[4],
Jan-Willem van der Rijt[5], Marc Roubens[6,7], and Harrie de Swart[8]

[1] Radboud University Nijmegen, Nijmegen School of Management, P.O. Box 9108
6500 HK Nijmegen, The Netherlands
a.rusinowska@fm.ru.nl
[2] Warsaw School of Economics, Department of Mathematical Economics
Al. Niepodleglosci 162, 02-554 Warsaw, Poland
arusin@sgh.waw.pl
[3] Institute of Computer Science, University of Kiel
Olshausenstraße 40, 24098 Kiel, Germany
rub@informatik.uni-kiel.de
[4] Umeå University, Department of Computing Science
SE-90187 Umeå, Sweden
peklund@cs.umu.se
[5] University of Groningen, Faculty of Philosophy, Oude Boteringestraat 52
9712 GL Groningen, The Netherlands
J.W.van.der.Rijt@rug.nl
[6] University of Liège, Institute of Mathematics, B37
B-4000 Liège, Belgium
[7] Faculté Polytechnique de Mons
Rue de Houdain 9, B-7000 Mons, Belgium
m.roubens@ulg.ac.be
[8] Tilburg University, Faculty of Philosophy, P.O. Box 90153
5000 LE Tilburg, The Netherlands
H.C.M.deSwart@uvt.nl

Abstract. This paper concerns an interdisciplinary approach to coalition formation. We apply the MacBeth software, relational algebra, the RelView tool, graph theory, bargaining theory, social choice theory, and consensus reaching to a model of coalition formation. A feasible government is a pair consisting of a coalition of parties and a policy supported by this coalition. A feasible government is stable if it is not dominated by any other feasible government. Each party evaluates each government with respect to certain criteria. MacBeth helps to quantify the importance of the criteria and the attractiveness and repulsiveness of governments to parties with respect to the given criteria. Feasibility, dominance, and stability are formulated in relation-algebraic terms. The RelView tool is used to compute the dominance relation and the set of all stable governments. In case there is no stable government, i.e., in case the dominance relation is cyclic, we apply graph-theoretical techniques for breaking the cycles. If the solution is not unique, we select

---

[*] Co-operation for this paper was supported by European COST Action 274 "Theory and Applications of Relational Structures as Knowledge Instruments" (TARSKI). We thank Gunther Schmidt for his most valuable contributions to this paper.

H. de Swart et al. (Eds.): TARSKI II, LNAI 4342, pp. 1–30, 2006.

the final government by applying bargaining or appropriate social choice rules. We describe how a coalition may form a government by reaching consensus about a policy.

Keywords: stable government, MacBeth, relational algebra, RELVIEW, graph theory, bargaining, social choice rule, consensus.

1   Introduction

This paper presents an overview of the results on coalition formation obtained from cooperation within the European COST Action 274: TARSKI (Theory and Applications of Relational Structures as Knowledge Instruments). The authors were connected to two different Work Areas of the COST Action, namely Work Area WA2 (Mechanization and Relational Reasoning) and Work Area WA3 (Relational Scaling and Preferences). This cooperation, which was not foreseen but gradually evolved over the years, resulted in an interdisciplinary approach to coalition formation. The MacBeth technique, relational algebra, the RELVIEW tool, graph theory, bargaining theory, social choice theory, and consensus reaching were applied to the basic model of coalition formation described in Rusinowska et al. [44].

Coalition formation is one of the more interesting and at the same time more popular topics, and consequently a lot of work has already been done in this field. There are several ways to distinguish different coalition formation theories: one may talk, for instance, about power-oriented versus policy-oriented theories, one-dimensional versus multi-dimensional models, or actor-oriented versus non-actor oriented theories. The power-oriented theories, where the motivation for political parties to join a coalition is based only on their personal gains, are the earliest theories of coalition formation. One may mention here the theory of minimal winning coalitions (von Neuman and Morgenstern [55]), the minimum size theory (Riker [40]), and the bargaining proposition (Leiserson [35]). In policy-oriented theories, the process of coalition formation is determined by both policy and power motivations. Some of the most important early policy-oriented theories were the minimal range theory (Leiserson [34]), conflict of interest theory (Axelrod [2]), and the policy distance theory (de Swaan [21]). Actor-oriented theories, like the dominant player theory (Peleg [38], [39]) and the center player theory (van Deemen [53]), select an actor that has a more powerful position in the process of coalition formation. Also a lot of work has been done on spatial coalition formation theories, especially with respect to multi-dimensional policy-oriented theories. A main assumption in such models is that policy positions of parties are very important in the coalition formation process. One must mention here the political heart solution (Schofield [48], [49], [50]), the protocoalition formation (Grofman [29]), the winset theory (Laver and Shepsle [32], [33]), and the competitive solution (McKelvey, Ordeshook and Winer [36]). Many authors also considered institutional theories of coalition formation. One of the first theorists who acknowledged the important role of institutions was Shepsle [52], followed, in particular, by Austen-Smith and Banks [1], Laver and Schofield [31], and

Baron [6]. For an overview of coalition formation models we also like to refer to van Deemen [54], de Vries [24], Kahan and Rapoport [30].

The point of departure in this paper is a multi-dimensional model of coalition formation (see Rusinowska et al. [44]) in which the notion of stable government is central. In the model, the approach we use to represent party preferences allows us to include both rent-seeking and idealistic (policy-seeking) motivations. Moreover, a policy space does not have to be a Euclidean space, as is assumed frequently in coalition formation models, but may be any kind of space. The policy space is assumed to be multi-dimensional, which allows us to consider many political issues at the same time.

A government is defined as a pair consisting of a coalition and a policy supported by that coalition. It has a value (utility) to each party with respect to every given issue. In order to determine these values in practice, we propose to use the MacBeth approach; see also Roubens et al. [41]. MacBeth, which stands for *Measuring Attractiveness by a Categorical Based Evaluation Technique*, is an interactive approach to quantify the attractiveness of each alternative, such that the measurement scale constructed is an interval scale. For an overview and some applications of the software, we refer to the web site (www.m-macbeth.com), Bana e Costa and Vansnick [3]; Bana e Costa et al. [5]. The notion of absolute judgement has also been used in Saaty's Analytical Hierarchy Process (AHP); see Saaty [45], [46]. In the MacBeth technique, the absolute judgements concern differences of attractiveness, while in Saaty's method they concern ratios of priority, or of importance. One of the advantages of using the MacBeth approach is related to ensuring consistency. In case of any inconsistency of the initial evaluations, the MacBeth software indicates to the user what is the cause of the inconsistency and how to reach consistency. For a critical analysis of the AHP, see Bana e Costa and Vansnick [4].

Another application to the coalition formation model we propose here concerns Relational Algebra and the RELVIEW tool which helps us to calculate stable governments; see also Berghammer et al. [11]. The RELVIEW system, which has been developed at Kiel University, is a computer system for the visualization and manipulation of relations and for relational prototyping and programming. The tool is written in the C programming language, uses reduced ordered binary decision diagrams for implementing relations, and makes full use of the X-windows graphical user interface. For details and applications see, for instance, Berghammer et al. [14], Behnke et al. [7], Berghammer et al. [10], and Berghammer et al. [13].

In this paper, we also present an application of Graph Theory to the model of coalition formation in question; see Berghammer et al. [12]. We present a graph-theoretical procedure for choosing a government in case there is no stable government. If, on the other hand, more than one stable government exists, we may apply Social Choice Theory to choose one government. For an overview and comparison of social choice rules see, for instance, Brams and Fishburn [16], and de Swart et al. [23]. Another natural application is based on Bargaining Theory. We use a strategic approach to bargaining; see Rubinstein [42], Fishburn and

Rubinstein [27], Osborne and Rubinstein [37]. We formulate several bargaining games in which parties bargain over the choice of one stable government, and next we look for refinements of Nash equilibria called subgame perfect equilibria (Selten [51]) of these games; see also Rusinowska and de Swart [43].

We describe a procedure for a coalition to choose a policy in order to propose a government, based on consensus reaching, by combining some ideas from Carlsson et al. [18] and Rusinowska et al. [44]. It has been first proposed in Eklund et al. [25], where the authors consider consensus reaching in a committee, and next in Eklund et al. [26], where a more complicated model, i.e., consensus reaching in coalition formation, is presented.

The paper is structured as follows. Section 2 introduces the model of coalition formation. In Section 4, the basic notions of relational algebra are presented. In Sections 3 and 5, we present applications of the MacBeth and RELVIEW tools, respectively, to the model in question. Section 6 concerns applications of Social Choice Theory and Bargaining Theory to the model, in order to choose a stable government in the case there exists more than one. Next, an application of Graph Theory to the model of coalition formation is proposed in Section 7, in order to choose a 'rather stable' government in the case that there exists no stable one. Section 8 describes how a coalition may reach consensus about a policy in order to propose a government. In Section 9, we present our conclusions.

2 The Model of Coalition Formation

In this section we recapitulate a model of coalition formation, first introduced in Rusinowska et al. [44], and further refined, in particular, in Eklund et al. [26].

2.1 Description of the Model

Let $N = \{1, \ldots, n\}$ be the set of political parties in a parliament, and let w_i denote the number of seats received by party $i \in N$. Moreover, let W denote the *set of all winning coalitions*. The model concerns the creation of a government by a winning coalition. It is assumed that there are some independent policy issues on which a government has to decide. Let P denote the *set of all policies*.

A *government* is defined as a pair $g = (S, p)$, where S is a winning coalition and p is a policy. Hence, the *set G of all governments* is defined as

$$G := \{(S, p) \mid S \in W \ \wedge \ p \in P\}. \tag{1}$$

Each party has preferences concerning all policies and all (winning) coalitions. A coalition is called *feasible* if it is acceptable to all its members. A policy is *feasible for a given coalition* if it is acceptable to all members of that coalition. A government (S, p) is *feasible* if both, S and p, are acceptable to each party belonging to S. By G^* we denote the *set of all feasible governments*, and by G_i^* the *set of all feasible governments containing party i*, i.e., for each $i \in N$,

$$G_i^* := \{(S, p) \in G^* \mid i \in S\}. \tag{2}$$

A *decision maker* is a party involved in at least one feasible government, i.e., the *set DM of all decision makers* is equal to

$$DM := \{i \in N \mid G_i^* \neq \emptyset\}. \tag{3}$$

Moreover, let the subset W^* of W be defined as

$$W^* := \{S \in W \mid \exists p \in P : (S, p) \in G^*\}. \tag{4}$$

A feasible government is evaluated by each decision maker with respect to the given policy issues and with respect to the issue concerning the coalition. Let C^* be the finite *set of criteria*. The criteria do not have to be equally important to a party, and consequently, each decision maker evaluates the importance of the criteria. Formally, for each $i \in DM$, we assume a function $\alpha_i : C^* \to [0, 1]$, such that the following property holds:

$$\forall i \in DM : \sum_{c \in C^*} \alpha_i(c) = 1. \tag{5}$$

The number $\alpha_i(c)$ is i's evaluation of criterion c. Moreover, each decision maker evaluates each feasible government with respect to all the criteria. Hence, for each $i \in DM$, we assume $u_i : C^* \times G^* \to \mathbb{R}$ where the real number $u_i(c, g)$ is called the *value of government $g \in G^*$ to party $i \in DM$ with respect to criterion $c \in C^*$*. Moreover, for each $i \in DM$, we define $U_i : G^* \to \mathbb{R}$ such that

$$(U_i(g))_{g \in G^*} = (\alpha_i(c))_{c \in C^*} \cdot (u_i(c, g))_{c \in C^*, g \in G^*}, \tag{6}$$

where $(\alpha_i(c))_{c \in C^*}$ is the $1 \times |C^*|$ matrix representing the evaluation (comparison) of the criteria by party i, $(u_i(c, g))_{c \in C^*, g \in G^*}$ is the $|C^*| \times |G^*|$ matrix containing party i's evaluation of all governments in G^* with respect to each criterion in C^*, and $(U_i(g))_{g \in G^*}$ is the $1 \times |G^*|$ matrix containing party i's evaluation of each government in G^*.

In order to determine in practice the values of $\alpha_i(c)$ and $u_i(c, g)$ for all parties $i \in DM$, criteria $c \in C^*$ and governments $g \in G^*$, we can use the MacBeth technique. We do so in Section 3.

The central notion of the model introduced in Rusinowska et al. [44] is the notion of *stability*. A feasible government $h = (S, p) \in G^*$ *dominates* a feasible government $g \in G^*$ (denoted as $h \succ g$) if the property

$$(\forall i \in S : U_i(h) \geq U_i(g)) \wedge (\exists i \in S : U_i(h) > U_i(g)) \tag{7}$$

holds. A feasible government is said to be *stable* if it is dominated by no feasible government. By

$$SG^* := \{g \in G^* \mid \neg \exists h \in G^* : h \succ g\} \tag{8}$$

we denote the *set of all (feasible) stable governments*. In Rusinowska et al. [44], necessary and sufficient conditions for the existence and the uniqueness of a stable government are investigated. Moreover, the authors introduce some alternative definitions of 'stability', and establish the relations between the new notions of 'stability' and the chosen one. In the present paper, we decide for the definition of a stable government given by (8), which we find the most natural definition of stability.

2.2   A Running Example

Let us consider a very small parliament consisting of only three parties. We assume each coalition consisting of at least two parties is winning and there are only two policy issues and four policies, i.e., we have

$$N = \{A, B, C\}, \quad W = \{AB, AC, BC, ABC\}, \quad P = \{p_1, p_2, p_3, p_4\}.$$

As a consequence, we have 16 governments. Assume that the grand coalition is not feasible, but all two-party coalitions are feasible. Further, assume both policies p_1 and p_2 are acceptable to all three parties, policy p_3 is not acceptable to party C, while policy p_4 is not acceptable to party B. Hence, policies p_1 and p_2 are feasible for coalitions AB, AC, and BC, policy p_3 is feasible for coalition AB, and p_4 is feasible for coalition AC.

Consequently, there are eight feasible governments, i.e.,

$$G^* = \{g_1, g_2, g_3, g_4, g_5, g_6, g_7, g_8\},$$

which are given as

$$g_1 = (AB, p_1), \quad g_2 = (AC, p_1), \quad g_3 = (BC, p_1), \quad g_4 = (AB, p_2),$$
$$g_5 = (AC, p_2), \quad g_6 = (BC, p_2), \quad g_7 = (AB, p_3), \quad g_8 = (AC, p_4)$$

and therefore obtain the governments containing the parties as

$$G_A^* = \{g_1, g_2, g_4, g_5, g_7, g_8\},$$
$$G_B^* = \{g_1, g_3, g_4, g_6, g_7\},$$
$$G_C^* = \{g_2, g_3, g_5, g_6, g_8\}.$$

Moreover, we have

$$DM = N, \quad W^* = \{AB, AC, BC\}, \quad C^* = \{1, 2, 3\},$$

where the criteria 1 and 2 refer to the first and the second policy issue, while criterion 3 concerns the (attractiveness of the) 'coalition'. In order to determine $\alpha_i(c)$ and $u_i(c, g)$ for each $i \in DM$, $c \in C^*$, and $g \in G^*$, we will use the MacBeth technique in the next section.

3   Applying MacBeth to Coalition Formation

When applying the coalition formation model described in Section 2 in practice, the question arises how to determine the $\alpha_i(c)$ and the $u_i(c, g)$ for $i \in DM$. The answer to this question will be given in this section, where we propose to use the MacBeth software to determine these values. In Subsection 3.1, we show how the utilities of governments to parties may be calculated using the MacBeth technique (see also [41]), while in Subsection 3.2 the application is illustrated by an example. It is assumed here that each party judges only a finite number of governments differently, even if there is an infinite number of possible governments.

3.1   Computing the Utilities by MacBeth

Given a party $i \in DM$ and a criterion $c \in C^*$, in order to determine the values $u_i(c, g)$ for each feasible government $g \in G^*$, we will use the MacBeth approach. For each criterion $c \in C^*$, each party ranks in a non-increasing order all feasible governments taking into account the attractiveness of these governments with respect to the given criterion. In particular, for each criterion $c \in C^*$, each party $i \in DM$ specifies two particular references:

- $neutral_i^c$ ('a for party i neutral government with respect to criterion c') defined as a for i neither satisfying nor unsatisfying government wrt. c,
- $good_i^c$ ('a for party i good government with respect to criterion c') defined as a for i undoubtedly satisfying government wrt. c.

These references may be fictitious. We need to add that $neutral_i^c$ and $good_i^c$ are only related to the component of the government concerning the given criterion c, which is either the policy on issue c or the coalition forming the government. For each $c \in C^*$ the remaining 'components' do not matter. Define for all $c \in C^*$ and $i \in DM$ the set

$$G_i^c = G^* \cup \{neutral_i^c, good_i^c\}.$$

For each $c \in C^*$, each party $i \in DM$ judges verbally the difference of attractiveness between each two governments $g, h \in G_i^c$, where g is at least as attractive to i as h. When judging, a party chooses one of the following categories:

D_0 : *no* difference of attractiveness,
D_1 : *very weak* difference of attractiveness,
D_2 : *weak* difference of attractiveness,
D_3 : *moderate* difference of attractiveness,
D_4 : *strong* difference of attractiveness,
D_5 : *very strong* difference of attractiveness,
D_6 : *extreme* difference of attractiveness.

(Formally, the categories are relations.) A party may also choose the union of several successive categories among these above or a *positive* difference of attractiveness in case the party is not sure about the difference of attractiveness.

Given a party $i \in DM$ and a criterion $c \in C^*$, a non-negative number $u_i(c, g)$ is associated to each $g \in G_i^c$. If there is no hesitation about the difference of attractiveness, the following rules are satisfied; see Bana e Costa and Vansnick [3], Bana e Costa et al. [5]. First, for all $g, h \in G_i^c$

$$u_i(c, g) > u_i(c, h) \iff g \text{ more attractive to } i \text{ wrt. } c \text{ than } h. \qquad (9)$$

Second, for all $k, k' \in \{1, 2, 3, 4, 5, 6\}$ with $k \geq k' + 1$ and all $g, g', h, h' \in G_i^c$ with $(g, g') \in D_k$ and $(h, h') \in D_{k'}$

$$u_i(c, g) - u_i(c, g') > u_i(c, h) - u_i(c, h'). \qquad (10)$$

The numerical scale, called the MacBeth basic scale, is obtained by linear programming, and it exists if and only if it is possible to satisfy rules (9) and (10).

In that case the matrix of judgements is called consistent. If it is impossible to satisfy rules (9) and (10), a message appears on the screen ('inconsistent judgements'), inviting the party to revise the judgements, and the MacBeth tool gives suggestions how to obtain a consistent matrix of judgements.

The basic MacBeth scale, which is still a *pre-cardinal scale*, is presented both in a numerical way and in a graphical way ('thermometer'). In order to obtain a *cardinal scale*, and the final utilities $u_i(c, g)$ for party i of the governments g with respect to the given criterion c, the party uses the thermometer. When a party selects with the mouse a government, an interval appears around this government. By moving the mouse, the position of the selected government within this interval is modified, by which the party obtains a new positioning of the governments such that both conditions (9) and (10) are satisfied. We obtain the *cardinal scale* and the (final, agreed) utilities of the governments with respect to the given criterion, when the party agrees that the scale adequately represents the relative difference of attractiveness with respect to the given criterion between any two governments.

Using the MacBeth software, we can also calculate the coefficients or weights $(\alpha_i(c))_{c \in C^*}$ of criterion c for party i. Let us assume that $C^* = \{1, 2, \ldots, m\}$. For each party $i \in N$, we consider the following reference profiles:

$$[neutral_i] = (neutral_i^1, neutral_i^2, \ldots, neutral_i^m)$$
$$[Crit._i^1] = (good_i^1, neutral_i^2, \ldots, neutral_i^m)$$
$$[Crit._i^2] = (neutral_i^1, good_i^2, \ldots, neutral_i^m)$$
$$\vdots$$
$$[Crit._i^m] = (neutral_i^1, neutral_i^2, \ldots, good_i^m)$$

For each $c \in C^*$, the difference in attractiveness between $[Crit._i^c]$ and $[neutral_i]$ corresponds to the added value of the 'swing' from $neutral_i^c$ to $good_i^c$. A party ranks the reference profiles in decreasing order of attractiveness and, using categories D_0 to D_6, judges the difference of attractiveness between each two reference profiles, where the first one is more attractive than the second one. After the adjustment of the MacBeth scale proposed by the software, an interval scale is obtained, which measures the overall attractiveness of the reference profiles, and leads to obtaining the coefficients $(\alpha_i(c))_{c \in C^*}$.

3.2   Example (Continued)

In order to determine for our running example (introduced in Subsection 2.2) the utilities to each party of all governments with respect to each criterion, and the coefficients concerning the importance of the criteria for each party, we will use the MacBeth approach. First, each party expresses its preferences. Note that since g_1, g_2, and g_3 have the same policy p_1, they must be equally attractive to each party with respect to the first and the second policy issue. The same holds for governments g_4, g_5, and g_6 which have the same policy p_2. Moreover, governments formed by the same coalition are equally attractive to each party with respect to the third issue, the one concerning the coalition.

In the following three tables we show for each party A, B, and C of our example the non-increasing order of all eight feasible governments $g_1, \ldots, g_8$ with respect to the first, the second, and the third (coalition) criterion. By the symbol $\sim_i$ we denote the equivalence relation for party $i \in DM$.

Table 1. Non-increasing order of all governments wrt. issue 1

party	order
A	$good_A^1 \quad g_1 \sim_A g_2 \sim_A g_3 \quad g_4 \sim_A g_5 \sim_A g_6 \quad g_8 \quad g_7 = neutral_A^1$
B	$good_B^1 \quad g_4 \sim_B g_5 \sim_B g_6 \quad g_1 \sim_B g_2 \sim_B g_3 \quad g_7 \quad g_8 = neutral_B^1$
C	$g_7 \quad good_C^1 = g_8 \quad g_1 \sim_C g_2 \sim_C g_3 \quad g_4 \sim_C g_5 \sim_C g_6 \quad neutral_C^1$

Table 2. Non-increasing order of all governments wrt. issue 2

party	order
A	$good_A^2 \quad g_1 \sim_A g_2 \sim_A g_3 \quad g_4 \sim_A g_5 \sim_A g_6 \quad g_7 \quad g_8 = neutral_A^2$
B	$good_B^2 \quad g_1 \sim_B g_2 \sim_B g_3 \quad g_4 \sim_B g_5 \sim_B g_6 \quad g_7 \quad g_8 = neutral_B^2$
C	$good_C^2 = g_8 \quad g_7 \quad g_1 \sim_C g_2 \sim_C g_3 \quad g_4 \sim_C g_5 \sim_C g_6 \quad neutral_C^2$

Table 3. Non-increasing order of all governments wrt. issue 3

party	order
A	$good_A^3 = g_1 \sim_A g_4 \sim_A g_7 \quad g_2 \sim_A g_5 \sim_A g_8 \quad g_3 \sim_A g_6 = neutral_A^3$
B	$good_B^3 = g_1 \sim_B g_4 \sim_B g_7 \quad g_3 \sim_B g_6 \quad g_2 \sim_B g_5 \sim_B g_8 = neutral_B^3$
C	$good_C^3 = g_2 \sim_C g_5 \sim_C g_8 \quad g_3 \sim_C g_6 \quad g_1 \sim_C g_4 \sim_C g_7 = neutral_C^3$

Each party $i \in DM$ also has to judge the difference of attractiveness between each two reference profiles. Here we obtain the following values:

$$[neutral_i] = (neutral_i^1, neutral_i^2, neutral_i^3)$$
$$[Crit._i^1] = (good_i^1, neutral_i^2, neutral_i^3)$$
$$[Crit._i^2] = (neutral_i^1, good_i^2, neutral_i^3)$$
$$[Crit._i^3] = (neutral_i^1, neutral_i^2, good_i^3)$$

Let us assume that Table 4 shows the decreasing orders of these reference profiles for all parties. Then we obtain:

Table 4. Decreasing order of the reference profiles

party	order of the profiles			
A	$[Crit._A^1]$	$[Crit._A^2]$	$[Crit._A^3]$	$[neutral_A]$
B	$[Crit._B^3]$	$[Crit._B^2]$	$[Crit._B^1]$	$[neutral_B]$
C	$[Crit._C^3]$	$[Crit._C^1]$	$[Crit._C^2]$	$[neutral_C]$

First, we consider party A which has to judge the difference of attractiveness for all the governments with respect to each issue. The following Tables 5, 6 and 7 show the matrices of judgements for this party.

Table 5. Judgements of the attractiveness for party A and issue 1

	$good_A^1$	g_1	g_4	g_8	$neutral_A^1$
$good_A^1$	**no**	very weak	weak	strong	extreme
g_1		**no**	weak	strong	extreme
g_4			**no**	strong	extreme
g_8				**no**	extreme
$neutral_A^1$					**no**

Table 6. Judgements of the attractiveness for party A and issue 2

	$good_A^2$	g_1	g_4	g_7	$neutral_A^2$
$good_A^2$	**no**	weak	moderate	strong	very strong
g_1		**no**	moderate	strong	very strong
g_4			**no**	strong	very strong
g_7				**no**	very strong
$neutral_A^2$					**no**

Table 7. Judgements of the attractiveness for party A and issue 3

	$good_A^3$	g_2	$neutral_A^3$
$good_A^3$	**no**	weak	extreme
g_2		**no**	extreme
$neutral_A^3$			**no**

Based on the above tables, next, the MacBeth tool proposes the basic scale for party A – using the thermometer – discusses the scale, and after that the final values (utilities) are calculated. The following Table 8 shows the results $u_A(c, g)$ for c ranging over the three issues and g ranging over all eight feasible governments $g_1, \ldots, g_8$ of our example:

Table 8. Values of the governments wrt. each issue for party A

$g =$	g_1	g_2	g_3	g_4	g_5	g_6	g_7	g_8
$u_A(1, g) =$	93.0	93.0	93.0	82.3	82.3	82.3	0.0	53.5
$u_A(2, g) =$	93.0	93.0	93.0	78.6	78.6	78.6	57.0	0.0
$u_A(3, g) =$	100.0	75.0	0.0	100.0	75.0	0.0	100.0	75.0

In a similar way, the values for parties B and C may be calculated. Tables 9 and 10 present the values for these parties.

Table 9. Values of the governments wrt. each issue for party B

$g =$	g_1	g_2	g_3	g_4	g_5	g_6	g_7	g_8
$u_B(1, g) =$	80.0	80.0	80.0	95.0	95.0	95.0	55.0	0.0
$u_B(2, g) =$	96.5	96.5	96.5	93.0	93.0	93.0	53.5	0.0
$u_B(3, g) =$	100.0	0.0	57.0	100.0	0.0	57.0	100.0	0.0

Table 10. Values of the governments wrt. each issue for party C

$g =$	g_1	g_2	g_3	g_4	g_5	g_6	g_7	g_8
$u_C(1, g) =$	90.0	90.0	90.0	60.0	60.0	60.0	110.0	100.0
$u_C(2, g) =$	64.2	64.2	64.2	53.5	53.5	53.5	92.8	100.0
$u_C(3, g) =$	0.0	100.0	96.5	0.0	100.0	96.5	0.0	100.0

Moreover, using the MacBeth technique, we can calculate the coefficients $\alpha_i(c)$ for all decision makers $i \in DM$ (in the case of the example, hence, for all parties A, B, and C) and all three issues $c \in C^*$. These numbers are summarized in the following Table 11.

Table 11. The scaling constants

$i \in DM$	$\alpha_i(1)$	$\alpha_i(2)$	$\alpha_i(3)$
A	0.6	0.3	0.1
B	0.1	0.3	0.6
C	0.3	0.1	0.6

Finally, based on all the values, the utilities of all governments are calculated by means of formula (6). The results are presented in Table 12. This table will be the base for obtaining the input of the RELVIEW tool in order to compute the stable governments, as described in the next section.

Table 12. The utilities of all feasible governments

$g \in G^*$	$U_A(g)$	$U_B(g)$	$U_C(g)$
$g_1 = (AB, p_1)$	93.7	97.0	33.4
$g_2 = (AC, p_1)$	91.2	37.0	93.4
$g_3 = (BC, p_1)$	83.7	71.2	91.3
$g_4 = (AB, p_2)$	82.7	97.4	23.4
$g_5 = (AC, p_2)$	80.5	37.4	83.4
$g_6 = (BC, p_2)$	73.0	71.6	81.3
$g_7 = (AB, p_3)$	27.1	81.6	42.3
$g_8 = (AC, p_4)$	39.6	0.0	100.0

4   Relational Algebraic Preliminaries

In this section we recall the basics of relational algebra and some further relational constructions, which are used in this paper later on. For more details on relations and relational algebra, see Schmidt et al. [47] or Brink et al. [17] for example.

4.1   Relational Algebra

If X and Y are sets, then a subset R of the Cartesian product $X \times Y$ is called a (binary) relation with *domain* X and *range* Y. We denote the set (in this context also called type) of all relations with domain X and range Y by $[X \leftrightarrow Y]$ and write $R : X \leftrightarrow Y$ instead of $R \in [X \leftrightarrow Y]$. If X and Y are finite sets of size m and n respectively, then we may consider a relation $R : X \leftrightarrow Y$ as a Boolean matrix with m rows and n columns. In particular, we write $R_{x,y}$ instead of $\langle x, y \rangle \in R$. The Boolean matrix interpretation of relations is used as one of the graphical representations of relations within the RELVIEW tool.

The basic operations on relations are R^{T} (*transposition*), $\overline{R}$ (*complement*), $R \cup S$ (*union*), $R \cap S$ (*intersection*), $R; S$ (*composition*), R^* (*reflexive-transitive closure*), and the special relations O (*empty relation*), L (*universal relation*), and I (*identity relation*). If R is included in S we write $R \subseteq S$, and equality of R and S is denoted as $R = S$.

A *vector* v is a relation v with $v = v; \mathsf{L}$. For v being of type $[X \leftrightarrow Y]$ this condition means: Whatever set Z and universal relation $\mathsf{L} : Y \leftrightarrow Z$ we choose, an element $x \in X$ is either in relationship $(v; \mathsf{L})_{x,z}$ to no element $z \in Z$ or to all elements $z \in Z$. As for a vector, therefore, the range is irrelevant, we consider in the following mostly vectors $v : X \leftrightarrow \mathbf{1}$ with a specific singleton set $\mathbf{1} := \{\bot\}$ as range and omit in such cases the second subscript, i.e., write v_x instead of $v_{x,\bot}$.

Analogously to linear algebra we use in the following lower-case letters to denote vectors. A vector $v : X \leftrightarrow \mathbf{1}$ can be considered as a Boolean matrix with exactly one column, i.e., as a Boolean column vector, and *describes* (or is a description of) the subset $\{x \in X \mid v_x\}$ of X.

As a second way to model sets we will use the relation-level equivalents of the set-theoretic symbol "$\in$", i.e., *membership-relations* $\varepsilon : X \leftrightarrow 2^X$. These specific relations are defined by $\varepsilon_{x,Y}$ if and only if $x \in Y$, for all $x \in X$ and $Y \in 2^X$. A Boolean matrix representation of the ε relation requires exponential space. However, in Berghammer et al. [10] an implementation of ε using reduced ordered binary decision diagrams is presented, the number of vertices of which is linear in the size of X.

4.2   Relational Products and Sums

Given a Cartesian product $X \times Y$ of two sets X and Y, there are two projection functions which decompose a pair $u = \langle u_1, u_2 \rangle$ into its first component u_1 and its second component u_2. For a relation-algebraic approach it is useful to consider instead of these functions the corresponding *projection relations* $\pi : X \times Y \leftrightarrow X$

and $\rho : X \times Y \leftrightarrow Y$ such that for all $u \in X \times Y$, $x \in X$, and $y \in Y$ we have $\pi_{u,x}$ if and only if $u_1 = x$ and $\rho_{u,y}$ if and only if $u_2 = y$. Projection relations enable us to describe the well-known pairing operation of functional programming relation-algebraically as follows. For relations $R : Z \leftrightarrow X$ and $S : Z \leftrightarrow Y$ we define their *pairing* (frequently also called *fork* or *tupling*) $[R, S] : Z \leftrightarrow X \times Y$ by

$$[R, S] := R; \pi^\mathsf{T} \cap S; \rho^\mathsf{T}. \tag{11}$$

Using (11), for all $z \in Z$ and $u \in X \times Y$ a simple reflection shows that $[R, S]_{z,u}$ if and only if R_{z,u_1} and S_{z,u_2}. As a consequence, the *exchange relation*

$$E := [\rho, \pi] = \rho; \pi^\mathsf{T} \cap \pi; \rho^\mathsf{T} \tag{12}$$

of type $[X \times Y \leftrightarrow X \times Y]$ exchanges the components of a pair. This means that for all $u, v \in X \times Y$ the relationship $E_{u,v}$ holds if and only if $u_1 = v_2$ and $u_2 = v_1$.

Analogously to the Cartesian product, the disjoint union (or direct sum) $X + Y := (X \times \{1\}) \cup (Y \times \{2\})$ of two sets X and Y leads to the two *injection relations* $\imath : X \leftrightarrow X+Y$ and $\kappa : Y \leftrightarrow X+Y$ such that for all $u \in X + Y$, $x \in X$, and $y \in Y$ we have $\imath_{x,u}$ if and only if $u = \langle x, 1 \rangle$ and $\kappa_{y,u}$ if and only if $u = \langle y, 2 \rangle$. In this case the counter-part of pairing is the *sum* $R + S : X+Y \leftrightarrow Z$ of two relations $R : X \leftrightarrow Z$ and $S : Y \leftrightarrow Z$, defined by

$$R + S := \imath^\mathsf{T}; R \cup \kappa^\mathsf{T}; S. \tag{13}$$

From specification (13) we obtain for all $u \in X + Y$ and $z \in Z$ that $(R + S)_{u,z}$ if and only if there exists $x \in X$ such that $u = \langle x, 1 \rangle$ and $R_{x,z}$ or there exists $y \in Y$ such that $u = \langle y, 2 \rangle$ and $S_{y,z}$.

The representation of a relation $R : X \leftrightarrow Y$ by a vector $vec(R) : X \times Y \leftrightarrow \mathbf{1}$ means that for all $x \in X$ and $y \in Y$ the properties $R_{x,y}$ and $vec(R)_{\langle x,y\rangle,\perp}$, or $vec(R)_{\langle x,y\rangle}$ for short, are equivalent. To obtain a relation-algebraic specification of $vec(R)$, i.e., an expression which does not use element relationships, but only the constants and operations of relational algebra, we assume $x \in X$ and $y \in Y$ and calculate as follows.

$$
\begin{aligned}
R_{x,y} &\iff \exists a : \pi_{\langle x,y\rangle,a} \wedge R_{a,y} && \pi : X \times Y \leftrightarrow X \text{ projection} \\
&\iff (\pi; R)_{\langle x,y\rangle,y} \\
&\iff \exists b : (\pi; R)_{\langle x,y\rangle,b} \wedge \rho_{\langle x,y\rangle,b} && \rho : X \times Y \leftrightarrow Y \text{ projection} \\
&\iff \exists b : (\pi; R \cap \rho)_{\langle x,y\rangle,b} \wedge \mathsf{L}_b && \mathsf{L} : Y \leftrightarrow \mathbf{1} \\
&\iff ((\pi; R \cap \rho); \mathsf{L})_{\langle x,y\rangle}.
\end{aligned}
$$

An immediate consequence of the last expression of this calculation and the equality of relations is the relation-algebraic specification

$$vec(R) = (\pi; R \cap \rho); \mathsf{L} \tag{14}$$

of the vector $vec(R) : X \times Y \leftrightarrow \mathbf{1}$; see also Schmidt et al. [47].

Later we also will consider a list $R^{(1)}, R^{(2)}, \ldots, R^{(n)}$ of relations $R^{(i)} : X \leftrightarrow Y$ and compute from these a new relation as follows. Let $N := \{1, \ldots, n\}$. If we identify this set with the disjoint union of n copies of $\mathbf{1}$, then

$$C := vec(R^{(1)})^\mathsf{T} + \ldots + vec(R^{(n)})^\mathsf{T} \tag{15}$$

defines a relation of type $[N \leftrightarrow X \times Y]$ such that, using Boolean matrix terminology, for all $i \in N$ the ith row of C equals the transpose of the vector $vec(R^{(i)})$. Hence, from the above considerations we obtain for all $i \in N, x \in X$, and $y \in Y$ the equivalence of $C_{i,\langle x,y \rangle}$ and $R^{(i)}_{x,y}$.

5   Applying RELVIEW to Coalition Formation

In this section we recapitulate the application of the RELVIEW tool to the model of a stable government (see Berghammer et al. [11]). The main purpose of the RELVIEW tool is the evaluation of relation-algebraic expressions. These are constructed from the relations of its workspace using pre-defined operations and tests, user-defined relational functions, and user-defined relational programs. A relational program is much like a function procedure in the programming languages Pascal or Modula 2, except that it only uses relations as data type. It starts with a head line containing the program name and the formal parameters. Then the declaration of the local relational domains, functions, and variables follows. Domain declarations can be used to introduce projection relations and pairings of relations in the case of Cartesian products, and injection relations and sums of relations in the case of disjoint unions, respectively. The third part of a program is the body, a while-program over relations. As a program computes a value, finally, its last part consists of a return-clause, which is a relation-algebraic expression whose value after the execution of the body is the result. RELVIEW makes the results visible in the form of graphs or matrices.

5.1   Computing the Dominance Relation by RELVIEW

In the following we step-wisely develop relation-algebraic specifications of the notions presented in Section 2, such as feasible governments, the dominance relationship, and stable governments. As we will demonstrate, these can be translated immediately into the programming language of the RELVIEW tool and, hence, the tool can be applied to deal with concrete examples.

In order to develop a relation-algebraic specification of feasible governments, we need two 'acceptability' relations A and B. We assume $A : DM \leftrightarrow P$ such that for all $i \in DM$ and $p \in P$

$$A_{i,p} \iff \text{party } i \text{ accepts policy } p,$$

and $B : DM \leftrightarrow W$ such that for all $i \in DM$ and $S \in W$

$$B_{i,S} \iff \text{party } i \text{ accepts coalition } S.$$

Next we consider the following three relations:

- A relation $isFea(A) : W \leftrightarrow P$ such that a coalition $S \in W$ and a policy $p \in P$ are in relationship $isFea(A)_{S,p}$ if and only if p is feasible for S. A formal predicate logic definition of this is

$$isFea(A)_{S,p} \iff \forall i : i \in S \to A_{i,p}. \tag{16}$$

- A vector $feaC(B) : W \leftrightarrow \mathbf{1}$ which describes the set of all feasible coalitions. For all $S \in W$ the predicate logic definition is

$$feaC(B)_S \iff \forall\, i : i \in S \to B_{i,S}. \tag{17}$$

- A relation $feaG(A, B) : W \leftrightarrow P$ which coincides with the set G^* of feasible governments. Here we have for all coalitions $S \in W$ and policies $p \in P$ the predicate logic description

$$feaG(A, B)_{S,p} \iff feaC(B)_S \,\wedge\, isFea(A)_{S,p}\,. \tag{18}$$

Our goal is to obtain from the predicate logic definitions (16), (17), and (18) of the relations $isFea(A)$, $feaC(B)$, and $feaG(A, B)$ equivalent relation-algebraic specifications. In Berghammer et al. [11] it is shown that

$$isFea(A) = \overline{\varepsilon^{\mathsf{T}} ; \overline{A}}\,, \tag{19}$$

$$feaC(B) = \overline{(\varepsilon \cap \overline{B})^{\mathsf{T}} ; \mathsf{L}}\,, \tag{20}$$

$$feaG(A, B) = \overline{\varepsilon^{\mathsf{T}} ; \overline{A}} \cap \overline{(\varepsilon \cap \overline{B})^{\mathsf{T}} ; \mathsf{L}} ; \mathsf{L}\,, \tag{21}$$

where $\varepsilon : DM \leftrightarrow W$ is the membership-relation between decision makers and winning coalitions. Note that $W \subseteq 2^{DM}$. Using matrix terminology, the relation ε is obtained from the ordinary membership-relation of type $[DM \leftrightarrow 2^{DM}]$ by removing from the latter all columns not corresponding to a set of W.

Next, we develop a relation-algebraic specification of the dominance relationship between feasible governments. To this end, we suppose a relational description of government membership to be given, that is, a relation $M : DM \leftrightarrow G^*$ such that for all $i \in DM$ and $g \in G^*$ the equivalence

$$M_{i,g} \iff \text{party } i \text{ is a member of government } g$$

holds. Moreover, for each party $i \in DM$, we introduce a utility (or comparison) relation $R^{(i)} : G^* \leftrightarrow G^*$ such that for all $g, h \in G^*$

$$R^{(i)}_{g,h} \iff U_i(g) \geq U_i(h).$$

Based on these relations, we introduce a global utility (or comparison) relation $C : DM \leftrightarrow G^* \times G^*$ as follows. For all $i \in DM$ and $g, h \in G^*$ we define

$$C_{i,\langle g,h \rangle} \iff R^{(i)}_{g,h}.$$

An immediate consequence of (15) is the equation

$$C = vec(R^{(1)})^{\mathsf{T}} + \ldots + vec(R^{(n)})^{\mathsf{T}}.$$

Next, we consider the dominance relationship, and we get for all $g, h \in G^*$

$$g \succ h \iff (\forall\, i : M_{i,g} \to C_{i,\langle g,h \rangle}) \wedge (\exists\, i : M_{i,g} \wedge \overline{C}_{i,\langle h,g \rangle}). \tag{22}$$

Since $\overline{C}_{i,\langle h,g \rangle} \iff (\overline{C}; E)_{i,\langle g,h \rangle}$, where $E : G^* \times G^* \leftrightarrow G^* \times G^*$ is the exchange relation $[\rho, \pi]$, we have the following description of dominance:

$$g \succ h \iff (\forall i : M_{i,g} \to C_{i,\langle g,h \rangle}) \wedge (\exists i : M_{i,g} \wedge (\overline{C}; E)_{i,\langle g,h \rangle}). \qquad (23)$$

In Berghammer et al. [11], the following fact is proved: Let $\pi : G^* \times G^* \leftrightarrow G^*$ and $\rho : G^* \times G^* \leftrightarrow G^*$ be the projection relations and $E : G^* \times G^* \leftrightarrow G^* \times G^*$ the exchange relation. If we define

$$dominance(M, C) = \overline{(\pi; M^{\mathsf{T}} \cap \overline{C}^{\mathsf{T}}); \mathsf{L}} \cap (\pi; M^{\mathsf{T}} \cap E; \overline{C}^{\mathsf{T}}); \mathsf{L}, \qquad (24)$$

then we have for all $u = \langle g, h \rangle \in G^* \times G^*$ that $dominance(M, C)_u$ if and only if $g \succ h$, i.e., g dominates h.

The relation-algebraic specification $dominance(M, C)$ of the vector describing the dominance relationship between feasible governments immediately leads to the following RELVIEW-program.

```
dominance(M,C)
  DECL Prod = PROD(M^*M,M^*M);
       pi, rho, E
  BEG  pi = p-1(Prod);
       rho = p-2(Prod);
       E = [rho,pi]
       RETURN -dom(pi*M^ & -C^) & dom(pi*M^ & E*-C^)
  END.
```

In this program the first declaration introduces `Prod` as a name for the direct product $G^* \times G^*$. Using the relational product domain `Prod`, the two projection relations and the exchange relation are then computed by the three assignments of the body and stored as `pi`, `rho`, and `E`, respectively. The return-clause of the program consists of a direct translation of (24) into RELVIEW-notation, where `^`, `-`, `&`, and `*` denote transposition, complement, intersection, and composition, and, furthermore, the pre-defined operation `dom` computes for a relation $R : X \leftrightarrow Y$ the vector $R; \mathsf{L} : X \leftrightarrow \mathbf{1}$.

Finally, we consider stability of feasible governments. Due to the original definition of stability and the above result concerning dominance we have for all $g \in G^*$ the equivalence

$$stable(M, C)_g \iff \neg \exists h : dominance(M, C)_{\langle h,g \rangle}. \qquad (25)$$

In Berghammer et al. [11], it is shown how to transform this specification into the relation-algebraic specification

$$stable(M, C) = \overline{\rho^{\mathsf{T}}; dominance(M, C)}. \qquad (26)$$

Also a translation of the relation-algebraic specification of $stable(M, C)$ into RELVIEW-code is straightforward.

5.2   Example (Continued)

The above RELVIEW-program `dominance` expects two relations as inputs. In the following, we show for our running example how these can be obtained from the hitherto results, and also how then the dominance relation can be computed and visualized with the aid of the RELVIEW tool.

The first input M of the RELVIEW-program `dominance` is a description of government membership in the form of a relation of type $[DM \leftrightarrow G^*]$ that column-wisely enumerates the governments. In the case of our running example, it immediately is obtained from the list of governments of Subsection 2.2. Its RELVIEW-representation as 3×8 Boolean matrix is shown in Figure 1, where we additionally have labeled the rows and columns of the matrix with the parties and governments, respectively, for explanatory purposes.

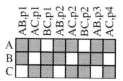

Fig. 1. Relational description of government membership

The second input is the global utility relation of type $[DM \leftrightarrow G^* \times G^*]$. It is constructed from the three utility relations $R^{(A)}, R^{(B)}, R^{(C)} : G^* \leftrightarrow G^*$ of the parties A, B, and C, respectively. The latter three relations are obtained immediately from Table 12 and the labeled 8×8 Boolean matrices representations look in RELVIEW as given in the following Figure 2.

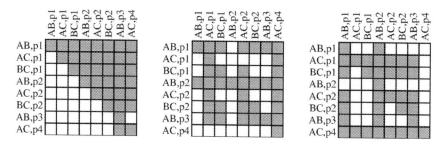

Fig. 2. The parties' Utility Relations

We renounce the RELVIEW-picture for the global utility relation, since the explanatory power of this 3×64 Boolean matrix is rather small. The same holds for the vector description (24) of the dominance relation. Instead we show in the following Figure 3 the dominance relation of the example as a labeled 8×8 Boolean matrix. For obtaining this matrix we used that the relation

$$R := \pi^{\mathsf{T}}; (\rho \cap v; \mathsf{L})$$

describes a vector $v : X \times Y \leftrightarrow \mathbf{1}$ as relation of type $[X \leftrightarrow Y]$, i.e., $v_{\langle x,y \rangle}$ and $R_{x,y}$ are equivalent for all $x \in X$ and $y \subset Y$ (where π and ρ are the projection relations of the direct product $X \times Y$). See Schmidt et al. [47] for details.

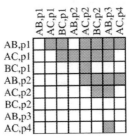

Fig. 3. The Dominance Relation

A representation of this relation as directed graph is shown in the following Figure 4. For drawing this graph, the RELVIEW tool used the specific layout algorithm of Gansner et al. [28].

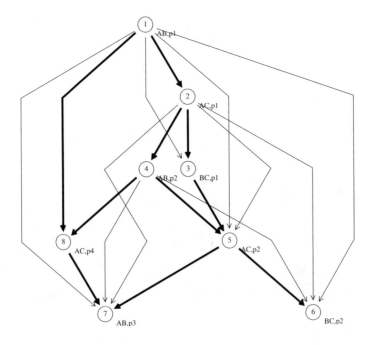

Fig. 4. Graphical Representation of the Dominance Relation

In this *dominance graph* we additionally have marked the immediate neighbourhood relationships as boldface arcs to make things more clear. From Figure 4 we immediately obtain $g_1 = (AB, p_1)$ as the only stable government of our example, since, by definition, a government is stable if and only if it is a source of the dominance graph.

6 Applying Social Choice Theory and Bargaining Theory

We will now address the question of how to proceed in cases where multiple stable governments exist. In such cases, social choice rules or bargaining theory may be applied.

6.1 Selection of Governments Via Social Choice Rules

The input for an application of social choice theory consists of: (at least two) selected governments (from which we have to choose one), parties forming these governments, and preferences of the parties over the governments. Moreover, for each government each party either accepts (approves of) or does not accept (disapproves of) it. We consider the following rules; see Subsection 7.2 for an illustration.

- *Plurality Rule*: Under this rule only the first preference of a party is considered. A government g is collectively preferred to a government h if the number of parties that prefer g most is greater than the number of parties that prefer h most. The government chosen under the plurality rule is the government which is put first by most parties.
- *Majority Rule*: This rule is based on the majority principle. A government g is collectively preferred to h if g defeats h, i.e., the number of parties that prefer g to h is greater than the number of parties that prefer h to g. If there is a government that defeats every other government in a pairwise comparison, this government is chosen, and it is called a Condorcet winner; see also Condorcet [19].
- *Borda Rule*: Here weights are given to all the positions of the governments in the individual preferences. For n governments, every party gives n points to its most preferred government, $n - 1$ points to its second preference, etc., and 1 point to its least preferred government. A decision is made based on the total score of every government in a given party profile; see also [20] for more details.
- *Approval Voting Rule*: Under Approval Voting (Brams and Fishburn [15]), each party divides the governments into two classes: the governments it approves of and the ones it disapproves of. Each time a government is approved of by a party is good for one point. The government chosen is the one that receives most points.

6.2 Selection of Governments Via Bargaining

Apart from the application of social choice rules, we may propose an alternative method for choosing a government. If there is more than one stable government, bargaining theory may be applied in order to choose one government. In Rusinowska and de Swart [43] (see also Berghammer et al. [12]), the authors define several bargaining games in which parties belonging to stable governments (assuming that there are at least two stable ones) bargain over the choice of one stable government. Subgame perfect equilibria of the games are investigated. Also a procedure for choosing the order of parties for a given game is proposed.

We define three kinds of bargaining games, denoted here as Games I, II, and III, in which parties involved in at least one stable government bargain over the choice of one government. The order of the parties in which they bargain is according to the number of seats in the parliament. The common assumptions for the bargaining games are as follows.

- A party, when submitting an offer, may propose only one government.
- The same offers are not repeated: a party cannot propose a government which has already been proposed before.
- It is assumed that choosing no government is the worst outcome for each party.

The differences between the three bargaining games are specified by the following four rules.

- In Game I, a party, when submitting an offer, may propose only a government the party belongs to. Each party involved in a proposed government either accepts of rejects the proposal. The acceptance of the offer by all parties involved causes the government to be formed. Rejection leads to proposing a government by the rejecting party.
- In Game II, a party does not have to belong to the government it proposes, and all parties have to react to each offer.
- In Game III, only the strongest party may submit an offer, and the other parties forming the proposed government have to react.

Our bargaining games differ from each other with respect to the bargaining procedures. We consider games in which a party prefers to form a government it likes most with a delay, rather than to form immediately (with no delay) a less preferred government. We refer to Subsection 7.2 for an illustration.

7   Applying Graph Theory to Coalition Formation

In this section we consider the case that there exists no stable government. Using graph-theoretical terminology this means that the computed dominance graph has no source. As we will show in the following, a combination of social choice rules, bargaining, and techniques from graph theory can be applied to select a government that can be considered as 'rather stable'.

7.1   Graph-Theoretical Procedure for Choosing a Government

First, we use *strongly connected components* (SCCs). A SCC of a directed graph is defined as a maximal set of vertices such that each pair of vertices is mutually reachable. In particular, we are interested in SCCs without arcs leading from outside into them. These SCCs are said to be *initial*. We also apply the concept of a minimum feedback vertex set, where a *feedback vertex set* (FVS) is a set of vertices that contains at least one vertex from every cycle of the graph. For computing the initial strongly connected components and minimum feedback vertex sets one may use the RELVIEW tool again, see Berghammer and Fronk [8], [9], and Berghammer et al. [12] for details.

We propose the following procedure for choosing a government in case there is no stable government (see also Berghammer et al. [12]):

(1) Compute the set $\mathfrak{I}$ of all initial SCCs of the dominance graph.
(2) For each SCC C from $\mathfrak{I}$ do:
    (a) Compute the set $\mathfrak{F}$ of all minimum FVSs of the subgraph generated by the vertices of C.
    (b) Select from all sets of $\mathfrak{F}$ with a maximal number of ingoing arcs one with a minimal number of outgoing arcs. We denote this one by F.
    (c) Break all cycles of C by removing the vertices of F from the dominance graph.
    (d) Select an un-dominated government from the remaining graph. If there is more than one candidate, use social choice rules or bargaining in order to choose one.
(3) If there is more than one set in $\mathfrak{I}$, select the final stable government from the results of the second step by applying social choice rules or bargaining again.

An outgoing arc of the dominance graph denotes that a government dominates another one and an ingoing arc denotes that a government is dominated by another one. Hence, the governments of an initial SCC can be seen as a cluster which is not dominated from outside. The application of the second step to such a set of 'candidates' corresponds to a removal of those candidates which are 'least attractive', because they are most frequently dominated and they dominate other governments least frequently.

According to the procedure just mentioned, if the application of steps (1) and (2) does not give a unique solution, we select the final government from among the 'graph-theoretical' results by applying again social choice rules or bargaining games.

7.2   Example (Continued)

The computation of the dominance graph of Figure 4 is based upon the values of columns 2 to 4 of Table 12. By changing our running example a little bit (viz. by rounding each value to the next natural number being a multiple of 5) the situation changes drastically. We obtain the dominance graph of Figure 5, that does not possess a source. In this RELVIEW-picture the subgraph induced by the only initial SCC (corresponding to the set $\{g_1, g_2, g_3, g_4, g_5\}$ of governments) is emphasized by black vertices and boldface arcs.

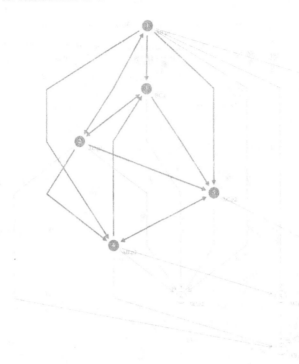

Fig. 5. Dominance relation and initial SCC after rounding

We have applied the procedure of Subsection 7.1 to obtain a government that can be considered as an approximation of a stable one. The next figure shows the two minimum FVSs of the initial SCC as presented on the RELVIEW screen:

AB,p1		
AC,p1	▨	▨
BC,p1		
AB,p2		▨
AC,p2	▨	
BC,p2		
AB,p3		
AC,p4		

Fig. 6. Minimum feedback vertex sets of the initial SCC

Each of the initial components possesses 3 ingoing arcs and the number of their outgoing arcs is also 3. If we select the minimum FVS represented by the first column of the matrix of Figure 6 in step (b) of our procedurs, then step (c) yields vertex 1 as source. A selection of the second column yields the same result. This shows that the stable government g_1 of the original example is rather 'robust' with respect to modifying the parties' utilities to a certain extent.

To demonstrate an application of the concepts of Section 6, we have changed our example again and used a still coarser scale for the utilities. It divides the values of Table 12 into four categories, viz. small (0 to 25), medium (26 to 50), large (51 to 75), and very large (76 to 100). Such a quatrigrade scale leads to the dominance graph depicted in Figure 7; in this RELVIEW-drawing again the only initial SCC is emphasized.

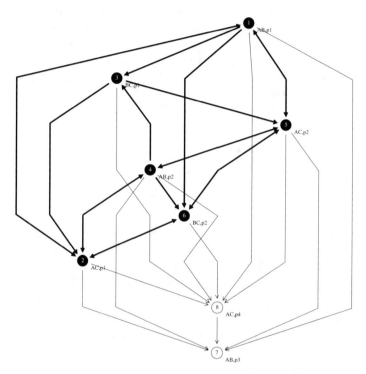

Fig. 7. Dominance relation and initial SCC after a coarser rounding

If we apply the procedure of Subsection 7.1 to this graph, we obtain vertices 2 and 5 as the only minimum FVS and their removal converts vertices 1 and 4 to sources. Hence, besides government g_1 now also government g_4 is a candidate for being selected as rather stable.

Let us apply the Plurality rule for the final selection. From the utility relations $R^{(A)}, R^{(B)}$, and $R^{(C)}$ of Figure 2 we obtain for the three parties A, B, and C the following preferences:

$$A : g_1 \text{ before } g_4, \qquad B : g_4 \text{ before } g_1, \qquad C : g_1 \text{ before } g_4.$$

Hence, government g_1 is put first by two parties whereas government g_4 is put first by one party only. This means that again g_1 is selected.

Alternatively, we can apply the bargaining games to this example. Since both governments g_1 and g_4 are formed by coalition AB, only parties A and B are involved in bargaining. Consequently, both Games I and II are the same. In Game I/II, with the order of parties (A, B), there is only one subgame perfect equilibrium, and it leads to the choice of government g_4 already in the first period of the game. Game I/II with the order of parties (B, A) has also one subgame perfect equilibrium, but it leads to the choice of government g_1 in the first period of the game. Let us note that being the first proposer in bargaining may be disadvantageous: when party A is the first proposer, the subgame perfect equilibrium gives g_4 which is worse for party A than government g_1. The same holds for party B being the first proposer: the subgame perfect equilibrium leads to government g_1 which is less attractive for party B than government g_4. When applying Game III, if party A is stronger than B (i.e., for instance, A has more seats in parliament than B), we get the same result as in Games I and II with the order (A, B). If party B is stronger than A, Game III gives the same result as in Games I and II with the order (B, A).

8   Consensus Reaching

In this section, we describe a procedure for a winning coalition to reach consensus on a policy in order to form a feasible government.

8.1   Consensus Reaching Within a Coalition

In what follows, we assume a kind of mediator, called the *chairman*, who does not belong to any party and is indifferent between all the parties. First of all, this chairman chooses the parties that should adjust their preferences if needed, and gives suggestions to the parties how they should change their preferences. Moreover, in case of any non-uniqueness, the chairman chooses one solution. Also, if a coalition seems to be unable to reach consensus, the chairman decides when to stop the process of consensus reaching within that coalition. If the attempts to reach consensus within a coalition fail, this means that the given coalition does not propose any government.

We propose the following procedure for consensus reaching within a winning coalition S; see also Eklund et al. [26]. Let G_S^* denote the *set of all feasible governments with $S \in W^*$ as coalition.* Each party $i \in S$ evaluates each government from G_S^* with respect to all the criteria. The notations here are similar to the ones presented in Subsection 2.1, except that we add the lower index S, since now the parties of coalition S only consider the governments formed by S. For each $i \in S$, we assume $u_{i,S} : C^* \times G_S^* \to [0, 1]$ such that

$$\forall c \in C^* : \sum_{g \in G_S^*} u_{i,S}(c, g) = 1. \tag{27}$$

The real number $u_{i,S}(c, g)$ is called the *value of government $g \in G_S^*$ to party $i \in S$ with respect to criterion $c \in C^*$.* Moreover, for each $i \in S$, we define $U_{i,S} : G_S^* \to [0, 1]$ such that

$$(U_{i,S}(g))_{g \in G_S^*} = (\alpha_i(c))_{c \in C^*} \cdot (u_{i,S}(c, g))_{c \in C^*, g \in G_S^*}, \tag{28}$$

where $(\alpha_i(c))_{c \in C^*}$ is the $1 \times |C^*|$ matrix representing the evaluation (comparison) of the criteria by party i, $(u_{i,S}(c, g))_{c \in C^*, g \in G_S^*}$ is the $|C^*| \times |G_S^*|$ matrix containing party i's evaluation (comparison) of all governments in G_S^* with respect to each criterion in C^*, and $(U_{i,S}(g))_{g \in G_S^*}$ is the $1 \times |G_S^*|$ matrix containing party i's evaluation of each government in G_S^*. Because of property (5) (with the set DM replaced by S) and (27) we have that

$$\sum_{g \in G_S^*} U_{i,S}(g) = 1. \tag{29}$$

Reaching consensus within a coalition means that the preferences of the parties from this coalition, as well as their evaluation of the importance of all criteria from C^*, should be relatively 'close' to each other. We specify this in detail. We define an assessment or 'distance' function $d_S : S \times S \to [0,1]$ satisfying the conditions $d_S(i,i) = 0$ and $d_S(i,j) = d_S(j,i)$ for all $i,j \in S$. In Eklund et al. [26], the authors consider the specific assessment function

$$d_S(i,j) = \sqrt{\frac{1}{|G_S^*|} \sum_{g \in G_S^*} (U_{i,S}(g) - U_{j,S}(g))^2}$$

but one may apply other assessment functions as well. Moreover, the *consensus degree between decision makers i and j in coalition S* is given by

$$\delta_S(i,j) = 1 - d_S(i,j). \tag{30}$$

The higher the consensus (degree), the smaller the 'distance' between pairs of decision makers, i.e., between i and j. In particular, if $d_S(i,j) = 0$, then we say that i and j are in *complete consensus in coalition S*. If $d_S(i,j) = 1$, then we say that i and j are in *complete disagreement in coalition S*. Moreover, we define

$$d_S^* = \max\{d_S(i,j) \mid i,j \in S\}, \tag{31}$$

and a *generalized consensus degree for coalition S* as

$$\delta_S^* = 1 - d_S^*, \tag{32}$$

which concerns the consensus reached by all the decision makers from S.

A certain consensus degree $0 < \tilde{\delta} < 1$ is required in the model. We say that coalition S reaches consensus if $\delta_S^* \geq \tilde{\delta}$. If $\delta_S^* < \tilde{\delta}$, then the chairman will ask at least one party to adjust its preferences. Any change of preferences leads to a new generalized consensus degree for the coalition.

Now, let D_S^* denote the set of all parties from S with most different preferences, that is, we have

$$D_S^* = \{i \in S \mid \exists j \in S \, [d_S(i,j) = d_S^*]\}. \tag{33}$$

The chairman decides which party from D_S^* will be advised to change its evaluation(s) regarding some government(s) and/or the importance of the criteria. The party $i_S^D \in D_S^*$ asked to adjust its preferences is a party such that

$$i_S^D = \arg\max_{i \in D_S^*} \sum_{j \in S} d_S(i,j). \tag{34}$$

If this party does not agree to adjust its evaluations according to the chairman's advice, the chairman may propose another change to the same party or a change to another party. Of course, this procedure of consensus reaching may consist of several steps.

Assuming that w_i is the weight of decision maker $i \in S$, we define the weighted value $U_S(g)$ of government $g \in G_S^*$ as

$$U_S(g) = \sum_{i \in S} w_i' \cdot U_{i,S}(g), \tag{35}$$

where

$$w_i' = \frac{w_i}{\sum_{j \in S} w_j}. \tag{36}$$

Finally, if the generalized (final) consensus degree is not smaller than $\tilde{\delta}$, the consensus government g_S^* formed by coalition S is chosen such that

$$g_S^* = \arg \max_{g \in G_S^*} U_S(g), \tag{37}$$

Of course, there may be more than one such government g_S^*. As noticed in Eklund et al. [26], any government g_S^* chosen by consensus reaching within coalition S is stable in G_S^*.

8.2   Example (Continued)

Consider coalition AB which has to choose from three policies p_1, p_2, p_3; p_4 is not acceptable to B. So AB has to choose from governments $\{g_1, g_4, g_7\}$; see Subsection 2.2. Suppose the weights of the three criteria for A are $\alpha_A = (1/3, 1/3, 1/3)$ and for B, $\alpha_B = (1/2, 1/4, 1/4)$ respectively. Also suppose that the matrices u_A and u_B of the utilities for A, respectively B, of the different governments with respect to the three criteria look as follows:

$$u_A = \begin{pmatrix} 1/2 & 1/4 & 1/4 \\ 1/4 & 1/2 & 1/4 \\ 1/4 & 1/4 & 1/2 \end{pmatrix} \text{ and } u_B = \begin{pmatrix} 1/4 & 1/2 & 1/4 \\ 1/4 & 1/2 & 1/4 \\ 1/4 & 1/4 & 1/2 \end{pmatrix}$$

Then $U_A = \alpha_A \cdot u_A = (1/3, 1/3, 1/3)$ and $U_B = \alpha_B \cdot u_B = (4/16, 7/16, 5/16)$. Hence,

$$d_{AB}^* = d_{AB}'(A, B) = \sqrt{\frac{1}{3}[(\frac{1}{3} - \frac{1}{4})^2 + (\frac{1}{3} - \frac{7}{16})^2 + (\frac{1}{3} - \frac{5}{16})^2]} = \frac{1}{48}\sqrt{2}.$$

Supposing that the required (generalized) consensus degree is $\frac{15}{16}$, the (generalized) consensus degree δ_{AB}^* for coalition AB, being $1 - \frac{1}{48}\sqrt{14}$, is too small. So, the chairman comes into play and suppose that after discussion he is able to convince party B to adjust its utilities as follows:

$$u_B' = \begin{pmatrix} 1/4 & 1/2 & 1/4 \\ 1/2 & 1/4 & 1/4 \\ 1/4 & 1/4 & 1/2 \end{pmatrix}$$

Then $U'_B = \alpha_B \cdot u'_B = (5/16, 6/16, 5/16)$ and consequently

$$d^*_{AB} = d'_{AB}(A, B) = \sqrt{\frac{1}{3}[(\frac{1}{3} - \frac{5}{16})^2 + (\frac{1}{3} - \frac{6}{16})^2 + (\frac{1}{3} - \frac{5}{16})^2]} = \frac{1}{48}\sqrt{2}.$$

Hence, the generalized consensus degree δ^*_{AB} becomes $1 - \frac{1}{48}\sqrt{2}$, which is larger than the required consensus degree of $\frac{15}{16}$. So, coalition AB reaches consensus. Assuming that each party has equal weight, we compute the utilities $U_{AB}(g)$ for coalition AB of each government $g \in \{g_1, g_4, g_7\}$ and we find that $U_{AB}(g_1) = \frac{1}{2}U_A(g_1) + \frac{1}{2}U_B(g_1) = \frac{1}{2}(1/3 + 5/16) = 31/96$, $U_{AB}(g_4) = \frac{1}{2}(1/3 + 6/16) = 34/96$ and $U_{AB}(g_7) = \frac{1}{2}(1/3 + 5/16) = 31/96$. Consequently, coalition AB will propose government g_4. Of course, it may happen that there is more than one government with a maximal utility for a given coalition, in which case the coalition may propose all these governments with maximal utility.

9 Conclusions

We used the MacBeth software in order to determine the utilities of policies to parties. Based on these utilities one can determine the feasible governments. Next we used the RELVIEW tool in order to calculate the stable governments. If there is more than one stable government we showed how social choice rules or bargaining may result in a particular choice. In case there is no stable government we used techniques from graph theory in order to choose a government which is as close as possible to being stable. We also indicated a procedure for a coalition to reach consensus about a policy, in order to propose a government.

Due to the MacBeth and RELVIEW software, our model of coalition formation seems to be applicable in practice. It could be helpful in the real world in order to form a stable government after elections in a rational way. It would be interesting to test the model in practice and to compare the outcome of the model with the actual outcome.

References

1. Austen-Smith D, Banks J (1988) Elections, coalitions, and legislative outcomes. *American Political Science Review* 82: 405-422
2. Axelrod R (1970) *Conflict of Interest; A Theory of Divergent Goals with Applications to Politics.* Chicago: Markham
3. Bana e Costa CA, Vansnick JC (1999) The MACBETH approach: basic ideas, software and an application. In: Meskens N, Roubens M (Eds.), *Advances in Decision Analysis*, Kluwer Academic Publishers, Dordrecht, pp. 131-157
4. Bana e Costa CA, Vansnick JC (2001) A fundamental criticism to Saaty's use of the eigenvalue procedure to derive priorities. LSE OR Working Paper 01.42, 2001. Downloadable via internet
5. Bana e Costa CA, De Corte JM, Vansnick JC (2003) MACBETH, LSE OR Working Paper 03.56

6. Baron D (1993) Government formation and endogenous parties. *American Political Science Review* 87: 34-47
7. Behnke R, Berghammer R, Meyer E, Schneider P (1998) RELVIEW – A system for calculation with relations and relational programming. In: Astesiano E (Ed.), Proc. Conf. "Fundamental Approaches to Software Engineering (FASE '98)", LNCS 1382, Springer, pp. 318-321
8. Berghammer R, Fronk A (2004) Considering design tasks in OO-software engineering using relations and relation-based tools. *Journal on Relational Methods in Computer Science* 1: 73-92
9. Berghammer R, Fronk A (2005) Exact computation of minimum feedback vertex sets with relational algebra. Fundamenta Informaticae 70 (4): 301-316
10. Berghammer R, Leoniuk B, Milanese U (2002) Implementation of relational algebra using binary decision diagrams. In: de Swart H (Ed.) Proc. 6th Int. Workshop "Relational Methods in Computer Science", LNCS 2561, Springer, pp. 241-257
11. Berghammer R, Rusinowska A, de Swart H (2005) Applying relational algebra and RELVIEW to coalition formation. *European Journal of Operational Research* (forthcoming)
12. Berghammer R, Rusinowska A, de Swart H (2005) Graph theory, RELVIEW and coalitions. Submitted
13. Berghammer R, Schmidt G, Winter M (2003) RELVIEW and RATH – Two systems for dealing with relations. In: [22], pp. 1-16
14. Berghammer R, von Karger B, Ulke C (1996) Relation-algebraic analysis of Petri nets with RELVIEW. In: Margaria T, Steffen B (Eds.) Proc. 2nd Workshop "Tools and Applications for the Construction and Analysis of Systems (TACAS '96)", LNCS 1055, Springer, pp. 49-69
15. Brams SJ, Fishburn PC (1983) *Approval Voting*. Birkhäuser, Boston
16. Brams SJ, Fishburn PC (2002) Voting procedures. In: Kenneth Arrow, Amartya Sen and Kotaro Suzumura (Eds.) *Handbook of Social Choice and Welfare*. Elsevier Science, Amsterdam
17. Brink C, Kahl W, Schmidt G (Eds.) (1997) *Relational Methods in Computer Science*, Advances in Computing Science, Springer
18. Carlsson C, Ehrenberg D, Eklund P, Fedrizzi M, Gustafsson P, Merkuryeva G, Riissanen T, Ventre A (1992) Consensus in distributed soft environments. *European Journal of Operational Research* 61: 165-185
19. Condorcet, *Sur les Elections et autres Textes*. Corpus des Oeuvres de Philosophie en Langue Française. Librairie Artheme Fayard, Paris, 1986
20. De Borda J-C (1781) Mémoire sur les Elections au Scrutin. English translation by De Grazia A, Mathematical Derivation of an Election System. *Isis*, 44 (1953) 42-51
21. de Swaan A (1973) *Coalition Theories and Cabinet Formations*. Elsevier, Amsterdam: North Holland
22. de Swart H, Orlowska E, Schmidt G, Roubens M (Eds.) (2003) *Theory and Applications of Relational Structures as Knowledge Instruments*. Springer's Lecture Notes in Computer Science, LNCS 2929, Springer, Heidelberg, Germany
23. de Swart H, van Deemen A, Van der Hout E, Kop P (2003) Categoric and ordinal voting: an overview. In: [22], pp. 147-195
24. De Vries M (1999) *Governing with your closest neighbour: An assessment of spatial coalition formation theories*, Ph.D. Thesis, Print Partners Ipskamp
25. Eklund P, Rusinowska A, de Swart H (2004) Consensus reaching in committees. *European Journal of Operational Research* (forthcoming)
26. Eklund P, Rusinowska A, de Swart H (2004) A consensus model of political decision-making. Submitted

27. Fishburn PC, Rubinstein A (1982) Time preferences. *International Economic Review* 23: 667-694
28. Gansner E, Koutsofios E, North C, Vo K (1993) *A technique for drawing directed graphs*, IEEE Trans. Software Eng. 19: 214-230
29. Grofman B (1982) A dynamic model of protocoalition formation in ideological n-space. *Behavioral Science* 27: 77-90
30. Kahan J, Rapoport A (1984) *Theories of Coalition Formation*. Lawrence Erlbauw Associates Publishers
31. Laver M, Schofield N (1990) *Multiparty Government; The Politics of Coalition in Europe*, Oxford: Oxford University Press
32. Laver M, Shepsle KA (1990) Coalitions and cabinet government. *American Political Science Review* 3: 873-890
33. Laver M, Shepsle K (1996) *Making and Breaking Governments; Cabinet and Legislatures in Parliamentary Democracies*. Cambridge: Cambridge University Press
34. Leiserson M (1966) *Coalitions in Politics: A Theoretical and Empirical Study*, Doctoral Dissertation, University Microfilms, Inc., Ann Arbor, Michigan
35. Leiserson M (1968) Factions and coalition in one-party Japan: an interpretation based on the theory of games, *American Political Science Review* 62: 770-787
36. McKelvey R, Ordeshook P, Winer M (1978) The competitive solution for n-person games without transferable utility with an application to committee games. *American Political Science Review* 72: 599-615
37. Osborne MJ, Rubinstein A (1990) *Bargaining and Markets*. Academic Press, San Diego
38. Peleg B (1980) A theory of coalition formation in committees. *Journal of Mathematical Economics* 7: 115-134
39. Peleg B (1981) Coalition formation in simple games with dominant players. *International Journal of Game Theory* 10: 11-33
40. Riker WH (1962) *The Theory of Political Coalitions*, New Haven/London: Yale University Press
41. Roubens M, Rusinowska A, de Swart H (2006) Using MACBETH to determine utilities of governments to parties in coalition formation. *European Journal of Operational Research 172/2:588-603*
42. Rubinstein A (1982) Perfect equilibrium in a bargaining model. *Econometrica* 50: 97-109
43. Rusinowska A, de Swart H (2004) Negotiating a stable government - an application of bargaining theory to a coalition formation model. Submitted
44. Rusinowska A, de Swart H, Van der Rijt JW (2005) A new model of coalition formation. *Social Choice and Welfare* 24: 129-154
45. Saaty TL (1977) A scaling method for priorities in hierarchical structures. *Journal of Mathematical Psychology* 15: 234-281
46. Saaty TL (1980) *The Analytic Hierarchy Process*, McGraw-Hill
47. Schmidt G, Ströhlein T (1993) Relations and Graphs. *Discrete Mathematics for Computer Scientists*, EATCS Monographs on Theoret. Comput. Sci., Springer
48. Schofield N (1993) Political competition and multiparty coalition governments. *European Journal of Political Research* 23: 1-33
49. Schofield N (1993) Party competition in a spatial model of coalition formation. In: Barnett W, Hinich M, Schofield N (Eds.) *Political Economy; Institutions, Competition and Representation*. Cambridge: Cambridge University Press
50. Schofield N (1995) Coalition politics; A formal model and empirical analysis. *Journal of Theoretical Politics* 7: 245-281

51. Selten R (1975) Reexamination of the perfectness concept for equilibrium points in extensive games. *International Journal of Game Theory* 4: 25-55
52. Shepsle KA (1979) Institutional arrangements and equilibrium in multidimensional voting models. *American Journal of Political Science* 23: 27-59
53. van Deemen A (1991) Coalition formation in centralized policy games. *Journal of Theoretical Politics* 3: 139-161
54. van Deemen A (1997) *Coalition Formation and Social Choice*, Kluwer
55. von Neumann J, Morgenstern O (1944) *Theory of Games and Economic Behavior*, Princeton: Princeton University Press

Investigating Finite Models of Non-classical Logics with Relation Algebra and RELVIEW[*]

Rudolf Berghammer[1] and Renate A. Schmidt[2]

[1] Institut für Informatik, Christian-Albrechts-Universität Kiel
Olshausenstraße 40, 24098 Kiel, Germany
`rub@informatik.uni-kiel.de`
[2] School of Computer Science, University of Manchester
Oxford Road, Manchester M13 9PL, United Kingdom
`Renate.Schmidt@manchester.ac.uk`

Abstract. In computer science, scenarios with interacting agents are often developed using modal logic. We show how to interpret modal logic of knowledge in relation algebra. This allows the use of the RELVIEW tool for the purpose of investigating finite models and for visualizing certain properties. Our approach is illustrated with the well-known 'muddy children' puzzle using modal logic of knowledge. We also sketch how to treat other non-classical logics in this way. In particular, we explore our approach for computational tree logic and illustrate it with the 'mutual exclusion' example.

.1 Introduction

For some time now researchers in computer science have been interested in reasoning about knowledge in multi-agent systems. Here a group of interacting agents is given and it is assumed that each agent takes into account not only facts that are true about the world, but also the knowledge of other agents. Applications of this scenario can be found in many domains of computer science, for instance in distributed computing, cryptography, and robotics.

The idea of using modal logic for reasoning about knowledge goes back to J. Hintikka and has been worked out in great detail, e.g. in the textbooks [9,12,19]. The standard semantics of modal logic is based on the agents' accessibility relations on a global set of possible worlds. In this paper we adopt an algebraic perspective. Relation algebra, and more generally Boolean algebras with operators, provide natural settings for studying modal logics and other kinds of non-classical logics, cf. [2,6,13,23] for example. A sufficient framework for interpreting modal logic of knowledge is dynamic algebra [16,24]. However in this paper we interpret modal logics in the more expressive setting of heterogeneous relation algebras with transitive closure (see [22,25,26]) and their

---

[*] The authors thank Harrie de Swart and the anonymous referees for their comments. The work was supported by EU COST Action 274 (Tarski).

H. de Swart et al. (Eds.): TARSKI II, LNAI 4342, pp. 31–49, 2006.

representation as Boolean matrices [25]. Representing sets (respectively, predicates on sets) by specific relations, viz. vectors, relation-algebraic specifications can be evaluated by calculations on Boolean matrices and vectors, and properties of relations can be verified in this way. Hence, the relation-algebraic manipulation and visualization system RELVIEW [1,20,4] can be applied for the purpose of model checking and similar tasks. It turns out that this can be achieved with very little effort and that the approach can be transferred to other important non-classical logics, which are embeddable into the programming language of RELVIEW, such as temporal logic which we consider in this paper but also Peirce logic and description logic.

The case study in this paper explores a novel application of the RELVIEW tool for which it was not originally designed. The application may be of interest to researchers working in the area of modal logics, since to our knowledge, there seem to be very few tools available for solving and visualizing computational problems of finite models in modal logic. One of the uses of RELVIEW we explore is its use as a finite model checker. However we do not claim any superiority of the system over existing implemented model checking systems such as MCMAS [18] and VERICS [14,15]). Sophisticated model checking tools which have been developed for computational tree logic, linear temporal logic, and the process algebra CSP include SPIN, SMV, KRONOS, UPPALL, and FDR2. Because of the global approach that RELVIEW takes, it cannot compete directly with systems based on local evaluations. Nevertheless, the underlying technology of RELVIEW is based on reduced, ordered BDDs which are fast [17,3,20]. Furthermore, the tool has a convenient graphical user interface and provides useful capabilities for manipulating and displaying relations and graphs. Particularly attractive in the context of modal logic is the presence of the operator **trans** for computing transitive closures in the tool's programming language. This is useful for performing finite model reasoning tasks for a modal logic with the common knowledge operator and also for dynamic logic. Such logics cannot be handled directly for example by first-order logic theorem provers since the transitive closure operator and the common knowledge operator are not first-order definable.

The remainder of the paper is organized as follows. Some basic notions of modal logic and modal logic of knowledge are recalled in Sections 2 and 3. Section 4 describes how to interpret modal logic of knowledge in relation algebra and how then the RELVIEW tool can be used for solving computational problems on finite models. The application of the approach to the well-known 'muddy children' puzzle is presented in Section 5. This example also demonstrates how RELVIEW can be used for visualizing models, and solutions of tasks. Our method can be extended to all non-classical logics, embeddable into the programming language of RELVIEW. Section 6 features the approach for computational tree logic and the 'mutual exclusion' example in more detail. In Section 7 some further applications of relation algebra and RELVIEW in the context of modal logic are considered. Finally, Section 8 concludes with some further remarks about the approach and the use of RELVIEW.

2 Modal Logic

The language of (propositional) modal logic with multiple modalities is defined over countably many propositional variables $p_1, p_2, p_3, \ldots$, and finitely many modalities $\Diamond_1, \ldots, \Diamond_n$, one for each agent $1, \ldots, n$. A *propositional atom* is a propositional variable or the constant $\top$ (the symbol for 'true') and a *modal formula* is either a propositional atom or a formula of the form $\neg\phi$, $\phi \wedge \psi$, and $\Diamond_i\phi$. We define the constant $\bot$ (the symbol for 'false') and the other propositional connectives $\vee$, $\rightarrow$, and $\leftrightarrow$ as usual, e.g. $\phi \rightarrow \psi := \neg\phi \vee \psi$. Furthermore, the dual operator of $\Diamond_i$ is defined by $\Box_i\phi := \neg\Diamond_i\neg\phi$.

The standard semantics of modal logic is given by the well-known *Kripke semantics* (or *possible world semantics*). A *frame* (or *relational structure*) for a modal logic is a pair $\mathcal{F} = (W, \{R_1, \ldots, R_n\})$, where W is a non-empty set of worlds and each R_i is a binary relation over W. W is the set of possible worlds (or states) in which the truth of formulae is evaluated. The R_i are the accessibility relations which determine the formulae deemed possible by an agent i in a given world $(1 \leq i \leq n)$. A *model* is a pair $M = (\mathcal{F}, \iota)$ of a frame $\mathcal{F}$ and a valuation function ι from the set of propositional variables to 2^W, where $\iota(p_i)$ is interpreted to be the set of worlds in which p_i is true. The *truth* of a modal formula in a world x of a model M is defined as follows (where the notation $R_i(x, y)$ means that the elements x and y are related via the relation R_i).

$$M, x \models \top$$
$$M, x \models p_i \;:\Longleftrightarrow\; x \in \iota(p_i)$$
$$M, x \models \neg\phi \;:\Longleftrightarrow\; M, x \not\models \phi$$
$$M, x \models \phi \wedge \psi \;:\Longleftrightarrow\; M, x \models \phi \text{ and } M, x \models \psi$$
$$M, x \models \Diamond_i\phi \;:\Longleftrightarrow\; \exists y \in W : R_i(x, y) \text{ and } M, y \models \phi$$

If $M, x \models \phi$ we also say that x satisfies ϕ. A modal formula is *valid* in a model M iff the formula is true in every world of M. It is valid in a frame $\mathcal{F}$ iff it is valid in all models based on the frame, i.e. in all models $(\mathcal{F}, \iota)$.

For the purposes of this paper it suffices to consider modal logic from a semantic perspective. (The reader interested in the axiomatizations of the considered logics should refer to standard textbooks, e.g. [5,7,10,11].) A modal logic L is said to be *sound* (respectively *complete*) *with respect to a class of frames* iff for any modal formula ϕ, any frame in the class validates ϕ if (respectively iff) ϕ is a theorem in L. A modal logic is said to be *complete* iff it is complete with respect to some class of frames.[1]

The basic multi-modal logic $K_{(m)}$ is complete with respect to the class of all frames. The table in Figure 1 lists the relation-algebraic correspondence properties satisfied by classes of frames for extensions of the basic logic $K_{(m)}$. This means, if L denotes an extension of the basic logic $K_{(m)}$ with a subset of the common axioms listed in the table then L is a logic (sound and) complete with

---

[1] Note in modal logic the notion of completeness is used differently than in other logical disciplines.

Axiom		Correspondence property	
T	$\Box_i p \rightarrow p$	reflexivity	$\mathsf{I} \subseteq R_i$
4	$\Box_i p \rightarrow \Box_i \Box_i p$	transitivity	$R_i; R_i \subseteq R_i$
B	$\Diamond_i \Box_i p \rightarrow p$	symmetry	$R_i \subseteq R_i{}^\mathsf{T}$
D	$\Box_i p \rightarrow \Diamond_i p$	seriality	$\mathsf{L} \subseteq R_i; \mathsf{L}$
alt_1	$\Diamond_i p \rightarrow \Box_i p$	functionality	$R_i{}^\mathsf{T}; R_i \subseteq \mathsf{I}$
5	$\Diamond_i \Box_i p \rightarrow \Box_i p$	Euclideanness	$R_i{}^\mathsf{T}; R_i \subseteq R_i$

Fig. 1. Modal axioms and their frame correspondence properties

respect to the class of all frames which satisfy each of the corresponding properties. In the table, I denotes the identity relation and L denotes the universal relation. Furthermore, $R; R$ denotes the composition of R with itself and R^T the transpose (converse) of R. Other relation-algebraic constructions used in this paper are the empty relation O, the Boolean constructs $R \cup S$ (union), $R \cap S$ (intersection), $\overline{R}$ (complement), and the transitive closure $R^+ := \bigcup_{k \geq 1} R^k$ of R. Here we assume powers are defined inductively by $R^0 := \mathsf{I}$ and $R^{k+1} := R; R^k$ for $k \geq 0$.

3   Modal Logic of Knowledge

Modal logic lends itself to formalize informational aspects of agent-based scenarios. Consider the language defined in Section 2 in which $\Box_i \phi$, from now on written $K_i \phi$, is interpreted as 'the agent i knows that property ϕ is the case'. For this reading it is usual to assume that the following axioms of the table in Figure 1 are valid: T (axiom of true knowledge), 4 (agents are positively introspective) and 5 (agents are negatively introspective). The accessibility relations R_i associated with the knowledge operators K_i are therefore equivalence relations on the set of worlds W (because each R_i is reflexive and transitive and $R_i^\mathsf{T} = R_i^\mathsf{T}; \mathsf{I} \subseteq R_i^\mathsf{T}; R_i \subseteq R_i$ shows symmetry).

In order to handle the common knowledge of a group of agents two additional modal operators, E_G and C_G, are required. Let G denote a finite set of agents. Then the modal formula $E_G \phi$ is read to mean that 'each of the agents in G knows that ϕ is the case', and the modal formula $C_G \phi$ is read to mean that 'it is common knowledge among the group G of agents that ϕ is the case'. Their semantics is defined by the following equivalences, where $E_G^k \phi$ is an abbreviation of the modal formula $E_G \ldots E_G \phi$ with k occurrences of the operator E_G.

$$M, x \models E_G \phi \quad :\Longleftrightarrow \quad \forall i \in G : M, x \models K_i \phi$$
$$M, x \models C_G \phi \quad :\Longleftrightarrow \quad \forall k \geq 1 : M, x \models E_G^k \phi$$

If $G = \{i_1, \ldots, i_m\}$, then we have the following equivalence.

$$M, x \models E_G \phi \quad \Longleftrightarrow \quad M, x \models K_{i_1} \phi \wedge \ldots \wedge K_{i_m} \phi$$

Thus, the formula $E_G\phi$ is true in a world of a model iff everyone in the group knows that ϕ is true. Furthermore, the formula $C_G\phi$ is true iff everyone in the group knows that ϕ is true and everyone in the group knows that everyone in the group knows that ϕ is true, and so on. The following three properties are not difficult to show for any model M and any world x of M. We assume that R is the union of the accessibility relations R_i for all $i \in G$, i.e. $R := \bigcup_{i \in G} R_i$.

$$M, x \models C_G\phi \iff M, x \models E_G(\phi \land E_G C_G \phi)$$
$$M, x \models E_G^k\phi \iff \forall\, y \in W : R^k(x, y) \text{ implies } M, y \models \phi$$
$$M, x \models C_G\phi \iff \forall\, y \in W : R^+(x, y) \text{ implies } M, y \models \phi$$

Distributed knowledge is another concept central to modal logics of knowledge. Here a group of agents can deduce a formula by pooling their knowledge together. Since this distributed knowledge is not used in the 'muddy children' puzzle of Section 5, we omit the technical details and refer to the textbooks cited in Section 2. Relation algebra does however allow us to model distributed knowledge by using the same techniques which we apply in the next section to model the modal logic of common knowledge.

4 Relational Model Checking

The term 'model checking' refers to automatic model-based verification approaches; see e.g., [21,8]. In the case of modal logic it involves solving tasks of the following kind. Suppose that $M = (\mathcal{F}, \iota)$ is a given *finite model*, where the frame is $\mathcal{F} = (W, \{R_1, \ldots, R_n\})$, and ϕ is a given modal formula.

(1) Determine whether ϕ is true in a given world of M (satisfiability in a given world of a model).

(2) Determine whether there is a world of M in which ϕ is true (satisfiability in a model).

(3) Determine whether ϕ is true in all worlds of M (global satisfiability/validity in a model).

(4) Determine the set of all worlds of M in which ϕ is true.

In this paper, we use relation algebra and the RELVIEW tool to compute the set of all worlds of M in which ϕ is true (i.e. to solve task (4)). This immediately leads to solutions of tasks (1)–(3), too.

Our solution is based on the representation of sets of worlds by so-called *vectors* over W. Such vectors are relations with W as the domain and a singleton set, $\{\bullet\}$ say, as the range. Since this specific range is irrelevant, in the following we omit for a vector v the second argument and write $v(x)$ instead of $v(x, \bullet)$. A vector v over W can be viewed as a Boolean column vector and *represents the set* $\{x \in W \mid v(x)\}$ of worlds.

Suppose we wish to describe an arbitrary modal formula ϕ via the vector of worlds in which it is true, that is, we want to compute the vector v_ϕ representing the set $\{x \in W \mid M, x \models \phi\}$. We start by defining for the constant $\top$ the vector $v_\top$ as the universal vector L over W (the universal relation with domain W and range $\{\bullet\}$). Then for each propositional variable p in ϕ we define a vector v_p representing the set $\iota(p)$. Using Boolean vector terminology, the latter means that we set the x-component of v_p to 1 if $x \in \iota(p)$ and we set it to 0 if $x \notin \iota(p)$. Due to the first two cases of the definition of truth in Section 2, the vector $v_\top$ represents the set $\{x \in W \mid M, x \models \top\}$ and the vector v_p represents the set $\{x \in W \mid M, x \models p\}$ for every propositional variable p in ϕ. Based on these facts, we then obtain the vector v_ϕ which we are looking for by recursively applying the following properties.

$$v_{\neg\psi} = \overline{v_\psi} \qquad v_{\psi \wedge \rho} = v_\psi \cap v_\rho \qquad v_{\Diamond_i\psi} = R_i; v_\psi$$

The proofs of these equations for arbitrary ψ and ρ use the remaining three cases of the definition of truth in Section 2 and the definition of relational complement, intersection, and composition. E.g., $v_{\Diamond_i\psi} = R_i; v_\psi$ holds since for all $x \in W$

$$\begin{aligned}
(R_i; v_\psi)(x) &\iff \exists y \in W : R_i(x, y) \text{ and } v_\psi(y) \\
&\iff \exists y \in W : R_i(x, y) \text{ and } M, y \models \psi \\
&\iff M, x \models \Diamond_i\psi.
\end{aligned}$$

It is obvious from the above equations, how to get the vectors for the constant $\bot$ and the other propositional connectives $\vee$, $\to$ and $\leftrightarrow$. A little reflection yields the vectors for the dual operators K_i (or $\Box_i$). With the help of the properties of Section 3 we, finally, obtain the vector-representation for the remaining modal operators E_G and C_G, too.

We present only the results for the dual operators K_i and the common knowledge operators E_G, and C_G. Here we have:

$$v_{K_i\psi} = \overline{R_i; \overline{v_\psi}} \qquad v_{E_G\psi} = \overline{(\bigcup_{i\in G} R_i); \overline{v_\psi}} \qquad v_{C_G\psi} = \overline{(\bigcup_{i\in G} R_i)^+; \overline{v_\psi}}$$

A proof of the first equation is

$$v_{K_i\psi} = v_{\neg\Diamond_i\neg\psi} = \overline{v_{\Diamond_i\neg\psi}} = \overline{R_i; v_{\neg\psi}} = \overline{R_i; \overline{v_\psi}} \ .$$

The second equation follows from the calculation

$$v_{E_G\psi} = v_{\bigwedge_{i\in G} K_i\psi} = \bigcap_{i\in G} v_{K_i\psi} = \bigcap_{i\in G} \overline{R_i; \overline{v_\psi}} = \overline{\bigcup_{i\in G} R_i; \overline{v_\psi}} = \overline{(\bigcup_{i\in G} R_i); \overline{v_\psi}} \ .$$

A simple induction shows that $v_{E_G^k\psi} = \overline{(\bigcup_{i\in G} R_i)^k; \overline{v_\psi}}$ for all $k \geq 1$. This property is used in the following proof of the third equation.

$$v_{C_G\psi} = \bigcap_{k\geq 1} v_{E_G^k\psi} = \bigcap_{k\geq 1} \overline{(\bigcup_{i\in G} R_i)^k; \overline{v_\psi}} = \overline{\bigcup_{k\geq 1}\bigcup_{i\in G} (R_i)^k; \overline{v_\psi}} = \overline{(\bigcup_{i\in G} R_i)^+; \overline{v_\psi}}$$

All the constructs of relation algebra we have used up to now are available in the programming language of the RELVIEW tool. More specifically, we have the

RELVIEW-operators – for complementation (prefix operator), ˆ for transposition (postfix operator), |, &, and * for union, intersection, and composition (infix operators), and **trans** for transitive closure (a pre-defined relational function). Furthermore, the tool allows for the definition of relational functions by the user. For instance, the box operators $\square_i$ can be modelled by the following binary RELVIEW-function **box**.

$$\texttt{box(S,v) = -(S * -v)}$$

Here S denotes a RELVIEW-relation (a Boolean matrix) and v a RELVIEW-vector. Consider the modal formula ϕ defined as follows.

$$K_1 p \wedge K_1 \neg K_2 K_1 p$$

In words, the formula ϕ says that agent 1 knows p and, furthermore, that agent 1 knows that agent 2 does not know agent 1 knows p. The vector-representation v_ϕ of the set of worlds in which ϕ is true is computed by RELVIEW as the result of the evaluation of the expression

$$\texttt{box(R1,p) \& box(R1,-box(R2,box(R1,p))).}$$

Here it is assumed that the accessibility relations R_1, R_2 of M and the vector v_p are stored in the tool's workspace under the names R1, R2, and p.

5   Example: The Muddy Children Puzzle

By way of the well-known 'muddy children' puzzle we now illustrate the support provided by the RELVIEW tool for solving certain problems on finite models of modal logic. Our description of the puzzle follows [9].

A group of n children play together. A number of them happen to get mud on their foreheads. Each child can see another child's forehead but it cannot see its own forehead. Since no child will tell another child whether it has mud on the forehead, the puzzle is the following. *Can a child know that it has mud on its own forehead?* Obviously, without any extra information the answer is no. But now the father comes onto the scene. He says for all to hear, that 'at least one of you has mud on your forehead'. He then asks the children over and over again: 'Do you know whether you have mud on your forehead?' with the instruction that the children have to answer the question simultaneously. Suppose the number of children with mud on their foreheads is k. Then in the first $k - 1$ rounds, the father asks the question all children will answer 'no'. However, in the kth round exactly the children with muddy foreheads will answer 'yes'; the remaining will answer 'no'.

This puzzle can be modelled and solved within the modal logic of knowledge defined in Section 3. The common knowledge operator C_G is particularly crucial for the solution.

As a concrete example of the 'muddy children' puzzle, in the following we elaborate an instance of the problem with three children. The possible states

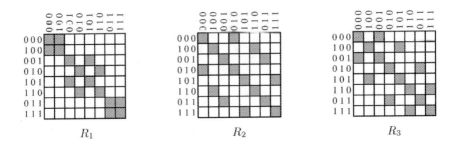

Fig. 2. The accessibility relations in the case of three children

(worlds) of the model are given by triples of 0's and 1's, where (s_1, s_2, s_3) is the state in which child i has mud on its forehead iff $s_i = 1$ and is clean iff $s_i = 0$ ($1 \leq i \leq 3$). The model, hence, consists of 8 states representing all combinations of associating 0 or 1 with the three children. Let us now consider what each child knows in a given state. For instance, in the state $(1, 0, 1)$ child 1 sees the foreheads of child 2 and child 3 but not its own, it therefore knows that child 2 does not have a muddy forehead but child 3 does. Initially the child does not know if its own head is muddy. Hence, $(0, 0, 1)$ and $(1, 0, 1)$ are the only possible successor states of the state $(1, 0, 1)$ with respect to the accessibility relation R_1. Similar considerations apply to the other children and states.

The three pictures in Figure 2 show the accessibility relations R_1, R_2, and R_3. This is how RELVIEW displays the relations as Boolean matrices (with labeled rows and columns). A black square in the matrix R_i means that the corresponding states are related via this relation and a white square means that they are not related. E.g., the above considerations on the knowledge of child 1 in the state $(1, 0, 1)$ correspond to the two black squares in the fifth row of R_1.

Suppose that the relation R is the union of the three accessibility relations R_1, R_2, and R_3. In Figure 3 it is shown how the RELVIEW tool displays the irreflexive part $R \cap \bar{I}$ of R as a labelled graph. This graph is the disjoint union of three subgraphs. These correspond to the possibilities of child 1 (boldface arcs), child 2 (dotted arcs), and child 3 (remaining arcs), but neglecting all self-loops. (We have omitted the self-loops in order to avoid cluttering in the graph.)

Now, we assume the propositional variable p_i, $1 \leq i \leq 3$, denotes that 'child i has mud on its forehead'. Then RELVIEW depicts the vector v_{p_i} representing the set $\iota(p_i)$ as a Boolean column vector as in Figure 4, where we have again used the tool's labeling mechanism to enhance understandability.

The three accessibility relations and these three vectors (Figures 2 and 4) provide a complete specification of the model M which we use as input to RELVIEW. We assume that these are stored in the tool's workspace under the names R1, R2, R3 and p1, p2, p3. Furthermore, we use the relational function box of Section 3.

In order to determine satisfiability of a formula ϕ in a state or set of states all that is required is to let RELVIEW evaluate the expression corresponding

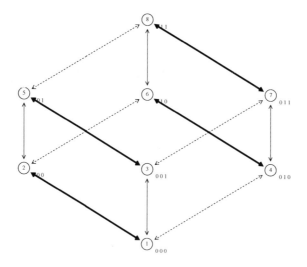

Fig. 3. Graphical representation of the accessibility relations

to ϕ, since this returns the set of worlds/states in which the formula is true as a vector. As first examples consider the following two statements.

$$M, x \models \neg K_1(p_1 \vee p_2) \qquad M, x \models K_1(p_2 \vee K_3 p_1)$$

The formula on the left says that child 1 does not know whether it or child 2 is muddy and the formula on the right says that child 1 knows that child 2 is muddy or that child 3 knows that child 1 is muddy. The RELVIEW-expressions representing the modal formulae $\neg K_1(p_1 \vee p_2)$ and $K_1(p_2 \vee K_3 p_1)$ are -box(R1,p1 | p2) respectively box(R1,p2 | box(R3,p1)). Evaluating these two expressions with the tool yields the vectors in Figure 5.

The labelling of the rows is as in Figure 4. Hence, the interpretation of the vectors is that $\neg K_1(p_1 \vee p_2)$ is true in the states $(0, 0, 0)$, $(1, 0, 0)$, $(0, 0, 1)$ and $(1, 0, 1)$ and $K_1(p_2 \vee K_3 p_1)$ is true in all states except $(0, 0, 0)$, $(1, 0, 0)$, $(0, 0, 1)$ and $(1, 0, 1)$. As a consequence, the statement

$$M, x \models K_1(p_1 \vee p_2) \leftrightarrow K_1(p_2 \vee K_3 p_1)$$

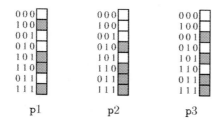

Fig. 4. The vectors for 'child i has mud on its forehead'

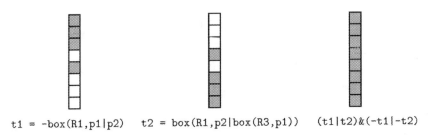

t1 = -box(R1,p1|p2)   t2 = box(R1,p2|box(R3,p1))   (t1|t2)&(-t1|-t2)

Fig. 5. Satisfiability of $t_1 = \neg K_1(p_1 \vee p_2)$, $t_2 = K_1(p_2 \vee K_3 p_1)$ and $\neg t_1 \leftrightarrow t_2$

is true for all states x of the model M, which menas that its formula is valid in M. This can be easily determined with the aid of RELVIEW by evaluating the expression (t1 | t2) & (-t1 | -t2) (i.e. $\neg$t1 $\leftrightarrow$ t2) where t1 and t2 denote $\neg K_1(p_1 \vee p_2)$ and $K_1(p_2 \vee K_3 p_1)$, respectively. This produces the universal vector which confirms that the equivalence is valid. In words, the equivalence says that child 1 knows that itself or child 2 is muddy iff it knows that either child 2 is muddy or that child 3 knows that child 1 is muddy.

The next example involves the common knowledge operator C_G. Consider the following statement.

$$M, x \models C_{\{1,2,3\}}(p_2 \rightarrow K_1 p_2)$$

Because $\mathsf{v}_{C_G \psi} = \overline{(\bigcup_{i \in G} R_i)^+; \overline{\mathsf{v}_\psi}}$ (see Section 4) and the definition of implication in terms of negation and disjunction, the RELVIEW-expression for the formula $C_{\{1,2,3\}}(p_2 \rightarrow K_1 p_2)$ is

box(trans(R1 | R2 | R3),-p2 | box(R1,p2)).

The RELVIEW result for this expression is the universal vector, which means that the formula $C_{\{1,2,3\}}(p_2 \rightarrow K_1 p_2)$ holds in all the worlds of the model M under consideration. Indeed, as is easy to verify, in this model it is common knowledge of all children that, if child 2 is muddy then child 1 knows this.

The above illustrates that RELVIEW has two modes for displaying relations: graph representations and matrix representations. Graph representations are particularly well suited for visualization. RELVIEW allows for the edges and nodes of graphs to be distinctively marked. For example, different edge styles can be used as in Figure 3 to specify designated (sub)relations and the nodes can be labelled. Matrix representations are in general less well-suited for visualization, but provide efficient representations of graphs and are easy to process by relation-algebraic (matrix) operations. In addition, certain properties have natural illustrations in matrices. E.g., it is easy to recognize at one glance from the matrices representing R_1, R_2 and R_3 that all three relations are reflexive and symmetric (because each matrix includes the diagonal, the identity relation, and is a mirror image in the diagonal). Also validity of a formula in the model is immediately recognizable when the evaluation returns a vector with all squares marked.

6   Treatment of Other Non-classical Logics

Until now, we have shown how to interpret modal logic of knowledge in relation algebra and how then the RELVIEW tool can be used for investigating finite models of this logic, for visualizing them and for computing solutions to certain computational tasks. This method can be extended to all non-classical logics, embeddable into the programming language of RELVIEW. Prominent examples are logics such as linear-time logic LTL, Hennessy-Milner logic HML, the modal μ-calculus, and the computational tree logic CTL. These are used in computer science for describing properties of computer systems, and model checking for these logics can then serve as a verification method.

In all the logics we have just mentioned some modalities are specified via fixed point constructions. This is no problem for RELVIEW. Far from it! Its programming language allows to formulation of while-loops. These can be used immediately to compute extremal fixed points of monotone functions f on finite lattices as limit of the finite ascending chain $0 \leq f(0) \leq f(f(0)) \leq \ldots$ in the case of the least fixed point (0 is the least element of the lattice) and of the finite descending chain $1 \geq f(1) \geq f(f(1)) \geq \ldots$ in the case of the greatest fixed point (1 is the greatest element of the lattice), respectively.

In the following, we consider computational tree logic CTL in more detail. Formulae of this logic are constructed using the propositional atoms and connectives of modal logic as introduced in Section 2 and the specific operators AX, EX, AU, EU, AF, EF, AG, and EG. The meaning of the operators AX (respectively EX) is the same as the meaning of the $\Box$-modality (respectively the $\Diamond$-modality) in classical modal logic. Hence, if we use again v_ϕ as vector representation of the set $\{M, x \models \phi\}$ we obtain the relation-algebraic specifications

$$\mathsf{v}_{AX(\phi)} = \overline{R; \overline{\mathsf{v}_\phi}} \qquad \mathsf{v}_{EX(\phi)} = R; \mathsf{v}_\phi,$$

where R is the transition relation of the model M. A formula of the form $AU(\phi, \psi)$ holds in a state x if for *all computation paths* $x_1, x_2, x_3, \ldots$ beginning with $x(= x_1)$ we have that ψ holds in some future state x_i and ϕ holds for all states x_j, $j < i$. Furthermore, a formula $EU(\phi, \psi)$ holds in a state x if there *exists a computation path* $x_1, x_2, x_3, \ldots$ beginning with $x(= x_1)$ such that ψ holds in some future state x_i and ϕ holds in all states x_j, $j < i$. Formally these properties can be described by least fixed point constructions (cf. [21]). These yield the following vector representations, where again R is the transition relation of the model M.

$$\mathsf{v}_{AU(\phi,\psi)} = \mu_f \text{ where } f(w) = \mathsf{v}_\psi \cup (\mathsf{v}_\phi \cap \overline{R; \overline{w}} \cap R; \mathsf{L})$$
$$\mathsf{v}_{EU(\phi,\psi)} = \mu_g \text{ where } g(w) = \mathsf{v}_\psi \cup (\mathsf{v}_\phi \cap R; w)$$

The remaining four operators can be reduced to AU and EU. We have $AF(\varphi) := AU(\top, \varphi)$, $EF(\varphi) := EU(\top, \varphi)$, $AG(\varphi) := \neg EF(\neg\varphi)$, and $EG(\varphi) := \neg AF(\neg\varphi)$ (see e.g., [21]). From these definitions we obtain the corresponding vector representations as follows:

$$\mathsf{v}_{AF(\varphi)} = \mathsf{v}_{AU(\top,\varphi)} \qquad \mathsf{v}_{AG(\varphi)} = \overline{\mathsf{v}_{EF(\neg\varphi)}}$$
$$\mathsf{v}_{EF(\varphi)} = \mathsf{v}_{EU(\top,\varphi)} \qquad \mathsf{v}_{EG(\varphi)} = \overline{\mathsf{v}_{AF(\neg\varphi)}}$$

```
AX(R,p) = -(R * -p).

AU(R,p,q)
    DECL w, v
    BEG  w = O(p);
         v = q | (p & -(R * -w) & R * L(p));
         WHILE -eq(w,v) DO
             w = v;
             v = q | (p & -(R * -v) & R * L(p)) OD
         RETURN w
    END.

AF(R,p) = AU(R,L(p),p).
```

Fig. 6. Programs to compute AX, AU, AF

A RELVIEW-implementation of CTL essentially consists of RELVIEW-programs for the operators of this logic. The code in Figure 6 shows the programs for the three operators AX, AU, and AF as they arise from the above vector representations. Guided by this code the reader should have no difficulties to obtain the RELVIEW-programs for the remaining five CTL-operators EX, EU, EF, AG, and EG from the corresponding vector representations.

We have experimented with a RELVIEW-implementation of CTL using standard examples from the literature. One of them is the 'mutual exclusion' of two processes P_1 and P_2. In the textbook [12] this example is modelled by a transition system in two ways and in each case some important properties (such as safety and liveness) are verified using CTL. The remainder of this section treats the first attempt of [12] with the aid of RELVIEW.

We assume six propositional variables. For $i \in \{1,2\}$ the variable n_i denotes that the process P_i is in a non-critical section, the variable t_i denotes that P_i tries to enter a critical section, and the variable c_i denotes that P_i is in a critical section. Based on these variables, a protocol for managing the admission to a critical section is given by a transition relation R on a set of states and a valuation of the propositional variables. A RELVIEW-description of the protocol is presented in the Figures 7 and 8. Figure 7 shows the transition relation R on the protocol's states as a Boolean matrix R and the valuation of the propositional

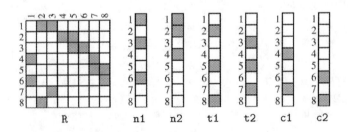

Fig. 7. Relational model of a mutual exclusion protocol

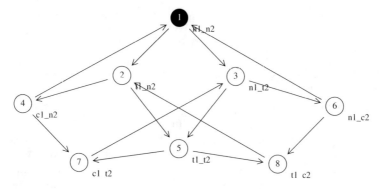

Fig. 8. Graphical representation of a mutual exclusion protocol

variables as six Boolean vectors n1, n2, t1, t2, c1, and c2. The graph representation of the model is shown in Figure 8. In this picture a node corresponds to a state and the labels of a node indicate which propositional variables are defined to be true in the corresponding state. E.g., the first node corresponds to the initial state where both processes are in a non-critical section and the second node corresponds to the state where P_1 tries to enter a critical section and P_2 remains in a non-critical section. Usually, the initial state of a transition system is indicated as a node with an incoming arrow without a source. Since in RELVIEW such 'partial arrows' are not possible, we have drawn the initial node as a black circle.

Having the RELVIEW-description of the protocol at hand, we have used the tool to verify fundamental properties of the protocol. For example, safety, liveness, and that a process can always request to enter a critical section are described by the following three CTL-formulae.

safety: $AG(\neg(c_1 \wedge c_2))$
liveness: $AG(t_1 \rightarrow AF(c_1))$
non-blocking: $AG(n_1 \rightarrow EX(t_1))$

If we evaluate the three corresponding RELVIEW-expressions AG(R,-(c1 & c2)), AG(R,-t1 | AF(R,c1)), and AG(R,-n1 | EX(R,t1)), we obtain in the first case and the third case the 8×1 universal vector and in the second case the 8×1 empty vector. This means that the properties of safety and non-blocking are satisfied in every state but liveness is satisfied in no state. This conclusion is in agreement with the results of [12].

7   Further Uses of RelView

Suppose q is a propositional variable in a modal formula ϕ and suppose the valuations $\iota(p)$ of all propositional variables p in ϕ with the exception of q are defined in M. A problem which we might be interested in is the following:

(5)   Compute a valuation $\iota(q)$ to q so that ϕ is satisfiable in a world of M.

Task (5) may be generalized to an optimization problem as follows:

(6)   Compute a valuation $\iota(q)$ for q so that ϕ is satisfiable in a maximal number of worlds of M.

A solution to the first problem is possible by applying the 'is-member-of' relation between W and the powerset 2^W. The 'is-member-of' relation ε relates a world x and a set of worlds X iff $x \in X$. It is available in RELVIEW via a pre-defined relational function called epsi. Problem (6), the generalization, can also be solved with RELVIEW. The solution uses besides the 'is-member-of' relation also the 'size-comparison' relation on 2^W, and the vector-representation of greatest elements with respect to a quasi-order. The 'size-comparison' relation relates two sets X and Y iff $|X| \leq |Y|$ and can be computed via a call of the pre-defined function cardrel.

In an array-like implementation of relations the memory consumption of the 'is-member-of' relation and the 'size-comparison' relation is exponential in the size of the base set. However, BDDs allow a very efficient implementation of these two relations. In [17] for the 'is-member-of' relation a BDD-implementation is developed that uses $O(n)$ BDD-nodes and [20] presents for the 'size-comparison' relation a BDD-implementation with $O(n^2)$ BDD-nodes. In both cases n is the number of elements of the base set, i.e., the cardinality of the set of worlds W in our case.

To give an impression of how to solve problem (5) by means of RELVIEW, we consider the formula $\neg K_1(p_1 \vee p_2)$ of Section 5 and replace the propositional variable p_2 by the (uninterpreted) propositional variable q. We assume again that the relation R_1 is as shown in Figure 2 and that the propositional variable p_1 denotes 'child 1 has mud on its forehead', i.e., the vector representation v_{p_1} of $\iota(p_1)$ is as shown in Figure 4. Then RELVIEW computes exactly 240 possible valuations $\iota(q)$ for q such that the modal formula

$$\neg K_1(p_1 \vee q)$$

becomes true in a world of the model M with relation R_1 and valuation function ι. The key to obtaining this result is the relation Q between the set of worlds W and the powerset 2^W, defined by

$$Q := R_1;\ \overline{\mathsf{v}_{p_1};\mathsf{L} \cup \varepsilon}.$$

This definition implies that for all $x \in W$ and $X \in 2^W$ we have that $Q(x, X)$ iff $X = \iota(q)$ implies $M, x \models \neg K_1(p_1 \vee q)$. In matrix terminology this means: If $\iota(q)$ is represented by column c of ε, then $\{x \in W \mid M, x \models \neg K_1(p_1 \vee q)\}$ is represented by the same column of Q. Hence, the vector $Q^\mathsf{T};\mathsf{L}$ (defined over 2^W) represents the 240 solutions of problem (5) with inputs $\neg K_1(p_1 \vee q)$, R_1, and $\iota(p_1)$. A column-wise description of these solutions is $\varepsilon; inj(Q^\mathsf{T};\mathsf{L})^\mathsf{T}$, where the relational function inj computes the injective mapping generated by a vector. (If the vector v over X represents the subset Y of X, then $inj(v)$ is the relation

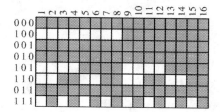

Fig. 9. Valuations not leading to satisfiability of $\neg K_1(p_1 \vee q)$

between Y and X such that $inj(v)(x,y)$ iff $x = y$.) This standard technique for representing sets of subsets is explained in, e.g. [3,4]. In our example it yields a 8×240 RELVIEW-matrix, which is too large to be presented here. Therefore, we show in Figure 9 a much smaller RELVIEW-matrix that column-wisely represents the non-solutions, i.e., the 16 valuations $\iota(q)$ for q which do not lead to satisfiability. For example, from the last column of this picture we see that no world of M satisfies the formula $\neg K_1(p_1 \vee q)$ if ι defines the variable q to be true in all worlds of M.

We have also used RELVIEW to solve problem (6) for the same three inputs $\neg K_1(p_1 \vee q)$, R_1, and $\iota(p_1)$. The system computes that exactly 16 of the 240 solutions of problem (5) maximize the number of worlds which satisfy the formula, there is only one such maximal set of worlds, and its cardinality is 4. The 16 solutions of problem (6) are column-wisely described by the 8×16 matrix of Figure 10. E.g., the last column of this matrix states that $\neg K_1(p_1 \vee q)$ is true in a maximal number of worlds if q is true in $\langle 1,0,0 \rangle$, $\langle 1,0,1 \rangle$, $\langle 1,1,0 \rangle$, and $\langle 1,1,1 \rangle$. The only 4 worlds which satisfy $\neg K_1(p_1 \vee q)$, if $\iota(q)$ is one of the 16 solutions of problem (6), are $\langle 1,0,0 \rangle$, $\langle 1,0,1 \rangle$, $\langle 1,1,0 \rangle$, and $\langle 1,1,1 \rangle$. This property follows from the RELVIEW-vector of Figure 10.

Like the solutions (respectively non-solutions) of problem (5) for the inputs $\neg K_1(p_1 \vee q)$, R_1, and $\iota(p_1)$, also the solutions of problem (6) can be specified by simple relation-algebraic expressions. Crucial to the solution is the vector

$$v := ge(C, syq(\varepsilon, Q); \mathsf{L})$$

over the powerset 2^W that represents the set of all maximal subsets X of W such that $M, x \models \neg K_1(p_1 \vee q)$ holds for all $x \in X$. In this definition the relations Q

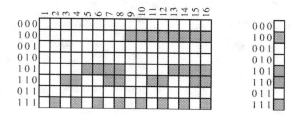

Fig. 10. Maximum satisfiability of $\neg K_1(p_1 \vee q)$

and ε are as above, C denotes the 'size-comparison' relation on 2^W, and the relational functions

$$ge(R, w) = w \cap \overline{\overline{R}^{\mathsf{T}}; w} \qquad syq(R, S) = \overline{R^{\mathsf{T}}; \overline{S}} \cap \overline{\overline{R}^{\mathsf{T}}; S}$$

compute the vector of the greatest elements of the vector w with respect to the quasi-order R and the symmetric quotient of R and S, respectively. In the present case the column-wise description $\varepsilon; inj(v)^{\mathsf{T}}$ of the maximal subsets consists of only one column and coincides with the vector of Figure 10. From it we obtain the vector representation of the set of 16 valuations leading to the only maximal subset[2] $\{\langle 1, 0, 0\rangle, \langle 1, 0, 1\rangle, \langle 1, 1, 0\rangle, \langle 1, 1, 1\rangle\}$ via the vector $syq(Q, \varepsilon; inj(v)^{\mathsf{T}})$ over 2^W, and the 8×16 matrix of Figure 10, finally, is exactly the column-wise description of this set of valuations.

So far we have used RELVIEW only for computing sets of worlds or for solving related tasks. But the application domain of the system is larger. For example, the tool can also be used for the following important task:

(7)   Determine whether a relation R in a given finite frame possesses certain properties.

The kinds of properties RELVIEW can express and handle are rather general. In particular, these are all properties which can be written as Boolean combinations of inclusions between relation-algebraic expressions. This includes all the correspondence properties of Section 2 (reflexivity of a relation R, transitivity or R, etc), and also properties such as irreflexivity ($\overline{\mathsf{I}} \subseteq R$) and acyclicity ($R^+ \subseteq \overline{\mathsf{I}}$) as well as Boolean combinations of these. For example, R is an equivalence relation iff it satisfies the conjunction of the first three correspondence properties of Section 2. In the syntax of the RELVIEW tool a corresponding evaluation test looks as follows:

```
incl(I(R),R) & incl(R*R,R) & incl(R,R^).
```

Let us consider a last application. For a given finite frame $\mathcal{F}$ with set of worlds W and a closure system[3] $\mathcal{C} \subseteq 2^{W \times W}$ of relations (like the Euclidean or the transitive relations), the RELVIEW tool very often allows us to solve the following task:

(8)   Compute the corresponding closure operator $cl : 2^{W \times W} \to 2^{W \times W}$, defined by $cl(R) = \bigcap \{S \in \mathcal{C} \mid R \subseteq S\}$.

The condition which is to be fulfilled is that the conjunction of $S \in \mathcal{C}$ and $R \subseteq S$ is equivalent to $f(S) \subseteq S$, with f being a monotone function on the set $2^{W \times W}$ of all relations over W. In this case $cl(R)$ coincides with the least fixed point μ_f of the function f, due to Tarski's fixed point theorem [27]. The

---

[2] In words, this vector marks exactly the 16 columns of Q each of which represent a set of worlds with the maximal cardinality 4.

[3] A subset $\mathcal{C}$ of a powerset 2^X is a closure system on X if $\bigcap Y \in \mathcal{C}$ for all $Y \subseteq \mathcal{C}$.

```
euclid(R)
  DECL S, fS
  BEG  S = O(R);
       fS = R;
       WHILE -eq(S,fS) DO
         S = fS;
         fS = R | fS^ * fS OD
       RETURN S
  END.
```

Fig. 11. Program to compute the Euclidean closure of a relation R

frame $\mathcal{F}$ is finite. Hence f is even $\cup$-continuous and we get the representation $\mu_f = \bigcup_{i \geq 0} f^i(O)$, where the chain $O \subseteq f(O) \subseteq f^2(O) \subseteq \ldots$ eventually becomes stationary. To give an example, the Euclidean closure of a relation R is computed by the RELVIEW-program euclid of Figure 11, because obviously a relation S is Euclidean (i.e., $S^{\mathsf{T}}; S \subseteq S$) and contains R iff $R \cup S^{\mathsf{T}}; S \subseteq S$.

Finally, it is worth mentioning that RELVIEW has some file input/output interfaces. Especially ASCII formatted files can be used to exchange data with other systems.

8   Concluding Remarks

Based on the interpretation of non-classical logics in relation algebra, in this paper we have shown how the RELVIEW tool can be used for investigating finite models of such logics and for visualizing them and solutions of certain computational tasks. Modal logic of knowledge and computational tree logic have been treated in detail and illustrated with two well-known examples, viz. the 'muddy children' puzzle and a 'mutual exclusion' protocol.

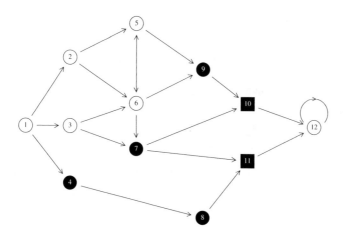

Fig. 12. Visualization of the meaning of the AF-operator

We believe that the attraction of RELVIEW in respect to the applications we have discussed in this paper lies in its flexibility, the concise form of its programs, and the various possibilities for manipulation, testing, and visualization. Because of these properties it is an excellent tool for prototyping, experimenting, and for university teaching. It can be programmed to handle different logics and perform typical tasks on them while avoiding unnecessary overhead. We found it very attractive to use RELVIEW also for producing good examples. Concerning teaching, its visualization possibilities can be used to demonstrate the meaning of logical operators and formulae for example.

To illustrate this point, consider the picture in Figure 12. It explains the meaning of the AF-operator of the logic CTL. The squares denote the states where a certain property, p say, holds and the black vertices (including the squares) denote the states x such that for all computation paths $x_1, x_2, \ldots$ beginning in x somewhere along the path p holds. Visualization is of particular importance when combined with the evaluation of RELVIEW-expressions in a stepwise fashion. All this can help students, and even be key to their fully understanding of an advanced concept.

References

1. R. Behnke, R. Berghammer, E. Meyer, and P. Schneider. RELVIEW - A system for calculating with relations and relational programming. In E. Astesiano, editor, *Fundamental Approaches to Software Engineering*, volume 1382 of *LNCS*, pages 318–321. Springer, Berlin, 1998.
2. R. Berghammer, P. Kempf, G. Schmidt, T. Ströhlein. Relational algebra and logic of programs. In H. Andréka, J.D. Monk, and I. Németi, editors, *Algebraic Logic*, volume 54 of *Colloq. Math. Soc. J. Bolyai*, pages 37–58. North-Holland, Amsterdam, 1991.
3. R. Berghammer, B. Leoniuk, and U. Milanese. Implementation of relational algebra using binary decision diagrams. In H. de Swart, editor, *Relational Methods in Computer Science*, volume 2561 of *LNCS*, pages 241–257. Springer, Berlin, 2002.
4. R. Berghammer, F. Neumann. RELVIEW – An OBDD-based Computer Algebra system for relations. In V.G. Gansha, E.W. Mayr, and E.V. Vorozhtsov, editors, *Computer Algebra in Scientific Computing*, volume 3718 of *LNCS*, pages 40–51. Springer, Berlin, 2005.
5. P. Blackburn, M. de Rijke, and V. Venema. *Modal Logic*. Cambridge Univ. Press, Cambridge, 2001.
6. C. Brink, K. Britz, and R. A. Schmidt. Peirce algebras. *Formal Aspects of Computing*, 6(3):339–358, 1994.
7. B. F. Chellas. *Modal Logic: An Introduction*. Cambridge Univ. Press, Cambridge, 1980.
8. E. Clarke, O. Grumberg and D. Peled. *Modal Checking*. MIT Press, Cambridge, 2000.
9. R. Fagin, J. Y. Halpern, Y. Moses, and M. Y. Vardi. *Reasoning about knowledge.* MIT Press, Cambridge, 1995.
10. R. Goldblatt. *Logics of Time and Computation*, volume 7 of *CSLI Lecture Notes*. Chicago Univ. Press, Chicago, 1987.

11. G. E. Hughes and M. J. Cresswell. *A New Introduction to Modal Logic*. Routledge, London, 1996.

12. M. Huth and M. Ryan. *Logic in Computer Science. Modelling and Reasoning About Systems*. Cambridge Univ. Press, Cambridge, 2000.

13. B. Jónsson and A. Tarski. Boolean algebras with operators, Part I. *American Journal of Mathematics*, 73:891–939, 1951.

14. W. Nabialek, A. Niewiadomski, W. Penczek, A. Polrola, and M. Szreter. VERICS 2004: A model checker for real time and multi-agent systems. In *Proc. International Workshop on Concurrency, Specification, and Programming*, volume 170 of Informatik-Berichte, pages 88-99, Humbold Universität Berlin, 2004.

15. M. Kacprzak, A. Lomusico, A. Niewiadomski, M. Szreter, W. Penczek, and F. Raimondi. Comparing BDD and SAT based techniques for model checking Chaum's Dining Cryptographers protocol. To appear in *Fundamenta Informaticae*, 2006.

16. D. Kozen. A representation theorem for models of *-free PDL. In J. de Bakker and J. van Leeuwen, editors, *Automata, Languages and Programming*, volume 85 of *LNCS*, pages 351–362. Springer, Berlin, 1980.

17. B. Leoniuk. ROBDD-based implementation of relational algebra with applications (in German). Ph.D. thesis, Institut für Informatik und Praktische Mathematik, Universität Kiel, 2001.

18. A. Lomuscio and F. Raimondi. MCMAS: a tool for verifying multi-agent systems. In H. Hermanns and J. Palsberg, editors, *Tools and Algorithms for the Aonstruction and Symposium of Systems*, volume 3920 of *LNCS*, pages 450–454. Springer, Berlin, 2006.

19. J.-J. Ch. Meyer and W. van der Hoek. *Epistemic Logic for AI and Computer Science*. Cambridge Univ. Press, Cambridge, 1995.

20. U. Milanese. On the implementation of a ROBDD-based tool for the manipulation and visualization of relations (in German). Ph.D. thesis, Institut für Informatik und Praktische Mathematik, Universität Kiel, 2003.

21. M. Müller-Olm, D. Schmidt, B. Steffen. Model-checking: A tutorial introduction. In A. Cortesi and G. Filé, editors, *Static Analysis Symposium 1999*, volume 1694 of *LNCS*, pages 330–354. Springer, Berlin, 1999.

22. K. C. Ng and A. Tarski. Relation algebras with transitive closure, abstract 742-02-09. *Notices Amer. Math. Soc.*, A29–A30, 1977.

23. E. Orlowska. Relational formalisation of nonclassical logics. In C. Brink, W. Kahl, and G. Schmidt, editors, *Relational Methods in Computer Science*, Advances in Computing, pages 90–105. Springer, Wien, 1997.

24. V. R. Pratt. Dynamic algebras: Examples, constructions, applications. Technical Report MIT/LCS/TM-138, MIT Laboratory for Computer Science, 1979.

25. G. Schmidt and T. Ströhlein. *Relations and Graphs. Discrete Mathematics for Computer Scientists*. Springer, Berlin, 1993.

26. A. Tarski. On the calculus of relations. *J. Symbolic Logic*, 6(3):73–89, 1941.

27. A. Tarski. A lattice-theoretical fixpoint theorem and its applications. *Pacific J. Math.*, 5:285–309, 1955.

On the Logic of Medical Decision Support

Patrik Eklund[1], Johan Karlsson[1], Jan Rauch[2], and Milan Šimůnek[3]

[1] Umeå University, Department of Computing Science,
SE-90187 Umeå, Sweden
peklund@cs.umu.se, johank@cs.umu.se
[2] University of Economics, Prague, Faculty of Informatics and Statistics,
Czech Republic
rauch@vse.cz
[3] Academy of Sciences of the Czech Republic, Institute of Computer Science
simunek@vse.cs

Abstract. Guideline development, implementation, utility and adherence require intelligence and multimedia to interact in decision support environments. However, efforts to combine all these aspects and to connect solutions into a effective, efficient and productive environment are rare. In this paper we use a regional health care perspective on maintenance and analysis of data, information and knowledge. Examples are drawn from cardiac diseases. Analysis and development is viewed from by-pass surgery point of view. Association rules are used for analysis, and we show how these rules take logical forms so as to prepare for development of guidelines.

1   Introduction

Clinical guidelines and evidence medicine are rather general concepts and therefore we provide some explanations so as to provide a language for our discussions. We will discuss (clinical) guidelines in the meaning of involving their development, implementation, utility and adherence. A clinical guideline is "systematically developed statements to assist practitioner and patient decisions about appropriate health care for specific clinical circumstances" [13]. By consensus among a large enough group of domain-experts, such guidelines can be said to represent, if not the only correct advice but at least given available research results, good enough advice [25].

Development typically involves definitions of measurement scopes, clinical trials with data collections, followed by data analysis using statistical tools. Data analysis results are converted to text based rules which constitute the so called guidelines. Indeed, representing clinical knowledge in a computer is difficult in practice since many clinical practice guidelines are still published simply as traditional text documents. Clearly, logic is very much missing at this stage.

The whole guideline development process is what traditionally is called an evidence based [6,7] approach. In essence, evidence based medicine proposes a shift from experience based care to acquiring knowledge through systematic reviews of the appropriate literature. Note that from a medical point of view, development does not include preparations for the next phase, namely the implementation of

H. de Swart et al. (Eds.): TARSKI II, LNAI 4342, pp. 50–59, 2006.

the text based guidelines into electronically readable forms. Further, the medical community is rather conservative concerning statistical tools, and there is no understanding of statistical tools that leads naturally into discussions on logical structures.

This is one of the major obstacles when aiming at computerized guidelines. As logic is missing, semantics is not available, ambiguities exist, and tools for program specifications are hard to invent. In fact, the problem is not only that logic is missing, but even more complicated as we need to decide on the most appropriate logic for that particular guidelines implementation task. The existence of a general-purpose logic covering representability of most guidelines seems very unlikely, even if related diagnosis and treatment problems can be expected to share some common logic features.

Selecting and establishing the underlying logic of guidelines reveals how statistics does not comply with logical inference, with logic on the other hand not providing language constructs that can handle statistical information. In such a situation, heuristics easily enters the scene, and a required synergy and even convergence of statistical and logical methods remains unseen. Alternatively, statistics and logic can be bound more tightly together. The goal is then to provide kind of an all-in-one computation that fulfills requirements for evidence-based statistics and reasoning, at the same time providing results represented more strictly within a corresponding logical machinery.

It is now more evident that the gap between development and implementation is where logic comes to the rescue. Clinical guidelines often provide the basis for logic in decision support systems. Such a logic can also be inferred more directly from patient data [3].

Use and utility of a particular system needs to conform with national and even regional requirements and needs. Further, where health care is at least partly public, it involves political decisions and considerations. Add to that attitudes among professionals, and we get installation procedures that can be rather complicated and very much dependent on organizational structures. Having these situations and processes in mind, it is obvious that plans for end-usage together guideline adherence studies should exist even before implementations can start.

No system is complete without its thorough evaluation. Once implemented, guidelines adherence must be investigated. To increase the adherence to clinical guidelines and thus evidence-based care, computer-based decision support tools are recognized to be important [36].

This paper is organized as follows. In Section 2 some comparisons are made to other guideline implementations. Section 3 provide background and motivation, in particular from a regional health care perspective, and presents briefly the Coronary Artery Bypass Grafting (CABG) medical framework. Section 4 presents the GUHA method with capability to produce rules that can be embedded into various computational schemas. GUHA rules are also presented as logical entities. In Section 5 we will see how association rules formally match with quantifiers in a certain extension of predicate calculus. Section 6 concludes the paper.

2   Related Work

Concerning implementation we should note that we cannot exclusively concentrate on rules of the guidelines but we need also consider system interfaces to other relevant sources of information used in the overall diagnosis and/or treatment process. In the case of hypertension treatment, the analysis step should be seen developed as manifested by the international guideline (JNC-VI [34]) for hypertension treatment. However, implementation of treatment suggestions cannot be isolated from the overall hypertension treatment context and other decisions required from that more general viewpoint. Pharmacological information with corresponding databases are typically required to be interconnected with the systems for treatment suggestions. It should here be immediately observed how pharmacological information and e.g. interaction analysis and identification in itself requires a deeper understanding of logical structures different from those of the treatment suggestions. Indeed, we are implementing a hybrid of logical systems in the overall support system for hypertension treatment as connected to pharmacological information systems. This system builds on a previous system [26,27]. In our developments on hypertension treatment, guideline adherence was also investigated, and has affected further developments of the system.

In [23], on diagnosis of cognitive disorder, dementia and dementia types, we go beyond the hypertension treatment approach and encode the DSM-IV guidelines [33], together with regional adaptations, in a probabilistic argumentation framework [24] as well as using a neural propositional logic [12,10].

3   Regional Experiences with Patient Data for Quality Assurance

The information management approach in this paper rests upon experiences within the County Council of Västerbotten in Northern Sweden. The population is small but the geographical area is large. There is one university hospital in the region, together with 13 regional hospitals. The region is unique in Sweden in that there is only one patient record system, which is used both within primary care as well throughout the hospitals. The potential for information flow between clinics is huge, even if not yet fully exploited. Further, the region maintains the responsibility for several national quality registers, where interactions with the patient record is highly prioritized. Various quality assurance programmes are on the agenda, and utility of data mining has been identified as having huge potentials.

In addition to quality registers, several clinics maintain their own research databases, such as for cardiac surgery, where data mining deployment often is more straightforward, but then usually at the expense of non-compliance, with respect to terminologies, with the electronic record structure. Screening is another field where careful maintenance of patient data over several years and decades provide valuable insight concerning trends within the population.

Computing facilities and software development capabilities within regions are important. The former usually builds upon traditions within biostatistics and the fact that curricula in medical and nursing education always involve at least some basic statistics. Computer science and the art of software development, on the other hand, is rarely, if ever, included in such curricula. Further, involving IT competence for the technical staff of hospitals comes with some lag of time.

The example for GUHA data analysis in this paper is coronary artery by-pass grafting (CABG)[1]. In Västerbotten, there are about 5-6 cardiac surgeries every day, most of which are coronary bypass operations. Several medical studies show relations between pre- and postoperative CABG data, see for example [28]. The research database in Umeå also involves intraoperative data. Outcome predictions are certainly needed, if possible, from preoperative data, but outcome predictions while operating is additionally useful.

Preoperative data includes information on diseases, heart conditions and function classes (typical follow-up parameter), number of injured vessels, character, if any, of angina pectoris, and so on. Important intraoperative information is e.g. time while aorta is closed and patient is in heart/lung machine, number of anastomoses, aorta quality and suitability for reoperation. Postoperative attributes include death within 30 days after operation, hours in intensive care, respirator time and postsurgical conditions of various kind.

4   GUHA

GUHA is an original Czech method of data mining. Its aim is to offer all interesting facts following from the analyzed data to the given problem. GUHA is realized by GUHA procedures. It is a computer program, the input of which consists of the analyzed data and of a simple definition of set of relevant (i.e. potentially interesting) patterns. GUHA procedure automatically generates each particular pattern and tests if it is true in the analysed data. The output of the procedure consists of all prime patterns. The pattern is prime if it is true in the analysed data and if it does not immediately follow from the other more simple output patterns [14].

The most important GUHA procedure is the procedure ASSOC [14] that mines for association rules. The association rules the procedure ASSOC mines for are more general than the classical association rules defined in [2]. This procedure deals among other things with association rules corresponding to statistical hypothesis tests. There are several implementations of the procedure ASSOC, see e.g. [15,16]. The latest one is the procedure *4ft-Miner*. It has various new important features and it mines also for conditional association rules [32].

There is academic software system LISp-Miner [32] that includes five new GUHA procedures in addition to the procedure 4ft-Miner. They mine for large variety of patterns. There are both simple patterns verified in one contingency

---

[1] Data has not been made public, but interested readers may contact one of the authors for enquiries concerning this particular data set.

table of two Boolean attributes and complex patterns corresponding to differences of two sets what concerns relation of two attributes. Such complex pattern is verified using a pair of contingency tables. Implementation of all GUHA procedures of the LISp-Miner system is based on *representation of analyzed data by strings of bits* [29,32].

There are important theoretical results related to the GUHA method. Observational calculi are defined and studied in [14] as a language in which statements concerning observed data are formulated. Logical calculi formulae of which correspond to generalized association rules are special case of observational calculi. Various theoretical results concerning observational calculi and namely association rules were achieved in [14]. Some new results concerning logic of association rules are e.g. in [30,31]. Theoretical results concerning association rules can play an important role when embedding association rules into various intelligent systems.

We show several examples of association rules concerning CABG data. These association rules were mined by the procedure 4ft-Miner.

The association rules is an expression of the form $\varphi \approx \psi$, where *antecedent* φ and *succedent* ψ are conjunctions of *literals*. Literal is a Boolean attribute (automatically) derived from the analyzed data. Boolean attributes such as `AnginaPectoris(STABLE)` and `Age⟨70; 80)` are examples of literals.

The symbol $\approx$ is called *4ft-quantifier*. It defines a relation of antecedent φ and succedent ψ. This relation can be true or false in a given data matrix $\mathcal{M}$. The association rule $\varphi \approx \psi$ is verified in the given data matrix $\mathcal{M}$ using the four-fold table 4ft(φ,ψ, $\mathcal{M}$) of φ and ψ in $\mathcal{M}$, see Table 1.

Table 1. 4ft table 4ft(φ,ψ, $\mathcal{M}$) of φ and ψ in $\mathcal{M}$

$\mathcal{M}$	ψ	$\neg\psi$
φ	a	b
$\neg\varphi$	c	d

The table should be given the interpretation that a is the number of objects satisfying both φ and ψ, b is the number of objects satisfying φ but not ψ, $a + b$ is the number of objects satisfying φ, and so on.

A condition concerning all 4ft tables is associated to each 4ft quantifier $\approx$. The association rule $\varphi \approx \psi$ is true in the analyzed data matrix $\mathcal{M}$ if and only if the condition associated to the 4ft quantifier $\approx$ is satisfied for the four-fold table $4ft(\varphi,\psi$, $\mathcal{M})$ of φ and ψ in $\mathcal{M}$. If this condition is not satisfied then the association rule $\varphi \approx \psi$ is false in the analyzed data matrix $\mathcal{M}$. There are various 4ft-quantifiers, see e.g. [14] and [32].

The 4ft-quantifier $\Rightarrow_{p, Base}$ of *founded implication* [14] is defined for $0 < p \leq 1$ and $Base > 0$ by the condition

$$\frac{a}{a + b} \geq p \wedge a \geq Base \ .$$

The association rule $\varphi \Rightarrow_{p;Base} \psi$ is interpreted as "100p % of objects satisfying φ also satisfy ψ" or "φ implies ψ on the level of 100p %".

The 4ft-quantifier $\sim^+_{p,Base}$ of *above average dependence* is defined for $0 < p$ and $Base > 0$ by the condition

$$\frac{a}{a+b} \geq (1+p)\frac{a+c}{a+b+c+d} \wedge a \geq Base .$$

This means that among the objects satisfying φ is at least $100p$ per cent more objects satisfying ψ than among all objects and that there are at least $Base$ objects satisfying both φ and ψ.

Analysis of CABG was done using the system LISp-Miner [32,35], and involved predictions, on one hand, from preoperative to postoperative conditions, on the other hand, from preoperative and intraoperative to postoperative conditions. Can we make useful and reliable preoperative-to-postoperative predictions without intraoperative information? Which are the most significant intraoperative variables used in addition to preoperative variables when predicting postoperative conditions? We have chosen to illuminate the possibilities of the 4ft-Miner procedure by looking at death after 30 days (no/yes) as an example of postoperative condition. The number of postoperative deaths in the data set is rather small, 44 cases which is less than 2% of the total number of records (2975 cases).

Tables 2, 3 and 4 present typical examples from analysis within GUHA and using LISp-Miner. In Table 2 we have an example of a rule that provides 100% survival 30 days after operation.

Table 2. $Reop(no) \wedge FunctClass(IIIA) \wedge LV - Funct(good) \Rightarrow_{1.0;436} Died30d(no)$

CABG	Died30d(no)	Died30d(yes)
$Reop(no) \wedge FunctClass(IIIA) \wedge LV - Funct(good)$	436	0
$\neg(Reop(no) \wedge FunctClass(IIIA) \wedge LV - Funct(good))$	2495	44

This is the strongest (founded) implication of the form

$$preop_1 \wedge \ldots \wedge preop_n \Rightarrow_{p;Base} Died30d(no).$$

There are several other strong implications, also of the form

$$Age \wedge preop_1 \wedge \ldots \wedge preop_n \Rightarrow_{p;Base} Died30d(no).$$

In the situation for non-survival after 30 days, the association rule for the above average relation turns out to be more suitable. The four-fold table for the strongest rule is shown in Table 3. The rule should be understood as patients satisfying $LV - Funct(bad) \wedge MainSten(no)$ are with 537% more likely to satisfy $Died30d(yes)$ as compared to all observed cases. There are 13 patients satisfying both $LV - Funct(bad) \wedge MainSten(no)$ as well as $Died30d(yes)$.

The weakest above average relation is

$$AngPect(unstable) \sim^+_{1.18;19} Died30d(yes).$$

Table 3. $LV - Funct(bad) \wedge MainSten(no) \sim^+_{5.37;13} Died30d(yes)$

$CABG$	$Died30d(yes)$	$Died30d(no)$
$LV - Funct(bad) \wedge MainSten(no)$	13	125
$\neg(LV - Funct(bad) \wedge MainSten(no))$	31	2806

Combinations with age, especially for patients in their later 60's and early 70's, show association rules where patients having $LV - Funct(bad)$ is worse than having $FunctClass(IV)$.

Finally, involving intraoperative information, Table 4 shows an example association rule with $ClampTime$.

Table 4. $ClampTime\langle 45; 90\rangle \wedge MainSten(no) \sim^+_{4.0;10} Died30d(yes)$

$CABG$	$Died30d(yes)$	$Died30d(no)$
$ClampTime\langle 45; 90\rangle \wedge MainSten(no)$	10	67
$\neg(ClampTime\langle 45; 90\rangle \wedge MainSten(no))$	34	2931

5   GUHA Logic

In the association rule $\varphi \approx \psi$, the symbol $\approx$ corresponds to a quantifier. We will now make this more precise.

The extended predicate language of GUHA [14] consists of predicates and variables. Further there are operators $0, 1, \neg, \wedge, \vee, \rightarrow, \leftrightarrow$. The extension is in inclusion of (a finite or infinite sequence of) quantifiers $q_1, \ldots$. Formulae are defined in the usual way. Further, $(qx)(\phi_1, \ldots, \phi_n)$ is a formula whenever q is a quantifier, x is a variable, and $\phi_1, \ldots, \phi_n$ are formulae. The association rule $\varphi \approx \psi$ would thus correspond to a quantification $(q_\approx x)(\varphi, \psi)$.

Before discussing semantics, observe that models M in our presentation can be viewed as matrices where columns correspond to properties and rows to observations.

The semantics of the operators is again as usual. In order to introduce the semantics of quantifiers, let us review the situation concerning $(\forall x)P(x)$, i.e. intuitively involving a one-column matrix in the case of P being atomic. Interpretations are relations on M, or equivalently, mappings f from M to $\{0, 1\}$. If P is interpreted in M by f we have $\|(\forall x)P(x)\|_M = 1$ if and only if f is 1 on M. The function $Asf_\forall$, given by $Asf_\forall(M, f) = 1$ if and only if f is 1 on M, defines the semantics of $\forall$.

The quantifier of implication $\Rightarrow$ ([5]) is defined by $Asf_\Rightarrow(M, f, g) = 1$ if and only if $g(o) = 1$ whenever $f(o) = 1$. Our examples in Section 4, such as the founded implication $\Rightarrow_{p,Base}$ and the above average dependence $\sim^+_{p,Base}$, can now be included into the list of possible quantifiers.

The deduction rule

$$\frac{\varphi \approx \psi}{\varphi' \approx \psi'}$$

means that $\varphi' \approx \psi'$ is true in M whenever $\varphi \approx \psi$ is true in M.

Quantifiers can be more or less implicational, and thus also more or less associational. Further, the soundness of certain deduction rules is connected to quantifiers being implicational. See [14,30,31] for more detail.

6 Conclusions

Traditionally, in evidence based medicine, logic is very sparsely seen as a computational discipline, even less understood as being a language for representation of rules within clinical guidelines. Evidence and belief is anchored in statistical computations, and consensus guidelines are documented as pieces of pure text. Arrival at guidelines is thus based on using statistical tools, where specification and implementation of rules in guidelines require languages of logic.

Knowledge representation using formal methods is very shallow, with guideline performance and adherence impossible to measure and evaluate.

GUHA provides a method for knowledge elicitation where rules are represented in a formal logic. GUHA data analysis on coronary artery by-pass grafting is shown to open up possibilities for computer supported production of guidelines. In particular for CABG, the GUHA approach turns out to be very suitable and providing useful insight related to concrete domain knowledge.

The logical understanding of association rules being quantifiers, and in the sense of being more or less implicational, makes guideline implementations feasible even if less trivial as compared to using a logic in a more clear clausal form. In the case of by-pass surgery, broader analysis with respect to prediction accuracy and guideline implementation is future work. Further, extensions of this paper in these directions also need to include end-user evaluations together with support for ensuring guideline adherence.

Regional and coherent approaches to information analysis, together with knowledge representation based on interaction between statistics and logic, provide impact on all levels of information management ranging from patient records, through a well-founded understanding of organization and workflow, all the way to guidelines based on computed evidence and implemented for the purpose of recommended or even enforced adherence.

Acknowledgement

We are grateful to Hjärtcentrum at the University Hospital of Northern Sweden for making the data set on by-pass surgery available to us. Further, valuable comments and suggestions for improvements of this paper has been provided by anonymous referees. This is gratefully acknowledged.

References

1. S. Achour, M. Dojat, C. Rieux, P. Bierling, E. Lepage, *A UMLS-Based Knowledge Acquisition Tool for Rule-Based Clinical Decision Support Systems Construction*, J. Amer. Med. Inform. Assoc., **8** No 4 (2001), 351-360.
2. R. Agrawal, H. Mannila, R. Srikant, H. Toivonen, A. I. Verkamo, *Fast discovery of association rules*, In: Fayyad UM, et al (eds). Advances in Knowledge Discovery and Data Mining. AAAI Press, Menlo Park California.
3. T. Anagnostou, M. Remzi, M. Lykourinas, B. Djavan, *Artificial neural networks for decision-making in urologic oncology*, European Urology **43** (2003), 596-603.
4. J. Bury, J. Fox, A. Seyfang, *The ProForma guideline specification language: Progress and prospects*, In: First European Workshop on Computer-based Support for Clinical Guidelines and Protocols, 2000, 1-14.
5. A. Church, *Introduction to mathematical logic, Volume I.*, Princeton, 1956.
6. A. Cohen, P. Stavri, W. Hersh, *A categorization and analysis of the criticisms of Evidence-Based Medicine* International Journal of Medical Informatics **73** (2004), 35-43.
7. F. Davidoff, B. Haynes, D. Sackett, R. Smith, *Evidence based medicine*, British Medical Journal **310** (1995), 1085-1086.
8. J. S. Einbinder, K. W. Scully, R. D. Pates, J. R. Schubart, R. E. Reynolds, *Case study: a data warehouse for an academic medical center*, J. Healthcare Information Management, **15** No 2 (2001), 165-175.
9. P. Eklund, *Network Size Versus Preprocessing*, in: Fuzzy Sets, Neural Networks and Soft Computing, ed. R. Yager, L. Zadeh, Van Nostrand Reinhold, New York, 1994, 250-264.
10. P. Eklund, F. Klawonn, *Neural Fuzzy Logic Programming*, IEEE Trans. Neural Networks, **3** No 5 (1992), 815-818.
11. P. Eklund, J. Karlsson, J. Rauch, M. Simunek, *Computational Coronary Artery Bypass Grafting*, Proc. 6th Int. Conf. on Computational intelligence and Multimedia Applications (Iccima'05), Volume 00 (August 16 - 18, 2005). ICCIMA. IEEE Computer Society, Washington, DC, 138-144.
12. P. Eklund, H. Lindgren, *Towards Dementia Diagnosis Logic*, In. Proc. 11th Int. Conf. Information Processing and Management of Uncertainty in Knowledge-based Systems (IPMU 2006), Paris, France, 2-7 July 2006.
13. M. Field, K. Lohr, *Clinical practice guidelines: directions for a new program*, Chapt. Attributes of good practice guidelines, National Academy Press, 1990, 53-77.
14. P. Hájek, T. Havránek, *Mechanising Hypothesis Formation - Mathematical Foundations for a General Theory*, Springer-Verlag, 1978.
15. P. Hájek (guest editor), Second special issue on GUHA, International Journal of Man-Machine Studies, 1981.
16. P. Hájek, A. Sochorová, J. Zvárová, *GUHA for personal computers*, Computational Statistics & Data Analysis, **19** (1995), 149-153.
17. G. Hripcsak, *Writing Arden syntax medical logic modules*, Comput. Biol. Med., **24** (1994), 331-363.
18. W. Inmon, *Building the data warehouse* (2nd Edition), 1996.
19. J. Karlsson, P. Eklund, *Data mining and structuring of executable data analysis reports: Guideline development and implementation in a narrow sense*, Medical Infobahn for Europe, Stud. Health Technol. Inform. **77** (2000), 790-794.
20. J. Karlsson, P. Eklund, C.-G. Hallgren, J. Sjödin, *Data warehousing as a basis for web-based documentation of data mining and analysis*, Medical Informatics Europe '99, Stud. Health Technol. Inform. **68** (1999), 423-427.

21. E. Kerkri, C. Quantin, K. Yetongnon, F. A. Allaert, L. Dusserre, *Application of the medical data warehousing architecture epidware to epidemiological follow-up*, Stud. Health Technol. Inform., **68** (1999), 414-418.
22. A. Kircher, J. Antonsson, A. Babic, H.Casimir-Ahn, *Quantitative data analysis for exploring outcomes in cardiac surgery*, Studies in Health Technology and Informatics, **68** (1990), 457-460.
23. H. Lindgren, P. Eklund, *Logic of dementia guidelines in a Probabilistic Argumentation Framework*, ECSQARU 2005 (Ed. L. Godo), LNAI 3571, 2005, 341-352.
24. S. Parsons, *A Proof Theoretic Approach to Qualitative Probabilistic Reasoning*, Int. J. Approximate Reasoning, **19** (1998), 265-297.
25. S. Pearson, C. Margolis, S. Davis, L. Schreier, H. Sokol, L. Gottlieb, *Is consensus reproducible? A study of an algorithmic guidelines development process. Medical Care* **33** (1995), 643-660.
26. M. Persson, J. Bohlin, P. Eklund, *Development and maintenance of guideline-based decision support for pharmacological treatment of hypertension*, Comp. Meth. Progr. Biomed., **61** (2000), 209-219.
27. M. Persson, T. Mjörndal, B. Carlberg, J. Bohlin, L. H. Lindholm, *Evaluation of a computer-based decision support system for treatment of hypertension with drugs: Retrospective, nonintervention testing of cost and guideline adherence*, Internal Medicine, **247** (2000), 87-93.
28. P-E. Puddu, G. Brancaccio, M. Leacche, F. Monti, M. Lanti, A. Menotti, C. Gaudio, U. Papalia, B. Marino, *Prediction of early and delayed post-operative deaths after coronary artery bypass surgery alone in Italy*, Italian Heart Journal, **3** (2002), 166-181.
29. J. Rauch, *Some Remarks on Computer Realisations of GUHA Procedures* International Journal of Man-Machine Studies **10** (1978), 23-28.
30. J. Rauch, *Logic of Association Rules*, Applied Intelligence **22** (2005), 9-28.
31. J. Rauch, *Definability of Association Rules in Predicate Calculus* In: T. Y. Lin, S. Ohsuga, C. J. Liau, X. Hu (eds), Foundations and Novel Approaches in Data Mining, Springer-Verlag, 2005, 23-40.
32. J. Rauch, M. Šimůnek, *Alternative Approach to Mining Association Rules* In: Lin T Y, Ohsuga S, Liau C J, and Tsumoto S (eds) *Data Mining: Foundations, Methods, and Applications*, Springer-Verlag, 2005, 219-238.
33. *Diagnostic and Statistical Manual of Mental Disorders*, Fourth Edition, Text Revision (DSM-IV-TR), American Psychiatric Association, 1994.
34. *The sixth report of the joint national committee on prevention detection, evaluation, and treatment of high blood pressure*, Technical Report 98-4080, National Institutes of Health, 1997.
35. *LISp-Miner*, http://lispminer.vse.cz
36. M. van Wijk, J. van der Lei, M. Mosseveld, A. Bohnen, J. van Bemmel, *Compliance of General Practitioners with a Guideline-based Decision Support System for Ordering Blood Tests*, Journal of Clinical Chemistry, **48** (2002), 55-60.

Generalizing and Modifying
the Hoede-Bakker Index

Agnieszka Rusinowska[1,2] and Harrie de Swart[3]

[1] Radboud University Nijmegen, Nijmegen School of Management, P.O. Box 9108
6500 HK Nijmegen, The Netherlands
a.rusinowska@fm.ru.nl
[2] Warsaw School of Economics, Department of Mathematical Economics
Al. Niepodleglosci 162, 02-554 Warsaw, Poland
arusin@sgh.waw.pl
[3] Tilburg University, Faculty of Philosophy, P.O. Box 90153
5000 LE Tilburg, The Netherlands
H.C.M.deSwart@uvt.nl

Abstract. In this paper, we generalize the Hoede-Bakker index, which is a measure for the power of agents in a network, taking into account the mutual influences of the agents. We adopt sets of axioms different from the one adopted in the original definition. In particular, we remove an original assumption according to which changing all inclinations of the players leads to the opposite group decision. Several examples showing the usefulness of this generalization are constructed. In particular, we may apply the generalized Hoede-Bakker index to a game with a vetoer. Next, the relation between the generalized Hoede-Bakker index and the Penrose measure is analysed. Moreover, we introduce several modifications of the Hoede-Bakker index which lead to the Coleman indices, the Rae index, and the König-Bräuninger index. In order to show the relation between the generalized or the modified Hoede-Bakker index and the other power indices, we use the probabilistic approach.

Keywords: Hoede-Bakker index, inclination vector, the Penrose measure, the Coleman indices, the Rae index, the König-Bräuninger index.

1 Introduction

In order to measure voting strength of actors in a voting situation, a number of power indices have been proposed in the course of more than fifty years (for instance, Penrose [38], Shapley and Shubik [45], see also Shapley [44], Banzhaf [1], Rae [39], Coleman [5], [6], Deegan and Packel [7], Johnston [26], Dubey and Shapley [8], Holler [17], Holler and Packel [18], König-Bräuninger [25]). For an extensive analysis of most of the power indices see, first of all, Felsenthal and Machover [10], but also, for instance, Lucas [32], Owen [37], and Straffin [49]. Also to be found in the literature are some values for games with a priori unions (Owen [35], [36]). Basically, there are two approaches to analyze power indices, that is, the axiomatic approach and the probabilistic one. Laruelle and Valenciano [29]

H. de Swart et al. (Eds.): TARSKI II, LNAI 4342, pp. 60–88, 2006.
© Springer-Verlag Berlin Heidelberg 2006

present a probabilistic model in which they re-examine the concepts of 'success' and 'decisiveness', and in which they also consider some conditional variants.

Apart from the theoretical analysis, one may find in the literature applications of power indices. So far, applications of power indices can be found especially in the field of decision-making in the European Union (see, for instance, Felsenthal and Machover [10], [11], [12], Hosli [19], [20], [21], [22], [23], [24], Laruelle [27], Laruelle and Widgren [30], Leech [31], Nurmi and Meskanen [33], Sutter [50], Widgren [52]). However, the power index approach can be applied equally well to national legislatures and parliaments (see, for instance, Sosnowska [46], [47], Rusinowska [40], Van Deemen and Rusinowska [51], Rusinowska and Van Deemen [42]).

Hoede and Bakker [15] introduced the concept of decisional power, which is still not widely known, although it surely deserves broader attention. This index takes the inclinations of the players into account, as well as the social structure in which players may influence each other. The essential point of the Hoede-Bakker index is the distinction between the inclination (to say 'yes' or 'no') and the final decision (apparent in a vote). Preliminary research on the Hoede-Bakker index has been initiated in Stokman and Willer [48], where an application of the Hoede-Bakker index to coalition formation has been presented, and in Rusinowska and De Swart [41]. In the latter paper, the authors investigate some properties of the Hoede-Bakker index. They check, in particular, whether the Hoede-Bakker index displays some voting power paradoxes and whether it satisfies some postulates for power indices. The paradoxes re-defined and checked for the Hoede-Bakker index were the redistribution paradox (Fischer and Schotter [14], see also Schotter [43]), the paradox of new members (Brams [2], Brams and Affuso [3]), and the paradox of large size (Brams [2]). The postulates re-defined for the Hoede-Bakker index were, in particular, the monotonicity postulate, the donation postulate, and the bloc postulate. An extensive theoretical analysis of these and some other postulates for power indices and voting power paradoxes is given, for instance, in Felsenthal and Machover [9], [10], Felsenthal, Machover and Zwicker [13], and in Laruelle [28].

The aim of this paper is to introduce and analyze a generalization and some modifications of the Hoede-Bakker index. The structure of this paper is as follows. In Section 2, using the probabilistic approach, we recapitulate the definitions of the Rae index, the Penrose measure (often called the absolute or non-normalized Banzhaf index), the Coleman indices, and the König-Bräuninger index. Section 3 concerns the original Hoede-Bakker index. We start with recapitulating the axioms and the definition of this index as adopted by Hoede and Bakker [15]. Next, in order to show the usefulness of generalizing the original Hoede-Bakker index, we present an example with a vetoer. In Section 4, the generalization of the Hoede-Bakker index is presented, and the relation between this generalized Hoede-Bakker index and the Penrose measure is established. Section 5 concerns some modifications of the Hoede-Bakker index. We show the relations between the modifications defined and the Coleman indices, the Rae index, and the König-Bräuninger index. In Section 6, several examples are constructed in which the removed axiom of Hoede and Bakker is violated. Finally, Section 7 contains conclusions.

2   Power Indices – Probabilistic Approach

In this Section, using the probabilistic approach, we recapitulate the definitions of several power indices. We present very briefly some main concepts described in Laruelle and Valenciano [29]. For a probabilistic approach to power indices and an extensive analysis, see also Felsenthal and Machover [10].

Once a proposal is submitted, voters cast votes, voting either 'yes' (abstention included) or 'no'. A vote configuration is a possible result of voting. Hence, for n voters, there are 2^n possible vote configurations. The vote configuration S refers to the result of voting where all voters in S vote 'yes', and all voters in $N \setminus S$ vote 'no', where $N = \{1, 2, ..., n\}$. The vote configurations leading to the passage of a proposal are called winning configurations. Let W be the set of winning configurations representing an N-voting rule. A voting rule is assumed to satisfy the following conditions: (i) $N \in W$; (ii) $\emptyset \notin W$; (iii) If $S \in W$, then $T \in W$ for any T containing S; (iv) If $S \in W$, then $N \setminus S \notin W$.

A probability distribution over all possible vote configurations is incorporated into the model. A probability distribution may be represented by a map $p : 2^N \to [0, 1]$, associating with each vote configuration S its probability $p(S)$ to occur. That is, $p(S)$ is the probability that all voters in S vote 'yes', and all voters in $N \setminus S$ vote 'no'. Laruelle and Valenciano [29] introduced the following definitions and formulae derived from the definitions:

Definition 2.1. *Let (W, p) be an N-voting situation, where W is the voting rule to be used and p is the probability distribution over vote configurations, and let $k \in N$. Then:*

$$\Omega_k(W, p) := Prob(k \text{ is successful}) = \sum_{S: k \in S \in W} p(S) + \sum_{S: k \notin S \notin W} p(S) \quad (1)$$

$$\Phi_k(W, p) := Prob(k \text{ is decisive}) = \sum_{\substack{S : k \in S \in W \\ S \setminus \{k\} \notin W}} p(S) + \sum_{\substack{S : k \notin S \notin W \\ S \cup \{k\} \in W}} p(S) \quad (2)$$

$$\Lambda_k(W, p) := Prob(k \text{ is lucky}) = \sum_{\substack{S : k \in S \\ S \setminus \{k\} \in W}} p(S) + \sum_{\substack{S : k \notin S \\ S \cup \{k\} \notin W}} p(S) \quad (3)$$

$$\Omega_k(W, p) = \Phi_k(W, p) + \Lambda_k(W, p) \quad (4)$$

$$\alpha(W, p) := Prob(acceptance) = \sum_{S: S \in W} p(S) \quad (5)$$

$$\gamma_k(p) := Prob(k \text{ votes 'yes'}) = \sum_{S: k \in S} p(S) \quad (6)$$

$$\Omega_k^{Acc}(W,p) := Prob(k \text{ is } successful \mid \text{the proposal is accepted}) =$$

$$= \frac{\sum_{S:k\in S\in W} p(S)}{\alpha(W,p)} \tag{7}$$

$$\Omega_k^{Rej}(W,p) := Prob(k \text{ is } successful \mid \text{the proposal is rejected}) =$$

$$= \frac{\sum_{S:k\notin S\notin W} p(S)}{1 - \alpha(W,p)} \tag{8}$$

$$\Omega_k^+(W,p) := Prob(k \text{ is } successful \mid k \text{ votes 'yes'}) = \frac{\sum_{S:k\in S\in W} p(S)}{\gamma_k(p)} \tag{9}$$

$$\Omega_k^-(W,p) := Prob(k \text{ is } successful \mid k \text{ votes 'no'}) = \frac{\sum_{S:k\notin S\notin W} p(S)}{1 - \gamma_k(p)} \tag{10}$$

$$\Phi_k^{Acc}(W,p) := Prob(k \text{ is } decisive \mid \text{the proposal is accepted}) =$$

$$= \frac{\sum_{\substack{S\,:\,k\,\in\,S\,\in\,W \\ S\setminus\{k\}\notin W}} p(S)}{\alpha(W,p)} \tag{11}$$

$$\Phi_k^{Rej}(W,p) := Prob(k \text{ is } decisive \mid \text{the proposal is rejected}) =$$

$$= \frac{\sum_{\substack{S\,:\,k\,\notin\,S\,\notin\,W \\ S\cup\{k\}\in W}} p(S)}{1 - \alpha(W,p)} \tag{12}$$

$$\Phi_k^+(W,p) := Prob(k \text{ is } decisive \mid k \text{ votes 'yes'}) = \frac{\sum_{\substack{S\,:\,k\,\in\,S\,\in\,W \\ S\setminus\{k\}\notin W}} p(S)}{\gamma_k(p)} \tag{13}$$

$$\Phi_k^-(W,p) := Prob(k \text{ is } decisive \mid k \text{ votes 'no'}) = \frac{\sum_{\substack{S\,:\,k\,\notin\,S\,\notin\,W \\ S\cup\{k\}\in W}} p(S)}{1 - \gamma_k(p)} \tag{14}$$

Laruelle and Valenciano [29] showed (see also Felsenthal and Machover [10]) that for a given probability distribution p, the three measures $\Phi(W,p)$, $\Phi^+(W,p)$, and $\Phi^-(W,p)$, coincide for every voting rule W, if and only if the vote of every voter is independent from the vote of the other voters.

Let us assume now that all vote configurations are equally probable, that is:

$$\forall S \subseteq N \; [p^*(S) := \frac{1}{2^n}]. \tag{15}$$

Some power indices can be seen as (unconditional or conditional) probabilities in the sense of Definition 2.1 for the probability distribution p^* assumed in (15). One may derive the following equalities (Laruelle and Valenciano [29]), for voting rule W and $k \in N$:

- *Rae index* (Rae [39], see also Dubey and Shapley [8])

$$Rae_k(W) = \Omega_k(W,p^*) = \sum_{S:k\in S\in W} \frac{1}{2^n} + \sum_{S:k\notin S\notin W} \frac{1}{2^n} \tag{16}$$

- *Penrose measure* (also called the *absolute Banzhaf index*, or the *non-normalized Banzhaf index*) (Penrose [38], Banzhaf [1], see also Owen [34])

$$PB_k(W) = \frac{number\ of\ winning\ configurations\ in\ which\ k\ is\ decisive}{total\ number\ of\ voting\ configurations\ containing\ k} \tag{17}$$

$$PB_k(W) = \Phi_k^+(W, p^*) = \Phi_k^-(W, p^*) = \Phi_k(W, p^*) =$$

$$= \sum_{\substack{S\ :\ k\ \in\ S\ \in\ W \\ S\ \setminus\ \{k\}\ \notin\ W}} \frac{1}{2^n} + \sum_{\substack{S\ :\ k\ \notin\ S\ \notin\ W \\ S\ \cup\ \{k\}\ \in\ W}} \frac{1}{2^n} \tag{18}$$

- *Coleman's 'power of a collectivity to act'* (Coleman [5], [6])

$$A(W) = \frac{number\ of\ winning\ configurations}{total\ number\ of\ voting\ configurations} \tag{19}$$

$$A(W) = \alpha(W, p^*) = \sum_{S:S\in W} \frac{1}{2^n} \tag{20}$$

- *Coleman's index 'to prevent action'* (Coleman [5], [6])

$$Col_k^P(W) = \frac{number\ of\ winning\ configurations\ in\ which\ k\ is\ decisive}{total\ number\ of\ winning\ configurations} \tag{21}$$

$$Col_k^P(W) = \Phi_k^{Acc}(W, p^*) = \frac{\sum_{\substack{S\ :\ k\ \in\ S\ \in\ W \\ S\ \setminus\ \{k\}\ \notin\ W}} \frac{1}{2^n}}{\sum_{S:S\in W} \frac{1}{2^n}} \tag{22}$$

- *Coleman's index 'to initiate action'* (Coleman [5], [6])

$$Col_k^I(W) = \frac{number\ of\ losing\ configurations\ in\ which\ k\ is\ decisive}{total\ number\ of\ losing\ configurations} \tag{23}$$

$$Col_k^I(W) = \Phi_k^{Rej}(W, p^*) = \frac{\sum_{\substack{S\ :\ k\ \notin\ S\ \notin\ W \\ S\ \cup\ \{k\}\ \in\ W}} \frac{1}{2^n}}{1 - \sum_{S:S\in W} \frac{1}{2^n}} \tag{24}$$

- *König-Bräuninger inclusiveness index* (König and Bräuninger [25])

$$KB_k(W) = \frac{number\ of\ winning\ configurations\ containing\ k}{total\ number\ of\ winning\ configurations} \tag{25}$$

$$KB_k(W) = \Omega_k^{Acc}(W, p^*) = \frac{\sum_{S:k\in S\in W} \frac{1}{2^n}}{\sum_{S:S\in W} \frac{1}{2^n}} \tag{26}$$

3   The Hoede-Bakker Index

Hoede and Bakker [15] introduced the concept of decisional power, the so called Hoede-Bakker index. In this section, we recapitulate the definition of this index. We consider the situation in which $n \geq 1$ players make a decision about a certain point at issue (for instance, to accept or to reject a bill, a candidate, etc). Let N denote the set of all players (actors, voters). Hence, $N = \{1, ..., n\}$. With respect to the point at issue, each player has an inclination either to say 'yes' (denoted by 1) or 'no' (denoted by -1)[1]. For n players, we have therefore 2^n possible *inclination vectors*, that is, n-vectors consisting of ones and minus ones. Let i denote an inclination vector, and let I be the set of all n-vectors. Due to the influences of other players in the network, each inclination vector $i \in I$ is transformed into a *decision vector*, denoted by b. Formally, such a transformation may be represented by an operator $B : I \rightarrow B(I)$, that is, $b = Bi$, where $B(I)$ denotes the set of all decision vectors. The decision vector b is an n-vector consisting of ones and minus ones and indicating the decisions made by all players. Due to influences of the other actors, the final decision of an actor may be different from his original inclination. Furthermore, the *group decision* $gd : B(I) \rightarrow \{+1, -1\}$ is introduced. It is a function defined on the decision vectors b, having the value $+1$ if the group decision is 'yes', and the value -1 if the group decision is 'no'.

Hoede and Bakker [15] adopted the following two axioms which have to be satisfied by B and gd:

AXIOM (A-0):

$$\forall i \in I \; [gd(Bi^c) = -gd(Bi)], \tag{27}$$

where $i^c = (i_1^c, ..., i_n^c)$ is the *complement* of inclination vector $i = (i_1, ..., i_n)$, that is, for each $k \in \{1, ..., n\}$

$$i_k^c = \begin{cases} +1 & if \quad i_k = -1 \\ -1 & if \quad i_k = +1 \end{cases}. \tag{28}$$

AXIOM (A-1):

$$\forall i \in I \; \forall i' \in I \; [i \leq i' \; \Rightarrow \; gd(Bi) \leq gd(Bi')], \tag{29}$$

where $i \leq i'$ is defined in the following way:

$$i \leq i' \iff \{k \in N \mid i_k = +1\} \subseteq \{k \in N \mid i_k' = +1\}. \tag{30}$$

Moreover, by $i < i'$ we mean: $i \leq i'$ and $i \neq i'$.

---

[1] In the original paper by Hoede and Bakker [15], the inclination 'no' is denoted by 0. In order to simplify some notations introduced later on, we use the symbol -1 instead of 0.

Definition 3.1. *Given B and gd, the decisional power index (the Hoede-Bakker index) of a player $k \in N$ is given by*

$$HB(k) = \frac{1}{2^{n-1}} \cdot \sum_{\{i:\ i_k = +1\}} gd(Bi). \tag{31}$$

Definition 3.1 assumes axiom (A-0) to be satisfied. According to this axiom, changing all inclinations leads to the opposite group decision. Hence, in Definition 3.1 the given player k is assumed to have an inclination 'yes', and then the group decisions for the 2^{n-1} inclination vectors with inclination 'yes' of the given player are considered. Since (A-0) is adopted, Hoede and Bakker do not consider all the remaining 2^{n-1} inclination vectors with an inclination 'no' of the given player.

We find axiom (A-0) too restrictive, since one may describe situations for which this axiom is not satisfied. Let us consider the following example:

Example 3.1. Suppose that there are 3 players, A, B, and C, and player A happens to be a vetoer. We may think of a weighted voting game with the following weights of the players: $w(A) = 2$, $w(B) = w(C) = 1$, and the quota $q = 3$. Hence, the sets of winning coalitions and minimal winning coalitions are equal to $\{AB, AC, ABC\}$ and $\{AB, AC\}$, respectively. Player A, belonging to each minimal winning coalition, is a vetoer. Table 3.1 shows the group decision for this example, assuming $Bi = i$, and the group decision is 'yes' iff player A with at least one of the other players says 'yes'.

Table 3.1. Group decision for Example 3.1

inclination i	$gd(Bi)$	inclination i	$gd(Bi)$
$(1,1,1)$	$+1$	$(-1,-1,-1)$	-1
$(1,1,-1)$	$+1$	$(-1,-1,1)$	-1
$(1,-1,1)$	$+1$	$(-1,1,-1)$	-1
$(-1,1,1)$	-1	$(1,-1,-1)$	-1

Note that axiom (A-0) is NOT satisfied in this example, since $gd(B(-1,1,1)) = \cdot gd(B(1,-1,-1)) = -1$. This example suggests that, when calculating the Hoede-Bakker index, both inclination vectors $(-1,1,1)$ and $(1,-1,-1)$ should be taken into consideration. Nevertheless, let us still use Definition 3.1, ignoring the violation of axiom (A-0). We find then $HB(A) = \frac{1}{2}$, and $HB(B) = HB(C) = 0$. The Hoede-Bakker indices of players B and C are both equal to 0, although none of these players is a dummy in this game.

4   Generalization of the Hoede-Bakker Index

Inspired by Example 3.1, we introduce a generalization of the Hoede-Bakker index, as recapitulated in Definition 3.1. We consider the same situation as described in Section 3, but we adopt a different set of axioms. Note that neither in the axioms adopted nor in the original definition of the Hoede-Bakker index, the operators B and gd are considered separately. When calculating the Hoede-Bakker index, only the relation between an inclination vector i and the group decision $gd(Bi)$ is taken into account. One may argue that the operators B and gd should be separated. We impose the following conditions on the operator B:

AXIOM (B-1):

$$\forall i \in I \ \forall i' \in I \ [i \leq i' \ \Rightarrow \ Bi \leq Bi']^2 \tag{32}$$

AXIOM (B-2):

$$B(+1, ..., +1) = (+1, ..., +1) \tag{33}$$

AXIOM (B-3):

$$B(-1, ..., -1) = (-1, ... -1), \tag{34}$$

and the following conditions on the operator gd:

AXIOM (G-1):

$$\forall i \in I \ \forall i' \in I \ [Bi \leq Bi' \ \Rightarrow \ gd(Bi) \leq gd(Bi')] \tag{35}$$

AXIOM (G-2):

$$gd(+1, ..., +1) = +1 \tag{36}$$

AXIOM (G-3):

$$gd(-1, ..., -1) = -1. \tag{37}$$

One may still adopt a different set of axioms, keeping the operators B and gd together, as in the original paper by Hoede and Bakker. We resign from axiom (A-0), keep axiom (A-1) and replace (A-0) by the two weaker axioms (A-2) and (A-3):

AXIOM (A-1):

$$\forall i \in I \ \forall i' \in I \ [i \leq i' \ \Rightarrow \ gd(Bi) \leq gd(Bi')] \tag{38}$$

AXIOM (A-2):

$$gd(B(+1, ..., +1)) = +1 \tag{39}$$

AXIOM (A-3):

$$gd(B(-1, ..., -1)) = -1. \tag{40}$$

Axiom (A-1) says that a group decision 'yes' cannot be changed into 'no' if the set of players with inclination 'yes' is enlarged. Axiom (A-2) means that if all players have the inclination 'yes', then the group decision is also 'yes'. According to axiom (A-3), if all actors have the inclination 'no', then the group decision will be 'no'. Note that

---

^{2} For the definition of $i \leq i'$ see equation (30).

Fact 4.1. *The set of axioms (B-1), (B-2), (B-3), (G-1), (G-2), (G-3) implies the set of axioms (A-1), (A-2), (A-3).*

Proof. Axioms (B-j) and (G-j) imply (A-j), for $j = 1, 2, 3$. □

Note, however, that Example 3.1 shows that there is no implication between the set of axioms (B-1), (B-2), (B-3), (G-1), (G-2), (G-3) and the set (A-0), (A-1). Of course, the two axioms (A-0) and (A-1) imply the three axioms (A-1), (A-2), and (A-3).

Definition 4.1. *Let I be the set of all inclination vectors. We introduce a bijection f from I to the power set of N (that is, a 1-1 map from I onto the set of all coalitions), $f : I \to 2^N$, such that*

$$\forall i \in I \ [f(i) = \{k \in N \mid i_k = +1\}]. \tag{41}$$

In particular, $f(i_1, ..., i_n) = N$ iff $i_k = +1$ for each $k = 1, 2, ..., n$, and $f(i_1, ..., i_n) = \emptyset$ iff $i_k = -1$ for each $k = 1, 2, ..., n$. Moreover, given B and gd:

- *a coalition $f(i)$, where $i \in I$, is said to be winning if $gd(Bi) = +1$,*
- *a coalition $f(i)$ is said to be losing if $gd(Bi) = -1$,*
- *a coalition $f(i)$ is said to be minimal winning if $gd(Bi) = +1$ and for each $i' < i$, $gd(Bi') = -1$,*
- *player $k \in N$ is a dummy if there is NO minimal winning coalition $f(i)$ such that $i_k = +1$,*
- *player $k \in N$ is a vetoer if for each minimal winning coalition $f(i)$, $i_k = +1$.*

Remark 4.1. In the model recapitulated in Section 2 (Laruelle and Valenciano [29]), four conditions (i)-(iv) have been imposed on a voting rule. In fact, axioms (A-1), (A-2), and (A-3) imposed on B and gd in our model, correspond to their conditions (iii), (i), and (ii), respectively. A condition corresponding to their condition (iv) would look like

$$\forall i \in I \ [gd(Bi) = +1 \ \Rightarrow \ gd(Bi^c) = -1],$$

where i^c is the complement of the inclination vector i defined by equation (28). We do not impose this condition in our model, what means that we allow the possibility that $gd(Bi) = gd(Bi^c) = +1$ for some $i \in I$. In other words, we do not like to exclude from our considerations games which are not *proper* (that is, games in which a coalition and its complement may be both winning). Nevertheless, even without this extra axiom corresponding to condition (iv), we can apply the probabilistic model as presented by Laruelle and Valenciano [29] to our model with the generalized Hoede-Bakker index, since all results recapitulated in Section 2 hold without condition (iv)[3]. Having only these three axioms (A-1), (A-2),

---

[3] We like to thank the authors Laruelle and Valenciano for confirming this fact.

and (A-3) adopted, we may consider situations such that $gd(Bi) = gd(Bi^c)$ for some $i \in I$, meaning that either $gd(Bi) = gd(Bi^c) = +1$ (as mentioned before) or $gd(Bi) = gd(Bi^c) = -1$ (as in Table 3.1).

Assuming all axioms (B-1), (B-2), (B-3), (G-1), (G-2), and (G-3) to be satisfied, we introduce the following definition:

Definition 4.2. *Given B and gd, the generalized Hoede-Bakker index of a player $k \in N$ is given by*

$$GHB(k) = \frac{1}{2^n} \cdot \left(\sum_{\{i:\ i_k=+1\}} gd(Bi) - \sum_{\{i:\ i_k=-1\}} gd(Bi) \right) = \frac{HB^+(k) + HB^-(k)}{2},$$

(42)

where HB^+ and HB^- are defined for each $k \in N$ in the following way:

$$HB^+(k) = \frac{1}{2^{n-1}} \cdot \sum_{\{i:\ i_k=+1\}} gd(Bi) = HB(k)$$

(43)

$$HB^-(k) = -\frac{1}{2^{n-1}} \cdot \sum_{\{i:\ i_k=-1\}} gd(Bi).$$

(44)

Remark 4.2. Note that HB^+ is simply the 'old' Hoede-Bakker index HB as defined by Hoede and Bakker [15] and recapitulated here in Definition 3.1. Without axiom (A-0), HB^+ does NOT have to be equal to GHB. But, of course, if axiom (A-0) is satisfied, then $GHB(k) = HB^+(k) = HB^-(k)$ for each $k \in N$. In Example 3.1, $GHB(A) = \frac{1}{8}(2+4) = \frac{3}{4}$ and $GHB(B) = GHB(C) = \frac{1}{8}(0+2) = \frac{1}{4}$, while we found earlier that $HB(A) = \frac{1}{2}$ and $HB(B) = HB(C) = 0$.

Fact 4.2. *We have:*

(a) for each $k \in N$, $HB^+(k) \leq 1$
(b) for each $k \in N$, $HB^-(k) \leq 1$
(c) for each $k \in N$, $0 \leq GHB(k) \leq 1$.

Proof. (a) Let us consider an arbitrary player $k \in N$. Since $gd(Bi) \leq 1$, and for a given player k there are 2^{n-1} inclination vectors i such that $i_k = +1$, from (43) we get $HB^+(k) \leq 1$.
(b) By analogy, since $gd(Bi) \geq -1$, and for a given player k there are 2^{n-1} inclination vectors i such that $i_k = -1$, (44) gives $HB^-(k) \leq 1$.
(c) Since $HB^+(k) \leq 1$ and $HB^-(k) \leq 1$, by virtue of (42) we get immediately that $GHB(k) \leq 1$ for each $k \in N$. Let us consider an arbitrary player $k \in N$. Note that for each inclination vector $i = (i_1, ..., i_n)$ such that $i_k = +1$, there is $i' = (i'_1, ..., i'_n)$ such that

$$i'_j = \begin{cases} i_j & for \quad j \neq k \\ -1 & for \quad j = k \end{cases}.$$

(45)

By virtue of (38), that is, axiom (A-1), since $i' \leq i$, $gd(Bi') \leq gd(Bi)$. Hence, $gd(Bi) - gd(Bi') \geq 0$. Note that the result of subtracting the two sums in (42) is, in fact, equal to the sum of 2^{n-1} non-negative expressions $gd(Bi) - gd(Bi') \geq 0$ with the property (45). Hence, $GHB(k) \geq 0$. $\qquad\square$

Note that we do not prove that $HB^+(k) \geq 0$ and $HB^-(k) \geq 0$ for each $k \in N$. In fact, one of these values may be negative what will be shown in examples in Section 6. Of course, since $GHB(k) \geq 0$ for each $k \in N$, it is impossible that both $HB^+(k)$ and $HB^-(k)$ will be negative for the same player k.

Given B and gd, we introduce the following notations for each player $k \in N$:

$$I_k^{++} = |\{i \in I \mid i_k = +1 \ \wedge \ gd(Bi) = +1\}| \tag{46}$$

I_k^{++} - number of inclination vectors with inclination 'yes' of player k that lead to the group decision 'yes'

$$I_k^{+-} = |\{i \in I \mid i_k = +1 \ \wedge \ gd(Bi) = -1\}| \tag{47}$$

I_k^{+-} - number of inclination vectors with inclination 'yes' of player k that lead to the group decision 'no'

$$I_k^{-+} = |\{i \in I \mid i_k = -1 \ \wedge \ gd(Bi) = +1\}| \tag{48}$$

I_k^{-+} - number of inclination vectors with inclination 'no' of player k that lead to the group decision 'yes'

$$I_k^{--} = |\{i \in I \mid i_k = -1 \ \wedge \ gd(Bi) = -1\}| \tag{49}$$

I_k^{--} - number of inclination vectors with inclination 'no' of player k that lead to the group decision 'no'

Next, we introduce the following definition:

Definition 4.3. *Given B and gd, we introduce for each player $k \in N$:*

$$GHB^+(k) = \frac{I_k^{++} - I_k^{-+}}{2^{n-1}} \tag{50}$$

$$GHB^-(k) = \frac{I_k^{--} - I_k^{+-}}{2^{n-1}}. \tag{51}$$

One may easily prove the following fact:

Fact 4.3. *Given B and gd, for each $k \in N$:*

$$GHB(k) = GHB^+(k) = GHB^-(k) \tag{52}$$

$$HB^+(k) = \frac{I_k^{++} - I_k^{+-}}{2^{n-1}} \tag{53}$$

$$HB^-(k) = \frac{I_k^{--} - I_k^{-+}}{2^{n-1}} \tag{54}$$

Proof. Let us consider an arbitrary player $k \in N$. Note that

$$\sum_{\{i:\ i_k=+1\}} gd(Bi) = I_k^{++} - I_k^{+-} \tag{55}$$

$$\sum_{\{i:\ i_k=-1\}} gd(Bi) = I_k^{-+} - I_k^{--}. \tag{56}$$

Hence, by virtue of (43) and (44), we get (53) and (54), respectively. Note that

$$\forall k \in N \ [I_k^{++} + I_k^{+-} = I_k^{--} + I_k^{-+} = 2^{n-1}]. \tag{57}$$

Hence, $I_k^{++} - I_k^{-+} = I_k^{--} - I_k^{+-}$ for each $k \in N$, and therefore, by virtue of (50) and (51), $GHB^+(k) = GHB^-(k)$ for each $k \in N$. Moreover,

$$GHB(k) = \frac{I_k^{++} - I_k^{+-} + I_k^{--} - I_k^{-+}}{2^n} = \frac{1}{2} \cdot \left(\frac{I_k^{++} - I_k^{-+}}{2^{n-1}} + \frac{I_k^{--} - I_k^{+-}}{2^{n-1}} \right) =$$

$$\frac{GHB^+(k) + GHB^-(k)}{2} = GHB^+(k) = GHB^-(k). \qquad \square$$

Remember that in Example 3.1 the Hoede-Bakker indices of players B and C are both equal to 0, although none of these players is a dummy in the game in question. However, for the generalized Hoede-Bakker index we have the following:

Fact 4.4. *Player $k \in N$ is a dummy if and only if $GHB(k) = 0$.*

Proof. Let us consider an arbitrary player $k \in N$. There are 2^{n-1} inclination vectors i such that $i_k = +1$, and 2^{n-1} inclination vectors i such that $i_k = -1$. Moreover, note that for each i such that $i_k = +1$, there is i' such that $i'_k = -1$ and $i'_j = i_j$ for each $j \neq k$. This means that $i' < i$, and hence, by virtue of axiom (A-1), $I_k^{++} \geq I_k^{-+}$.

($\Leftarrow$) Suppose that player $k \in N$ is not a dummy. Hence, there is a coalition $f(i)$ such that $i_k = +1$, $gd(Bi) = +1$, and $gd(Bi'') = -1$ for each $i'' < i$. Hence, in particular, $gd(Bi') = -1$ for i' such that $i'_k = -1$ and $i'_j = i_j$ for each $j \neq k$. But this means that $I_k^{++} > I_k^{-+}$, and therefore, by virtue of (50) and (52), $GHB(k) > 0$.

($\Rightarrow$) Suppose now that $GHB(k) > 0$. By virtue of (50) and (52), this means that $I_k^{++} > I_k^{-+}$. Hence, there is i such that $i_k = +1$, $gd(Bi) = +1$, and $gd(Bi') = -1$ for i' such that $i'_k = -1$ and $i'_j = i_j$ for each $j \neq k$. But this means that player k does affect the outcome, and therefore he is not a dummy. $\qquad \square$

In this paper, we assume all inclination vectors to be equally probable, that is, similarly as defined in (15):

$$\forall i \in I \ [p^*(i) := \frac{1}{2^n}]. \tag{58}$$

We will not mention assumption (58) explicitly when presenting our results. Nevertheless, condition (58) is assumed throughout this paper. In fact, since there is a bijection between coalitions and inclination vectors (see Definition 4.1), conditions (15) and (58) are expressing the same assumption.

Remark 4.3. As was mentioned before, if all vote configurations are assumed to be equally probable (condition (15)), then some power indices can be seen as (unconditional or conditional) probabilities. An inclination vector i in our model corresponds uniquely with a vote configuration $f(i) = \{k \in N \mid i_k = +1\}$, and an inclination vector i such that $gd(Bi) = +1$ corresponds with a winning coalition $f(i)$. All the inclination vectors in our model are assumed to be equally probable (condition (58)). A decision of an actor may depend on the inclinations of the others (and it frequently does), but the inclinations of the players are assumed to be independent of each other. With the interpretation of vote configurations and winning coalitions just mentioned, we get the following proposition:

Proposition 4.1. Let $\Phi_k(W,p)$, $\Phi_k^+(W,p)$, $\Phi_k^-(W,p)$, and p^* be as defined by equations (2), (13), (14), and (15), respectively. Then

$$\forall k \in N \ [\Phi_k^+(W,p^*) = GHB^+(k)] \tag{59}$$

$$\forall k \in N \ [\Phi_k^-(W,p^*) = GHB^-(k)] \tag{60}$$

$$\forall k \in N \ [\Phi_k(W,p^*) = GHB(k)]. \tag{61}$$

Proof. The notion of a winning coalition for our model has been introduced in Definition 4.1. Re-writing some notions introduced in Definition 2.1 for our model, we get for each $k \in N$:

$$\gamma_k(p^*) = \sum_{S:k \in S} p^*(S) = \frac{|\{i \in I \mid i_k = +1\}|}{2^n} = \frac{1}{2} \tag{62}$$

$$\sum_{S:k \in S \in W} p^*(S) = \frac{|\{i \in I \mid i_k = +1 \ \wedge \ gd(Bi) = +1\}|}{2^n} = \frac{I_k^{++}}{2^n} \tag{63}$$

$$\sum_{\substack{S:k \in S \in W \\ S \setminus \{k\} \in W}} p^*(S) = \frac{|\{i \in I \mid i_k = -1 \ \wedge \ gd(Bi) = +1\}|}{2^n} = \frac{I_k^{-+}}{2^n}. \tag{64}$$

Hence,

$$\sum_{\substack{S:k \in S \in W \\ S \setminus \{k\} \notin W}} p^*(S) = \frac{I_k^{++} - I_k^{-+}}{2^n}, \tag{65}$$

and therefore, applying (62) and (65) to (13), and comparing it with (50), we get $\Phi_k^+(W,p^*) = GHB^+(k)$.

By virtue of (62),

$$1 - \gamma_k(p^*) = \frac{1}{2}. \tag{66}$$

Moreover, we have

$$\sum_{S:k\notin S\notin W} p^*(S) = \frac{|\{i\in I \mid i_k = -1 \ \wedge \ gd(Bi) = -1\}|}{2^n} = \frac{I_k^{--}}{2^n} \qquad (67)$$

$$\sum_{\substack{S\,:\,k\notin S\notin W \\ S\cup\{k\}\notin W}} p^*(S) = \frac{|\{i\in I \mid i_k = +1 \ \wedge \ gd(Bi) = -1\}|}{2^n} = \frac{I_k^{+-}}{2^n}. \qquad (68)$$

Hence,

$$\sum_{\substack{S\,:\,k\notin S\notin W \\ S\cup\{k\}\in W}} p^*(S) = \frac{I_k^{--} - I_k^{+-}}{2^n}, \qquad (69)$$

and therefore, applying (66) and (69) to (14), and comparing it with (51), we have $\Phi_k^-(W,p^*) = GHB^-(k)$.

Finally, by virtue of (2), (50)-(52), (65), and (69), we have

$$\Phi_k(W,p^*) = \frac{I_k^{++} - I_k^{-+} + I_k^{--} - I_k^{+-}}{2^n} = \frac{GHB^+(k) + GHB^-(k)}{2} = GHB(k). \quad \square$$

Conclusion 4.1. *The generalized Hoede-Bakker index coincides with the Penrose measure, that is,*

$$\forall k \in N \ [GHB(k) = PB_k(W)]. \qquad (70)$$

Proof. This immediately results from (18) and (61), since for each $k \in N$, $GHB(k) = \Phi_k(W,p^*) = PB_k(W)$. $\qquad\square$

In Rusinowska and De Swart [41], it was shown that if there is no influence between players, and the number of players is odd, then the original Hoede-Bakker index (with axiom (A-0) imposed) coincides with the absolute Banzhaf index. The result given in Conclusion 4.1 is more general than the result presented in Rusinowska and De Swart [41].

Example 4.1. Let us calculate all the notions introduced for our Example 3.1. In this case, all axioms (B-1), (B-2), (B-3), (G-1), (G-2) and (G-3) are satisfied.

$$I_A^{++} = 3, \quad I_A^{+-} = 1, \quad I_A^{-+} = 0, \quad I_A^{--} = 4$$
$$I_B^{++} = I_C^{++} = 2, \quad I_B^{+-} = I_C^{+-} = 2, \quad I_B^{-+} = I_C^{-+} = 1, \quad I_B^{--} = I_C^{--} = 3.$$

Hence, we have

$$HB(A) = HB^+(A) = \tfrac{1}{2}, \quad HB(B) = HB^+(B) = HB^+(C) = HB(C) = 0$$
$$HB^-(A) = 1, \quad HB^-(B) = HB^-(C) = \tfrac{1}{2}$$
$$GHB(A) = GHB^+(A) = GHB^-(A) = \tfrac{3}{4}$$
$$GHB(B) = GHB^+(B) = GHB^-(B) = \tfrac{1}{4}$$
$$GHB(C) = GHB^+(C) = GHB^-(C) = \tfrac{1}{4}.$$

Note that $GHB(B) > 0$ (and $GHB(C) > 0$), which confirms our observation that there are situations in which player B (player C, respectively) is decisive.

5   Modifications of the Hoede-Bakker Index

In this Section, we define some modifications of the generalized Hoede-Bakker index as introduced in Section 4. As before, we impose the axioms (B-1), (B-2), (B-3), (G-1), (G-2), and (G-3). Let I_k^{++}, I_k^{+-}, I_k^{-+}, and I_k^{--}, for each $k \in N$, be as defined by equations (46)-(49), respectively. Moreover, given B and gd, we introduce two additional symbols:

$$I^+ = |\{i \in I \mid gd(Bi) = +1\}| \qquad (71)$$

I^+ - number of inclination vectors leading to the group decision 'yes'

$$I^- = |\{i \in I \mid gd(Bi) = -1\}| \qquad (72)$$

I^- - number of inclination vectors leading to the group decision 'no'

Next, we introduce several modifications of the generalized Hoede-Bakker index.

5.1   Modifications Leading to the Coleman Indices

Let us introduce the following definition:

Definition 5.1. *Given B and gd:*

$$for\ each\ k \in N,\ M_1 GHB(k) = \frac{I_k^{++} - I_k^{-+}}{I^+} \qquad (73)$$

$$for\ each\ k \in N,\ M_2 GHB(k) = \frac{I_k^{--} - I_k^{+-}}{I^-} \qquad (74)$$

$$M_3 GHB = \frac{I^+}{2^n} = \frac{|\{i \in I \mid gd(Bi) = +1\}|}{2^n} \qquad (75)$$

Remark 5.1. Note that by virtue of (39), that is, axiom (A-2), $I^+ \geq 1$, and by virtue of (40), that is, axiom (A-3), $I^- \geq 1$. Hence, $M_1 GHB$ and $M_2 GHB$ are well defined: the denominators given in (73) and (74) are never equal to 0.

Fact 5.1. *We have:*

$$\forall k \in N\ [1 \geq M_1 GHB(k) \geq 0] \qquad (76)$$

$$\forall k \in N\ [1 \geq M_2 GHB(k) \geq 0] \qquad (77)$$

$$1 > M_3 GHB \geq \frac{1}{2^n}. \qquad (78)$$

Proof. Since $I^+ \geq 1$, and $I^+ < 2^n$, we get immediately $1 > M_3GHB \geq \frac{1}{2^n}$. Note that

$$\forall k \in N \ [I^+ = I_k^{++} + I_k^{-+}], \tag{79}$$

and hence, $M_1GHB(k) \leq 1$ for each $k \in N$. Moreover,

$$\forall k \in N \ [I^- = I_k^{--} + I_k^{+-}], \tag{80}$$

and therefore, $M_2GHB(k) \leq 1$ for each $k \in N$.

Let us consider an arbitrary player $k \in N$. We take an arbitrary inclination vector $i = (i_1, ..., i_n) \in I_k^{-+}$. This means that $i_k = -1$ and $gd(Bi) = +1$. On the other hand, note that for each $i = (i_1, ..., i_n) \in I_k^{-+}$ there is $i' = (i'_1, ..., i'_n)$ such that

$$i'_j = \begin{cases} i_j & for \ j \neq k \\ +1 & for \ j = k \end{cases}. \tag{81}$$

Hence, $i \leq i'$, and by virtue of (A-1), $gd(Bi) \leq gd(Bi')$. Since $gd(Bi) = +1$, we get $gd(Bi') = +1$. Hence, $I_k^{++} - I_k^{-+} \geq 0$, and therefore $M_1GHB(k) \geq 0$ for each $k \in N$. Moreover, by virtue of (57),

$$\forall k \in N \ [I_k^{++} - I_k^{-+} = I_k^{--} - I_k^{+-}], \tag{82}$$

and hence $I_k^{--} - I_k^{+-} \geq 0$, which gives $M_2GHB(k) \geq 0$ for each $k \in N$. $\quad\square$

Fact 5.2. *We have:*
(a) Player $k \in N$ is a dummy if and only if $M_1GHB(k) = 0$.
(b) Player $k \in N$ is a dummy if and only if $M_2GHB(k) = 0$.

Proof. By virtue of (50), (52) and (73), for each $k \in N$, $GHB(k) = 0$ if and only if $M_1GHB(k) = 0$. By analogy, from (51), (52) and (74), for each $k \in N$, $GHB(k) = 0$ if and only if $M_2GHB(k) = 0$. Hence, by virtue of Fact 4.4, we get Fact 5.2. $\quad\square$

Proposition 5.1. *Let $\Phi_k^{Acc}(W,p)$, $\Phi_k^{Rej}(W,p)$, $\alpha(W,p)$, and p^* be as defined by equations (11), (12), (5) and (15), respectively. Then*

$$\forall k \in N \ [\Phi_k^{Acc}(W,p^*) = M_1GHB(k)] \tag{83}$$

$$\forall k \in N \ [\Phi_k^{Rej}(W,p^*) = M_2GHB(k)] \tag{84}$$

$$\alpha(W,p^*) = M_3GHB. \tag{85}$$

Proof. Let us apply again the probabilistic model recapitulated in Section 2 to our situation, interpreting a coalition as an inclination vector and a winning coalition as an inclination vector i with $gd(Bi) = +1$. Then

$$\alpha(W,p^*) = \sum_{S:S\in W} p^*(S) = \frac{|\{i \in I \mid gd(Bi) = +1\}|}{2^n} = \frac{I^+}{2^n}, \tag{86}$$

and since $I^+ + I^- = 2^n$, we have

$$1 - \alpha(W, p^*) = \frac{I^-}{2^n}. \qquad (87)$$

Comparing (75) and (86), we get immediately $\alpha(W, p^*) = M_3GHB$.

Applying (65) and (86) to (11), and comparing it with (73), we get for each $k \in N$

$$\Phi_k^{Acc}(W, p^*) = \frac{I_k^{++} - I_k^{-+}}{I^+} = M_1GHB(k). \qquad (88)$$

And finally, applying (69) and (87) to (12), and comparing it with (74), we get for each $k \in N$

$$\Phi_k^{Rej}(W, p^*) = \frac{I_k^{--} - I_k^{+-}}{I^-} = M_2GHB(k). \qquad (89)$$

$\square$

Conclusion 5.1. *The modified Hoede-Bakker indices M_1GHB, M_2GHB, and M_3GHB coincide with the Coleman indices, that is, Coleman's index 'to prevent action', Coleman's index 'to initiate action', and Coleman's 'power of a collectivity to act', respectively. We have*

$$\forall k \in N \; [M_1GHB(k) = Col_k^P(W)] \qquad (90)$$

$$\forall k \in N \; [M_2GHB(k) = Col_k^I(W)] \qquad (91)$$

$$M_3GHB = A(W). \qquad (92)$$

Proof. From (83) and (22) we have $M_1GHB(k) = \Phi_k^{Acc}(W, p^*) = Col_k^P(W)$ for each $k \in N$. By virtue of (84) and (24), $M_2GHB(k) = \Phi_k^{Rej}(W, p^*) = Col_k^I(W)$ for each $k \in N$. And finally, from (85) and (20), $M_3GHB = \alpha(W, p^*) = A(W)$. $\square$

Example 5.1. Let us calculate the new modifications introduced for Example 3.1. As before, we use the calculations done in Example 4.1. Moreover, we have (see Table 3.1) $I^+ = 3$ and $I^- = 5$. Hence,

$M_1GHB(A) = 1, \quad M_1GHB(B) = M_1GHB(C) = \frac{1}{3}$

$M_2GHB(A) = \frac{3}{5}, \quad M_2GHB(B) = M_2GHB(C) = \frac{1}{5}$

$M_3GHB = \frac{3}{8}$.

5.2   Modification Leading to the Rae Index

Next, we introduce a modification of the generalized Hoede-Bakker index which appears to lead to the Rae index.

Definition 5.2. *Given B and gd, for each player $k \in N$:*

$$M_4GHB^+(k) = \frac{I_k^{++}}{2^{n-1}} = \frac{|\{i \in I \mid i_k = +1 \; \wedge \; gd(Bi) = +1\}|}{2^{n-1}} \qquad (93)$$

$$M_4GHB^-(k) = \frac{I_k^{--}}{2^{n-1}} = \frac{|\{i \in I \mid i_k = -1 \ \wedge \ gd(Bi) = -1\}|}{2^{n-1}} \quad (94)$$

$$M_4GHB(k) = \frac{M_4GHB^+(k) + M_4GHB^-(k)}{2} = \frac{|\{i \in I \mid i_k = gd(Bi)\}|}{2^n}. \quad (95)$$

Fact 5.3. *We have for each $k \in N$:*

$$1 \geq M_4GHB^+(k) \geq \frac{1}{2^{n-1}} \quad (96)$$

$$1 \geq M_4GHB^-(k) \geq \frac{1}{2^{n-1}} \quad (97)$$

$$1 \geq M_4GHB(k) \geq \frac{1}{2^{n-1}}. \quad (98)$$

Proof. From axiom (A-2), $I_k^{++} \geq 1$ for each $k \in N$. Hence, we have $M_4GHB^+(k)$ $\geq \frac{1}{2^{n-1}}$. By virtue of axiom (A-3), $I_k^{--} \geq 1$ for each $k \in N$, and therefore $M_4GHB^-(k) \geq \frac{1}{2^{n-1}}$. Hence, also $M_4GHB(k) \geq \frac{1}{2^{n-1}}$. Moreover, $I_k^{++} \leq 2^{n-1}$, and $I_k^{--} \leq 2^{n-1}$ for each $k \in N$. Hence, $M_4GHB^+(k) \leq 1$, $M_4GHB^-(k) \leq 1$, and therefore also $M_4GHB(k) \leq 1$ for each $k \in N$. □

Proposition 5.2. *Let $\Omega_k(W,p)$, $\Omega_k^+(W,p)$, $\Omega_k^-(W,p)$, and p^* be as defined by equations (1), (9), (10), and (15), respectively. Then*

$$\forall k \in N \ [\Omega_k^+(W,p^*) = M_4GHB^+(k)] \quad (99)$$

$$\forall k \in N \ [\Omega_k^-(W,p^*) = M_4GHB^-(k)] \quad (100)$$

$$\forall k \in N \ [\Omega_k(W,p^*) = M_4GHB(k)]. \quad (101)$$

Proof. From (9), (62), (63), and (93),

$$\Omega_k^+(W,p^*) = \frac{I_k^{++}}{2^{n-1}} = M_4GHB^+(k). \quad (102)$$

Applying (66) and (67) to (10), and comparing it with (94), we have

$$\Omega_k^-(W,p^*) = \frac{I_k^{--}}{2^{n-1}} = M_4GHB^-(k). \quad (103)$$

Finally, using (1), (63), (67), and (95), we get

$$\Omega_k(W,p^*) = \frac{I_k^{++}}{2^n} + \frac{I_k^{--}}{2^n} = \frac{M_4GHB^+(k) + M_4GHB^-(k)}{2} = M_4GHB(k). \quad (104)$$

□

Conclusion 5.2. *The modified Hoede-Bakker index M_4GHB coincides with the Rae index, that is,*

$$\forall k \in N \ [M_4GHB(k) = Rae_k(W)] \tag{105}$$

Proof. By virtue of (16) and (101), we have for each $k \in N$, $M_4GHB(k) = \Omega_k(W, p^*) = Rae_k(W)$. □

Example 5.2. We will calculate the new notions introduced for Example 3.1. Using the calculations from Example 4.1, we get

$M_4GHB^+(A) = \frac{3}{4}$, $M_4GHB^+(B) = M_4GHB^+(C) = \frac{1}{2}$

$M_4GHB^-(A) = 1$, $M_4GHB^-(B) = M_4GHB^-(C) = \frac{3}{4}$

$M_4GHB(A) = \frac{7}{8}$, $M_4GHB(B) = M_4GHB(C) = \frac{5}{8}$.

5.3   Modification Leading to the König-Bräuninger Index

Finally, we like to introduce two new modifications of the generalized Hoede-Bakker index. One of them happens to coincide with the König-Bräuninger index.

Definition 5.3. *Given B and gd, for each $k \in N$:*

$$M_5GHB(k) = \frac{I_k^{++}}{I^+} = \frac{|\{i \in I \mid i_k = gd(Bi) = +1\}|}{|\{i \in I \mid gd(Bi) = +1\}|} \tag{106}$$

$$M_6GHB(k) = \frac{I_k^{--}}{I^-} = \frac{|\{i \in I \mid i_k = gd(Bi) = -1\}|}{|\{i \in I \mid gd(Bi) = -1\}|} \tag{107}$$

Remark 5.2. Note that by virtue of axioms (A-2) and (A-3), $I^+ \geq 1$, and $I^- \geq 1$, respectively. Hence, $M_5GHB(k)$ and $M_6GHB(k)$ are well defined, since the denominators I^+ and I^- are never equal to 0.

Fact 5.4. *We have for each $k \in N$:*

$$1 \geq M_5GHB(k) > \frac{1}{2^n} \tag{108}$$

$$1 \geq M_6GHB(k) > \frac{1}{2^n}. \tag{109}$$

Proof. By virtue of axioms (A-2) and (A-3), $I^+ < 2^n$, $I^- < 2^n$, and for each $k \in N$, $I_k^{++} \geq 1$, and $I_k^{--} \geq 1$. Hence, we get

$$M_5GHB(k) = \frac{I_k^{++}}{I^+} > \frac{I_k^{++}}{2^n} \geq \frac{1}{2^n}, \quad M_6GHB(k) = \frac{I_k^{--}}{I^-} > \frac{I_k^{--}}{2^n} \geq \frac{1}{2^n}. \tag{110}$$

By virtue of (79), $I^+ \geq I_k^{++}$ for each $k \in N$, and therefore, $M_5GHB(k) \leq 1$ for each $k \in N$. By analogy, from (80), $I^- \geq I_k^{--}$ for each $k \in N$, and hence, $M_6GHB(k) \leq 1$ for each $k \in N$. □

Proposition 5.3. *Let $\Omega_k^{Acc}(W, p)$, $\Omega_k^{Rej}(W, p)$, and p^* be as defined by equations (7), (8), and (15), respectively. Then*

$$\forall k \in N \; [\Omega_k^{Acc}(W, p^*) = M_5 GHB(k)] \tag{111}$$

$$\forall k \in N \; [\Omega_k^{Rej}(W, p^*) = M_6 GHB(k)]. \tag{112}$$

Proof. Applying the probabilistic model from Section 2 to our model, we get the following results. By virtue of (7), (63), (86), and (106), we have for each $k \in N$,

$$\Omega_k^{Acc}(W, p^*) = \frac{I_k^{++}}{I^+} = M_5 GHB(k). \tag{113}$$

By analogy, from (8), (67), (87), and (107), we get for each $k \in N$,

$$\Omega_k^{Rej}(W, p^*) = \frac{I_k^{--}}{I^-} = M_6 GHB(k). \tag{114}$$

$\square$

Conclusion 5.3. *The modified Hoede-Bakker index $M_5 GHB$ coincides with the König-Bräuninger index, that is,*

$$\forall k \in N \; [M_5 GHB(k) = KB_k(W)]. \tag{115}$$

Proof. This follows immediately from (26) and (111), since for each $k \in N$ $M_5 GHB(k) = \Omega_k^{Acc}(W, p^*) = KB_k(W)$. $\square$

Example 5.3. Let us calculate the new modifications introduced for Example 3.1. We find

$M_5 GHB(A) = 1, \quad M_5 GHB(B) = M_5 GHB(C) = \frac{2}{3}$
$M_6 GHB(A) = \frac{4}{5}, \quad M_6 GHB(B) = M_6 GHB(C) = \frac{3}{5}.$

We finish this section with the following fact:

Fact 5.5. *Given B and gd, if axioms (A-0) and (A-1) are satisfied, then:*

$$\forall k \in N \; [HB^+(k) = HB^-(k) = GHB^+(k) = GHB^-(k) = GHB(k) =$$

$$= M_1 GHB(k) = M_2 GHB(k)] \tag{116}$$

$$M_3 GHB = \frac{1}{2} \tag{117}$$

$$\forall k \in N \; [M_4 GHB^+(k) = M_4 GHB^-(k) = M_4 GHB(k) =$$

$$= M_5 GHB(k) = M_6 GHB(k)]. \tag{118}$$

Proof. If axiom (A-0) is additionally satisfied, then we have

$$I^+ = I^- = 2^{n-1} \tag{119}$$

$$\forall k \in N \ [I_k^{++} = I_k^{--}] \tag{120}$$

$$\forall k \in N \ [I_k^{+-} = I_k^{-+}]. \tag{121}$$

From (50)-(54), and (120)-(121), we get for each $k \in N$, $HB^+(k) = HB^-(k) = GHB^+(k) = GHB^-(k) = GHB(k)$.

From (50)-(52), (73)-(74), and (119)-(121), we have for each $k \in N$, $M_1GHB(k) = M_2GHB(k) = GHB(k)$.

By virtue of (75) and (119), $M_3GHB = \frac{1}{2}$.

Finally, from (93)-(95), (106), (107), and (119)-(121), we get for each $k \in N$, $M_4GHB^+(k) = M_4GHB^-(k) = M_4GHB(k) = M_5GHB(k) = M_6GHB(k)$. $\square$

6   Examples

The examples presented in this Section have been constructed in order to show some advantages of skipping axiom (A-0) as adopted by Hoede and Bakker [15]. In these examples, axiom (A-0) is not satisfied, and hence, we cannot apply the (original) Hoede-Bakker index. Having introduced the generalized version of the Hoede-Bakker index, we may calculate all the measures introduced in this paper.

In Hoede and Bakker [15], it was assumed that the ability to influence does not depend on the inclination. This means that if a player follows another actor who influences him, then this influenced player will always decide according to the inclination of his 'boss', no matter what the inclinations are. We find such a requirement too restrictive, since one may face situations in which the ability to influence does depend on the inclinations. Hence, in this paper, we do NOT adopt this assumption.

Example 6.1. Let us analyze the situation in which a married couple considers a proposal to have a holiday this month. Unfortunately, there are three players involved in this game: husband (player 1), wife (player 2), and wife's boss denoted as player 3. Hence, $N = \{1, 2, 3\}$. In fact, player 1 is fully influenced by player 2 (in particular, when considering this proposal), and he always does what his wife asks for. In the matter of going on holiday, player 2 is 'partially' influenced by her boss: if player 3 feels like 'yes' (that is, I like you to have a holiday), she will follow her own inclination, but if the boss has the inclination 'no' (I do not want you to have a holiday now), she will decide according to his wish and continue working. The couple will go on holiday only if all three actors involved will decide 'yes'. As was already mentioned, we face a kind of 'partial' influence here. Figure 6.1 illustrates this situation. There are two different arrows in Figure 6.1. The 'normal' arrow going from node 2 to node 1 means that player 1 always follows the inclination of player 2. The dashed line going from node 3 to node 2 denotes that player 2 follows the inclination 'no' of player 3, and otherwise, if actor 3 has the inclination 'yes', actor 2 will decide according to her own inclination.

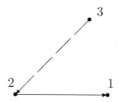

Fig. 6.1. Graph for Example 6.1

Table 6.1 presents the group decision for this situation.

Table 6.1. Group decision for Example 6.1

inclination i	Bi	$gd(Bi)$	inclination i	Bi	$gd(Bi)$
$(1,1,1)$	$(1,1,1)$	$+1$	$(-1,-1,-1)$	$(-1,-1,-1)$	-1
$(1,1,-1)$	$(1,-1,-1)$	-1	$(-1,-1,1)$	$(-1,-1,1)$	-1
$(1,-1,1)$	$(-1,-1,1)$	-1	$(-1,1,-1)$	$(1,-1,-1)$	-1
$(-1,1,1)$	$(1,1,1)$	$+1$	$(1,-1,-1)$	$(-1,-1,-1)$	-1

First of all, one may note that axiom (A-0) is indeed not satisfied here, but all axioms (B-1), (B-2), (B-3), (G-1), (G-2) and (G-3) are satisfied. There are two winning coalition, that is, $f(-1,1,1) = \{2,3\}$ which is the only one minimal winning coalition, and of course the grand coalition $f(1,1,1) = \{1,2,3\} = N$. Player 1 is a dummy, and players 2 and 3 are the vetoers in this game. Let us calculate the generalized Hoede-Bakker index and all its modifications introduced. We have

$$I^+ = 2, \quad I^- = 6$$
$$I_1^{++} = 1, \quad I_1^{+-} = 3, \quad I_1^{-+} = 1, \quad I_1^{--} = 3$$
$$I_2^{++} = I_3^{++} = 2, \quad I_2^{+-} = I_3^{+-} = 2, \quad I_2^{-+} = I_3^{-+} = 0, \quad I_2^{--} = I_3^{--} = 4$$

Hence, the final results of our calculations are as follows:

$$HB^+(1) = -\tfrac{1}{2} < 0, \quad HB^+(2) = HB^+(3) = 0$$

As one can see, HB^+ may be negative.

$$HB^-(1) = \tfrac{1}{2}, \quad HB^-(2) = HB^-(3) = 1$$
$$GHB(1) = GHB^+(1) = GHB^-(1) = 0$$
$$GHB(2) = GHB^+(2) = GHB^-(2) = \tfrac{1}{2}$$
$$GHB(3) = GHB^+(3) = GHB^-(3) = \tfrac{1}{2}$$
$$M_1 GHB(1) = 0, \quad M_1 GHB(2) = M_1 GHB(3) = 1$$
$$M_2 GHB(1) = 0, \quad M_2 GHB(2) = M_2 GHB(3) = \tfrac{1}{3}$$
$$M_3 GHB = \tfrac{1}{4}$$

$$M_4GHB^+(1) = \tfrac{1}{4}, \quad M_4GHB^+(2) = M_4GHB^+(3) = \tfrac{1}{2}$$
$$M_4GHB^-(1) = \tfrac{3}{4}, \quad M_4GHB^-(2) = M_4GHB^-(3) - 1$$
$$M_4GHB(1) = \tfrac{1}{2}, \quad M_4GHB(2) = M_4GHB(3) = \tfrac{3}{4}$$
$$M_5GHB(1) = \tfrac{1}{2}, \quad M_5GHB(2) = M_5GHB(3) = 1$$
$$M_6GHB(1) = \tfrac{1}{2}, \quad M_6GHB(2) = M_6GHB(3) = \tfrac{2}{3}.$$

When using the original definition of the Hoede-Bakker index (with axiom (A-0) adopted), the problem was faced that for some networks with an even number of players a draw might appear. One of the advantages of the generalized Hoede-Bakker index is that it can be calculated without any problem for an arbitrary number of players, in particular, if there is an even number of actors.

Example 6.2. Let us analyze the network presented in Figure 6.2.

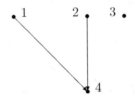

Fig. 6.2. Graph for Example 6.2

There are four players in this network. Players 1 and 2 influence player 4, and player 3 is independent. We apply the standard procedure to this network (see Rusinowska and De Swart [41]), according to which the players decide as follows:

- Players 1, 2, and 3 follow their own inclinations.
- If players 1 and 2 have different inclinations, then player 4 decides according to his own inclination, otherwise he follows the inclination of players 1 and 2.
- The group decision is 'yes' if and only if at least three players decide to say 'yes'.

The group decision for this example is shown in Table 6.2.

Table 6.2. Group decision for Example 6.2

inclination i	Bi	$gd(Bi)$	inclination i	Bi	$gd(Bi)$
$(1,1,1,1)$	$(1,1,1,1)$	$+1$	$(-1,-1,-1,-1)$	$(-1,-1,-1,-1)$	-1
$(1,1,1,-1)$	$(1,1,1,1)$	$+1$	$(-1,-1,-1,1)$	$(-1,-1,-1,-1)$	-1
$(1,1,-1,1)$	$(1,1,-1,1)$	$+1$	$(-1,-1,1,-1)$	$(-1,-1,1,-1)$	-1
$(1,-1,1,1)$	$(1,-1,1,1)$	$+1$	$(-1,1,-1,-1)$	$(-1,1,-1,-1)$	-1
$(-1,1,1,1)$	$(-1,1,1,1)$	$+1$	$(1,-1,-1,-1)$	$(1,-1,-1,-1)$	-1
$(1,1,-1,-1)$	$(1,1,-1,1)$	$+1$	$(-1,-1,1,1)$	$(-1,-1,1,-1)$	-1
$(1,-1,1,-1)$	$(1,-1,1,-1)$	-1	$(-1,1,-1,1)$	$(-1,1,-1,1)$	-1
$(1,-1,-1,1)$	$(1,-1,-1,1)$	-1	$(-1,1,1,-1)$	$(-1,1,1,-1)$	-1

In this case, all axioms (B-1), (B-2), (B-3), (G-1), (G-2) and (G-3) are also satisfied, but again axiom (A-0) is not. There are six winning coalitions here, that is, N, $\{1,2,3\}$, $\{1,2,4\}$, $\{1,3,4\}$, $\{2,3,4\}$, and $\{1,2\}$. The last three coalitions are minimal. There is no vetoer and no dummy. By virtue of Table 6.2 we have

$$I^+ = 6, \quad I^- = 10$$

$$I_1^{++} = I_2^{++} = 5, \quad I_1^{+-} = I_2^{+-} = 3, \quad I_1^{-+} = I_2^{-+} = 1, \quad I_1^{--} = I_2^{--} = 7$$

$$I_3^{++} = I_4^{++} = 4, \quad I_3^{+-} = I_4^{+-} = 4, \quad I_3^{-+} = I_4^{-+} = 2, \quad I_3^{--} = I_4^{--} = 6.$$

We get the following final results:

$$HB^+(1) = HB^+(2) = \tfrac{1}{4}, \quad HB^+(3) = HB^+(4) = 0$$

$$HB^-(1) = HB^-(2) = \tfrac{3}{4}, \quad HB^-(3) = HB^-(4) = \tfrac{1}{2}$$

$$GHB(1) = GHB^+(1) = GHB^-(1) = \tfrac{1}{2}$$

$$GHB(2) = GHB^+(2) = GHB^-(2) = \tfrac{1}{2}$$

$$GHB(3) = GHB^+(3) = GHB^-(3) = \tfrac{1}{4}$$

$$GHB(4) = GHB^+(4) = GHB^-(4) = \tfrac{1}{4}$$

$$M_1GHB(1) = M_1GHB(2) = \tfrac{2}{3}, \quad M_1GHB(3) = M_1GHB(4) = \tfrac{1}{3}$$

$$M_2GHB(1) = M_2GHB(2) = \tfrac{2}{5}, \quad M_2GHB(3) = M_2GHB(4) = \tfrac{1}{5}$$

$$M_3GHB = \tfrac{3}{8}$$

$$M_4GHB^+(1) = M_4GHB^+(2) = \tfrac{5}{8}, \quad M_4GHB^+(3) = M_4GHB^+(4) = \tfrac{1}{2}$$

$$M_4GHB^-(1) = M_4GHB^-(2) = \tfrac{7}{8}, \quad M_4GHB^-(3) = M_4GHB^-(4) = \tfrac{3}{4}$$

$$M_4GHB(1) = M_4GHB(2) = \tfrac{3}{4}, \quad M_4GHB(3) = M_4GHB(4) = \tfrac{5}{8}$$

$$M_5GHB(1) = M_5GHB(2) = \tfrac{5}{6}, \quad M_5GHB(3) = M_5GHB(4) = \tfrac{2}{3}$$

$$M_6GHB(1) = M_6GHB(2) = \tfrac{7}{10}, \quad M_6GHB(3) = M_6GHB(4) = \tfrac{3}{5}.$$

Example 6.3. Let us analyze the network presented in Figure 6.3.

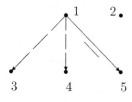

Fig. 6.3. Graph for Example 6.3

We may say that this is an example of 'a positive and opposite influence'. The network analyzed consists of five players. Players 1 and 2 always decide according to their own inclinations. Player 2 is fully independent: he neither influences nor is influenced. Player 1 influences players 3, 4, and 5. In the matter of his influence on players 3 and 4, we face a kind of 'partial' ('positive') influence. We mean by this that if player 1 has the inclination 'yes', then players 3 and 4 follow his inclination, but if his inclination is 'no', actors 3 and 4 decide according to their own inclinations. Such a partial influence is denoted on Figure 6.3 by the two dashed vectors going from node 1 to nodes 3 and 4. In the matter of the influence on player 5, we find a kind of 'opposite' influence, because player 5 will always make a decision opposite to the inclination of player 1: if player 1's inclination is 'yes', player 5 will decide for 'no', and if player 1 has the inclination 'no', player 5 will say 'yes'. In order to stress this 'opposite' influence, we marked additionally the vector going from node 1 to node 5 in Figure 6.3. The group decision is made

Table 6.3. Group decision for Example 6.3

inclination i	Bi	$gd(Bi)$	inclination i	Bi	$gd(Bi)$
$(1,1,1,1,1)$	$(1,1,1,1,-1)$	$+1$	$(-1,-1,-1,-1,-1)$	$(-1,-1,-1,-1,1)$	-1
$(1,1,1,1,-1)$	$(1,1,1,1,-1)$	$+1$	$(-1,-1,-1,-1,1)$	$(-1,-1,-1,-1,1)$	-1
$(1,1,1,-1,1)$	$(1,1,1,1,-1)$	$+1$	$(-1,-1,-1,1,-1)$	$(-1,-1,-1,1,1)$	-1
$(1,1,-1,1,1)$	$(1,1,1,1,-1)$	$+1$	$(-1,-1,1,-1,-1)$	$(-1,-1,1,-1,1)$	-1
$(1,-1,1,1,1)$	$(1,-1,1,1,-1)$	$+1$	$(-1,1,-1,-1,-1)$	$(-1,1,-1,-1,1)$	-1
$(-1,1,1,1,1)$	$(-1,1,1,1,1)$	$+1$	$(1,-1,-1,-1,-1)$	$(1,-1,1,1,-1)$	$+1$
$(1,1,1,-1,-1)$	$(1,1,1,1,-1)$	$+1$	$(-1,-1,-1,1,1)$	$(-1,-1,-1,1,1)$	-1
$(1,1,-1,1,-1)$	$(1,1,1,1,-1)$	$+1$	$(-1,-1,1,-1,1)$	$(-1,-1,1,-1,1)$	-1
$(1,-1,1,1,-1)$	$(1,-1,1,1,-1)$	$+1$	$(-1,1,-1,-1,1)$	$(-1,1,-1,-1,1)$	-1
$(-1,1,1,1,-1)$	$(-1,1,1,1,1)$	$+1$	$(1,-1,-1,-1,1)$	$(1,-1,1,1,-1)$	$+1$
$(1,1,-1,-1,1)$	$(1,1,1,1,-1)$	$+1$	$(-1,-1,1,1,-1)$	$(-1,-1,1,1,1)$	$+1$
$(1,-1,1,-1,1)$	$(1,-1,1,1,-1)$	$+1$	$(-1,1,-1,1,-1)$	$(-1,1,-1,1,1)$	$+1$
$(-1,1,1,-1,1)$	$(-1,1,1,-1,1)$	$+1$	$(-1,1,1,-1,-1)$	$(-1,1,1,-1,1)$	$+1$
$(1,-1,-1,1,1)$	$(1,-1,1,1,-1)$	$+1$	$(1,-1,-1,1,-1)$	$(1,-1,1,1,-1)$	$+1$
$(-1,1,-1,1,1)$	$(-1,1,-1,1,1)$	$+1$	$(1,-1,1,-1,-1)$	$(1,-1,1,1,-1)$	$+1$
$(-1,-1,1,1,1)$	$(-1,-1,1,1,1)$	$+1$	$(1,1,-1,-1,-1)$	$(1,1,1,1,-1)$	$+1$

according to the majority's decision: gd is equal to $+1$ if and only if at least three players will decide for 'yes'. Table 6.3 presents the group decision for this example.

Player 5 is a dummy in this example. There is no vetoer. In this case, none of the axioms (B-1), (B-2), (B-3) is satisfied, but all axioms (A-1), (A-2) and (A-3) are satisfied. By virtue of Table 6.3, we get the following results:

$$I^+ = 24, \quad I^- = 8$$
$$I_1^{++} = 16, \quad I_1^{+-} = 0, \quad I_1^{-+} = 8, \quad I_1^{--} = 8$$
$$I_2^{++} = I_3^{++} = I_4^{++} = 14, \quad I_2^{+-} = I_3^{+-} = I_4^{+-} = 2$$
$$I_2^{-+} = I_3^{-+} = I_4^{-+} = 10, \quad I_2^{--} = I_3^{--} = I_4^{--} = 6$$

$$I_5^{++} = 12, \quad I_5^{+-} = 4, \quad I_5^{-+} = 12, \quad I_5^{--} = 4$$

$$HB^+(1) = 1, \quad HB^+(2) = HB^+(3) = HB^+(4) = \tfrac{3}{4}, \quad HB^+(5) = \tfrac{1}{2}$$

$$HB^-(1) = 0, \quad HB^-(2) = HB^-(3) = HB^-(4) = -\tfrac{1}{4}, \quad HB^-(5) = -\tfrac{1}{2}$$

Note than HB^- is negative for all players but player 1.

$$GHB(1) = GHB^+(1) = GHB^-(1) = \tfrac{1}{2}$$

$$GHB(k) = GHB^+(k) = GHB^-(k) = \tfrac{1}{4} \text{ for } k = 2, 3, 4$$

$$GHB(5) = GHB^+(5) = GHB^-(5) = 0$$

$$M_1 GHB(1) = \tfrac{1}{3}, \quad M_1 GHB(k) = \tfrac{1}{6} \text{ for } k = 2, 3, 4, \quad M_1 GHB(5) = 0$$

$$M_2 GHB(1) = 1, \quad M_2 GHB(k) = \tfrac{1}{2} \text{ for } k = 2, 3, 4, \quad M_2 GHB(5) = 0$$

$$M_3 GHB = \tfrac{3}{4}$$

$$M_4 GHB^+(1) = 1, \quad M_4 GHB^+(k) = \tfrac{7}{8} \text{ for } k = 2, 3, 4, \quad M_4 GHB^+(5) = \tfrac{3}{4}$$

$$M_4 GHB^-(1) = \tfrac{1}{2}, \quad M_4 GHB^-(k) = \tfrac{3}{8} \text{ for } k = 2, 3, 4, \quad M_4 GHB^-(5) = \tfrac{1}{4}$$

$$M_4 GHB(1) = \tfrac{3}{4}, \quad M_4 GHB(k) = \tfrac{5}{8} \text{ for } k = 2, 3, 4, \quad M_4 GHB(5) = \tfrac{1}{2}$$

$$M_5 GHB(1) = \tfrac{2}{3}, \quad M_5 GHB(k) = \tfrac{7}{12} \text{ for } k = 2, 3, 4, \quad M_5 GHB(5) = \tfrac{1}{2}$$

$$M_6 GHB(1) = 1, \quad M_6 GHB(k) = \tfrac{3}{4} \text{ for } k = 2, 3, 4, \quad M_6 GHB(5) = \tfrac{1}{2}.$$

7   Conclusions

The Hoede-Bakker index was introduced more than twenty years ago, but, in our opinion, up till now, it did not get the attention it deserves, because it takes the mutual influences of the players in a social network into account. By resigning from the requirement, imposed in the original definition, that changing all inclinations of the players leads to an opposite group decision, the applicability of the index is extended considerably. We present several examples showing the usefulness of such a generalization. In particular, the generalized Hoede-Bakker index may be applied to a game with a vetoer.

Moreover, we allow the ability of influencing other players to depend on the inclination. It may happen, for instance, that a player will follow the positive inclination of another player, but not his negative inclination. It may also happen that a player will decide according to the inclination 'no' of the influencing player, but is not sensitive to the inclination 'yes' of that player. In such situations we cannot apply the original Hoede-Bakker index, since one of the axioms adopted by Hoede and Bakker [15], that is, the axiom mentioned above, is not satisfied. With the new and weaker set of axioms and the generalized definition of the Hoede-Bakker index, it is possible to analyze such situations.

Although it has a completely different motivation in terms of a social network with mutual influences among the agents, the generalized Hoede-Bakker index happens to coincide with the Penrose measure. This means that the generalized Hoede-Bakker index of a player is the probability that the given player is decisive, assuming that all inclination vectors are equally probable. This probability is equal to the two conditional probabilities: the probability that a player is decisive if he votes 'yes', and the probability that a player is decisive if he votes 'no'.

In this paper, also several modifications of the generalized Hoede-Bakker index are introduced. Given a group decision function, we may calculate all the inclination vectors leading to a positive group decision. This modification gives the probability of the acceptance of a proposal, and hence, it coincides with Coleman's 'power of a collectivity to act'. Two other modifications, being the conditional probability that a player is decisive if the proposal is accepted, and the probability that a player is decisive if the proposal is rejected, lead to Coleman's index 'to prevent action' and to Coleman's index 'to initiate action', respectively. In another modification, we calculate the probability that a player is successful. This modification leads to the Rae index. Of course, we may also calculate the four conditional probabilities that a player is successful if he votes 'yes', if he votes 'no', if the proposal is accepted, or if the proposal is rejected. The third one, that is, the conditional probability that a player is successful if the proposal is accepted, gives the König-Bräuninger index.

References

1. Banzhaf J (1965) Weighted voting doesn't work: a mathematical analysis, *Rutgers Law Review* 19: 317-343
2. Brams SJ (1975) Game Theory and Politics, New York, Free Press
3. Brams SJ, Affuso P (1976) Power and size: a new paradox, *Theory and Decision* 7: 29-56
4. Brams SJ, Lucas WF, Straffin PD (eds) (1982) Political and related models, volume 2 in series *Models in Applied Mathematics*, New York, Springer
5. Coleman JS (1971) Control of collectivities and the power of a collectivity to act, in: *Social Choice*, edited by Lieberman, Gordon and Breach, London
6. Coleman JS (1986) Individual Interests and Collective Action: Selected Essays, Cambridge University Press
7. Deegan J, Packel EW (1978) A new index of power for simple n-person games, *International Journal of Game Theory* 7: 113-123
8. Dubey P, Shapley LS (1979) Mathematical properties of the Banzhaf power index, *Mathematics of Operations Research* 4: 99-131
9. Felsenthal DS, Machover M (1995) Postulates and paradoxes of relative voting power - a critical reappraisal, *Theory and Decision* 38: 195-229
10. Felsenthal DS, Machover M (1998) The Measurement of Voting Power: Theory and Practice, Problems and Paradoxes, London: Edward Elgar Publishers
11. Felsenthal DS, Machover M (2000) Enlargement of the EU and weighted voting in its Council of Ministers, London, London School of Economics and Political Science, Centre for Philosophy of Natural and Social Science, VPP 01/00
12. Felsenthal DS, Machover M (2001) The Treaty of Nice and qualified majority voting, *Social Choice and Welfare* 19 (3): 465-483
13. Felsenthal DS, Machover M, Zwicker WS (1998) The bicameral postulates and indices of a priori voting power, *Theory and Decision* 44: 83-116
14. Fischer D, Schotter A (1978) The inevitability of the paradox of redistribution in the allocation of voting weights, *Public Choice* 33: 49-67
15. Hoede C, Bakker R (1982) A theory of decisional power, *Journal of Mathematical Sociology* 8: 309-322
16. Holler MJ (ed.) (1981) Power, voting, and voting power, Physica Verlag, Würzburg

17. Holler MJ (1982) Forming coalitions and measuring voting power, *Political Studies* 30: 262-271
18. Holler MJ, Packel EW (1983) Power, luck and the right index, *Journal of Economics* 43: 21-29
19. Hosli M (1993) Admission of European Free Trade Association States to the European Community: Effects on voting power in the European Community's Council of Ministers, *International Organization* 47: 629-643
20. Hosli M (1995) The balance between small and large: Effects of a double-majority system on voting power in the European Union, *International Studies Quarterly* 39: 351-370
21. Hosli M (1996) Coalitions and power: Effects of qualified majority voting in the Council of the European Union, *Journal of Common Market Studies* 34(2): 255-273
22. Hosli M (1997) Voting strength in the European Parliament: The influence of national and partisan actors, *European Journal of Political Research* 31: 351-366
23. Hosli M (1998) An institution's capacity to act: What are the effects of majority voting in the Council of the EU and the European Parliament?, Current European Issues, Maastricht, European Institute of Public Administration
24. Hosli M (2002) Preferences and power in the European Union, *Homo Oeconomicus* 19: 311-326
25. König T, Bräuninger (1998) The inclusiveness of European decision rules, *Journal of Theoretical Politics* 10: 125-142
26. Johnston RJ (1978) On the measurement of power: Some reactions to Laver, *Environment and Planning A* 10: 907-914
27. Laruelle A (1998) Game Theoretical Analysis of Decision-Making Processes with Applications to the European Union, PhD dissertation, University of Louvain, Belgium
28. Laruelle A (2000) On the choice of power indices, University of Alicante, Spain
29. Laruelle A, Valenciano F (2005) Assessing success and decisiveness in voting situations, *Social Choice and Welfare* 24: 171-197
30. Laruelle A, Widgren M (1998) Is the allocation of power among EU states fair?, *Public Choice* 94: 317-339
31. Leech D (2002) Designing the voting system for the EU Council of Ministers, *Public Choice* 113: 437-464
32. Lucas WF (1982) Measuring power in weighted voting games, in: Brams SJ, Lucas WF and Straffin PD, o.c.
33. Nurmi H, Meskanen T (1999) A priori power measures and the institutions of the European Union, *European Journal of Political Research* 35(2): 161-179
34. Owen G (1975) Multilinear extensions and the Banzhaf value, *Naval Research Logistic Quarterly* 22: 741-750
35. Owen G (1977) Values of games with a priori unions, in: *Essays in Mathematical Economics and Game Theory*, edited by Hein and Moeschlin, New York, Springer-Verlag, pp. 77-88
36. Owen G (1981) Modification of the Banzhaf-Coleman index for games with a priori unions, in Holler, pp. 232-238
37. Owen G (1995) Game Theory, San Diego: Academic Press
38. Penrose LS (1946) The elementary statistics of majority voting, *Journal of the Royal Statistical Society* 109: 53-57
39. Rae D (1969) Decision rules and individual values in constitutional choice, *American Political Science Review* 63: 40-56
40. Rusinowska A (2001) Paradox of redistribution in Polish politics, *Annals of the Marie Curie Fellowships*, Vol. 2, pp. 46-54, European Commission, Brussels

41. Rusinowska A, De Swart H (2006) On some properties of the Hoede-Bakker index, forthcoming in *Journal of Mathematical Sociology*
42. Rusinowska A, Van Deemen AMA (2005), The redistribution paradox and the paradox of new members in the German parliament, in: L.A. Petrosjan and V.V. Mazalov (eds.) *Game Theory and Applications*, Volume X, pp. 153-174, Nova Science Publishers, Inc. New York
43. Schotter A (1981) The paradox of redistribution: Some theoretical and empirical results, in Holler, o.c.
44. Shapley LS (1953) A value for n-person games, *Annals of Mathematics Studies* 28: 307-317
45. Shapley LS, Shubik M (1954) A method for evaluating the distribution of power in a committee system, *American Political Science Review* 48: 787-792
46. Sosnowska H (1996) Shapley values of games with a priori unions as a method of analysis of elections in Poland 1989-1994, discussion paper, series Mathematical Economics, No. 1/EM/96, Warsaw School of Economics, Warsaw, Poland
47. Sosnowska H (1999) Generalized paradox of redistribution and size. Application to Polish elections in 1997, discussion paper, series Mathematical Economics, No. 2/EM/99, Warsaw School of Economics, Warsaw, Poland
48. Stokman F, Willer D (1999) Exchange networks, power indices, and cabinet formation, mimeo
49. Straffin PD (1982) Power indices in politics, in: Brams SJ, Lucas WF and Straffin PD, o.c.
50. Sutter M (2000) Fair allocation and re-weighting of votes and voting power in the EU before and after the next enlargement, *Journal of Theoretical Politics* 12(4): 433-449
51. Van Deemen AMA, Rusinowska A (2003) Paradoxes of voting power in Dutch politics, *Public Choice* 115: 109-137
52. Widgren M (1994) Voting power in the EU and the consequences of two different enlargements, *European Economic Review* 38: 1153-1170

An Environment for Specifying Properties of Dyadic Relations and Reasoning About Them II: Relational Presentation of Non-classical Logics*

Andrea Formisano[1], Eugenio G. Omodeo[2], and Ewa Orłowska[3]

[1] Dipartimento di Informatica, Università di L'Aquila, Italy
formisano@di.univaq.it
[2] Dipartimento di Matematica e Informatica, Università di Trieste, Italy
eomodeo@units.it
[3] National Institute of Telecommunications, Warsaw, Poland
orlowska@itl.waw.pl

Abstract. This paper contributes to the vast literature on relational renderings of non-classical logics providing a general schema for automatic translation. The translation process is supported by a flexible Prolog tool. Many specific translations are already implemented, typically leading from an unquantified logic into the calculus of binary relations. Thanks to the uniformity of the translation pattern, additional source languages (and, though less commonly, new target languages) can be installed very easily into this Prolog-based translator. The system also integrates an elementary graphical proof assistant based on Rasiowa-Sikorski dual-tableau rules.

Keywords: Relational systems, translation methods, modal logic.

Introduction

Common approaches to the automation of modal inferences often exploit *ad hoc*, direct inference methods (cf., e.g., [23, 33]). An alternative approach, discussed in the ongoing and aimed at developing a uniform relational platform for modal reasoning, is intended to benefit from relational renderings of non-classical logics (cf. [27] among others).

The envisaged framework covers a full-fledged inferential apparatus, where the inferential activity is viewed as consisting of two phases. First, a translation phase carries a (propositional) modal formalization φ of a problem into its relational counterpart. Then, within the relational context, a deductive method is exploited to seek a proof of the translated formula φ (cf. Fig. 1).

There are several kinds of proof systems for relational reasoning, such as tableaux [17], Gentzen-style systems [34, 22], systems à la Rasiowa-Sikorski

---

* Research partially funded by INTAS project *Algebraic and deduction methods in non-classical logic and their applications to Computer Science*, and by the European Concerted Research Action COST 274, *TARSKI: Theory and Applications of Relational Structures as Knowledge Instruments*.

H. de Swart et al. (Eds.): TARSKI II, LNAI 4342, pp. 89–104, 2006.

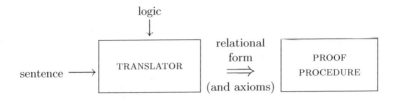

Fig. 1. General scheme of the inferential framework

[25, 30, 14], display calculus [16], and of course equational proof systems based
on relation algebras [11, 12]. The system we have in mind should be seen as providing a convenient input for any of those proof systems. Specifically, the input
for a tableaux-based system, a Gentzen system, or a Rasiowa-Sikorski system
will be an expression of the form $x\, t(\varphi)\, y$, where x and y stand for individual variables and $t(\varphi)$ for a relational term translating the given formula φ, obtainable
e.g. by means of a system which we have implemented in Prolog along the lines
that will be expounded below. On the other hand, our input for an equational
proof system will be an equation $t(\varphi) = \mathbf{1}$, where $\mathbf{1}$ denotes the top element of
a relation algebra.

This paper focuses on the translation phase: we describe a prototypical,
Prolog-based implementation of a tool, named *transIt*, which uniformly carries
out translations from various modal logics to the relational formalism [35]. As
an aside, we give some details about possible approaches towards the inter-
action/integration between the translator and a deductive engine. The development of an efficient relational deductive system (actually, in the Rasiowa-Sikorski
style) is the theme of [8].

We verified that this approach offers indeed a high degree of uniformity: *transIt* is able to treat varied modal logics, all by the very same machinery. Moreover,
extensions to further families of logics can easily be obtained by routine application of their declarative Prolog specifications.

Moreover, the adoption of an approach based on declarative programming allows us to develop the system in an incremental way and ensures high modularity
and extensibility of the application. As a matter of fact, in the same easy routine
fashion in which source languages can be added, the system can also be extended
to encompass other target languages, so as to "drive" different (relational) proof
systems. We exemplify this adaptability by extending *transIt* in order to use
it as a front-end for two deductive frameworks for relation algebras which are
rather different in nature (Section 4). One of the two consists in a minimal implementation of a proof-assistant (with some form of automated capabilities)
based on Rasiowa-Sikorski rewriting rules [29]. Actually, this proof-assistant has
been easily integrated in *transIt* by means of a common graphical user interface.
As a second approach to relational reasoning, we show how *transIt* can be used
as a front-end for a first-order theorem-prover which is exploited as relational
inference engine very much in the spirit of [11, 12].

The paper is organized as follows. In Section 1 we describe source and target languages. For most of the modal logics, we provide the corresponding translation rules. Section 2 illustrates the architecture of *transIt* and the successive phases of the translation process, while an outline of the input/output formats is given in Section 3. Finally, Sections 4 and 5 describe the interface to the built-in proof-assistant and speculate on improvements to the overall inferential framework one can envisage.

1   Source and Target Languages

The main target language which our translation supports is the algebra of binary relations. For this target, given a formula φ the system produces a relational term $t(\varphi)$ belonging to an algebraic language encompassing the usual constructs of Boolean algebra plus further operators specific to the realm of relations. To be more specific, following the work of Alfred Tarski [35], let us recall the basic notions on such formalism. The intended *universe of discourse* is a collection $\Re$ of binary relations over a non-null domain $\mathcal{U}$. We assume that the *top* relation $\bigcup \Re$, and the *diagonal* relation consisting of all pairs $\langle u, u \rangle$ with u in $\mathcal{U}$, belong to this universe, which is also closed under the *intersection* ($\cap$), *union* ($\cup$), *complement* ($^{-}$) relative to $\bigcup \Re$, *composition* (;), and *conversion* ($^{\smile}$) operations. Within such a system, two primitive constants $\mathbf{1}$ and $\boldsymbol{I}$ designate the top and the diagonal relation, while the operations are interpreted as one expects (here, for any relational expression R we are indicating by $R^{\Im}$ the relation over $\mathcal{U}$ designated by R), for instance:

- $P^{\smile}$ designates the relation consisting of all pairs $\langle v, u \rangle$ with $\langle u, v \rangle$ in $P^{\Im}$;
- $P; Q$ designates the relation consisting of all pairs $\langle u, w \rangle$ such that there is at least one v for which $\langle u, v \rangle$ and $\langle v, w \rangle$ belong to $P^{\Im}$ and to $Q^{\Im}$, respectively;
- $P \cap Q$ designates the relation consisting of all pairs $\langle u, v \rangle$ which simultaneously belong to $P^{\Im}$ and to $Q^{\Im}$;

and similarly for the other constructs.

Designations for further constants, operations over relations, or equations of a special kind, can be introduced through definitions, e.g.:

$$\mathbf{0} =_{\text{Def}} \overline{\mathbf{1}}, \qquad\qquad\qquad \boldsymbol{D} =_{\text{Def}} \overline{\boldsymbol{I}},$$
$$P{-}Q =_{\text{Def}} P \cap \overline{Q}, \qquad\qquad P{+}Q =_{\text{Def}} (Q \cup P){-}(Q \cap P),$$
$$P \sqsubseteq Q \leftrightarrow_{\text{Def}} P{-}Q = \mathbf{0}.$$

Another target language currently supported is the binary first-order predicate calculus with three variables, namely $\mathcal{L}_3$ [35]. For this target, the translation is obtained by first performing the translation into the algebra of relations, and then exploiting first-order characterizations of the relational operators. Clearly, in order to limit the overall number of first-order variables to three, in doing the latter transformation we must rely on a suitable variable-recycling mechanism.

It should be noted that a first-order sentence is logically equivalent to a sentence of $\mathcal{L}_3$ if and only if it is expressible in the algebra of relations. This is because $\mathcal{L}_3$ is equipollent to the arithmetic of binary relations [35, Chap. 3]. On the other hand, this is no more the case if we consider sentences of the full first-order predicate calculus. Actually, it is known (cf. [35, 21]) that the collection of all first-order sentences expressible with three variables (and hence having a relational rendering) is undecidable. As a consequence, the translation from first-order predicate calculus into the algebra of relations is not always doable and it can only be achieved (in favorable cases) by means of conservative techniques. Therefore, our Prolog-based translator may fail in translating a sentence. Anyway, the translation process terminates in every case, and a diagnostic message is issued when the translation is not carried through. Notice that the translation process could be improved by resorting to conservative refinements such as those proposed in [2].

Similar enhancements can be applied in order to build more target languages into the tool. One could easily achieve this goal by describing such languages in terms of suitable rewriting rules. As an example we mention another currently available translation for modal formulas (see below, for a description of the source languages), having a set-theoretical language as target. This approach is described in [4, 1, 31], where it is shown that even a very weak set theory can offer adequate means for expressing the semantics of modal systems of propositional logic. In this context, a modal formula is translated into a formula of a very weak set theory. Then, in order to perform (semi-)automated modal inference, the result of the translation could be fed into a deduction system for theory-based reasoning [13] or, alternatively, into a Rasiowa-Sikorski proof system for set-theory, as described in [31]. Another possibility could consist in performing one further translation step, from the set theoretical framework into the relational calculus, as suggested in [9], to then exploit any deductive system for relational reasoning.

Let us now briefly highlight most of the source languages currently accepted by the translator. We characterize the languages of the logics which employ binary accessibility relations in terms of their Kripke-style models. Our translator does not, as yet (although we plan extensions of this kind), deal with the languages of relevant logics or the logics with binary modalities—requiring ternary relations in their models. The translation functions for many of these languages are known, see [26] for a translation of languages of relevant logics.

The main idea of the translation is to assign relational terms to formulas of non-classical logics so that validity is preserved. These terms must represent *right ideal* relations, a binary relation R on the domain $\mathcal{U}$ being called right ideal when it meets the condition $R; \mathbf{1} = R$. In other words, a right ideal relation is of the form $X \times \mathcal{U}$ for some $X \subseteq \mathcal{U}$. Intuitively speaking, if a formula is replaced by a right ideal relation, then its domain represents the set of states where the formula is true, and its range represents the universe of all states. For atomic formulas the property of being right ideal can be enforced by postulating that a propositional variable, say p, is translated into a relational term $P; \mathbf{1}$, where

P is a relation variable uniquely associated with p. It follows that, given a language, a relational translation of its formulas can be defined provided that, first, the propositional operations of the language can be mapped into the relational operations which preserve the property of being right ideal and, second, the translation will preserve validity. It is known that Boolean operations preserve the property of being right ideal and the composition of any relation with a right ideal relation results in a right ideal relation. So if a logic is based on a classical logic whose propositional connectives are Boolean, or if a logic has a lattice as a basis, then the only problem is to appropriately translate the remaining intensional propositional operations of the logic. Since their semantics depends on the accessibility relation(s) which usually are not right ideal, the translation should use these relations only as first arguments of the composition operator, making use of the property stated above. If this can be done with preservation of validity, then the translation process is successful.

In the following we present definitions of the translation functions of languages for several families of logics whose accessibility relations are binary. In all the listed cases the validity-preserving theorems are known and can be found in the cited references.

Mono-Modal Logics. This is the basic translation of (propositional) modal formulas into relational terms originated in [25]. The source language involves usual propositional connectives together with necessity and possibility operators (here ψ and χ stand for propositional sentences):

- $t(p_i) \ =_{\text{Def}} \ P_i \, ; \, \mathbf{1}$, where P_i is a relational variable uniquely corresponding to the propositional variable p_i;
- $t(\neg \psi) \ =_{\text{Def}} \ \overline{t(\psi)}$;
- $t(\psi \,\&\, \chi) \ =_{\text{Def}} \ t(\psi) \cap t(\chi)$;
- $t(\Diamond \, \psi) \ =_{\text{Def}} \ \mathsf{R} \, ; \, t(\psi)$, where R is a constant relation designating the *accessibility relation* between possible worlds;

and similarly for the other customary propositional connectives (see also [25], for a very detailed treatment).

Lattice-Based Modal Logics. Lattice-based modal logics have the operations of disjunction and conjunction and, moreover, each of them includes a modal operator which can be either a possibility or necessity or sufficiency or dual sufficiency operator. Since negation is not available in these logics, both in the possibility–necessity and in the sufficiency–dual-sufficiency pair neither operator is expressible in terms of the other. We can also consider mixed languages with any subset of these operators. The target relational language for all of these lattice-based logics includes the following specific accessibility relations: binary relations $\leqslant_1$ and $\leqslant_2$, which are assumed to be reflexive and transitive and to satisfy the condition $\leqslant_1 \cap \leqslant_2 = \boldsymbol{I}$. Such relations are needed in order to provide semantics for the operation of disjunction which, in the case of lattice-based logics, does not necessarily distribute over conjunction. All of these logics have been deeply investigated in [32, 7, 20]. The translation of disjunction and conjunction is:

- $t(\psi \vee \chi) \quad =_{\text{Def}} \overline{\leqslant_1; \overline{\leqslant_2; (t(\psi) \cup t(\chi))}};$
- $t(\psi \mathbin{\&} \chi) \quad =_{\text{Def}} t(\psi) \cap t(\chi).$

Considering a source language with a possibility operator $\Diamond$, the target language includes two relations $R_\Diamond$ and $S_\Diamond$ subject to the following conditions:

$$\leqslant_1^{\smile}; R_\Diamond; \leqslant_1^{\smile} \sqsubseteq R_\Diamond, \qquad R_\Diamond \sqsubseteq S_\Diamond; \leqslant_1^{\smile},$$
$$\leqslant_2; S_\Diamond; \leqslant_2 \sqsubseteq S_\Diamond, \qquad S_\Diamond \sqsubseteq \leqslant_2; R_\Diamond.$$

The translation of a formula involving the modal operator is

$$t(\Diamond\chi) \quad =_{\text{Def}} \overline{\leqslant_1; \overline{S_\Diamond; \leqslant_2; t(\chi)}}.$$

Also in the case of a language involving the necessity operator $\Box$, the target language includes two relations $R_\Box$ and $S_\Box$ subject to:

$$\leqslant_1; R_\Box; \leqslant_1 \sqsubseteq R_\Box, \qquad R_\Box \sqsubseteq \leqslant_1; S_\Box,$$
$$\leqslant_2^{\smile}; S_\Box; \leqslant_2^{\smile} \sqsubseteq S_\Box, \qquad S_\Box \sqsubseteq R_\Box; \leqslant_2^{\smile}.$$

The translation of a formula involving the modal operator is

$$t(\Box\chi) \quad =_{\text{Def}} \overline{R_\Box; \overline{t(\chi)}}.$$

Formulas involving the sufficiency operator $\boxdot$ are translated into relational expressions by introducing two relations $R_\boxdot$ and $S_\boxdot$ subject to the following conditions:

$$\leqslant_1; R_\boxdot; \leqslant_2 \sqsubseteq R_\boxdot, \qquad R_\boxdot \sqsubseteq \leqslant_1; S_\boxdot,$$
$$\leqslant_2^{\smile}; S_\boxdot; \leqslant_1^{\smile} \sqsubseteq S_\boxdot, \qquad S_\boxdot \sqsubseteq R_\boxdot; \leqslant_1^{\smile}.$$

Within such a framework, the translation of a formula involving the sufficiency operator is

$$t(\boxdot\chi) \quad =_{\text{Def}} \overline{R_\boxdot; \overline{\leqslant_2; t(\chi)}}.$$

Finally, the translation of formulas involving the dual sufficiency operator $\lozenge\!\!\!\lozenge$, has as its target a relational language with two relations, $R_\lozenge$ and $S_\lozenge$, subject to the following conditions:

$$\leqslant_1^{\smile}; R_\lozenge; \leqslant_2^{\smile} \sqsubseteq R_\lozenge, \qquad R_\lozenge \sqsubseteq S_\lozenge; \leqslant_2^{\smile},$$
$$\leqslant_2; S_\lozenge; \leqslant_1 \sqsubseteq S_\lozenge, \qquad S_\lozenge \sqsubseteq \leqslant_2; R_\lozenge.$$

Then, the translation of a formula involving $\lozenge\!\!\!\lozenge$ is

$$t(\lozenge\!\!\!\lozenge\chi) \quad =_{\text{Def}} \overline{\leqslant_1; \overline{\overline{S_\lozenge; t(\chi)}}}.$$

Logics of Knowledge and Information. These modal logics come from [5]:

$\star$ Logic with knowledge operator $\mathbf{K}$, subject to the following translation rule:

$$t(\mathbf{K}\varphi) \quad =_{\text{Def}} \overline{R; \overline{t(\varphi)}} \cup \overline{R; t(\varphi)}.$$

⋆ Logic of non-deterministic information (NIL) [5, Sect. 7.2]. A multi-modal logic with three modalities, determined by the relations of informational inclusions ($\leqslant$ and $\geqslant$) and similarity (σ) subject to the following conditions:

- $\leqslant$ is reflexive and transitive and such that $\leqslant \; = \; \geqslant^{\smile}$,
- σ is reflexive and symmetric,
- $\geqslant ; \sigma ; \leqslant \; \sqsubseteq \sigma$.

⋆ Information logic (IL) [5, Sect. 7.3]. A modal logic with three modal operators corresponding to the relations of indiscernibility ($\equiv$), forward inclusion ($\leqslant$), and similarity (σ) subject to the following conditions:

- $\equiv$ is an equivalence relation,
- $\leqslant$ is reflexive and transitive,
- σ is reflexive and symmetric,
- $\leqslant^{\smile} ; \sigma \sqsubseteq \sigma$ and $\leqslant \cap \leqslant^{\smile} \; = \; \equiv$.

Intuitionistic Logic. The translation of intuitionistic logic is based on the following rules:

$$t(\psi \rightarrow \chi) =_{\text{Def}} \overline{\leqslant ; (t(\psi) \cap \overline{t(\chi)})}, \qquad t(\psi \,\&\, \chi) =_{\text{Def}} t(\psi) \cap t(\chi),$$
$$t(\neg \psi) =_{\text{Def}} \overline{\leqslant ; t(\psi)}, \qquad t(\psi \vee \chi) =_{\text{Def}} t(\psi) \cup t(\chi),$$

where $\leqslant$ is a reflexive and transitive relation.

Multi-modal Logic. These logics correspond to multi-modal frames consisting of a relational system (W, Rel) where Rel is a family of accessibility relations (enjoying closure properties with respect to relational constructs). Modalities are then of the form $[R]$ and $\langle R \rangle$, where R is any relational term of Rel (cf. [27]).

The translation of modal operators is the same as in the case of mono-modal logic. The differences between operators are articulated in terms of the properties of the corresponding accessibility relations.

Temporal Logics. By taking the relational formalization of temporal logics given in [28], we extended the translator in order to deal with temporal formulas. The basic modal operators (referring to states in the future or in the past) are:

- $G\varphi$ interpreted as "always, in the future, φ will be true";
- $F\varphi$ interpreted as "sometimes, in the future, φ will be true ";
- $H\varphi$ interpreted as "φ was always true in the past";
- $P\varphi$ interpreted as "φ was true in some past time";
- $\varphi \, U \, \chi$ interpreted as "at some moment χ will be true and from now till then φ will be true";
- $\varphi \, S \, \chi$ interpreted as "there was a moment when χ was true and such that φ has ever since been true";
- $X\varphi$ interpreted as "φ will be true in the next moment in time".

In this context, relational representations of temporal formulas are expressed by considering an accessibility relation R that (together with its converse $R^\smile$) links time instants. The relational translations of the modalities G, F, H, and P are as follows:

$$t(G\varphi) =_{\text{Def}} \overline{R; \overline{t(\varphi)}}, \qquad\qquad t(H\varphi) =_{\text{Def}} \overline{R^\smile; \overline{t(\varphi)}},$$
$$t(F\varphi) =_{\text{Def}} R; t(\varphi), \qquad\qquad t(P\varphi) =_{\text{Def}} R^\smile; t(\varphi),$$
$$t(\varphi U \chi) =_{\text{Def}} t(\varphi) U t(\chi), \qquad\qquad t(\varphi S \chi) =_{\text{Def}} t(\varphi) S t(\chi),$$
$$t(X\varphi) =_{\text{Def}} t((\varphi \,\&\, \neg\varphi) U \varphi),$$

where, in the translations of the modal operators U and S we use the same symbols to denote two newly introduced relational constructs. These new constructs cannot be defined in terms of the primitive relational constructs (page 91). The intended interpretation of U is as follows: PUQ designates the binary relation consisting of all pairs $\langle u, v \rangle$ such that there exists t such that $\langle u, t \rangle$ belongs to the accessibility relation $R^\Im$, $\langle t, v \rangle$ belongs to $Q^\Im$, and for all w, if $\langle u, w \rangle \in R^\Im$ and $\langle w, t \rangle \in R^\Im$ then $\langle w, v \rangle \in P^\Im$. (The interpretation of S is analogous, with respect of $R^\smile$.)

Other Modal Logics. Other modal logics currently accepted by the translator involve: logics with specification operators [18, 24], logics with Humberstone operators [19], logics with sufficiency operators [15, 6].

Following the semantics developed by Hoare and Jifeng [18], the operators of the weakest prespecification (\\) and the weakest postspecification (/) are modeled with residuals of the relational composition which are definable with composition, converse and complement:

$$Q\backslash R =_{\text{Def}} \overline{\overline{R}; Q^\smile} \qquad \text{and} \qquad R/P =_{\text{Def}} \overline{P^\smile; \overline{R}}.$$

Consequently, $P; Q \sqsubseteq R$ if and only if $P \sqsubseteq Q\backslash R$ if and only if $Q \sqsubseteq R/P$.

The Humberstone operators are the modal operators of possibility and necessity determined by the complement of an accessibility relation. It follows that their translation can easily be derived from the translation of the mono-modal operators.

The sufficiency ($\boxdot$) and dual sufficiency ($\diamondsuit$) operators receive the following relational translation:

$$t(\boxdot \varphi) =_{\text{Def}} \overline{\overline{R}; t(\varphi)},$$
$$t(\diamondsuit \varphi) =_{\text{Def}} \overline{R}; \overline{t(\varphi)},$$

where R is a relational constant representing an accessibility relation of the models of a logic under considerations.

It follows that our translation tool is able to translate the formulas of any of the information logics presented in [5], as they involve, together with Boolean or lattice operators, intensional operators that are either modal or sufficiency or knowledge operators.

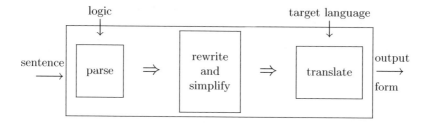

Fig. 2. The translator architecture

2 The Translation Process

The translator takes as input a formula of a specific source language (see Section 1). As shown in Fig. 2, the first of these transformations yields an internal representation of the formula, while the last step generates its final rendering. Then, a sequence of rewritings and simplifications is performed. Finally, the desired translation is produced.

More specifically, here is the sequence of the salient phases which usually form the translation (some of them being skipped in specific cases, for instance double-negation removal in intuitionistic logic):

Lexical and syntactical analyses. This phase accepts a formula only if it is syntactically correct and its constructs belong to the specific language at hand. The syntax-directed translation implemented through this stage is described by an attributed definite clause grammar. Hence, any extension to further logics can be achieved by simply adding a suitable set of grammar rules which characterize the (new) well-formed formulas. The outcome of this stage is an intermediate representation of the abstract syntax tree (AST) of the input formula.

Generation of an internal representation. By means of a rewriting process which acts in a bottom-up recursive fashion, the outcome of the preceding phase is turned into an internal representation of the AST (in form of a Prolog term), independent of the source language.

Abstract propositional evaluation. The internal representation of the given formula is analyzed in order to extract its propositional schema. The schema so obtained is then (possibly) simplified through replacements of some of its sub-formulas by tautologically equivalent ones.

Reduction to primitive constructs. In this phase the formula is rewritten in terms of a small repertoire of constructs and connectives, to be regarded as being "primitive". For instance, biimplication $\leftrightarrow$ is rewritten as a conjunction of two implications, and so on. Notice that some of these rewritings must be inhibited at times, insofar as unsound with respect to the specific logic at hand. The aim of this transformation is to make the next phase easier.

Propositional simplifications. Through this phase the internal representation of the formula is simplified by applying a number of propositional

```
rewrite(Rules,From,To)  :- transl(Rules,From,M),     % rewrite until
          (From==M, To=M ; rewrite(Rules,M,To)).      % fix-point

transl(_,T,T)  :- var(T).
transl(R,T,S)  :- T =.. [F|Argg], translArgg(R,Argg,Brgg),
                  M =.. [F|Brgg], transl0(R,M,S).
translArgg(_,[],[]).
translArgg(R,[H|B],[SH|SB])  :- transl(R,H,SH),
                                translArgg(R,B,SB).
transl0(R,T,S)  :- Goal =.. [R,T,S],   (Goal ; S=T).

rewrite1(R,T,S)  :-  once(transl(R,T,S)).             % rewrite once
```

Fig. 3. A simple and powerful post-order rewriting procedure

simplifications to it, mainly aimed at reducing the size of the formula (for instance, elimination of tautological sub-formulas and of double negations).

Relational translation. This is the main step of the translation process: the internal representation of the given formula is translated into the calculus of binary relations. The kind of rewriting rules employed may depend on the source language of the input formula (see Section 1). The outcome of this phase is a relational term.

Relational simplifications. The overall translation process ends with a series of relational simplifications applied to the relational term produced by the preceding step. The simplest among these rewritings take care of the idempotency, absorption or involution properties of (some of) the relational constructs. The process can easily be extended to perform more complex simplifications.

It should be noticed that most of the above steps are all uniformly performed by exploiting the same simple meta-rewriter. Fig. 3 displays the basic Prolog specification of this post-order rewriting procedure. The main predicate is `rewrite/3`. Intuitively speaking, it accepts as its first parameter (`Rules`) a Prolog predicate describing one of the possible translation steps. Then it recursively processes the term `From` in order to produce its translation `To`.

Further phases could be added, for instance in order to apply semantical transformations to the relational term, possibly with respect to a set of axiomatic assumptions characterizing a particular class of relational structures as constituting the target framework.

Example 1. As an example we provide here the textual output produced by the various steps of the translation into the calculus of relations of the multi-modal formula:

$$[R \cup Q] < Q > p \rightarrow q.$$

Here is a tracing of the translation process (where p1, p2, and R3 are internal names corresponding to the external names p, q, and Q, respectively):

```
?- enu2tg(A,polyModal).
|: [R+Q]<Q>p -> q.
...i(nn(pp(1,-3),u(0,-3)),2)...
from intermediate to internal representation:
    NEC(POSS(p1,R3),u(R,R3))imp p2...
in primitive connectives:
    NEC(NEC(p1 imp f,R3)imp f,u(R,R3))imp p2...
after propositional simplification:
    NEC(NEC(p1 imp f,R3)imp f,u(R,R3))imp p2...
after translation to calculus of relations:
    u(c(c(k(u(R,R3),c(u(c(c(k(R3,c(u(c(k(p1,U)),Z)))),Z)))),k(p2,U))...
after relational simplifications:
    u(k(p2,U),k(u(R,R3),c(k(k(R3,p1),U))))...

A = (u(k(p2,'U'),k(u('R','R3'),c(k(k('R3',p1),'U'))))='U')
```

The Prolog term produced is the representation of the relational equality

$$q \,;\, 1 \cup (R \cup Q)\,;\, \overline{Q\,;\, p\,;\, 1} = 1.$$

Proving that the initial modal formula is a theorem amounts to deriving this equation within the calculus of relations.

3   Input and Output Formats

When rawly used, our Prolog-based translation tool system reads a pure-text input typed in by the user (cf. Example 1). The output is then written, again in pure-text format, to the standard output stream (usually, the screen). This kind of interaction is, however, quite unsatisfactory, because the ASCII character set is rather poor. In order to overcome this disadvantage and ease the input/output of complex formulas and expressions, a user-friendly interface has been developed. Such a graphical interface allows the user to type in formulas using graphical LATEX-generated symbols. In doing this, we exploited the useful integration facilities offered by SICStus Prolog [37] with respect to other programming languages, in particular to the Tcl/Tk toolkit [36]. Hence, the input of formulas is achieved through dialogues that are generated at run-time depending on the specific language chosen by the user. For instance, Fig. 4 displays the input dialogue generated for multi-modal formulas.

The system also provides the possibility of processing a text file, as well as to generate a text file as output. Through this feature it is possible to produce input files for different deduction tools (see Section 4).

Let us briefly illustrate the use of the graphical interface with a simple example. Consider the multi-modal formula $[R \cup Q] < Q > p \to q$. This formula can be input to the translator easily, as shown in Fig. 4. The relational equation obtained can also be displayed graphically, as in Fig. 5.

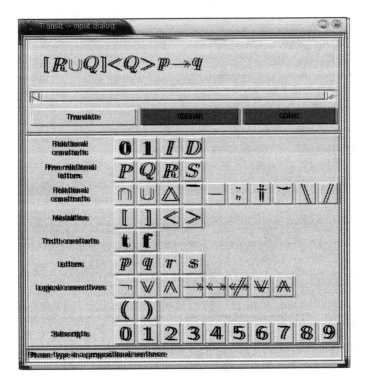

Fig. 4. Input dialogue for multi-modal formulas

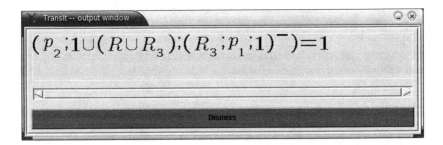

Fig. 5. Output of a translation process

4   Driving a Deductive Tool

As mentioned at the outset, the main purpose of *transIt* is to provide an extensible front-end for (relational) deductive systems.

To exemplify how well this goal is approached, in what follows we report on two extensions of *transIt*, designed in order to use it as a front-end for two deductive frameworks for relation algebras which are rather different in nature. One

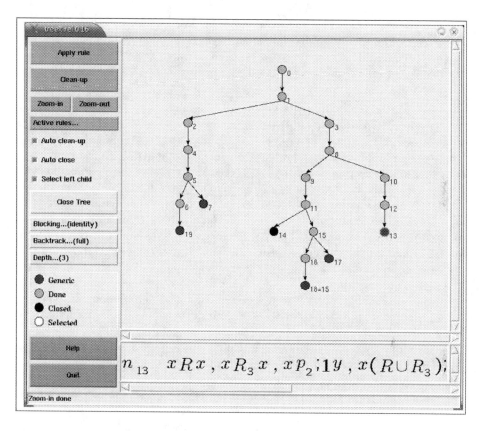

Fig. 6. Assisted development of a proof tree

of the two consists in a minimal implementation of a proof-assistant (showing some degree of autonomy) based on Rasiowa-Sikorski rewriting rules [29]. Such proof-assistant is accessible through *transIt*'s graphical interface. Once the user has obtained a relational rendering of a theorem, (s)he can proceed to try building a proof-tree of the relational translation. Fig. 6 shows a simple example of a derivation tree. The user interacts with the system by simply choosing a node of the tree in order to apply one of the rewriting rules. The system takes care of verifying applicability of rules, performing the extension of the tree, and checking whether, as a consequence of rule applications, any branch becomes closed. Some form of (semi-)automated reasoning capabilities are also implemented: it is possible to ask the system to try, autonomously, to close all branches of a (sub-)tree.

Another viable approach to relational reasoning consists in using *transIt* as a front-end for a first-order theorem-prover: Otter, in our choice. This is achieved by extending the translation process: a new set of rewriting rules is used to implement automated generation of an input file to be fed into Otter. Once the input file is available, Otter can be used as described in [11, 12] to search for a proof of the theorem within the relational framework. Obviously, the very same approach can be used with other theorem provers.

Currently, *transIt* can be downloaded from the site `http://www.di.univaq.it/TARSKI/transIt/` and easily installed. It is developed under Linux, but we also provide a porting for Windows XP.

5   Improving the System

The modular approach we adopted both in developing the translator and in extending the collection of source and target languages, plainly permits steady improvements to and extensions of the system. At the moment, most of the phases of the translation process are carried out by means of syntactical rewritings. Nevertheless, the translation process could benefit from improvements to its ability to exploit semantic properties of connectives and constructs. As a matter of fact, in the current implementation this ability lies exclusively in the abstract propositional evaluation phase (see page 97).

Another amelioration, in the same frame of mind, is the exploitation, in the derivation process (both for the assisted and for the autonomous functioning mode), of specific rewriting rules depending on the particular logic of the theorem being proved.

As mentioned, the system can deal with different target languages (see page 92). As a further example, we mention here a particularly interesting future development: extend the collection of target languages, so as to permit the translation into languages of ternary relations needed for handling relevant logics and other substructural logics whose translations are presented in [26].

Further challenging themes for long-term activities regard exploring the possibilities offered by

- the integration with/within existing tools for translation and deduction. In particular, a fruitful synergy could develop from the integration/interaction with the "Anamorpho system", an environment for describing relational specifications which is based on definitional extension mechanisms (see [3]).
- the integration with visual-oriented tools for manipulation of relational formulas (based, for instance, on graphical representation of relational expressions and on graph-rewriting techniques [10]).

Acknowledgments

Thanks are due to Marianna Nicolosi Asmundo and to Hui Wang for fruitful discussions on the topics of this paper.

References

[1] J. F. A. K. van Benthem, G. D'Agostino, A. Montanari, and A. Policriti. Modal deduction in second-order logic and set theory-I, *Journal of Logic and Computation*, **7**(2):251–265, 1997.

[2] D. Cantone, A. Formisano, E. G. Omodeo, C. G. Zarba. Compiling dyadic first-order specifications into map algebra. *Theoretical Computer Science*, **293**(2):447–475, Elsevier, 2003.

[3] P. Caianiello, S. Costantini, E. G. Omodeo. An Environment for Specifying Properties of Dyadic Relations and Reasoning about Them. I: Language Extension Mechanisms. In H. de Swart, E. Orłowska, G. Schmidt, and M. Roubens eds., *Theory and Applications of Relational Structures as Knowledge Instruments*, LNCS **2929**, pp.87–106, Springer, 2004.

[4] G. D'Agostino, A. Montanari, and A. Policriti. A set-theoretic translation method for polymodal logics, *Journal of Automated Reasoning*, **3**(15):317–337, Kluwer, 1995.

[5] S. Demri, E. Orłowska. Incomplete Information: Structure, Inference, Complexity. EATCS Monographs in Theoretical Computer Science, Springer, 2002.

[6] I. Düntsch and E. Orłowska. Beyond modalities: sufficiency and mixed algebras. In E. Orłowska and A. Szalas eds., *Relational Methods in Computer Science Applications*. pp.277–299. Physica-Verlag, Heidelberg, 2000.

[7] I. Düntsch, E. Orłowska, A. M. Radzikowska, and D. Vakarelov. Relational Representation Theorems for Some Lattice-Based Structures. *Journal on Relational Methods in Computer Science*, **1**:132–160, 2004.

[8] A. Formisano and M. Nicolosi Asmundo. An efficient relational deductive system for propositional non-classical logics. *Journal of Applied Non-Classical Logics*, to appear. (A draft version is available as report 8/06 of the Dipartimento di Informatica, Università di L'Aquila, 2006.)

[9] A. Formisano, E. Omodeo, E. Orłowska, and A. Policriti. Uniform relational frameworks for modal inferences. In I. Düntsch and M. Winter eds., Proceedings of the 8th International Conference on Relational Methods in Computer Science (RelMiCS 8), 2005.

[10] A. Formisano, E. Omodeo, M. Simeoni, A graphical approach to relational reasoning. *Electronic Notes in Theoretical Computer Science*, **44**(3), Elsevier, 2001.

[11] A. Formisano, E. G. Omodeo, M. Temperini. Instructing Equational Set-Reasoning with Otter. In R. Goré, A. Leitsch, and T. Nipkow eds., *Automated Reasoning*, Proceedings of IJCAR 2001. LNCS **2083**, pp.152–167, Springer, 2001.

[12] A. Formisano, E. G. Omodeo, M. Temperini. Layered map reasoning: An experimental approach put to trial on sets. *Electronic Notes in Theoretical Computer Science*, **48**, Elsevier, 2001.

[13] A. Formisano and A. Policriti. T-Resolution: Refinements and Model Elimination. *Journal of Automated Reasoning*, **22**(4):433–483, Kluwer, 1999.

[14] M. Frias and E. Orłowska. Equational reasoning in nonclassical logics. *Journal of Applied Non-Classical Logics*, **8**(1-2):27–66, 1998.

[15] V. Goranko. Modal definability in enriched languages. *Notre Dame Journal of Formal Logic*, **31**:81–105, 1990.

[16] R. Goré. Cut-free display calculi for relation algebras. In D. van Dalen and M. Bezem eds., *CSL96: Selected Papers of the Annual Conference of the Association for Computer Science Logic*, LNCS **1258**, pp.198–210. Springer, 1996.

[17] M. C. B. Hennessy. A proof system for the first order relational calculus. *Journal of Computer and System Sciences*, **20**:96–110, 1980.

[18] C. A. R. Hoare and H. Jifeng. The weakest prespecification. Part I *Fundamenta Informaticae*, **IX**:51–84; Part II ibidem **IX**:217-252, 1986.

[19] I. L. Humberstone. Inaccessible worlds. *Notre Dame Journal of Formal Logic*, **24**:346–352, 1983.

[20] J. Järvinen and E. Orłowska. Relational correspondences for lattices with operators. In I. Düntsch and M. Winter eds., Proceedings of the 8th International Conference on Relational Methods in Computer Science (RelMiCS 8), pp.111–118, 2005.

[21] M. K. Kwatinetz. *Problems of expressibility in finite languages.* PhD thesis, University of California, Berkeley, 1981.

[22] R. Maddux. A sequent calculus for relation algebras. *Annals of Pure and Applied Logic,* **25**:73–101, 1983.

[23] H. J. Ohlbach, A. Nonnengart, M. de Rijke, and D. Gabbay. Encoding Two-Valued Nonclassical Logics in Classical Logic, In *Handbook of Automated Reasoning,* vol. II, pp.1403–1486, Elsevier, 2001.

[24] E. Orłowska. Proof system for weakest prespecification. *Information Processing Letters,* **27**:309–313, 1988.

[25] E. Orłowska. Relational interpretation of modal logics, In H. Andreka, D. Monk, and I. Nemeti eds., *Algebraic Logic.* Colloquia Mathematica Societatis Janos Bolyai, vol. 54, pp.443–471, North Holland, 1988.

[26] E. Orłowska. Relational proof systems for relevant logics, *Journal of Symbolic Logic,* **57**, pp.167–186. 1992.

[27] E. Orłowska. Relational semantics for nonclassical logics: formulas are relations, Philosophical Logic in Poland (J. Wolenski, ed.), pp.167–186. Kluwer, 1994.

[28] E. Orłowska. Temporal logics in a relational framework. Time and Logic—A computational Approach. pp.249–227. University College London Press, 1995.

[29] E. Orłowska. Relational proof systems for modal logics, In H. Wansing ed., *Proof theory of modal logic,* Applied logic series, vol.2, pp.55–78. Kluwer, 1996.

[30] E. Orłowska. Relational formalization of nonclassical logics. In C. Brink, W. Kahl, and G. Schmidt eds., *Relational Methods in Computer Science,* pp.90-105. Springer, 1997.

[31] E. G. Omodeo, E. Orłowska, and A. Policriti. Rasiowa-Sikorski style relational elementary set theory. In R. Berghammer and B. Moeller eds., Proceedings 7th International Conference on Relational Methods in Computer Science (RelMiCS 7), LNCS **3051**, pp. 215-226, Springer, 2003.

[32] E. Orłowska and D. Vakarelov. Lattice-based modal algebras and modal logics. In P. Hajek, L. M. Valdés-Villanueva, D. Westerstahl eds., Logic, Methodology and Philosophy of Science. Proceedings of the 12th International Congress, pp.147–170. KCL Puplications, 2005.

[33] R. Schmidt and U. Hustadt. Mechanized reasoning and model generation for extended modal logics. In H. de Swart, E. Orłowska, G. Schmidt, and M. Roubens, eds., *Theory and Applications of Relational Structures as Knowledge Instruments,* LNCS **2929**, pp.38–67, Springer, 2004.

[34] W. Schoenfeld. Upper bounds for a proof search in a sequent calculus for relational equations. *Zeitschrift fuer Mathematische Logic und Grundlagen der Mathematik* **28**:239–246, 1982.

[35] A. Tarski and S. Givant. A formalization of Set Theory without variables, Colloquium Publications, vol. 41, American Mathematical Society, 1987.

[36] Web resources for the Tcl/Tk toolkit: http://tcl.sourceforge.net.

[37] Web reference for SICStus Prolog: http://www.sics.se/sicstus.

Relational Approach to Order-of-Magnitude Reasoning*

Alfredo Burrieza[1], Manuel Ojeda-Aciego[2], and Ewa Orłowska[3]

[1] Dept. Filosofía. Univ. Málaga, Spain
burrieza@uma.es
[2] Dept. Matemática Aplicada. Univ. Málaga, Spain
aciego@ctima.uma.es
[3] National Institute of Telecommunications, Warsaw, Poland
orlowska@itl.waw.pl

Abstract. This work concentrates on the automated deduction of logics of order-of-magnitude reasoning. Specifically, a translation of the multimodal logic of qualitative order-of-magnitude reasoning into relational logics is provided; then, a sound and complete Rasiowa-Sikorski proof system is presented for the relational version of the language.

1 Introduction

Qualitative order-of-magnitude reasoning has received considerable attention in the recent years; however, the analogous development of a logical approach has received little attention. Various multimodal approaches have been promulgated, for example, for qualitative spatial and temporal reasoning but, as far as we know, the only logic approaches to order-of-magnitude reasoning (OMR) are [1, 2, 3].

These first approaches to the logics of qualitative order-of-magnitude reasoning are based on a system with two landmarks, which is both simple enough to keep under control the complexity of the system and rich enough so as to permit the representation of a subset of the usual language of qualitative order-of-magnitude reasoning. The intuitive representation of our underlying frames is given below, in which two landmarks $-\alpha$ and $+\alpha$ are considered

In the picture, $-\alpha$ and $+\alpha$ represent, respectively, the greatest negative observable and the least positive observable, partitioning the real line in classes of positive observable OBS^+, negative observable OBS^- and non-observable (also called infinitesimal) numbers INF. This choice makes sense, in particular, when considering physical metric spaces in which we always have a smallest unit which can be measured; however, it is not possible to identify a least or greatest non-observable number.

---

* The first two authors are partially supported by projects TIC2003-9001-C02-01 and TIN2006-15455-C03-01.

H. de Swart et al. (Eds.): TARSKI II, LNAI 4342, pp. 105–124, 2006.
© Springer-Verlag Berlin Heidelberg 2006

In order to introduce a few intuitive explanations about the practical use of the OMR-relations of comparability and negligibility, firstly, assume one aims at specifying the behaviour of a device to automatically control the speed of a car; assume the system has, ideally, to maintain the speed close to some speed limit v. For practical purposes, any value in an interval $[v - \varepsilon, v + \varepsilon]$ for small ε is admissible. The extreme points of this interval can then be considered as the milestones α^- and α^+; on the other hand, the sets OBS$^-$, INF, and OBS$^+$ can be interpreted as SLOW, OK and HIGH speed.

Regarding negligibility, the representation capabilities of a pocket calculator provides an illustrative example of this type of relation. In such a device, it is not possible to represent any number whose absolute value is less than 10^{-99}. Therefore, it makes sense to consider $-\alpha = -10^{-99}$ and $+\alpha = +10^{-99}$ since any number between -10^{-99} and 10^{-99} cannot be observed/represented.[1] On the other hand, a number x can be said to be negligible with respect to y provided that the difference $y - x$ cannot be distinguished from y. Numerically, and assuming an 8+2 (digits and mantissa) display, this amounts to state that x is negligible wrt y iff $y - x > 10^8$. Furthermore, this example above suggests a real-life model in which, for instance -1000 is negligible with respect to -1. Simply, interpret the numbers above as exponents, since 10^{-1000} can be considered negligible with respect to 10^{-1}.

In this paper the paradigm 'formulas are relations' formulated in [11] is applied to the modal logic for order-of-magnitude reasoning of [3]. A relational formalisation of logics is based on an observation that a standard relational structure (a Boolean algebra with a monoid) constitutes a common core of a great variety of nonclassical logics. Exhibiting this common core on all the three levels of syntax, semantics and deduction, enables us to create a general framework for representation, investigation and implementation of nonclassical logics. Relational formalization of nonclassical logics is performed on the following methodological levels:

Syntax: With the formal language of a logic L there is associated a language of relational terms.

Semantics and model theory: With logic L there is associated a class of relational models for L and in these models the formulas from L are interpreted as relations.

Proof theory: With logic L there is associated a relational logic $Re(L)$ for L such that its proof system provides a deduction method for L.

In relational representation of logical systems we articulate explicitly information about both their syntax and semantics. Generally speaking, formulas are represented as terms over some appropriate algebras of relations. Each of the propositional connectives becomes a relational operation and in this way an original syntactic form of formulas is coded. Semantic information about a formula which normally is included in a satisfiability condition for that formula, consists of the two basic parts: first, we say which states satisfy the subformulas

---

[1] Of course, there are much more numbers which cannot be represented, but this is irrelevant for this example.

of the given formula, and second, how those states are related to each other via an accessibility relation. Those two ingredients of semantic information are of course interrelated and unseparable. In relational representation of formulas the terms representing accessibility relations are included explicitly in the respective relational terms corresponding to the formulas. They become the arguments of the relational operations in a term in the same way as the other of its subterms, obtained from subformulas of the given formula. In this way semantic information is provided explicitly on the same level as syntactic information. Thus the relational term corresponding to a formula encodes both syntactic and semantic information about the formula.

In the paper we develop a relational logic $Re(OM)$ based on algebras of relations generated by some relations specific to the frames of OM-logics. We define a translantion from the language of OM-logics to the language of $Re(OM)$. Next, we construct a deduction system for $Re(OM)$ in the Rasiowa-Sikorski style [14]. The Rasiowa-Sikorski systems are dual to the Tableaux systems, as shown in [15,8]. The system includes the rules of the classical relational logics and the rules specific to the relations from the frames of OM-logics. We present the basic steps of the proof of completeness theorem for this system The modular structure of the system enables us to use the existing implementation of relational proof systems [5] and to include to it the specific rules of $Re(OM)$ logic.

The structure of the paper is the following: The syntax and semantics of the language OM is given in Section 2, then a relational language for order-of-magnitude reasoning, $Re(OM)$, is presented in Section 3. Next, in Section 4 a translation function is given, which transforms a multimodal formula in OM into a relational formula in $Re(OM)$. Then, Section 5 introduces the relational proof system for the logic $Re(OM)$, together with proofs of some axioms of the proof system MQ^N of [3]. The next two sections are devoted to the soundness and completeness of the relational proof system. Finally, some conclusions are presented, together with prospects of future work.

2   The Modal Language OM

In our syntax we consider three types of modal connectives, each one associated to certain order relation: $\overrightarrow{\Box}$ and $\overleftarrow{\Box}$ to deal with an ordering $<$, the connectives $\overrightarrow{\blacksquare}$ and $\overleftarrow{\blacksquare}$ to deal with a second ordering $\sqsubset$ and the connectives $\overrightarrow{\boxdot}$ and $\overleftarrow{\boxdot}$ to deal with a third order relation $\prec$ (the specific conditions required on comparability and negligibility relations, $\sqsubset$ and $\prec$, will be stated later).

The intuitive meanings of each modal connective is as follows:

$\overrightarrow{\Box}A$ means A *is true for all numbers which are greater than the current one.*

$\overrightarrow{\blacksquare}A$ is read A *is true for all numbers which are greater than and comparable with the current one.*[2]

$\overleftarrow{\Box}A$ means A *is true for all numbers which are less than the current one.*

---

[2] Note that the use of "comparable" has to be understood as the comparability relation in OMR, hence it is related to the ordering $\sqsubset$ introduced in Definition 1.

$\overleftarrow{\blacksquare}A$ means A is true for all numbers which are less than and comparable with the current one.

$\overrightarrow{\boxed{n}}A$ means A is true for all numbers with respect to which the current one is negligible.

$\overleftarrow{\boxed{n}}A$ means A is true for all numbers which are negligible with respect to the current one.

The intuitive description of the meaning of the negligibility-related modalities deserves some explanatory comments. Depending on the particular context in which we are using the concept of negligibility, several possible definitions can arise. We have chosen to use an intrinsically directional notion of negligibility, in that negligible numbers are always to the left. There are other approaches in which the negligibility relation is bi-directional, so a point x can be negligible wrt points smaller than x and also wrt points greater than x, for instance, in [4,17] it is the absolute value of an element that is considered before considering the negligibility relation, whereas in [1] yet another definition of bidirectional neglibility is presented.

The syntax of our initial language for qualitative reasoning with comparability and negligibility is introduced below:

The alphabet of the language OM is defined by using:

- A stock of atoms or propositional variables, $\mathcal{V}$.
- The classical connectives $\neg, \wedge, \vee$ and $\rightarrow$ and the constants $\top$ and $\bot$.
- The unary modal connectives $\overrightarrow{\Box}, \overleftarrow{\Box}, \overrightarrow{\blacksquare}, \overleftarrow{\blacksquare}, \overrightarrow{\boxed{n}}$ and $\overleftarrow{\boxed{n}}$.
- The constants α^+ and α^-.
- The auxiliary symbols: (,).

Formulas are generated from $\mathcal{V} \cup \{\alpha^+, \alpha^-, \top, \bot\}$ by the construction rules of classical propositional logic adding the following rule: If A is a formula, then so are $\overrightarrow{\Box}A$, $\overleftarrow{\Box}A$, $\overrightarrow{\blacksquare}A$, $\overleftarrow{\blacksquare}A$, $\overrightarrow{\boxed{n}}A$ and $\overleftarrow{\boxed{n}}A$.

The *mirror image* of A is the result of replacing in A each occurrence of $\overrightarrow{\Box}$, $\overleftarrow{\Box}$, $\overrightarrow{\blacksquare}$, $\overleftarrow{\blacksquare}$, $\overrightarrow{\boxed{n}}$, $\overleftarrow{\boxed{n}}$, α^+, α^- by $\overleftarrow{\Box}$, $\overrightarrow{\Box}$, $\overleftarrow{\blacksquare}$, $\overrightarrow{\blacksquare}$, $\overleftarrow{\boxed{n}}$, $\overrightarrow{\boxed{n}}$, α^-, α^+, respectively. We shall use the symbols $\overrightarrow{\Diamond}, \overleftarrow{\Diamond}, \overrightarrow{\blacklozenge}, \overleftarrow{\blacklozenge}, \overrightarrow{\diamondsuit}$ and $\overleftarrow{\diamondsuit}$ as abbreviations respectively of $\neg\overrightarrow{\Box}\neg, \neg\overleftarrow{\Box}\neg, \neg\overrightarrow{\blacksquare}\neg, \neg\overleftarrow{\blacksquare}\neg, \neg\overrightarrow{\boxed{n}}\neg$ and $\neg\overleftarrow{\boxed{n}}\neg$.

Observe that due to the presence of constants α^- and α^+ in the language, the logic OM belongs to the family of logics with nominals. The use of nominals in modal logic originated in the papers [12,13,7]. Since then the use of nominals in modal languages is a usual practice, which increases their expressibility as it was already shown in [13]. More recently, hybrid logics make also an extensive use of nominals.

The intended meaning of our language is based on a multi-modal approach, therefore the semantics is given by using the concept of frame. Intuitively, the carrier of our frames can be seen as the real line, although in our approach we will only require it to be a linearly ordered set.

Definition 1. *A multimodal qualitative frame for OM (or, simply, a frame) is a tuple $\Sigma = (\mathbb{S}, +\alpha, -\alpha, <, \prec)$, where*

1. $(\mathbb{S}, <)$ *is a linearly ordered set.*
2. $+\alpha$ *and* $-\alpha$ *are designated points in* $\mathbb{S}$ *(called* frame constants*) which allow to form the sets* OBS^+, INF, *and* OBS^- *that are defined as follows:*

$$\text{OBS}^- = \{x \in \mathbb{S} \mid x \le -\alpha\};$$
$$\text{INF} = \{x \in \mathbb{S} \mid -\alpha < x < +\alpha\};$$
$$\text{OBS}^+ = \{x \in \mathbb{S} \mid +\alpha \le x\}$$

3. *The negligibility relation* $\prec$ *is a restriction of* $<$, *i.e.* $\prec \subseteq <$, *and satisfies:*
   (i) *If* $x \prec y < z$, *then* $x \prec z$
   (ii) *If* $x < y \prec z$, *then* $x \prec z$
   (iii) *If* $x \prec y$, *then either* $x \notin \text{INF}$ *or* $y \notin \text{INF}$

The comparability relation $x \sqsubset y$ *is used an abbreviation of "$x < y$ and $x, y \in \text{EQ}$, where* $\text{EQ} \in \{\text{INF}, \text{OBS}^+, \text{OBS}^-\}$".*

It is worth noticing that as a consequence of items (i) and (ii) we have the transitivity of $\prec$; on the other hand, item (iii) states that two non-observable elements cannot be compared by the negligibility relation.

Definition 2. *Let* Σ *be a multimodal qualitative frame, a* multimodal qualitative model on Σ *is an ordered pair* $\mathcal{M} = (\Sigma, h)$, *where* h *is a* meaning function *(or,* interpretation*)* $h: \mathcal{V} \longrightarrow 2^{\mathbb{S}}$.

Any interpretation can be uniquely extended to the set of all formulas in OM (also denoted by h) by means of the usual conditions for the classical boolean connectives and the constants $\top$ and $\bot$, and the following conditions for the modal operators and frame constants:

$$h(\overrightarrow{\Box} A) = \{x \in \mathbb{S} \mid y \in h(A) \text{ for all } y \text{ such that } x < y\}$$
$$h(\overrightarrow{\blacksquare} A) = \{x \in \mathbb{S} \mid y \in h(A) \text{ for all } y \text{ such that } x \sqsubset y\}$$
$$h(\overrightarrow{\boxdot} A) = \{x \in \mathbb{S} \mid y \in h(A) \text{ for all } y \text{ such that } x \prec y\}$$
$$h(\overleftarrow{\Box} A) = \{x \in \mathbb{S} \mid y \in h(A) \text{ for all } y \text{ such that } y < x\}$$
$$h(\overleftarrow{\blacksquare} A) = \{x \in \mathbb{S} \mid y \in h(A) \text{ for all } y \text{ such that } y \sqsubset x\}$$
$$h(\overleftarrow{\boxdot} A) = \{x \in \mathbb{S} \mid y \in h(A) \text{ for all } y \text{ such that } y \prec x\}$$
$$h(\alpha^+) = \{+\alpha\}$$
$$h(\alpha^-) = \{-\alpha\}$$

The concepts of truth and validity are defined in a straightforward manner.

3   The Relational Language $Re(OM)$

Syntax of $Re(OM)$
The alphabet of the language $Re(OM)$ consists of the disjoint sets listed below:

- A (nonempty) set $\mathbb{OV} = \{x, y, z, \dots\}$ of object variables.
- A set $\mathbb{OC} = \{\alpha^-, \alpha^+\}$ of object constants.

- A (nonempty) set $\mathbb{RV} = \{P, Q, R, \ldots\}$ of binary relation variables.
- A set $\mathbb{RC} = \{1, 1', \aleph^-, \aleph^+, <, \sqsubset, \prec\}$ of relation constants.
- A set $\mathbb{OP} = \{-, \cup, \cap, ; , ^{-1}\}$ of relational operation symbols.

The relational constants 1 and 1′ are intended to represent the universal relation and the identity relation, respectively. We use here the traditional notation for these constants originated in [16], and commonly used in the field of relation algebras.

Definition 3

- *The set of* relation terms $\mathbb{RT}$ *is the smallest set of expressions that includes all the relational variables and relational constants and is closed with respect to the operation symbols from* $\mathbb{OP}$.
- *The set* $\mathbb{FR}$ *of formulas, consists of expressions of the form* xRy *where* x, y *denote individual (or object) variables or constants and* R *is a relational term built from the relational variables and the relational operators.*

The defined relations $>, \leq$ and $\geq$ will be used hereafter in order to simplify some relational formulas. The definition of these relations is given as follows:

$$> := <^{-1} \qquad \leq := < \cup 1' \qquad \geq := <^{-1} \cup 1'$$

Semantics of $Re(OM)$

A model for $Re(OM)$ is a pair $M = (W, m)$ where $W = W' \cup \{-\alpha, +\alpha\}$ for a nonempty set W', and m is a meaning function such that:

1. Assigns elements of W to object constants as follows:
   (a) $m(\alpha^-) = -\alpha$
   (b) $m(\alpha^+) = +\alpha$
2. Assigns binary relations on W to relation constants as follows:
   For relation constants we should have:
   (a) $m(1) = W \times W$
   (b) $m(1') = \{(w, w) \mid w \in W\}$
   (c) $m(\aleph^-) = \{-\alpha\} \times W$
   (d) $m(\aleph^+) = \{+\alpha\} \times W$
   (e) $m(<)$ is a strict linear relation in W satisfying that $(-\alpha, +\alpha) \in m(<)$. Notice that the linearity of $m(<)$ allows to divide W into the classes OBS$^-$, OBS$^+$ and INF, defined as in the previous section.
   (f) $m(\sqsubset) = m(<) \cap ((\text{OBS}^- \times \text{OBS}^-) \cup (\text{INF} \times \text{INF}) \cup (\text{OBS}^+ \times \text{OBS}^+))$ Notice that, as a consequence of this requirement, $m(\sqsubset)$ turns out to inherit irreflexivity, left and right linearity and transitivity from $m(<)$.
   (g) $m(\prec)$ is a restriction of $m(<)$, i.e. $m(\prec) \subseteq m(<)$, which satisfies the following frame conditions:

$$\forall x, \forall y \text{ if } (x, y) \in m(\prec) \text{ and } (y, z) \in m(<), \text{ then } (x, z) \in m(\prec) \quad \textit{(fc-i)}$$
$$\forall x, \forall y \text{ if } (x, y) \in m(<) \text{ and } (y, z) \in m(\prec), \text{ then } (x, z) \in m(\prec) \quad \textit{(fc-ii)}$$
$$\forall x, \forall y \text{ if } x \in \text{INF and } (x, y) \in m(\prec), \text{ then } (+\alpha, y) \in m(< \cup 1') \quad \textit{(fc-iii)}$$
$$\forall x, \forall y \text{ if } x \in \text{INF and } (y, x) \in m(\prec), \text{ then } (y, -\alpha) \in m(< \cup 1') \quad \textit{(fc-iv)}$$

Notice that these conditions mimic the requirements (3.i)–(3.iii) in the definition of frame for OM. The required conditions ensure that $m(\prec)$ is irreflexive and transitive.

3. Assigns binary relations on W to relation variables.
4. Assigns operations on binary relations to the relational operation symbols in $\mathbb{OP}$.
5. Extends homomorphically to the set of terms in the usual manner:
   (a) $m(R \cup S) = m(R) \cup m(S)$ (union of relations)
   (b) $m(R \cap S) = m(R) \cap m(S)$ (intersection of relations)
   (c) $m(R; S) = m(R); m(S)$ (composition of relations)
   (d) $m(-R) = -m(R)$ (opposite relation)
   (e) $m(R^{-1}) = m(R)^{-1}$ (inverse relation)

We list below a set of frame conditions which are entailed by the requirements on the function m and will be used later:

$$\forall x \forall y, (x,y) \in m(\aleph^-) \text{ if and only if } (x,-\alpha) \in m(1') \quad \text{(fc-1)}$$

$$\forall x \forall y, (x,y) \in m(\aleph^+) \text{ if and only if } (x,+\alpha) \in m(1') \quad \text{(fc-2)}$$

$$\forall x, \text{ if } (x,-\alpha) \in m(1') \text{ then } (x,+\alpha) \in m(<) \quad \text{(fc-3)}$$

$$\forall x, \forall y \text{ if } (x,-\alpha) \in m(1') \text{ then } (x,y) \notin m(\sqsubset) \quad \text{(fc-4)}$$

$$\forall x, \forall y \text{ if } (y,+\alpha) \in m(1') \text{ then } (x,y) \notin m(\sqsubset) \quad \text{(fc-5)}$$

$$\forall x \forall y, \text{ if } x \in \text{INF} \text{ and } (x,y) \in m(\sqsubset), \text{ then } (-\alpha,y) \in m(<) \quad \text{(fc-6)}$$

$$\forall x \forall y, \text{ if } x \in \text{INF} \text{ and } (x,y) \in m(\sqsubset), \text{ then } (y,+\alpha) \in m(<) \quad \text{(fc-7)}$$

$$\forall x \forall y, \text{ if } (x,-\alpha) \in m(<) \text{ and } (x,y) \in m(\sqsubset), \text{ then } (y,-\alpha) \in m(< \cup 1') \quad \text{(fc-8)}$$

$$\forall x \forall y, \text{ if } (x,y) \in m(<) \text{ and } (y,-\alpha) \in m(< \cup 1'), \text{ then } (x,y) \in m(\sqsubset) \quad \text{(fc-9)}$$

$$\forall x \forall y, \text{ if } (x,y) \in m(<) \text{ and } (+\alpha,x) \in m(< \cup 1'), \text{ then } (x,y) \in m(\sqsubset) \quad \text{(fc-10)}$$

$$\forall x \forall y, \text{ if } (x,y) \in m(<) \text{ and } x \in \text{INF} \text{ and } y \in \text{INF}, \text{ then } (x,y) \in m(\sqsubset) \quad \text{(fc-11)}$$

$$\forall x, \forall y \text{ if } (x,y) \in m(\sqsubset), \text{ then } (x,y) \in m(<) \quad \text{(fc-12)}$$

Furthermore, it can be proved that the fulfillment of all the frame conditions, plus the requirements of $<$ being strict and linear entail the properties from 2.c to 2.f in the definition of model. This fact will be used later during the proof of completeness.

Finally, the notions of satisfiability and validity in the relational logic are introduced as follows:

Definition 4

– A valuation *in a model* $M = (W, m)$ *is a function* $v \colon \mathbb{OV} \cup \mathbb{OC} \to W$ *such that* $v(c) = m(c)$ *for all constant symbols.*[3] *We say that* v satisfies *a relational formula* xRy *if* $(v(x), v(y)) \in m(R)$.
– A *relational formula* xRy *is* true *in* M *if every valuation in* M *satisfies* xRy. *Moreover, if* xRy *is true in every model, we say that* xRy *is* valid *in the relational logic.*

---

[3] Notice the use of $\mathbb{OS}$ to denote the union of $\mathbb{OV}$ and $\mathbb{OC}$.

4 Translation from OM to $Re(OM)$

A translation function t from the language of OM to the language of $Re(OM)$ is introduced in this section.

The translation function $t\colon \Pi \to \mathbb{RV}$ from the set of propositional variables to the set of relational variables is defined for propositional connectives as follows:

$$t(p) := P; 1 \qquad\qquad t(\neg A) := -t(A)$$
$$t(A \vee B) := t(A) \cup t(B) \qquad t(A \wedge B) := t(A) \cap t(B)$$
$$t(A \to B) := -t(A) \cup t(B)$$

For the modal connectives, the translation is based on the general schema, which translates a modality based on a relation R as follows:

$$t(\langle R \rangle A) := R; t(A) \qquad t([R]A) := -(R; -t(A))$$

Specifically, in our case we have the following for the future connectives (for the past connectives the translation is similar):

- $t(\overrightarrow{\Diamond} A) := <; t(A)$
- $t(\overrightarrow{\Box} A) := -(<; -t(A))$
- $t(\overrightarrow{\blacklozenge} A) := \sqsubset; t(A)$

- $t(\overrightarrow{\blacksquare} A) := -(\sqsubset; -t(A))$
- $t(\overrightarrow{\diamondsuit} A) := \prec; t(A)$
- $t(\overrightarrow{\boxdot} A) := -(\prec; -t(A))$

Finally, the α-constants are translated, as expected, into the $\aleph$-relational constants:

$$t(\alpha^-) = \aleph^- \qquad\qquad t(\alpha^+) = \aleph^+$$

Proposition 1. *In relational logic $Re(OM)$ we can verify both validity and entailment of logic OM, namely*

1. *A formula A of logic OM is valid iff a formula $x\,t(A)\,y$ of the corresponding logic $Re(OM)$ is valid, where x, y are any object variables such that $x \neq y$,*
2. *$A_1, \ldots, A_n \models A$ in OM iff $x\left(1; -(t(A_1) \cap \cdots \cap t(A_n)); 1 \cup t(A)\right) y$ is valid in $Re(OM)$.*

Notice that this proposition states that a deduction system of the relational logic can serve as a theorem prover for the logic OM.

5 Relational Proof Systems for Modal $Re(OM)$

Relational proofs have the form of finitely branching trees whose nodes are finite sets of formulas. Given a relational formula xAy, where A may be a compound relational expression, we successively apply decomposition or specific rules. In this way we form a tree whose root consists of xAy and each node (except the root) is obtained by an application of a rule to its predecessor node. We stop applying rules to formulas in a node after obtaining an axiomatic set, or when none of the rules is applicable to the formulas in this node. Such a tree is referred to as a proof tree for the formula xAy. A branch of a proof tree is said to be closed whenever it contains a node with an axiomatic set of formulas. A tree is closed iff all of its branches are closed.

5.1   Rules for the Calculus of Binary Relations with Equality

In the present section we, first, recall the deduction rules for the classical relational logic [10], that is the logic whose formulas xAy are built from the terms A generated by relation variables and constants 1 and $1'$ with the standard relational operations of union, intersection, complement, composition and converse. Second, we define the specific rules that characterise the specific relations of $Re(OM)$. The rules apply to finite sets of relational formulas. As usual, we omit the set brackets when presenting the rules. The rules that refer to relational operations are decomposition rules. They enable us to decompose a formula in a set into some simpler formulas. As a result of decomposition we usually obtain finitely many new sets of formulas. The rules that encode properties of relational or object constants are referred to as specific rules. They enable us to modify a set to which they are applied, they have a status of structural rules. The role of axioms is played by what is called axiomatic sets.

A rule is said to be correct in $Re(OM)$ whenever the following holds: the upper set of formulas in the rule is valid iff all the lower sets are valid, where the validity of a finite set of formulas is understood as a validity of the (metalevel) disjunction of its elements. It follows that the branching in a rule is interpreted as conjunction.

As usual, we present the rules in a form of schemes. A scheme of the form A/B, where A and B are finite sets of formulas represents a family of rules $\Gamma \cup A / \Gamma \cup B$ for any finite set Γ of formulas, and similarly for the branching rules.

In order to introduce here the standard rules for the calculus of binary relations, note that the comma is interpreted disjunctively, whereas the vertical bar is interpreted conjunctively and that a variable is declared *new* in a rule whenever we require that it does not appear in any formula above the line in the rule.

Firstly, we consider the rules for $\cup$:

$$\frac{x(R \cup S)y}{xRy, xSy} \; (\cup) \qquad \frac{x-(R \cup S)y}{x-Ry \mid x-Sy} \; (-\cup)$$

Rules for $\cap$

$$\frac{x(R \cap S)y}{xRy \mid xSy} \; (\cap) \qquad \frac{x-(R \cap S)y}{x-Ry, x-Sy} \; (-\cap)$$

Rules for double complement and inverse relation

$$\frac{x--Ry}{xRy} \; (--) \qquad \frac{xR^{-1}y}{yRx} \; (^{-1}) \qquad \frac{x-R^{-1}y}{y-Rx} \; (-^{-1})$$

Now, we state the rules for the composition

$$\frac{x(R; S)y}{xRz, x(R; S)y \mid zSy, x(R; S)y} \; z \text{ any variable} \quad (;)$$

$$\frac{x-(R;S)y}{x-Rz,\,z-Sy} \quad z \text{ new variable} \quad (-;)$$

Finally, the rules for equality are introduced, where z is any variable

$$\frac{xRy}{xRz,xRy \mid y1'z,xRy} \ (1'\text{-}1) \qquad \frac{xRy}{x1'z,xRy \mid zRy,xRy} \ (1'\text{-}2)$$

5.2   Specific Rules for $Re(OM)$

Here we introduce the rules for handling the specific object constants and relation symbols $<, \sqsubset$ and $\prec$ of the language $Re(OM)$.

The rules below interpret adequately the behaviour of the relation constants $\aleph^-$ and $\aleph^+$:

$$\frac{x\aleph^-y}{x1'\alpha^-,\,x\aleph^-y} \ (\textbf{c1a}) \qquad\qquad \frac{x-\aleph^-y}{x-1'\alpha^-,\,x-\aleph^-y} \ (\textbf{c1b})$$

$$\frac{x\aleph^+y}{x1'\alpha^+,\,x\aleph^+y} \ (\textbf{c2a}) \qquad\qquad \frac{x-\aleph^+y}{x-1'\alpha^+,\,x-\aleph^+y} \ (\textbf{c2b})$$

The following rule expresses that α^- precedes α^+

$$\frac{x<\alpha^+}{x1'\alpha^-,\,x<\alpha^+} \ (\textbf{c3})$$

The remaining rules are stated below. The numbering of the rules reflects their relationship with the corresponding frame conditions:

$$\frac{x-\sqsubset y}{x1'\alpha^-,\,x-\sqsubset y} \ (\textbf{c4}) \qquad\qquad \frac{x-\sqsubset y}{y1'\alpha^+,\,x-\sqsubset y} \ (\textbf{c5})$$

$$\frac{x\le\alpha^-,\alpha^+\le x,x-\sqsubset y}{x\le\alpha^-,\alpha^+\le x,x-\sqsubset y,y\le\alpha^-} \ (\textbf{c6}) \qquad \frac{x\le\alpha^-,\alpha^+\le x,x-\sqsubset y}{x\le\alpha^-,\alpha^+\le x,x-\sqsubset y,\alpha^+\le y} \ (\textbf{c7})$$

$$\frac{\alpha^-\le x,x-\sqsubset y}{\alpha^-\le x,x-\sqsubset y,\alpha^-<y} \ (\textbf{c8})$$

$$\frac{x-<y,\alpha^-<y}{x-<y,\alpha^-<y,x-\sqsubset y} \ (\textbf{c9}) \qquad\qquad \frac{x-<y,x<\alpha^+}{x-<y,x<\alpha^+,x-\sqsubset y} \ (\textbf{c10})$$

$$\frac{x\le\alpha^-,\alpha^+\le x,y\le\alpha^-,\alpha^+\le y,x\sqsubset y}{x\le\alpha^-,\alpha^+\le x,y\le\alpha^-,\alpha^+\le y,x\sqsubset y,x<y} \ (\textbf{c11}) \qquad \frac{x-\sqsubset y}{x-\sqsubset y,x-<y} \ (\textbf{c12})$$

We include below the rules for irreflexivity and linearity properties of the relation constant $<$.

$$\frac{}{x < x}\ (\textbf{Iref}) \qquad \frac{}{y - {<} x \mid x - {<} y \mid x - 1' y}\ (\textbf{Lin})$$

The transitivity for the three relation constants is stated by the rule below, where $R \in \{<, \sqsubset, \prec\}$

$$\frac{xRy}{xRy, xRz, \mid xRy, zRy}\ z \text{ any var}\quad (\textbf{Tran})$$

The following cut-like rule will be needed later in the proof of completeness

$$\frac{}{x \sqsubset y \mid x - \sqsubset y}\ (\textbf{cut-}\ \sqsubset)$$

Finally, the following rules for $\prec$ reflect the frame conditions for negligibility:

$$\frac{x < y}{x \prec y, x < y}\ (\textbf{n-0})$$

$$\frac{x \prec z}{x \prec y, x \prec z \mid y < z, x \prec z}\ y \text{ any var}\quad (\textbf{n-i})$$

$$\frac{x \prec z}{x < y, x \prec z \mid y \prec z, x \prec z}\ y \text{ any var}\quad (\textbf{n-ii})$$

$$\frac{\alpha^+ \le y}{\alpha^- < x, \alpha^+ \le y \mid x < \alpha^+, \alpha^+ \le y \mid x \prec y, \alpha^+ \le y}\ (\textbf{n-iii})$$

$$\frac{y \le \alpha^-}{\alpha^- < x, y \le \alpha^- \mid x < \alpha^+, y \le \alpha^- \mid y \prec x, y \le \alpha^-}\ (\textbf{n-iv})$$

Axiomatic Sets

An axiomatic set is any finite set of formulas which includes a subset of either of the following forms for a relational term R and x, y are any object variables.

We have to introduce schemas of axiomatic sets for the universal relation, the identity relation and linearity, together with others which allow us to adequately interpret the constant relation symbols ℵ, together with the symbols $\pm\alpha$.

The axiomatic sets of $Re(OM)$ state valid formulas of the system, the following postulate the behaviour of the universal relation 1 and the equality relation $1'$, the *tertium non datur*, and the conditions related to the constant symbols α^- and α^+ are expressed by

$$\{x1y\} \qquad \{x1'x\} \qquad \{x - Ry, xRy\} \qquad \{\alpha^- < \alpha^+\}$$

where $x, y \in \mathbb{OS}$ and $R \in \mathbb{RT}$.

5.3   Proving Some Axioms of MQ^N

In this section we show the relational proof system at work, and prove some of the axioms of the system MQ^N of qualitative order-of-magnitude reasoning presented in [3].

Example 1. Axiom (c4): $\alpha^- \rightarrow \overrightarrow{\blacksquare} A$

The translated version of the axiom in the relational language is

$$-\aleph^- \cup -(\sqsubset; -(A; 1))$$

We consider $x(-\aleph^- \cup -(\sqsubset; -(A; 1)))y$, apply the rule $(\cup)$, and then, the following tree is generated:

$$
\cfrac{
\cfrac{
\cfrac{
\cfrac{
\boxed{x - \aleph^- y}, x - (\sqsubset; -(A; 1))y
}{
x - \aleph^- y, x - 1'\alpha^-, \boxed{x - (\sqsubset; -(A; 1))y}
} \ (c1b)
}{
x - \aleph^- y, x - 1'\alpha^-, x - \sqsubset z, \boxed{z - -(A; 1)y}
} \ (-;) \ z \text{ new}
}{
x - \aleph^- y, x - 1'\alpha^-, x - \sqsubset z, \boxed{z(A; 1)y}
} \ (--)
}{
\Gamma, zAw \mid \Gamma, \boxed{w1y}
} \ (;) \ \text{any } w
$$

where $\Gamma = x - \aleph^- y, x - 1'\alpha^-, x - \sqsubset z$.

The right branch closes because of $w1y$, whereas rule (c4) applies to the left branch against $x - 1'\alpha^-$, obtaining

$$\boxed{x - 1'\alpha^-}, \boxed{x1'\alpha^-}, x - \sqsubset z, zAw$$

which closes.

Example 2. Axiom (c1): $\overleftarrow{\Diamond} \alpha^- \vee \alpha^- \vee \overrightarrow{\Diamond} \alpha^-$

$$
\cfrac{
\boxed{x(>; \aleph^-)y}, x\aleph^- y, x(<; \aleph^-)y
}{
x < \alpha^-, x\aleph^- y, x(>; \aleph^-)y, x(<; \aleph^-)y \mid \alpha^- \aleph^- y, x\aleph^- y, x(>; \aleph^-)y, x(<; \aleph^-)y
} \ (;)
$$

where variable z has been instantiated to α^- in the application of the rule.

Note that the right branch closes, since it contains an axiomatic set for $\aleph^-$. On the other hand, the left branch continues as follows, where we use Γ to denote the pair of formulas $x(>; \aleph^-)y, x(<; \aleph^-)y$

$$
\cfrac{
x < \alpha^-, x\aleph^- y, x(>; \aleph^-)y, \boxed{x(<; \aleph^-)y}
}{
x < \alpha^-, \boxed{x\aleph^- y}, x > \alpha^-, \Gamma \mid x < \alpha^-, x\aleph^- y, \boxed{\alpha^- \aleph^- y}, \Gamma
} \ (;)[z/\alpha^-]
$$

the left branch closes after applying (c1a) and linearity, whereas the right branch closes because of the axiomatic set for $\aleph$.

6   Soundness of the Relational Proof System

Recall that a rule is said to be correct if the validity of the upper set entails the validity of the lower set and vice versa.

The frame conditions stated in Section 3 are used here in order to prove soundness of the relational proof system: we will show the equivalence between the correctness of the specific rules of $Re(OM)$ and the validity of the corresponding frame conditions. As a result, since all the frame conditions hold in every model of $Re(OM)$, we get that the specific rules of $Re(OM)$ are all correct.

Proposition 2

1. *For $k \in \{1, 2\}$, rules $(ck\, a)$ and $(ck\, b)$ are correct for a deduction system of $Re(OM)$ iff in every model of $Re(OM)$ condition $(fc\text{-}k)$ is satisfied.*
2. *For $k \in \{3, \ldots, 12\}$, rule $(c\, k)$ is correct for a deduction system of $Re(OM)$ iff in every model of $Re(OM)$ condition $(fc\text{-}k)$ is satisfied.*
3. *For $j \in \{i, ii, iii, iv\}$, rule $(n\, j)$ is correct for a deduction system of $Re(OM)$ iff in every model of $Re(OM)$ condition $(fc\text{-}j)$ is satisfied.*

Proof. 1. Let us prove the case of (fc-2), since the other is similar:

Assume that the rules are correct and, and let us prove the two implications which form the frame condition. We proceed by contradiction and consider that the frame condition

$$\forall x \forall y, (x, y) \in m(\aleph^+) \text{ if and only if } (x, +\alpha) \in m(1') \qquad (\text{fc-2})$$

does not hold.

Reasoning by cases, on the one hand, suppose that for some objects a, b we have $(a, +\alpha) \in m(1')$ and $(a, b) \notin m(\aleph^+)$. Consider the following instance of rule $(c2a)$, in which we add the context $\Gamma = x{-}1'\alpha^+$ to both the upper and lower sets:

$$\frac{x\aleph^+ y, x{-}1'\alpha^+}{x\aleph^+ y, x1'\alpha^+, x{-}1'\alpha^+}$$

The lower set is valid, so since the rule is correct, the upper set must be valid, that is, the formula $\forall x \forall y (x\aleph^+ y \vee x{-}1'\alpha^+)$ is valid in first order logic. But the valuation v such that $v(x) = a$ and $v(y) = b$ is a counterexample.

On the other hand, suppose conversely that for some objects a, b we have $(a, b) \in m(\aleph^+)$ and $(a, +\alpha) \notin m(1')$. Consider the following instance of rule $(c2b)$, in which we add the context $\Gamma = x1'\alpha^+$ to both the upper and lower sets:

$$\frac{x1'\alpha^+, x{-}\aleph^+ y}{x1'\alpha^+, x{-}1'\alpha^+, x{-}\aleph^+ y}$$

The lower set is valid, so since the rule is correct, the upper set must be valid, that is, the formula $\forall x \forall y (x1'\alpha^+ \vee x{-}\aleph^+ y)$ is valid in first order logic. But the valuation v such that $v(x) = a$ and $v(y) = b$ is a counterexample.

Reciprocally, assume the validity of the frame condition (fc-1$^+$) and let us prove that both rules (c1$^+$a) and (c1$^+$b) are correct. Clearly, validity of the upper set of the rules implies validity of the lower set. Now, assuming validity of the lower set, validity of the upper set follows easily from the frame condition.

2. For $k = 3$.

Assume that the rule is correct and suppose that $(fc\text{-}3)$ does not hold, i.e., for some object a we have $(a, -\alpha) \in 1'$ and $(a, +\alpha) \notin <$. Consider the following instance of rule $(c3)$

$$\frac{x < \alpha^+, x - 1'\alpha^-}{x1'\alpha^-, x < \alpha^+, x - 1'\alpha^-}$$

Clearly, the lower set is valid, so since the rule is correct, the upper set must be valid. This means that the formula $\forall x (x < \alpha^+ \vee x - 1'\alpha^-)$ is valid in first order logic. But the valuation v such that $v(x) = a$ does not satisfy that formula, a contradiction.

Reciprocally, assume $(fc\text{-}3)$. Validity of the upper set of the rule implies validity of the lower set. Assuming validity of the lower set, validity of the upper set follows from the frame condition.

The proof for the rest of the cases is similar, we just introduce the context to be used when considering the instance for the corresponding rule.

For $k = 4$, assume $\Gamma = x(-1')\alpha^-$.
For $k = 5$, assume $\Gamma = y(-1')\alpha^+$.
For $k = 6$, assume $\Gamma = \alpha^- < y$.
For $k = 7$, assume $\Gamma = y < \alpha^+$.
For $k = 8$, assume $\Gamma = y \leq \alpha^-$.
For $k = 9, 10, 11$, assume $\Gamma = x \sqsubset y$.
For $k = 12$, assume $\Gamma = x < y$.

3. For $j = 0$, the context $\Gamma = x - \prec y$ proves that the rule (n-0) is correct if and only if $\prec$ is a restriction of $<$.

For $j = i$, take the context $x - \prec y, y - < z$.
For $j = ii$, consider $\Gamma = x - < y, y - \prec z$
For $j = iii$, assume $\Gamma = x \leq \alpha^-, \alpha^+ < x, x - \prec y$.
For $j = iv$, assume $\Gamma = x \leq \alpha^-, \alpha^+ < x, y - \prec x$.   □

The rest of the rules are the standard ones for defining properties related of order relations and the equality. As a result, we have the following proposition:

Proposition 3

1. *All the rules of the deduction system for $Re(OM)$ are correct.*
2. *All the axiomatic sets are valid.*

The soundness theorem follows from the correctness of the rules and from validity of the axiomatic sets of the system.

Proposition 4 (Soundness). *If there is a closed proof tree for a formula xAy, then xAy is valid.*

7   Completeness of the Relational Proof System

A completeness proof for dual tableaux systems involves a notion of a complete proof tree. Intuitively, a proof tree is complete if all the rules that can be applied to its nodes have been applied. A non-closed branch b of a proof tree is complete whenever it satisfies some appropriate completion conditions. The conditions say that given a rule applicable to a node of b, there is a node on b which contains a set of formulas resulting from an application of that rule.

Completion Conditions. A non-closed branch b of a proof tree is said to be *complete* whenever for all $x, y \in \mathbb{OS}$ it satisfies the completion conditions on Table 1.

It is known that any proof tree can be extended to a complete proof tree. A complete and non-closed branch is said to be open.

Let b be an open branch of a proof tree. We define a branch structure $M^b = (W^b, m^b)$:

$$W^b = \mathbb{OV} \cup \mathbb{OC}$$
$$m^b(R) = \{(x, y) \in W^b \times W^b \mid xRy \notin b\} \text{ for } R \in \mathbb{RV} \cup \mathbb{RC}$$
$$m^b(\alpha^+) = \alpha^+, \qquad m^b(\alpha^-) = \alpha^-$$

and m^b extends homomorphically to all the relation terms.

Let $v^b \colon \mathbb{OV} \to W^b \setminus \mathbb{OC}$ be an identity valuation, i.e., $v^b(x) = x$ for every object variable x.

Throughout the rest of the paper we shall often write R^b for $m^b(R)$.

Note that, as in the case of first order logic with equality, the relation $1'^b$ can only be proved to be an equivalence relation.

Lemma 1. *The relation $1'^b$ is an equivalence relation.*

Proof. $1'^b$ is reflexive: We have $x1'x \notin b$ (otherwise b would be closed) which means, by definition of m^b, that $(x, x) \in 1'^b$.

$1'^b$ is symmetric: In order to reach a contradiction, consider $x, y \in W^b$ such that $(x, y) \in 1'^b$ but $(y, x) \notin 1'^b$, then by definition of m^b we have both $x1'y \notin b$ and $y1'x \in b$. Now from the completion condition (cpl $1'$-1), we have either $y1'y \in b$ or $x1'y \in b$. Since b is open, we obtain $x1'y \in b$, a contradiction.

$1'^b$ is transitive: Consider $x, y, z \in W^b$ such that $(x, y) \in 1'^b, (y, z) \in 1'^b$ and $(x, z) \notin 1'^b$, which means, by definition of m^b, that $x1'y \notin b$, $y1'z \notin b$ and $x1'z \in b$. Given $x1'z \in b$, from the completion condition (cpl $Tran$) we have either $x1'y \in b$ or $y1'z \in b$ and we reach a contradiction in both cases.   □

In order to obtain the expected behaviour of $1'^b$ as the equality relation, we consider a quotient model $[M^b]_{1'^b} = ([W^b]_{1'^b}, n)$ where:

- $[W^b]_{1'^b}$ is the set of equivalence classes of W^b wrt $1'^b$.
- $n(R) = \{([x]_{1'^b}, [y]_{1'^b}) \mid (x, y) \in R^b\}$ for $R \in \mathbb{RT}$.
- Valuation u in $[M^b]_{1'^b}$ is such that $u(x) = [x]_{1'^b}$.

Table 1. Completion conditions

(cpl $\cup$**)** If $x(R \cup S)y \in b$, then both $xRy \in b$ and $xSy \in b$
(cpl $-\cup$**)** If $x - (R \cup S)y \in b$, then either $x - Ry \in b$ or $x - Sy \in b$
(cpl $\cap$**)** If $x(R \cap S)y \in b$, then either $xRy \in b$ or $xSy \in b$
(cpl $-\cap$**)** If $x - (R \cap S)y \in b$, then both $x - Ry \in b$ and $x - Sy \in b$
(cpl $--$**)** If $x --Ry \in b$, then $xRy \in b$
(cpl $^{-1}$**)** If $xR^{-1}y \in b$, then $yRx \in b$
(cpl $-^{-1}$**)** If $x - R^{-1}y \in b$, then $y - Rx \in b$
(cpl ;**)** If $x(R;S)y \in b$, then for every $z \in \mathbb{OS}$, either $xRz \in b$ or $zSy \in b$
(cpl $-$;**)** If $x - (R;S) \in b$, then for some $z \in \mathbb{OS}$ both $x - Rz \in b$ and $z - Sy \in b$
(cpl $1'$-1**)** If $xRy \in b$, then for every $z \in \mathbb{OS}$ either $xRz \in b$ or $y1'z \in b$
(cpl $1'$-2**)** If $xRy \in b$, then for every $z \in \mathbb{OS}$ either $x1'z \in b$ or $zRy \in b$
(cpl $c1a$**)** If $x\aleph^- y \in b$ then $x1'\alpha^- \in b$
(cpl $c1b$**)** If $x-\aleph^- y \in b$, then $x-1'\alpha^- \in b$
(cpl $c2a$**)** If $x\aleph^+ y \in b$ then $x1'\alpha^+ \in b$
(cpl $c2b$**)** If $x-\aleph^+ y \in b$, then $x-1'\alpha^+ \in b$
(cpl $c3$**)** If $x < \alpha^+ \in b$ then $x1'\alpha^- \in b$
(cpl $c4$**)** If $x -\sqsubset y \in b$ then $x1'\alpha^- \in b$
(cpl $c5$**)** If $x -\sqsubset y \in b$ then $y1'\alpha^+ \in b$
(cpl $c6$**)** If $x \leq \alpha^- \in b$, $\alpha^+ \leq x \in b$ and $x -\sqsubset y \in b$, then $y \leq \alpha^- \in b$
(cpl $c7$**)** If $x \leq \alpha^- \in b$, $\alpha^+ \leq x \in b$ and $x -\sqsubset y \in b$, then $\alpha^+ \leq y \in b$
(cpl $c8$**)** If $\alpha^- \leq x \in b$ and $x -\sqsubset y \in b$, then $\alpha^- < y \in b$
(cpl $c9$**)** If both $x -< y \in b$ and $\alpha^- < y \in b$, then $x -\sqsubset y \in b$
(cpl $c10$**)** If both $x -< y \in b$ and $x < \alpha^+ \in b$, then $x -\sqsubset y \in b$
(cpl $c11$**)** If $x \leq \alpha^- \in b, \alpha^+ \leq x \in b, y \leq \alpha^- \in b, \alpha^+ \leq y \in b$ and $x \sqsubset y \in b$ then $x < y \in b$,
(cpl $c12$**)** If $x -\sqsubset y \in b$, then $x -< y \in b$
(cpl cut-$\sqsubset$**)** Either $x \sqsubset y \in b$ or $x -\sqsubset y \in b$
(cpl n-0**)** If $x < y \in b$, then $x \prec y \in b$
(cpl n-i**)** If $x \prec z \in b$, then for every $y \in \mathbb{OS}$ either $x \prec y \in b$ or $y < z \in b$
(cpl n-ii**)** If $x \prec z \in b$, then for every $y \in \mathbb{OS}$ either $x < y \in b$ or $y \prec z \in b$
(cpl n-iii**)** If $\alpha^+ \leq y \in b$, then $\alpha^- < x \in b$ or $x < \alpha^+ \in b$ or $x \prec y \in b$
(cpl n-iv**)** If $y \leq \alpha^- \in b$, then $\alpha^- < x \in b$ or $x < \alpha^+ \in b$ or $y \prec x \in b$
(cpl $Iref$**)** $x < x \in b$
(cpl $Tran$**)** If $xRy \in b$, then for every $z \in \mathbb{OS}$, either $xRz \in b$ or $zRy \in b$ (where $R \in \{<, \sqsubset, \prec\}$).
(cpl Lin**)** Either $x -< y \in b$ or $x - 1'y \in b$ or $y -< x \in b$

Now, we have the following proposition:

Proposition 5

1. *For every formula* xAy, $[M^b]_{1'b}, u \models xAy$ *iff* $M^b, v^b \models xAy$.
2. $[M^b]_{1'b}$ *is a model of* $Re(OM)$.

Proof.

1. This condition is easily verified using the corresponding definitions.

2. We only give the proofs for some conditions on the model; the proofs of the remaining conditions are similar.

(a) $n(1) = [W^b]_{1'^b} \times [W^b]_{1'^b}$

Since b is open, $x1y \notin b$ for all $x, y \in \mathbb{OS}$. So, by definition of m^b, we get $(x, y) \in m^b(1)$; note that this means that $M^b, v^b \models x1y$. Now, by the item 1 above, we have $[M^b]_{1'^b}, u \models x1y$. Hence $([x]_{1'^b}, [y]_{1'^b}) \in n(1)$.

(c) $n(\aleph^-) = \{[\alpha^-]_{1'^b}\} \times [W^b]_{1'^b}$

We have that

$$([x]_{1'^b}, [y]_{1'^b}) \in n(\aleph^-) \quad \text{if and only if} \quad [M^b]_{1'^b}, u \models x\aleph^- y$$
$$\text{if and only if} \quad M^b, v^b \models x\aleph^- y \text{ (by item 1 above)}$$
$$\text{if and only if} \quad (x, y) \in m^b(\aleph^-)$$
$$\text{if and only if} \quad x\aleph^- y \notin b.$$

On the other hand, we have

$$[x]_{1'^b} \neq [\alpha^-]_{1'^b} \quad \text{if and only if} \quad ([x]_{1'^b}, [\alpha^-]_{1'^b}) \notin n(1)$$
$$\text{if and only if} \quad [M^b]_{1'^b}, u \not\models x1'\alpha^- \text{ (by item 1 above)}$$
$$\text{if and only if} \quad M^b, v^b \not\models x1'\alpha^-$$
$$\text{if and only if} \quad x1'\alpha^- \in b.$$

If either $n(\aleph^-) \subset \{[\alpha^-]_{1'^b}\} \times [W^b]_{1'^b}$ or $n(\aleph^-) \supset \{[\alpha^-]_{1'^b}\} \times [W^b]_{1'^b}$ would not hold, then completion conditions (cpl c1a) and (cpl c1b) would generate a contradiction.

In the proofs of the remaining conditions we shall abuse the notation and the symbols of quotient classes will not be written, and moreover, we shall write A^b instead of $n(A)$, and W^b instead of $[W^b]_{1'^b}$.

fc-3 Let us show that $\forall x \in W^b$, if $(x, \alpha^-) \in 1'^b$ then $(x, \alpha^+) \in <^b$.

Assume that $(x, \alpha^-) \in 1'^b$ and suppose that $(x, \alpha^+) \notin <^b$. By definition of m^b we get $x1'\alpha^- \notin b$ and $x < \alpha^+ \in b$. From the completion condition (cpl c2) we get $x1'\alpha^- \in b$. Hence $(x, \alpha^-) \notin 1'^b$, a contradiction.

fc-6 $\forall x, y \in W^b$ if $x \in \text{INF}^b$ and $(x, y) \in \sqsubset^b$, then $(\alpha^-, y) \in <^b$.

Assume that $(\alpha^-, x) \in <^b, (x, \alpha^+) \in <^b$ (that is, $x \in \text{INF}^b$) and $(x, y) \in \sqsubset^b$. Suppose also that $(\alpha^-, y) \notin <^b$. By definition of m^b, we get $\alpha^- < x \notin b$, $x < \alpha^+ \notin b, x \sqsubset y \notin b$ and $\alpha^- < y \in b$. Now we have $y \leq \alpha^- \notin b$ (otherwise b should be closed). From the completion condition (clp c6) we obtain $x \leq \alpha^- \notin b$ or $\alpha^+ \leq x \notin b$ or $x- \sqsubset y \notin b$. From the completion condition (cpl cut-$\sqsubset$) we get $x \sqsubset y \notin b$ and, by definition of m^b, we have that $(x, \alpha^-) \in \leq^b$ or $(\alpha^+, x) \in \leq^b$ or $(x, y) \notin \sqsubset^b$. In any case we easily reach a contradiction.

fc-i $\forall x, y \in W^b$, if both $(x, y) \in \prec^b$ and $(y, z) \in <^b$, then $(x, z) \in \prec^b$.

Assume that $(x, y) \in \prec^b$ and $(y, z) \in <^b$. Then $x \prec y \notin b$ and $y < z \notin b$. If it were $(x, z) \notin \prec^b$ also, then $x \prec z \in b$. By the completion condition (clp $n\text{-}i$) we obtain either $x \prec y \notin b$ or $y < z \notin b$. In both cases we get a contradiction.

fc-iii $\forall x, y \in W^b$, if $x \in \text{INF}^b$ and $(x, y) \in \prec^b$, then $(\alpha^+, y) \in \leq^b$.

Assume that $(\alpha^-, x) \in <^b, (x, \alpha^+) \in <^b$ and $(x, y) \in \prec^b$ and also $(\alpha^+, y) \notin \leq^b$. Then, by definition of m^b, we obtain $\alpha^- < x \notin b, x < \alpha^+ \notin b, x \prec y \notin b$ and $\alpha^+ \leq y \in b$. Given $\alpha^+ \leq y \in b$, by the completion condition (cpl $n\text{-}iii$) we have that $\alpha^- < x \in b$ or $x < \alpha^+ \in b$ or $x \prec y \in b$, which lead us a contradiction in any case.   □

The following proposition has a standard proof by induction on the structure of term A.

Proposition 6. *For every open branch b of a proof tree and for every formula xAy the following holds: $M^b, v^b \models xAy$ implies $xAy \notin b$.*

Now the completeness theorem follows.

Theorem 1 (Completeness). *If a formula is valid, then there is a closed proof tree for it.*

Proof. Assume that a formula xAy is valid. Suppose all the proof trees for xAy are not closed. Take any of those trees. We may assume that it is complete. Let b be one of its open branches (which exists by the König's lemma). Since $xAy \in b$, by the previous proposition we know that v^b does not satisfy xAy in M^b. Therefore also the valuation u in the quotient model $[M^b]_{1^{\prime b}}$ does not satisfy xAy, a contradiction.   □

8   Conclusions

A relational deduction system for the logic OM of order of magnitude reasoning has been presented. OM is a propositional logic with modal operators determined by three accessibility relations related to each other and their converses. We defined a translation from the language of OM to a target relational language such that both accessibility relations from the frames of OM and the formulas of OM became the relational terms. All the frame conditions on the accessibility relations were postulated as the axioms in the target language.

Two groups of deduction rules have been defined: those that characterize the relational operators of the target language which corresponded to the propositional operators of OM, and those that reflect the axioms imposed on the accesibility relations.

We proved completeness of our proof system adjusting a standard method to the specific features of OM. The key steps of the proof include a development of the completion conditions associated with every rule which enable us to express the notion of a complete proof tree (or equivalently a saturated proof tree) and a development of a branch structure determined by a branch of a proof tree which must then be proved to be a model of the target relational language.

An implementation of translation procedures from the languages of nonclassical logics to relational languages is presented in [6]. Another recent implementation of the core rules of relational proof systems is described in [5]; further work is planned on relational proof systems for variants of OM logic and on implementation of the specific rules for OM within the latter system.

References

1. A. Burrieza, E. Muñoz, and M. Ojeda-Aciego. Order-of-magnitude qualitative reasoning with bidirectional negligibility. *Lect. Notes in Artificial Intelligence*, 4177:370–378, 2006.

2. A. Burrieza and M. Ojeda-Aciego. A multimodal logic approach to order of magnitude qualitative reasoning. *Lect. Notes in Artificial Intelligence*, 3040:431–440, 2004.

3. A. Burrieza and M. Ojeda-Aciego. A multimodal logic approach to order of magnitude qualitative reasoning with comparability and negligibility relations. *Fundamenta Informaticae*, 68:21–46, 2005.

4. P. Dague. Symbolic reasoning with relative orders of magnitude. In *Proc. 13th Intl. Joint Conference on Artificial Intelligence*, pages 1509–1515. Morgan Kaufmann, 1993.

5. J. Dallien and W. MacCaull. RelDT—a dual tableaux system for relational logics. In press, 2005. Available from `http://logic.stfx.ca/reldt/`

6. A.Formisano, E. Omodeo, and E. Orłowska. A PROLOG tool for relational translation of modal logics: A front-end for relational proof systems. In: B. Beckert (ed) Tableaux'05 Position Papers and Tutorial Descriptions. Universität Koblenz-Landau, Fachberichte Informatik No 12, 2005, 1-10. System available from `http://www.di.univaq.it/TARSKI/transIt/`

7. G. Gargov and V. Goranko. Modal logic with names. *Journal of Philosophical Logic*, 22:607–636, 1993.

8. J. Golińska-Pilarek and E. Orłowska. Tableaux and dual tableaux: Transformation of proofs. Submitted, 2005.

9. W. MacCaull and E. Orłowska. Correspondence results for relational proof systems with application to the Lambek calculus. *Studia Logica*, 71:279–304, 2002.

10. E. Orłowska. Relational interpretation of modal logics. In H. Andreka, D. Monk, and I. Nemeti, editors, *Algebraic Logic*, volume 54 of *Colloquia Mathematica Societatis Janos Bolyai*, pages 443–471. North Holland, 1988.

11. E. Orłowska. Relational semantics for nonclassical logics: Formulas are relations. In J. Wolenski, editor, *Philosophical Logic in Poland*, page 167–186. Kluwer, 1994.

12. S. Passy and T. Tinchev. PDL with data constants. *Information Processing Letters*, 20:35–41, 1985.

13. S. Passy and T. Tinchev. An essay in combinatory dynamic logic. *Information and Computation*, 93:263–332, 1991.

14. H. Rasiowa and R. Sikorski. *Mathematics of Metamathematics*. Polish Scientific Publishers, 1963.
15. R. Schmidt, E. Orłowska, and U. Hustadt. Two proof systems for Peirce algebras. *Lecture Notes in Computer Science*, 3051:235–248, 2004.
16. A. Tarski. On the calculus of relations. *Journal of Symbolic Logic* 6:73–89, 1941.
17. L. Travé-Massuyès, F. Prats, M. Sánchez, and N. Agell. Consistent relative and absolute order-of-magnitude models. In *Proc. Qualitative Reasoning 2002 Conference*, 2002.

Relational Logics and Their Applications

Joanna Golińska-Pilarek and Ewa Orłowska

National Institute of Telecommunications, Szachowa 1, 04-894 Warsaw, Poland
J.Golinska-Pilarek@itl.waw.pl,
E.Orlowska@itl.waw.pl

Abstract. Logics of binary relations corresponding, among others, to the class RRA of representable relation algebras and the class FRA of full relation algebras are presented together with the proof systems in the style of dual tableaux. Next, the logics are extended with relational constants interpreted as point relations. Applications of these logics to reasoning in non-classical logics are recalled. An example is given of a dual tableau proof of an equation which is RRA-valid, while not RA-valid.

1 Introduction

We present a survey of relational logics which provide a general framework for specification and reasoning (verification of validity, model checking and entailment) in non-classical logics. They also provide a common background for a broad class of relational structures used in computer science. We present the logics step by step, starting with a logic of binary relations with basic relational operations of relation algebras (RL-logic), then expanding the language with the constant 1 (RL(1)-logic), next with the constant $1'$ (RL($1'$)-logic), then with the constants 1 and $1'$ put together (RL($1, 1'$)-logic), and finally adding relational constants interpreted as point relations (RL$_{ax}(C)$-logic and RL$_{df}(C)$-logic). The logics are based on various classes of models which differ in the interpretation of relational constants, for example, 1 may be interpreted as a universal relation or as an equivalence relation, $1'$ may be interpreted as an equivalence relation or an identity. We present completeness theorems with respect to all those classes of models. We also show which classes of models of RL($1, 1'$)-language enable us to simulate the RRA-validity and FRA-validity. Logic RL($1, 1'$) with the class of models corresponding to full relation algebras plays the role of a generic logic within which many non-classical logics can be expressed. Its applications to modal logics originated in [15]. Then, after few more examples of logics treated in a relational framework (see e.g., [16], [17]), a paradigm 'formulas are relations' has been formulated in [18]. Since then relational proof systems have been developed for several theories, see e.g., [3], [4], [8], [11], [12], [13], [19], [20], [21], [10] and [9]. Any particular relational proof system consists of the deduction system for RL($1, 1'$) augmented with the specific rules which reflect properties of accessibility relations from the models of a non-classical logic in question. An important feature of RL($1, 1'$)-logic is that it is expressive enough for performing the major

H. de Swart et al. (Eds.): TARSKI II, LNAI 4342, pp. 125–161, 2006.

logical tasks, namely verification of validity, entailment, model checking and satisfiability, as it is shown in Sections 10, 11, 12, and 13. A correspondence theory for relational proof systems is considered in [14]. A general method of defining deduction rules reflecting various constraints imposed on relations in the models of $\mathsf{RL}(1, 1')$-logic is presented in that paper.

Recent implementations of the proof system for $\mathsf{RL}(1, 1')$-logic are described in [2] and [6]. The first one is available at `http://logic.stfx.ca/reldt`. In [5] an implementation of translation procedures from the languages of non-classical logics to relational languages is presented. The system can be downloaded from `http://www.di.univaq.it/TARSKI/transIt/`. For the algebraic background of the relational logics see [24], [25] and [23].

2   A General Scheme of Relational Logics

Each relational logic L is determined by its language and its class of models. In this paper we consider logics of binary relations. There are two kinds of expressions of relational languages: terms and formulas. Terms represent relations and formulas express the facts that a pair of objects stands in a relation.

The vocabulary $\mathcal{V}_\mathsf{L}$ of L-language consists of the symbols from the following pairwise disjoint sets:

- a countable infinite set of object variables $\mathbb{OV}_\mathsf{L}$;
- a countable (possibly empty) set of object constants $\mathbb{OC}_\mathsf{L}$;
- a countable (possibly empty) set of relational variables $\mathbb{RV}_\mathsf{L}$;
- a countable (possibly empty) set of relational constants $\mathbb{RC}_\mathsf{L}$;
- a set of relational operation symbols $\mathbb{OP}_\mathsf{L} = \{-, \cup, \cap, ; ,^{-1}\}$, where $-, \cup, \cap$ are Boolean operations, $;$ is a relative product, and $^{-1}$ is the operation of converse;
- a set of parentheses $\{(,)\}$.

The set $\mathbb{RA}_\mathsf{L} = \mathbb{RV}_\mathsf{L} \cup \mathbb{RC}_\mathsf{L}$ is called the *set of atomic relational terms*. The set $\mathbb{OS}_\mathsf{L} = \mathbb{OV}_\mathsf{L} \cup \mathbb{OC}_\mathsf{L}$ is called the *set of objects symbols*. The set $\mathbb{RT}_\mathsf{L}$ of *relational terms* is the smallest (wrt inclusion) set of expressions that includes all atomic relational terms and is closed with respect to all relational operation symbols. L-*formulas* are of the form xRy, where $x, y \in \mathbb{OS}_\mathsf{L}$ and $R \in \mathbb{RT}_\mathsf{L}$. An L-formula xRy is said to be *atomic* whenever $R \in \mathbb{RA}_\mathsf{L}$.

With an L-language a class of L-models is associated. An L-*model* is a structure $\mathcal{M} = (U, m)$, where U is a non-empty set and m is a meaning function which assigns:

- elements of U to object constants, that is $m(c) \in U$, for every $c \in \mathbb{OC}_\mathsf{L}$;
- binary relations on U to atomic relational terms, that is $m(R) \subseteq U \times U$, for every $R \in \mathbb{RA}_\mathsf{L}$;

and extends to compound relational terms as follows:

- some condition about $m(-R)$ is assumed (see Sections 4 and 5 for the examples of the definitions of the complement operations);

- $m(R \cup S) = m(R) \cup m(S)$;
- $m(R \cap S) = m(R) \cap m(S)$;
- $m(R^{-1}) = (m(R))^{-1} = \{(x, y) \in U \times U : (y, x) \in m(R)\}$;
- $m(R; S) = m(R); m(S) = \{(x, y) \in U \times U : \exists z((x, z) \in m(R) \wedge (z, y) \in m(S))\}$;
- some additional conditions about m may be assumed (see Sections 5 and 6).

Let $\mathcal{M} = (U, m)$ be an L-model. An L-*valuation* in $\mathcal{M}$ is any function $v : \mathbb{OS}_L \rightarrow U$ such that $v(c) = m(c)$, for every $c \in \mathbb{OC}_L$. Let $\mathcal{M}$ be an L-model, let v be an L-valuation in $\mathcal{M}$ and let xRy be an L-formula. *Satisfiability* of xRy by v in $\mathcal{M}$ is defined as follows:

- If $1 \notin \mathbb{RC}_L$, then $\mathcal{M}, v \models xRy$ iff $(v(x), v(y)) \in m(R)$.
- If $1 \in \mathbb{RC}_L$, then $\mathcal{M}, v \models xRy$ iff $(v(x), v(y)) \notin m(1)$ or $(v(x), v(y)) \in m(1) \cap m(R)$.

Note that in the latter case, satisfiability is defined in a non-standard way. This is because we want to relativize satisfiability to the interpretation of the relational constant 1. In the general case, this interpretation need not be the universal relation. In the case it is, clearly the two definitions are equivalent.

An L-formula xRy is *true* in $\mathcal{M}$ whenever it is satisfied in $\mathcal{M}$ by all L-valuations. An L-formula xRy is L-*valid* whenever it is true in all L-models.

Fact 1
Let L and L$'$ be relational logics such that every L-model is an L$'$-model. Then for any relational formula xRy, if xRy is L$'$-valid, then it is L-valid.

3   A General Scheme of Relational Proof Systems

Relational proof systems in the style of dual tableaux are founded on the Rasiowa-Sikorski system for the first order logic [22]. They are powerful tools for performing the major reasoning tasks: verification of validity, verification of entailment, model checking, and verification of satisfiability. Every relational proof system is determined by its axiomatic sets of formulas and rules which most often apply to finite sets of relational formulas. Some relational proof systems with infinitary rules are known in the literature, but in the present paper we confine ourselves to finitary rules only. The axiomatic sets take the place of axioms. The rules are intended to reflect properties of relational operations and constants. There are two groups of rules: decomposition rules and specific rules. Given a formula, the decomposition rules of the system enable us to transform it into simpler formulas, or the specific rules enable us to replace a formula by some other formulas. The rules have the following general form:

$$(*) \qquad \frac{\Phi}{\Phi_1 \mid \ldots \mid \Phi_n}$$

where $\Phi_1, \ldots, \Phi_n$ are finite non-empty sets of formulas, $n \geq 1$, and Φ is a finite (possibly empty) set of formulas. A rule of the form $(*)$ is said to be *applicable* to a set X of formulas whenever $\Phi \subseteq X$. As a result of an application of a rule of the form $(*)$ to a set X, we obtain the sets $(X \setminus \Phi) \cup \Phi_i$, $i = 1, \ldots, n$. A set to which a rule has been applied is called the *premise* of the rule, and the sets obtained by an application of the rule are called its *conclusions*. As usual, any concrete rule will always be presented in a short form, that is we will indicate only the formulas which are essential for a transformation to be performed by the rule and also we will omit set brackets. Given a formula, successive applications of the rules result in a tree whose nodes consist of finite sets of formulas. Each node includes all the formulas of its predecessor node, possibly except for those which have been transformed. A node of the tree does not have successors whenever its set of formulas includes an axiomatic subset or none of the rules is applicable to it. We say that a variable in a rule is *new* whenever it appears in a conclusion of the rule and does not appear in its premise.

Let L be a relational logic. A relational proof system for L (L-system for short) contains a set $\mathcal{DR}_L$ of L-*decomposition rules* and a set $\mathcal{SR}_L$ of L-*specific rules*, where in each particular logic L the terms and the object symbols range over the corresponding sets of L.

The set of decomposition rules $\mathcal{DR}_L$ includes the set $\mathcal{DR}_0$ of rules of the following forms:

Let $x, y, \in \mathbb{OS}_L$ and $R, S \in \mathbb{RT}_L$.

$(\cup)$ $\quad \dfrac{x(R \cup S)y}{x Ry, x Sy}$ $\qquad (-\cup)$ $\quad \dfrac{x-(R \cup S)y}{x-Ry \mid x-Sy}$

$(\cap)$ $\quad \dfrac{x(R \cap S)y}{x Ry \mid x Sy}$ $\qquad (-\cap)$ $\quad \dfrac{x-(R \cap S)y}{x-Ry, x-Sy}$

$(--)$ $\quad \dfrac{x--Ry}{x Ry}$

$(^{-1})$ $\quad \dfrac{x R^{-1} y}{y Rx}$ $\qquad (-^{-1})$ $\quad \dfrac{x-R^{-1}y}{y-Rx}$

$(;)$ $\quad \dfrac{x(R;S)y}{x Rz, x(R;S)y \mid z Sy, x(R;S)y}$ $\qquad z \in \mathbb{OS}_L$

$(-;)$ $\quad \dfrac{x-(R;S)y}{x-Rz, z-Sy}$ $\qquad z \in \mathbb{OV}_L$ and z is new

The set of specific rules includes the rules that reflect the properties of constants assumed in an L-language in question.

In all the systems considered in this paper the sets containing a subset $\{xRy, x-Ry\}$, for $x, y \in \mathbb{OS}_L, R \in \mathbb{RT}_L$, are assumed to be L-*axiomatic sets*. A finite set of formulas $\{\varphi_1, \ldots, \varphi_n\}$ is said to be an L-*set* whenever for every L-model $\mathcal{M}$ and for every L-valuation v in $\mathcal{M}$ there exists $i \in \{1, \ldots, n\}$ such that φ_i is satisfied by v in $\mathcal{M}$. Let Φ be a non-empty set of L-formulas. A rule $\dfrac{\Phi}{\Phi_1 \mid \ldots \mid \Phi_n}$ is L-*correct* whenever it holds: Φ is an L-set if and only if Φ_i is an L-set, for every $i \in \{1, \ldots, n\}$. In the case when Φ is empty, L-correctness can be expressed as follows: a rule $\dfrac{}{\Phi_1| \ldots |\Phi_n}$ is L-*correct* whenever there exists $i \in \{1, \ldots, n\}$ such that Φ_i is not an L-set. It follows that the rules are semantically invertible. It is a characteristic feature of all Rasiowa-Sikorski style deduction systems (see [22] and [9]). A transfer of validity from the bottom sets of a rule to the upper set is needed for soundness of the system. The other direction is used in a proof of completeness. Observe that the classical tableau system for first-order logic has in fact the analogous property of preserving and reflecting unsatisfiability. Although this fact is not provable directly from the definition of tableau rules, it can be proved under the additional assumptions on repetition of some formulas in the process of application of the rules. In tableau system this assumption is hidden, it is shifted to a strategy of building the proof trees. In our systems the required repetitions are explicitly indicated in the rules.

Let xRy be an L-formula. An L-*proof tree for* xRy is a tree with the following properties:

- the formula xRy is at the root of this tree;
- each node except the root is obtained by an application of an L-rule to its predecessor node;
- a node does not have successors whenever it is an L-axiomatic set.

A branch of an L-proof tree is said to be L-*closed* whenever it contains a node with an L-axiomatic set of formulas. A tree is L-*closed* iff all of its branches are L-closed.

Due to the forms of decomposition rules of $\mathcal{DR}_0$ we obtain the following:

Fact 2
Let L-*system consists of decomposition rules from* $\mathcal{DR}_0$. *If a node of an* L-*proof tree does not contain an* L-*axiomatic subset and contains an* L-*formula* xRy *or* $x-Ry$, *for atomic* R, *then all of its successors contain this formula as well.*

An L-formula xRy is L-*provable* whenever there is a closed L-proof tree for it.

Fact 3
For every relational logic L, *if we show that:*

1. *All* L-*rules are* L-*correct.*
2. *All* L-*axiomatic sets are* L-*sets.*

then we obtain the soundness theorem for L-*logic: if an* L-*formula* xRy *is* L-*provable, then it is* L-*valid.*

As usual in proof theory a concept of completeness of a non-closed proof tree is needed. Intuitively, completeness of a non-closed tree means that all the rules that can be applied have been applied. By abusing the notation, for any branch b and a formula xRy, we write $xRy \in b$, if xRy belongs to a set of formulas of a node of branch b.

A non-closed branch b of an L-proof tree is said to be L-*complete* whenever it satisfies L-completion conditions. L-completion conditions determined by the rules of $\mathcal{DR}_0$ are the following:

For all $x, y \in \mathbb{OS}_L$ and for all $R, S \in \mathbb{RT}_L$:

Cpl($\cup$) (resp. Cpl($-\cap$)) If $x(R \cup S)y \in b$ (resp. $x-(R \cap S)y \in b$), then both $xRy \in b$ (resp. $x-Ry \in b$) and $xSy \in b$ (resp. $x-Sy \in b$).

Cpl($\cap$) (resp. Cpl($-\cup$)) If $x(R \cap S)y \in b$ (resp. $x-(R \cup S)y \in b$), then either $xRy \in b$ (resp. $x-Ry \in b$) or $xSy \in b$ (resp. $x-Sy \in b$).

Cpl($-$) If $x(--R)y \in b$, then $xRy \in b$.

Cpl($^{-1}$) If $xR^{-1}y \in b$, then $yRx \in b$.

Cpl($-^{-1}$) If $x-R^{-1}y \in b$, then $y-Rx \in b$.

Cpl($;$) If $x(R; S)y \in b$, then for every $z \in \mathbb{OS}_L$, either $xRz \in b$ or $zSy \in b$.

Cpl($-;$) If $x-(R; S)y \in b$, then for some $z \in \mathbb{OV}_L$, both $x-Rz \in b$ and $z-Sy \in b$.

An L-proof tree is said to be L-*complete* iff all of its non-closed branches are L-complete. An L-complete non-closed branch is said to be L-*open*.

By Fact 2 and since the set containing a subset $\{xRy, x-Ry\}$ is L-axiomatic, in every L-system containing only decomposition rules of $\mathcal{DR}_0$ the following holds:

Fact 4
Let L-system be a system with decomposition rules of $\mathcal{DR}_0$ as the only rules and let b be an L-open branch of an L-proof tree. Then there is no atomic L-formula xRy such that $xRy \in b$ and $x-Ry \in b$.

Due to Facts 2 and 4 it is easy to prove the following proposition:

Proposition 1
Let L-system be a system with decomposition rules of $\mathcal{DR}_0$ as the only rules and let b be a branch of an L-proof tree. If there are $x, y \in \mathbb{OS}_L$ and $R \in \mathbb{RT}_L$ such that $xRy \in b$ and $x-Ry \in b$, then b is closed.

Sometimes if the logic L is clear from the context we will omit the index L.

4   Basic Relational Logic RL

The logic presented in this section is a common core of all the logics relevant for binary relations. The vocabulary of the language of RL-logic is defined as in Section 2 where:

- $\mathbb{RC}_{RL} = \emptyset$.

An RL-*model* is a structure $\mathcal{M} = (U, m)$, where U is a non-empty set and $m : \mathbb{RV}_{RL} \cup \mathbb{OC}_{RL} \to \mathcal{P}(U \times U) \cup U$ is a meaning function such that m extends to all compound relational terms as defined in Section 2 with the condition:

$$m(-R) = (U \times U) \setminus m(R)$$

where on the right hand side '$\setminus$' denotes the set difference.

The decomposition rules $\mathcal{DR}_{\mathsf{RL}}$ of the RL-system are the rules of $\mathcal{DR}_0$ presented in Section 3 adjusted to the RL-language. There are no specific rules in this system. RL-*axiomatic* set is any set containing $\{xRy, x{-}Ry\}$, as defined in Section 3, where $x, y \in \mathbb{OS}_{\mathsf{RL}}$ and R is a relational term of $\mathbb{RT}_{\mathsf{RL}}$.

For each rule $(\#) \in \mathcal{DR}_{\mathsf{RL}}$ its correctness follows directly from semantics of relational terms built with the operator $\#$.

Proposition 2

1. *All* RL-*rules are* RL-*correct.*
2. *All* RL-*axiomatic sets are* RL-*sets.*

Due to the above proposition and Fact 3 we obtain:

Theorem 1 (Soundness of RL)
Let xRy be an RL-*formula. If xRy is* RL-*provable, then it is* RL-*valid.*

A non-closed branch b of a proof tree is said to be RL-*complete* whenever it satisfies RL-completion conditions of Section 3 determined by the rules from $\mathcal{DR}_{\mathsf{RL}}$.

Let b be an RL-open branch of an RL-proof tree. We define a branch structure $\mathcal{M}^b = (U^b, m^b)$ as follows:

- $U^b = \mathbb{OS}_{\mathsf{RL}}$;
- $m^b(c) = c$, for every $c \in \mathbb{OC}_{\mathsf{RL}}$
- $m^b(R) = \{(x, y) \in U^b \times U^b : xRy \notin b\}$, for every relational variable R;
- m^b extends homomorphically to all compound relational terms as in the RL-models.

Fact 5
For every RL-*open branch b, $\mathcal{M}^b$ is an* RL-*model.*

Any structure $\mathcal{M}^b$ is referred to as an RL-*branch model*. Let $v^b : \mathbb{OS}_{\mathsf{RL}} \to U^b$ be an RL-valuation in $\mathcal{M}^b$ such that $v^b(x) = x$ for every $x \in \mathbb{OS}_{\mathsf{RL}}$.

Proposition 3
For every open branch b of an RL-*proof tree, and for every* RL-*formula xRy:*

$$(*) \quad if \quad \mathcal{M}^b, v^b \models xRy, \quad then \quad xRy \notin b.$$

Proof. The proof is by induction on the complexity of formulas.

Let xRy be an atomic RL-formula. Assume $\mathcal{M}^b, v^b \models xRy$, that is $(x, y) \in m^b(R)$. By the definition of a branch model $xRy \notin b$. Let $R \in \mathbb{RV}$ and $\mathcal{M}^b, v^b \models x{-}Ry$, that is $(x, y) \notin m^b(R)$. Therefore $xRy \in b$. By Fact 4, $x{-}Ry \notin b$.

By way of example we prove $(*)$ for $R = S;T$ and $R = -(S;T)$.

Let $\mathcal{M}^b, v^b \models xRy$, for $R = S;T$. Then $(x,y) \in m^b(S;T)$, that is there exists $z \in \mathbb{OS}_{\mathsf{RL}}$ such that $xSz \notin b$ and $zTy \notin b$. Suppose $x(S;T)y \in b$. By the completion condition $\mathrm{Cpl}(;)$, for every $z \in \mathbb{OS}_{\mathsf{RL}}$ either $xSz \in b$ or $zTy \in b$, a contradiction.

Let $\mathcal{M}^b, v^b \models xRy$, for $R = -(S;T)$. Then $(x,y) \notin m^b(S;T)$, that is for every $z \in \mathbb{OS}_{\mathsf{RL}}$ either $xSz \in b$ or $zTy \in b$. Suppose $x-(S;T)y \in b$. By the completion condition $\mathrm{Cpl}(-;)$, for some $z \in \mathbb{OV}_{\mathsf{RL}}$ both $x-Sz \in b$ and $z-Ty \in b$. By Proposition 1, b is closed, a contradiction. □

The above proposition enables us to prove the following completeness theorem:

Theorem 2 (Completeness of RL)
Let xRy be an RL-formula. If xRy is RL-valid, then xRy is RL-provable.

Proof. Assume xRy is RL-valid. Suppose there is no any closed RL-proof tree for xRy. Consider a non-closed RL-proof tree for xRy. We may assume that this tree is complete. Let b be an open branch of the complete RL-proof tree for xRy. Since $xRy \in b$, by Proposition 3 in the branch model $\mathcal{M}^b$ valuation v^b does not satisfy xRy. Hence xRy is not RL-valid, a contradiction. □

5   Relational Logics with the Constant 1

In this section we present a relational logic RL(1) obtained from RL by expanding its language with a relational constant 1. There are two classes of models associated with the logic RL(1): in the first one the relational constant 1 is interpreted as an equivalence relation on a non-empty set U, while in the second 1 is interpreted as a universal relation. The vocabulary of the language of RL(1)-logic is defined as in Section 2 with

- $\mathbb{RC}_{\mathsf{RL}(1)} = \{1\}$.

An RLN(1)-*model* is a structure $\mathcal{M} = (U, m)$, where U is a non-empty set and $m: \mathbb{RA}_{\mathsf{RL}(1)} \cup \mathbb{OC}_{\mathsf{RL}(1)} \to \mathcal{P}(U \times U) \cup U$ is a meaning function such that:

- $m(1)$ is an equivalence relation on U;
- m extends to all compound relational terms as defined in Section 2 with the following additional condition: $m(-R) = m(1) \cap (U \times U \setminus m(R))$.

An RLN(1)-model is said to be RL(1)-model whenever 1 is interpreted as an universal relation, that is $m(1) = U \times U$. It follows that if $\mathcal{M} = (U, m)$ is RLN(1)-model or RL(1)-model, then truth of a formula xRy in $\mathcal{M}$ is equivalent to $m(1) \subseteq m(R)$.

Due to the definitions of RLN(1)-models and RL(1)-models we obtain the following:

Fact 6
For every RL(1)*-formula* xRy*, if* xRy *is* RLN(1)*-valid, then it is* RL(1)*-valid.*

RL(1)-decomposition rules are precisely the rules of $\mathcal{DR}_{\mathsf{RL}}$, that is $\mathcal{DR}_{\mathsf{RL}(1)} = \mathcal{DR}_{\mathsf{RL}}$. Moreover, the relational proof system for RL(1)-logic (RL(1)-system for short) contains RL(1)-axiomatic sets defined below. A set is an RL(1)-axiomatic whenever it includes any of the subsets (Ax1) or (Ax2), where:

(Ax1) $\{x1y\}$, where $x, y \in \mathbb{OS}_{\mathsf{RL}(1)}$;
(Ax2) $\{xRy, x{-}Ry\}$, where $x, y \in \mathbb{OS}_{\mathsf{RL}(1)}$ and $R \in \mathbb{RT}_{\mathsf{RL}(1)}$.

As in the case of RL-logic, it is easy to prove the following:

Proposition 4

1. *All* RL(1)*-rules are* RLN(1)*-correct.*
2. *All* RL(1)*-axiomatic sets are* RLN(1)*-sets.*

Due to the above proposition and Fact 3 we have the following:

Proposition 5
Let xRy *be an* RL(1)*-formula. If* xRy *is* RL(1)*-provable, then it is* RLN(1)*-valid.*

Due to Fact 6 the following holds:

Corollary 1
Let xRy *be an* RL(1)*-formula. If* xRy *is* RL(1)*-provable, then it is* RL(1)*-valid.*

RL(1)-completion conditions are the same as the completion conditions defined in Section 3 determined by the rules from $\mathcal{DR}_{\mathsf{RL}(1)}$ and adapted to the RL(1)-language.

Let b be an open branch of an RL(1)-proof tree. A branch structure $\mathcal{M}^b = (U^b, m^b)$ is defined as for RL-logic, taking the object symbols of RL(1) as the elements of U^b, defining m^b for atomic RL(1)-terms and for object constants as in RL-branch model and defining m^b for all RL(1)-terms as in RL(1)-models.

Proposition 6
For every RL(1)*-open branch* b*, a branch structure* $\mathcal{M}^b$ *is an* RL(1)*-model.*

Proof. For all $x, y \in \mathbb{OS}_{\mathsf{RL}(1)}$ $x1y \notin b$, since otherwise b would be closed. So $m^b(1) = U^b \times U^b$. Therefore by the definition, $\mathcal{M}^b$ is an RL(1)-model. □

Let $v^b \colon \mathbb{OS}_{\mathsf{RL}(1)} \to U^b$ be an RL(1)-valuation in $\mathcal{M}^b$ such that $v^b(x) = x$ for every $x \in \mathbb{OS}_{\mathsf{RL}(1)}$.

Proposition 7
For every open branch b *of an* RL(1)*-proof tree, and for every* RL(1)*-formula* xRy:

$$(*) \quad if \quad \mathcal{M}^b, v^b \models xRy, \quad then \quad xRy \notin b.$$

Since $m^b(1)$ is the universal relation, the proof is similar to the proof of Proposition 3. Due to Proposition 7 we obtain the following:

Proposition 8
Let xRy be an RL(1)*-formula. If xRy is* RL(1)*-valid, then xRy is* RL(1)*-provable.*

Finally, due to Corollary 1 and Propositions 5 and 8 we obtain the following theorem:

Theorem 3 (Soundness and Completeness of RL(1)**)**
Let xRy be an RL(1)*-formula. The the following conditions are equivalent:*

- *xRy is* RL(1)*-provable;*
- *xRy is* RL(1)*-valid;*
- *xRy is* RLN(1)*-valid.*

The above theorem confirms the known fact that the classes of equations provable in algebras of relations with 1 being the universal relation and with 1 being an equivalence relation are the same. It will be discussed in more details in Section 14.

6   Relational Logics with Constant $1'$

A logic considered in this section is obtained from RL-logic by expanding its language with a constant $1'$. The vocabulary of the language of RL($1'$)-logic is defined as in Section 2 with

- $\mathbb{RC}_{\text{RL}(1')} = \{1'\}$.

An RL($1'$)-*model* is a structure $\mathcal{M} = (U, m)$, where U is a non-empty set and $m \colon \mathbb{RA}_{\text{RL}(1')} \cup \mathbb{OC}_{\text{RL}(1')} \to \mathcal{P}(U \times U) \cup U$ is a meaning function such that the following conditions are satisfied:

- $m(1')$ is an equivalence relation on U;
- $m(1'); m(R) = m(R); m(1') = m(R)$ for every $R \in \mathbb{RA}_{\text{RL}(1')}$ (extensionality);
- m extends to all compound relational terms as in the RL-models.

By an easy induction the following can be proved :

Proposition 9
Let $\mathcal{M} = (U, m)$ be an RL($1'$)*-model. Then for every relational term R of* RL($1'$)*-language, the following extensionality property holds:*

$$m(1'); m(R) = m(R); m(1') = m(R).$$

Proof
By way of example we show that the extensionality property holds for $R = -S$ and $R = (S; T)$.

Proof of $m(-S) = m(1'); m(-S)$

Assume $(x, y) \in m(-S)$. Since $m(1')$ is reflexive, $(x, x) \in m(1')$ and $(x, y) \in m(-S)$. Hence there exists $z \in U$ such that $(x, z) \in m(1')$ and $(z, y) \in m(-S)$. Therefore $(x, y) \in m(1'); m(-S)$.

Assume $(x, y) \in m(1'); m(-S)$, that is there exists $z \in U$ such that $(x, z) \in m(1')$ and $(z, y) \notin m(S)$. By the induction hypothesis, for all $u \in U$ $((z, u) \notin m(1')$ or $(u, y) \notin m(S))$. Let $u := x$. It follows that $(z, x) \notin m(1')$ or $(x, y) \notin m(S)$. Since $m(1')$ is symmetric, it must be $(x, y) \notin m(S)$. Therefore $(x, y) \in m(-S)$.

Proof of $m(S; T) = m(1'); m(S; T)$

Since $m(1')$ is reflexive, $m(S; T) \subseteq m(1'); m(S; T)$.

Assume $(x, y) \in m(1'); m(S; T)$, that is there exist $z, u \in U$ such that $(x, z) \in m(1')$, $(z, u) \in m(S)$ and $(u, y) \in m(T)$. By the induction hypothesis we get $(x, u) \in m(S)$. Therefore $(x, y) \in m(S; T)$. □

Proposition 10
Let $\mathcal{M} = (U, m)$ be a structure such that U is a non-empty set and $m\colon \mathbb{RA}_{\mathsf{RL}(1')} \cup \mathbb{OC}_{\mathsf{RL}(1')} \to \mathcal{P}(U \times U) \cup U$ is a meaning function satisfying the following conditions:

- *$m(1')$ is reflexive;*
- *m extends to all compound relational terms as in the RL-models;*
- *$m(1'); m(R) = m(R); m(1') = m(R)$ for every $R \in \mathbb{RT}_{\mathsf{RL}(1')}$.*

Then $\mathcal{M}$ is an RL(1')-model.

Proof
It suffices to show that $m(1')$ is symmetric and transitive. Let $R = (1')^{-1}$. Then $m(1')^{-1}; m(1') = m(1')^{-1} = m(1'); m(1')^{-1}$, thus $(m(1'); m(1')^{-1}); m(1') = m(1')^{-1}$. It implies that: (*) $(y, x) \in m(1')$ iff there exist $z, u \in U$ such that $(x, z) \in m(1')$, $(u, z) \in m(1')$ and $(u, y) \in m(1')$. Assume $(x, y) \in m(1')$, for some $x, y \in U$. Then $z := y$ and $u := x$ satisfy the right side of condition (*), so $(y, x) \in m(1')$. Therefore $m(1')$ is symmetric. Assume $(x, y) \in m(1')$ and $(y, z) \in m(1')$. Since $m(1'); m(1') \subseteq m(1')$, $(x, z) \in m(1')$. Therefore $m(1')$ is transitive, hence it is an equivalence relation on U. □

It follows that the equivalent set of conditions on the RL(1')-models could be reflexivity of $m(1')$ and the extensionality property for all the relational terms.

An RL(1')-model $\mathcal{M} = (U, m)$ is said to be *standard* whenever $m(1')$ is the identity on U, that is $m(1') = \{(x, x) : x \in U\}$. Any standard RL(1')-model will be referred to as RL*(1')-model. A formula xRy is said to be RL*(1')-valid iff it is true in all standard RL(1')-models.

Fact 7
If xRy is RL$(1')$ *valid, then it is* RL$^*(1')$-*valid.*

The decomposition rules of the RL$(1')$-system are the rules obtained from the rules in $\mathcal{DR}_0$ presented in Section 3 by adjusting them to the RL$(1')$-language. The specific rules of RL$(1')$-system have the following forms:

Let $x, y \in \mathbb{OS}_{\mathsf{RL}(1')}$ and $R \in \mathbb{RA}_{\mathsf{RL}(1')}$.

$(1'1)$ $\dfrac{xRy}{xRz, xRy \mid y1'z, xRy}$ $z \in \mathbb{OS}_{\mathsf{RL}(1')}$

$(1'2)$ $\dfrac{xRy}{x1'z, xRy \mid zRy, xRy}$ $z \in \mathbb{OS}_{\mathsf{RL}(1')}$

A finite set of formulas is RL$(1')$-*axiomatic* whenever it includes (Ax1) or (Ax2), where:

(Ax1) $\{x1'x\}$, where $x \in \mathbb{OS}_{\mathsf{RL}(1')}$
(Ax2) $\{xRy, x-Ry\}$,   where $x, y \in \mathbb{OS}_{\mathsf{RL}(1')}$ and $R \in \mathbb{RT}_{\mathsf{RL}(1')}$

It is easy to see that the properties of Facts 2, 4 and Proposition 1 are satisfied in RL$(1')$, that is in the RL$(1')$-system the following holds:

Proposition 11
Let b be a branch of an RL$(1')$-*proof tree. If $xRy \in b$ and $x-Ry \in b$, for some relational term R and for some $x, y \in \mathbb{OS}_{\mathsf{RL}(1')}$, then b is closed.*

Proposition 12

1. *All* RL$(1')$-*rules are* RL$(1')$-*correct.*
2. *All* RL$(1')$-*axiomatic sets are* RL$(1')$-*sets.*

Proof
Since $m(1')$ is reflexive, $\{x1'x\}$ is an RL$(1')$-set. To prove 1. it suffices to show correctness of the specific rules, correctness of the decomposition rules follows from the definitions of the relational operations. Let us prove that the rule $(1'1)_{\mathsf{RL}(1')}$ is correct, for any atomic relational term R. It is easy to see that if $\{xRy\}$ is an RL$(1')$-set, then $\{xRy, xRz\}$ and $\{y1'z, xRy\}$ are RL$(1')$-sets. Assume $\{xRy, xRz\}$ and $\{y1'z, xRy\}$ are RL$(1')$-sets, that is, by symmetry of $m(1')$, for every RL$(1')$-model $\mathcal{M}$ and for every RL$(1')$-valuation v:

$\mathcal{M}, v \models xRz$ or $\mathcal{M}, v \models xRy$ and $\mathcal{M}, v \models z1'y$ or $\mathcal{M}, v \models xRy$

Let $\mathcal{M}$ be an RL$(1')$-model and v be an RL$(1')$-valuation in $\mathcal{M}$. Suppose $\mathcal{M}, v \models xRz$ and $\mathcal{M}, v \models z1'y$. Then $(v(x), v(z)) \in m(R)$ and $(v(z), v(y)) \in m(1')$. Since $m(R); m(1') \subseteq m(R)$, $(v(x), v(y)) \in m(R)$. Hence $\mathcal{M}, v \models xRy$. In the remaining cases the proofs are obvious. The proof for the rule $(1'2)$ is similar.   □

Due to the above proposition and Fact 3 we obtain the following:

Proposition 13
Let xRy be an RL$(1')$-*formula. If xRy is* RL$(1')$-*provable, then it is* RL$(1')$-*valid.*

Corollary 2
Let xRy be an RL(1′)*-formula. If xRy is* RL(1′)*-provable, then it is* RL*(1′)*-valid.*

A non-closed branch b of an RL(1′)-proof tree is said to be RL(1′)-*complete* whenever it satisfies RL(1′)-completion conditions which consist of the completion conditions determined by decomposition rules of $\mathcal{DR}_{\mathsf{RL}(1′)}$ and the following:

For every $R \in \mathbb{RA}_{\mathsf{RL}(1′)}$ and for all $x, y \in \mathbb{OS}_{\mathsf{RL}(1′)}$:

Cpl(1′1) If $xRy \in b$, then for every $z \in \mathbb{OS}_{\mathsf{RL}(1′)}$, either $xRz \in b$ or $y1′z \in b$.

Cpl(1′2) If $xRy \in b$, then for every $z \in \mathbb{OS}_{\mathsf{RL}(1′)}$, either $x1′z \in b$ or $zRy \in b$.

Let b be an open branch of an RL(1′)-proof tree. We define a branch structure $\mathcal{M}^b = (U^b, m^b)$ similarly as for RL-logic adapted to the RL(1′)-language. In particular, $m^b(1′) = \{(x, y) \in U^b \times U^b : x1′y \notin b\}$.

Proposition 14
For every RL(1′)*-open branch b, a branch structure $\mathcal{M}^b$ is an* RL(1′)*-model.*

Proof
We need to prove that (1) $m^b(1′)$ is an equivalence relation on U^b and (2) $m^b(1′); m^b(R) = m^b(R); m^b(1′) = m^b(R)$ for every $R \in \mathbb{RA}_{\mathsf{RL}(1′)}$.

Proof of (1)
For every $x \in U^b$, $x1′x \notin b$, since otherwise b would be closed. Therefore $(x, x) \in m^b(1′)$, hence $m^b(1′)$ is reflexive. Assume $(x, y) \in m^b(1′)$, that is $x1′y \notin b$. Suppose $(y, x) \notin m^b(1′)$. Then $y1′x \in b$. By the completion condition Cpl(1′1), either $y1′y \in b$ or $x1′y \in b$, a contradiction. Therefore $m^b(1′)$ is symmetric. To prove transitivity, assume $(x, y) \in m^b(1′)$ and $(y, z) \in m^b(1′)$, that is $x1′y \notin b$ and $y1′z \notin b$. Suppose $(x, z) \notin m^b(1′)$. Then $x1′z \in b$. By the completion condition Cpl(1′1), either $x1′y \in b$ or $z1′y \in b$. In the first case we get a contradiction, so $z1′y \in b$. By the completion condition Cpl(1′1) applied to $z1′y$, either $z1′z \in b$ or $y1′z \in b$, a contradiction. Therefore $m^b(1′)$ is transitive.

Proof of (2)
Since $m^b(1′)$ is reflexive, $m^b(R) \subseteq m^b(1′); m^b(R)$ and $m^b(R) \subseteq m^b(R); m^b(1′)$.

Now assume $(x, y) \in m^b(1′); m^b(R)$, that is there exists $z \in U^b$ such that $x1′z \notin b$ and $zRy \notin b$. Suppose $(x, y) \notin m^b(R)$. Then $xRy \in b$. By the completion condition Cpl((1′2), for every $z \in U^b$, either $x1′z \in b$ or $zRy \in b$, a contradiction.

Assume $(x, y) \in m^b(R); m^b(1′)$, that is, by symmetry of $m^b(1′)$, there exists $z \in U^b$ such that $xRz \notin b$ and $y1′z \notin b$. Suppose $(x, y) \notin m^b(R)$. Then $xRy \in b$. By the completion condition Cpl(1′1), for every $z \in U^b$, either $xRz \in b$ or $y1′z \in b$, a contradiction. $\qquad\square$

Any structure $\mathcal{M}^b$ is referred to as an RL(1′)-*branch model*. Let $v^b \colon \mathbb{OS}_{\mathsf{RL}(1′)} \to U^b$ be an RL(1′)-valuation in $\mathcal{M}^b$ such that $v^b(x) = x$ for every $x \in \mathbb{OS}_{\mathsf{RL}(1′)}$.

Proposition 15
For every open branch b of an RL(1′)*-proof tree, and for every* RL(1′)*-formula xRy:*

$$(*) \quad if \quad \mathcal{M}^b, v^b \models xRy, \quad then \quad xRy \notin b.$$

The proof is similar to the proof of Proposition 3.

Since $m^b(1')$ is an equivalence relation on U^b, given an RL(1')-branch model $\mathcal{M}^b$, we may define the quotient model $\mathcal{M}_q^b = (U_q^b, m_q^b)$ as follows:

- $U_q^b = \{\|x\| : x \in U^b\}$, where $\|x\|$ is the equivalence class of $m^b(1')$ generated by x;
- $m_b^q(c) = \|c\|$, for every $c \in \mathbb{OC}_{\mathsf{RL}(1')}$;
- $m_q^b(R) = \{(\|x\|, \|y\|)) \in U_q^b \times U_q^b : (x,y) \in m^b(R)\}$, for every $R \in \mathbb{RA}_{\mathsf{RL}(1')}$;
- m_q^b extends for all compound relational terms as in the RL(1')-models.

Since a branch model satisfies the extensionality property, the definition of $m_q^b(R)$ is correct, that is the following condition is satisfied:

$$\text{if } (x,y) \in m^b(R) \text{ and } (x,z),(y,t) \in m^b(1'), \text{ then } (z,t) \in m^b(R).$$

Let v_q^b be an RL(1')-valuation in $\mathcal{M}_q^b$ such that $v_q^b(x) = \|x\|$, for every $x \in \mathbb{OS}_{\mathsf{RL}(1')}$.

Proposition 16

1. The model $\mathcal{M}_q^b$ is a standard RL(1')-model,
2. For every RL(1')-formula xRy:

$$(*) \quad \mathcal{M}^b, v^b \models xRy \quad \text{iff} \quad \mathcal{M}_q^b, v_q^b \models xRy$$

Proof

1. We have to show that $m_q^b(1')$ is the identity on U_q^b. Indeed, we have:

$$(\|x\|, \|y\|) \in m_q^b(1') \text{ iff } (x,y) \in m^b(1') \text{ iff } \|x\| = \|y\|$$

2. The proof is by an easy induction on the complexity of formulas. □

Proposition 17
Let xRy be an RL(1')-formula. If xRy is RL*(1')-valid, then xRy is RL(1')-provable.

Proof
Assume xRy is RL*(1')-valid. Suppose there is no closed RL(1')-proof tree for xRy. Consider a non-closed RL(1')-proof tree for xRy. We may assume that this tree is complete. Let b be an open branch of the complete RL(1')-proof tree for xRy. Since $xRy \in b$, so by Proposition 15, the branch model $\mathcal{M}^b$ does not satisfy xRy. By Proposition 16 condition 2. also the quotient model $\mathcal{M}_q^b$ does not satisfy xRy. Since $\mathcal{M}_q^b$ is a standard RL(1')-model, so xRy is not RL(1')-valid, a contradiction. □

From Fact 7 and Propositions 13, and 17 we obtain:

Theorem 4 (Soundness and Completeness of RL(1'))
Let xRy be an RL(1')-formula. Then the following conditions are equivalent:

- xRy is RL(1')-provable;
- xRy is RL(1')-valid;
- xRy is RL*(1')-valid.

7 Relational Logics with Constants $1'$ and 1

The vocabulary of the language of $RL(1, 1')$ is such that:

- $\mathbb{RC}_{RL(1,1')} = \{1', 1\}$.

An $RL(1, 1')$-*model* is a structure $\mathcal{M} = (U, m)$, where U is a non-empty set and $m: \mathbb{RA}_{RL(1,1')} \cup \mathbb{OC}_{RL(1,1')} \to \mathcal{P}(U \times U) \cup U$ is a meaning function such that $\mathcal{M}$ is an $RL(1')$-model and $\mathcal{M}$ is an $RL(1)$-model.

An $RLN(1, 1')$-*model* is a structure $\mathcal{M} = (U, m)$, where U is a non-empty set and $m: \mathbb{RA}_{RL(1,1')} \cup \mathbb{OC}_{RL(1,1')} \to \mathcal{P}(U \times U) \cup U$ is a meaning function such that

- $\mathcal{M}$ is an $RLN(1)$-model;
- $m(1')$ is an equivalence relation on U;
- $m(1'); m(R) = m(R); m(1') = m(R)$ for every atomic R.

An $RL(1, 1')$-model (resp. $RLN(1, 1')$-model) $\mathcal{M} = (U, m)$ is said to be *standard* whenever $m(1')$ is the identity on U. Standard $RL(1, 1')$-models (resp. $RLN(1, 1')$-models) are referred to as $RL^*(1, 1')$-models (resp. $RLN^*(1, 1')$-models).

$RL(1, 1')$-system consists of $RL(1')$-rules, $RL(1')$-axiomatic sets, and $RL(1)$-axiomatic sets adjusted to the language of $RL(1, 1')$-logic.

Note that in order to prove completeness we construct, as usual, the branch model. $m^b(1)$ is the universal relation in a branch model. It follows that completeness and soundness can be proved in a similar way as in $RL(1')$-logic and then by using Theorems 3 and 4 we obtain the following:

Theorem 5 (Soundness and Completeness of $RL(1, 1')$)
For any $RL(1, 1')$-formula xRy the following conditions are equivalent:

- xRy *is* $RL(1, 1')$-*provable;*
- xRy *is* $RL(1, 1')$-*valid;*
- xRy *is* $RL^*(1, 1')$-*valid;*
- xRy *is* $RLN(1, 1')$-*valid;*
- xRy *is* $RLN^*(1, 1')$-*valid.*

The class of $RLN(1, 1')$-models is closely related to the class RRA of representable relation algebras, while the class of $RL(1, 1')$-models corresponds to the class FRA of full relation algebras, as it will be proved in Section 14.

8 Relational Logics with Point Relations Introduced with Axioms

In the present section and in the subsequent Section 9 we consider the logics intended for providing a means of relational reasoning in the theories which refer to objects of their domains. There are two relational formalisms for coping with individual objects. A logic $RL_{ax}(C)$ presented in this section is a purely relational formalism where objects are introduced through point relations which, in turn are presented axiomatically with a well known set of axioms. The axioms say that a binary relation is a point relation whenever it is non-empty, right ideal relation with one-element domain. A binary relation R on a set U is right ideal whenever $R; 1 = R$, where $1 = U \times U$. In other words such an R is of the form $X \times U$, for some $X \subseteq U$. We may think of right ideal relations as representing sets, they are sometimes referred to as vectors (see [23]). Therefore if the domain of a right ideal relation is a singleton set, the relation may be seen as a representation of an individual object. A logic $\mathsf{RL}_{df}(C)$ presented in Section 9 includes object constants in its language interpreted as singletons, and moreover, with each object constant c there is associated a relation R_c such that its meaning in every model is defined as a right ideal relation with the domain consisting of the single element being a meaning of c.

The language of the logics considered in this section includes, apart from the relational constants 1 and $1'$, a family of relational constants interpreted as point relations determined axiomatically by the conditions 1, 2, and 3 below. The vocabulary of the language of $\mathsf{RL}_{ax}(C)$-logic is such that:

- $\mathbb{RC}_{\mathsf{RL}_{ax}(C)} = \{1', 1\} \cup C$, where $C = \{R_i : i \in I\}$ for some fixed set I.

An $\mathsf{RL}_{ax}(C)$-*model* is a structure $\mathcal{M} = (U, m)$, where U is a non-empty set and $m \colon \mathbb{RA}_{\mathsf{RL}_{ax}(C)} \cup \mathbb{OC}_{\mathsf{RL}_{ax}(C)} \to \mathcal{P}(U \times U) \cup U$ is a meaning function such that $\mathcal{M}$ is an $\mathsf{RL}(1, 1')$-model and the following hold:

- for every $R_i \in C$
    1. $m(R_i) \neq \emptyset$;
    2. $m(R_i) = m(R_i); m(1)$;
    3. $m(R_i); m(R_i)^{-1} \subseteq m(1')$;
- m extends to all compound relational terms as in RL-logic.

An $\mathsf{RL}_{ax}(C)$-model $\mathcal{M} = (U, m)$ is said to be *standard* ($\mathsf{RL}^*_{ax}(C)$-model for short) whenever $m(1')$ is the identity on U.

The above conditions 1., 2., and 3. say that relations R_i are point relations. Condition 2. guarantees that R_i is a right ideal relation, and condition 3. says that in the standard models the domains of relations R_i are singleton sets.

$\mathsf{RL}_{ax}(C)$-system consists of decomposition rules and specific rules of $\mathsf{RL}(1')$-system adjusted to the $\mathsf{RL}_{ax}(C)$-language and additional specific rules of the following forms that characterize relational constants R_i:

Let $x, y \in \mathbb{OS}_{\mathsf{RL}_{ax}(C)}$ and $R_i \in C$.

$(C1) \quad \dfrac{}{z - R_i t} \qquad\qquad z, t \in \mathbb{OV}_{\mathsf{RL}_{ax}(C)}$ are new

$(C2) \quad \dfrac{x R_i y}{x R_i y, \, x R_i z} \qquad\qquad z \in \mathbb{OS}_{\mathsf{RL}_{ax}(C)}$

$(C3) \quad \dfrac{x 1' y}{x R_i z, x 1' y \mid y R_i z, x 1' y} \qquad z \in \mathbb{OS}_{\mathsf{RL}_{ax}(C)}$

$RL_{ax}(C)$-*axiomatic* sets are those of $RL(1, 1')$ adapted to the $RL_{ax}(C)$-language. As in the previous cases, the conditions of Facts 2, 4, and Proposition 1 are satisfied in $RL_{ax}(C)$, that is the $RL_{ax}(C)$-system satisfies the property of Proposition 11. Therefore the following can be proved easily:

Proposition 18

1. *All* $RL_{ax}(C)$-*rules are* $RL_{ax}(C)$-*correct.*
2. *All* $RL_{ax}(C)$-*axiomatic sets are* $RL_{ax}(C)$-*sets.*

It is easy to see that correctness of the rules $(C1)$, $(C2)$, and $(C3)$ follows directly from the semantic conditions 1., 2., and 3., respectively.

Due to the above proposition and Fact 3 we obtain:

Proposition 19
Let xRy *be an* $RL_{ax}(C)$-*formula. If* xRy *is* $RL_{ax}(C)$-*provable, then it is* $RL_{ax}(C)$-*valid.*

Corollary 3
Let xRy *be an* $RL_{ax}(C)$-*formula. If* xRy *is* $RL_{ax}(C)$-*provable, then it is* $RL_{ax}^*(C)$-*valid.*

To prove completeness of $RL_{ax}(C)$-system it suffices to define the branch structure so that it will be an $RL_{ax}(C)$-model and the usual property will hold: if a formula is satisfied in a branch model determined by an open branch b, then it does not belong to b.

A non-closed branch b of an $RL_{ax}(C)$-proof tree is said to be $RL_{ax}(C)$-*complete* whenever it satisfies $RL_{ax}(C)$-completion conditions which consist of the completion conditions determined by the decomposition rules of $\mathcal{DR}_{RL_{ax}(C)}$, the specific rules for $1'$, and additionally the following:

For every $R_i \in C$ and for all $x, y \in \mathbb{OS}_{RL_{ax}(C)}$:

Cpl($C1$) There exist $z, t \in \mathbb{OV}_{RL_{ax}(C)}$ such that $z - R_i t \in b$.
Cpl($C2$) If $xR_i y \in b$, then for every $z \in \mathbb{OS}_{RL_{ax}(C)}$ $xR_i z \in b$.
Cpl($C3$) If $x1'y \in b$, then for every $z \in \mathbb{OS}_{RL_{ax}(C)}$ either $xR_i z \in b$ or $yR_i z \in b$.

Let b be an open branch of an $RL_{ax}(C)$-proof tree. We define a branch structure $\mathcal{M}^b = (U^b, m^b)$ with $U^b = \mathbb{OS}_{RL_{ax}(C)}$ similarly as in RL-logic by adjusting it to the $RL_{ax}(C)$-language.

Proposition 20
For every open branch b, *the branch structure* $\mathcal{M}^b$ *is an* $RL_{ax}(C)$-*model.*

Proof
It suffices to prove that for every $R_i \in C$, (1) $m^b(R_i) \neq \emptyset$, (2) $m^b(R_i) = m^b(R_i); m^b(1)$, and (3) $m^b(R_i); m^b(R_i)^{-1} \subseteq m^b(1')$.

Proof of (1)
By the completion condition Cpl(C1) there exist $z, t \in U^b$ such that $z - R_i t \in b$. Hence $z R_i t \notin b$, since otherwise b would be closed. Therefore there exist $z, t \in U^b$ such that $(z, t) \in m^b(R_i)$.

Proof of (2)
Since $m^b(1) = U^b \times U^b$, so $m^b(R_i) \subseteq m^b(R_i); m^b(1)$. Assume there exists $z \in U^b$ such that $(x, z) \in m^b(R_i)$ and $(z, y) \in m^b(1)$, that is $x R_i z \notin b$ and $z 1 y \notin b$. Suppose $(x, y) \notin m^b(R_i)$. Then $x R_i y \in b$. By the completion condition Cpl(C2) for every $z \in U^b$, $x R_i z \in b$, a contradiction.

The proof of (3) is similar.   □

Note that $m^b(R)$ is defined for all atomic relational terms R. Therefore due to the above proposition, the proof of completeness is similar to that of RL(1')-logic.

Proposition 21
Let xRy be an $\mathsf{RL}_{ax}(C)$-formula. If xRy is $\mathsf{RL}^*_{ax}(C)$-valid, then it is $\mathsf{RL}_{ax}(C)$-provable.

Corollary 4
Let xRy be an $\mathsf{RL}_{ax}(C)$-formula. If xRy is $\mathsf{RL}_{ax}(C)$-valid, then it is $\mathsf{RL}_{ax}(C)$-provable.

Due to Fact 1 and Propositions 19, and 21 we obtain the following:

Theorem 6 (Soundness and Completeness of $\mathsf{RL}_{ax}(C)$)
Let xRy be an $\mathsf{RL}_{ax}(C)$-formula. Then the following conditions are equivalent:

- *xRy is $\mathsf{RL}_{ax}(C)$-provable;*
- *xRy is $\mathsf{RL}_{ax}(C)$-valid;*
- *xRy is $\mathsf{RL}^*_{ax}(C)$-valid.*

9   Relational Logics with Point Relations Introduced with Definitions

The vocabulary of the language of $\mathsf{RL}_{df}(C)$-logic is such that:

- $\mathbb{OC}^0_{\mathsf{RL}_{df}(C)} \subseteq \mathbb{OC}_{\mathsf{RL}_{df}(C)}$, where $\mathbb{OC}^0_{\mathsf{RL}_{df}(C)} = \{c_i : i \in I\}$ for a fixed set I;
- $\mathbb{RC}_{\mathsf{RL}_{df}(C)} = \{1', 1\} \cup C$, where $C = \{R_i : i \in I\}$.

An $\mathsf{RL}_{df}(C)$-*model* is a structure $\mathcal{M} = (U, m)$, where U is a non-empty set and $m \colon \mathbb{RA}_{\mathsf{RL}_{df}(C)} \cup \mathbb{OC}_{\mathsf{RL}_{df}(C)} \to \mathcal{P}(U \times U) \cup U$ is a meaning function such that $\mathcal{M}$ is an RL(1, 1')-model and the following holds:

- $m(R_i) = \{(x, y) \in U \times U : (x, m(c_i)) \in m(1')\}$, for every $R_i \in C$;
- m extends to all compound relational terms as in RL-models.

An $RL_{df}(C)$-model $\mathcal{M} = (U, m)$ is said to be *standard* ($RL^*_{df}(C)$-model for short) whenever $m(1')$ is the identity on U. In the standard models relations R_i are right ideal relations with singleton domains.

$RL_{df}(C)$-system consists of decomposition rules $\mathcal{DR}_{RL_{df}(C)}$ obtained from $\mathcal{DR}_L$ by adjusting them to the $RL_{df}(C)$-language, the specific rules for $1'$ of $RL(1')$-system adapted to $RL_{df}(C)$-language, and the specific rules that characterize relational constants R_i:

Let $x, y \in \mathbb{OS}_{RL_{df}(C)}$, $c_i \in \mathbb{OC}^0_{RL_{df}(C)}$ and $R_i \in C$.

$(CD1)$ $\qquad \dfrac{xR_iy}{xR_iy, x1'c_i}$

$(CD2)$ $\qquad \dfrac{x-R_iy}{x-R_iy, x-1'c_i}$

$RL_{df}(C)$-*axiomatic* sets are those of $RL(1, 1')$ adjusted to the $RL_{df}(C)$-language. As in the previous cases, the $RL_{df}(C)$-system satisfies the property of Proposition 11. Therefore the following holds:

Proposition 22

1. *All* $RL_{df}(C)$-*rules are* $RL_{df}(C)$-*correct.*
2. *All* $RL_{df}(C)$-*axiomatic sets are* $RL_{df}(C)$-*sets.*

Proof
It suffices to show correctness of the new specific rules. It is easy to see that correctness of the rule $(CD1)$ follows from the property: if $(x, m(c_i)) \in m(1')$, then for every $y \in U$, $(x, y) \in m(R_i)$. The correctness of the rule $(CD2)$ follows from the property: if $(x, m(c_i)) \notin m(1')$, then for every $y \in U$, $(x, y) \notin m(R_i)$.

$\square$

Due to the above proposition and Fact 3 we obtain the following:

Proposition 23
Let xRy be an $RL_{df}(C)$-formula. If xRy is $RL_{df}(C)$-provable, then it is $RL_{df}(C)$-valid.

Corollary 5
Let xRy be an $RL_{df}(C)$-formula. If xRy is $RL_{df}(C)$-provable, then it is $RL^*_{df}(C)$-valid.

To prove completeness of $RL_{df}(C)$-system we define as usual the branch structure satisfying the appropriate conditions.

A non-closed branch b of an $RL_{df}(C)$-proof tree is said to be $RL_{df}(C)$-*complete* whenever it satisfies $RL_{df}(C)$-completion conditions which consist of the completion conditions determined by the decomposition rules, the completion conditions determined by the specific rules for $1'$, and additionally the following completion conditions determined by the specific rules for relational constants R_i:

For every $R_i \in C$ and for all $x, y \in \mathbb{OS}_{\mathsf{RL}_{df}(C)}$:

Cpl($CD1$) If $xR_iy \in b$, then $x1'c_i \in b$.

Cpl($CD2$) If $x-R_iy \in b$, then $x-1'c_i \in b$.

Let b be an open branch of an $\mathsf{RL}_{df}(C)$-proof tree. We define a branch structure $\mathcal{M}^b = (U^b, m^b)$ as follows:

- $U^b = \mathbb{OS}_{\mathsf{RL}_{df}(C)}$;
- $m^b(c) = c$, for every $c \in \mathbb{OC}_{\mathsf{RL}_{df}(C)}$;
- $m^b(R) = \{(x, y) \in U^b \times U^b : xRy \notin b\}$, for every $R \in \mathbb{RV}_{\mathsf{RL}_{df}(C)} \cup \{1, 1'\}$;
- $m^b(R_i) = \{x \in U^b : (x, c_i) \in m^b(1')\} \times U^b$, for every $R_i \in C$;
- m extends to all compound relational terms as in $\mathsf{RL}_{df}(C)$-models.

Fact 8
For every open branch b, $\mathcal{M}^b$ defined above is an $\mathsf{RL}_{df}(C)$-model.

Proposition 24
Let b be an open branch of an $\mathsf{RL}_{df}(C)$-proof tree and xRy be an $\mathsf{RL}_{df}(C)$-formula. Then

$$(*) \quad if \quad \mathcal{M}^b, v^b \models xRy, \quad then \ xRy \notin b.$$

Proof
It suffices to prove that $(*)$ holds for R being R_i or $-R_i$, where $R_i \in C$.

Let $R = R_i$ for some $R_i \in C$. Assume $(x, y) \in m^b(R_i)$, that is $(x, c_i) \in m^b(1')$. Then $x1'c_i \notin b$. Suppose $xR_iy \in b$. By the completion condition determined by the rule $(CD1)$, $x1'c_i \in b$, a contradiction.

Let $R = -R_i$ for some $R_i \in C$. Assume $(x, y) \in m^b(-R_i)$, that is $(x, c_i) \notin m^b(1')$. Then $x1'c_i \in b$. Suppose $x-R_iy \in b$. By the completion condition Cpl($CD2$), $x-1'c_i \in b$, hence b is closed a contradiction.   □

Due to the above proposition, the proof of completeness is similar to that of RL($1'$)-logic.

Proposition 25
Let xRy be an $\mathsf{RL}_{df}(C)$-formula. If xRy is $\mathsf{RL}^*_{df}(C)$-valid, then it is $\mathsf{RL}_{df}(C)$-provable.

Corollary 6
Let xRy be an $\mathsf{RL}_{df}(C)$-formula. If xRy is $\mathsf{RL}_{df}(C)$-valid, then it is $\mathsf{RL}_{df}(C)$-provable.

Due to Fact 1 and propositions 23, and 25 we obtain the following:

Theorem 7 (Soundness and Completeness of $\mathsf{RL}_{df}(C)$)
Let xRy be an $\mathsf{RL}_{df}(C)$-formula. Then the following conditions are equivalent:

- *xRy is $\mathsf{RL}_{df}(C)$-provable;*
- *xRy is $\mathsf{RL}_{df}(C)$-valid;*
- *xRy is $\mathsf{RL}^*_{df}(C)$-valid.*

10   Applications to Verification of Validity in Non-classical Logics

The logic $RL(1, 1')$ serves as a basis for the relational formalisms for non-classical logics whose Kripke-style semantics is determined by frames with binary accessibility relations. Let L be a modal logic with classical modal operators of possibility ($\langle R \rangle$) and necessity ($[R]$). The relational logic appropriate for expressing L-formulas is $RL_L(1, 1')$ obtained from $RL(1, 1')$ by expanding its language with a relational constant R representing the accessibility relation from the models of L-language and by assuming all the properties of R from these models in the $RL_L(1, 1')$-models. For example, if a relation R in a modal frame of a logic L is assumed to satisfy some conditions, e.g., reflexivity (logic T), symmetry (logic B), transitivity (logic S4) etc., then in the models of the corresponding logic $RL_L(1, 1')$ we add the respective conditions as the axioms of its models. The translation of a modal formula into a relational term starts with an assignment of relational variables to the propositional variables of the formula. Let τ' be such an assignment. Then the translation τ of the modal formulas is defined inductively as follows:

- $\tau(p) := \tau'(p); 1$ for propositional variable p;
- $\tau(\neg \alpha) := -\tau(\alpha)$;
- $\tau(\alpha \vee \beta) := \tau(\alpha) \cup \tau(\beta)$;
- $\tau(\alpha \wedge \beta) := \tau(\alpha) \cap \tau(\beta)$;
- $\tau(\langle R \rangle \alpha) := R; \tau(\alpha)$;
- $\tau([R]\alpha) := -(R; -\tau(\alpha))$.

The translation is defined so that it preserves validity of formulas.

Proposition 26
For every L-formula φ and for every L-model $\mathcal{M}$ there exists $RL_L^*(1, 1')$-model $\mathcal{M}'$ such that

$$\mathcal{M} \models \varphi \quad \text{iff} \quad \mathcal{M}' \models x\tau(\varphi)y$$

where x and y are object variables such that $x \neq y$.

Proof
Let φ be an L-formula and let $\mathcal{M} = (U, m)$ be an L-model. We define the corresponding $RL_L^*(1, 1')$-model $\mathcal{M}' = (U', m')$ as follows:

- $U' = U$;
- $m'(1) = U' \times U'$;
- $m'(1')$ is an identity on U';
- $m'(\tau'(p)) = \{(x, y) \in U' \times U' : x \in m(p)\}$, for any propositional variable p;
- $m'(R) = m(R)$;
- m' extends to all compound relational terms as in $RL(1, 1')$-models.

Given a valuation $v : \mathbb{OS}_{RL_L(1,1')} \to U$ we show by induction on the complexity of φ that the following property holds:

$$\mathcal{M}, v(x) \models \varphi \quad \text{iff} \quad \mathcal{M}', v \models x\tau(\varphi)y.$$

From that, we can conclude that φ is true in $\mathcal{M}$ iff $x\tau(\varphi)y$ is true in $\mathcal{M}'$. By way of example we prove the required condition for the formulas of the form: $\psi_1 \vee \psi_2$ and $\langle R \rangle \psi$.

- If $\varphi = \psi_1 \vee \psi_2$ then $\mathcal{M}, v(x) \models \psi_1 \vee \psi_2$ iff $\mathcal{M}, v(x) \models \psi_1$ or $\mathcal{M}, v(x) \models \psi_2$, iff, by inductive hypothesis, $\mathcal{M}', v \models x\tau(\psi_1)y$ or $\mathcal{M}', v \models x\tau(\psi_2)y$, iff $\mathcal{M}', v \models x(\tau(\psi_1) \cup \tau(\psi_2))y$ iff $\mathcal{M}', v \models x\tau(\psi_1 \cup \psi_2)y$.
- If $\varphi = \langle R \rangle \psi$ then $\mathcal{M}, v(x) \models \langle R \rangle \psi$ iff there exists $s \in U$ such that $(v(x), s) \in m(R)$ and $\mathcal{M}, s \models \psi$ iff, by inductive hypothesis, there exists $s \in U'$ such that $(v(x), s) \in m'(R)$ and $(s, v(y)) \in m'(\tau(\psi))$, iff $(v(x), v(y)) \in m'(R; \tau(\psi))$ iff $\mathcal{M}', v(x) \models x\tau(\langle R \rangle \psi)y$. $\qquad\square$

Proposition 27
For every L-formula φ and for every $RL_L^*(1, 1')$-model $\mathcal{M}'$ there exists L-model $\mathcal{M}$ such that

$$\mathcal{M} \models \varphi \quad iff \quad \mathcal{M}' \models x\tau(\varphi)y$$

where x and y are object variables such that $x \neq y$.

Proof
Let φ be an L-formula and let $\mathcal{M}' = (U', m')$ be an $RL_L^*(1, 1')$-model. We define the corresponding L-model $\mathcal{M} = (U, m)$ as follows:

- $U = U'$;
- for every propositional variable p, $s \in m(p)$ iff $(s, s') \in m'(\tau'(p))$ for some $s' \in U'$;
- $m(R) = m'(R)$.

The rest of the proof is similar to the proof of Proposition 26. $\qquad\square$

From Theorem 5 and Propositions 26, and 27 we obtain the following:

Theorem 8
For every formula φ of a logic L, φ is valid in L iff $xt(\varphi)y$ is valid in $RL_L(1, 1')$, where x and y are object variables such that $x \neq y$.

Once a translation from a non-classical logic L into an appropriate relational logic $RL_L(1, 1')$ is defined, we develop a dual tableau proof system for $RL_L(1, 1')$ which by the above theorem is a validity checker for L. The core of such a system is the $RL(1, 1')$-system. For each particular logic L the rules and/or axiomatic sets must be added reflecting the properties of the constant R. Defining these rules we follow the general principles presented in [14].

For example, a relational formalism for the modal logic K is the logic $RL_K(1, 1')$ obtained from $RL(1, 1')$ by assuming that the set of relational constants includes additionally a relational constant, say R, representing the accessibility relation from the frames of K. Since in the K-models there is no any specific assumption about R, $RL_K(1, 1')$-proof system can be obtained from that of $RL(1, 1')$ by adjusting it to the $RL_K(1, 1')$-language, in particular by postulating $\mathbb{RC}_{RL_K(1,1')} = \{1', 1, R\}$.

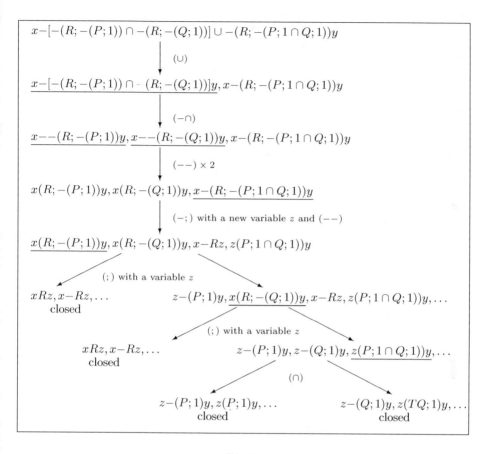

Fig. 1.

Let us consider the following formula φ of modal logic K:

$$\neg([R]p \wedge [R]q) \vee [R](p \wedge q)$$

Let $\tau'(p) = P$ and let $\tau'(q) = Q$. The translation $\tau(\varphi)$ of the above formula into a relational term of $\mathsf{RL_K}(1,1')$ is:

$$-[-(R;-(P;1)) \cap -(R;-(Q;1))] \cup -(R;-(P;1\cap Q;1))$$

We show that the formula φ is K-valid, that is $x\tau(\varphi)y$ is $\mathsf{RL_K}(1,1')$-valid. In each node of the proof tree we underline a formula which determines an applicable rule. Figure 1 presents a closed $\mathsf{RL_K}(1,1')$-proof tree for the formula $x\tau(\varphi)y$.

The method of relational formalization of non-classical logics is applicable to a great variety of logics, see e.g., [1], [10], [11], [15], [16] and [7].

11   Applications to Verification of Entailment in Non-classical Logics

The logic $\mathsf{RL}(1,1')$ can be also used to verify the entailment of formulas of non-classical logics, provided that they can be translated into a relational logic. The method is based on the following fact. Let $R_1, \ldots, R_n, R$ be binary relations on a set U and let $1 = U \times U$. It is known that $R_1 = 1, \ldots, R_n = 1$ imply $R = 1$ iff $(1; -(R_1 \cap \ldots \cap R_n); 1) \cup R = 1$. It follows that for every $\mathsf{RL}(1,1')$-model $\mathcal{M}$, $\mathcal{M} \models xR_1y, \ldots, \mathcal{M} \models xR_ny$ imply $\mathcal{M} \models xRy$ iff $\mathcal{M} \models x(1; -(R_1 \cap \ldots \cap R_n); 1) \cup R)y$ which means that entailment in $\mathsf{RL}(1,1')$ can be expressed in its language.

For example, in K-logic the formulas $[R]p$ and $[R](p \to q)$ imply $[R]q$. That is in $\mathsf{RL_K}(1,1')$-logic, relations $-(R; -(P;1))$ and $-(R; -(-(P;1) \cup (Q;1)))$ imply $-(R; -(Q;1))$. To prove this we need to show that the formula

$$x[(1; -(-(R; -(P;1)) \cap -(R; -(-(P;1) \cup (Q;1)))); 1) \cup -(R; -(Q;1))]y$$

is $\mathsf{RL_K}(1,1')$-provable. Figure 2 presents a closed $\mathsf{RL_K}(1,1')$-proof tree for this formula.

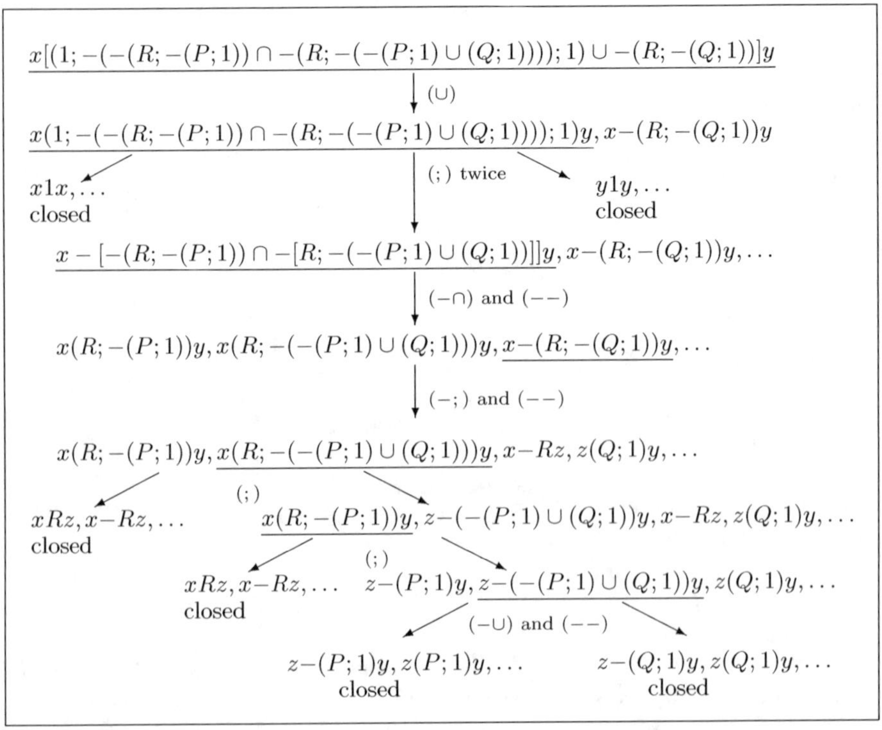

Fig. 2.

12 Applications to Model Checking in Non-classical Logics

The logic $RL(1, 1')$ is used in the formalisms of relational logics whose model checking problem is in question. Let $\mathcal{M} = (U, m)$ be a fixed $RL^*(1, 1')$-model and let $\varphi = xRy$ be an $RL(1, 1')$-formula, where R is a relational term and x, y are any object symbols. In order to obtain the relational formalism for the problem '$\mathcal{M} \models \varphi$?', we consider an instance $RL_{\mathcal{M}, \varphi}$ of the logic $RL(1, 1')$. Its language provides a code of model $\mathcal{M}$ and formula φ, and in its models the syntactic elements of φ are interpreted as in the model $\mathcal{M}$. The vocabulary of the logic $RL_{\mathcal{M}, \varphi}$ consists of the following pairwise disjoint sets:

- $\mathbb{OV}_{RL_{\mathcal{M}, \varphi}}$ a countable infinite set of object variables;
- $\mathbb{OC}_{RL_{\mathcal{M}, \varphi}} = \mathbb{OC}^0_{RL_{\mathcal{M}, \varphi}} \cup \mathbb{OC}^1_{RL_{\mathcal{M}, \varphi}}$, where $\mathbb{OC}^0_{RL_{\mathcal{M}, \varphi}} = \{c_a : a \in U\}$ and $\mathbb{OC}^1_{RL_{\mathcal{M}, \varphi}} = \{c \in \mathbb{OC}_{RL(1,1')} : c \text{ occurs in } \varphi\}$;
- $\mathbb{RC}_{RL_{\mathcal{M}, \varphi}} = \{S : S \text{ is an atomic subterm of } R\} \cup \{1, 1'\}$;
- $\mathbb{OP}_{RL_{\mathcal{M}, \varphi}} = \{-, \cup, \cap, ;, ^{-1}\}$;
- a set of parentheses $\{(,)\}$.

Note that the language of $RL_{\mathcal{M}, \varphi}$ does not contain relational variables.

An $RL_{\mathcal{M}, \varphi}$-model is a pair $\mathcal{N} = (W, n)$ where

- $W = U$;
- $n(c) = m(c)$, for every $c \in \mathbb{OC}^1_{RL_{\mathcal{M}, \varphi}}$;
- $n(c_a) = a$, for any $c_a \in \mathbb{OC}^0_{RL_{\mathcal{M}, \varphi}}$;
- $n(S) = m(S)$, for any atomic subterm S of R;
- $n(1), n(1')$ are defined as in $RL^*(1, 1')$-models;
- n extends to compound terms as in $RL^*(1, 1')$-models.

Observe that the above definition implies: for every atomic subterm S of R, $\mathcal{N}, v \models xSy$ iff there exist $a, b \in U$ such that $(a, b) \in m(S)$ and $v(x) = a$ and $v(y) = b$. Moreover, it is easy to prove that $n(R) = m(R)$. Note also that the class of $RL_{\mathcal{M}, \varphi}$-models has exactly one element up to isomorphism. Therefore, $RL_{\mathcal{M}, \varphi}$-validity is equivalent to the truth in a single $RL_{\mathcal{M}, \varphi}$-model $\mathcal{N}$, that is the following holds:

Proposition 28
The following statements are equivalent:

- $\mathcal{M} \models xRy$
- xRy is $RL_{\mathcal{M}, \varphi}$-valid

The relational proof system for $RL_{\mathcal{M}, \varphi}$ consists of the rules and axiomatic sets of $RL(1, 1')$-system adapted to the language of $RL_{\mathcal{M}, \varphi}$, and additionally:

- for every atomic subterm S of R and for any $x, y \in \mathbb{OS}_{\mathsf{RL}_{\mathcal{M},\varphi}}$ we add the rules of the following form:

$$(-S) \quad \frac{x-Sy}{x-1'c_{\mathsf{a}}, y-1'c_{\mathsf{b}}, c_{\mathsf{a}}-Sc_{\mathsf{b}}, x-Sy} \qquad c_{\mathsf{a}}, c_{\mathsf{b}} \in \mathbb{OC}^0_{\mathsf{RL}_{\mathcal{M},\varphi}} \text{ are new}$$

$$(1') \quad \frac{}{x-1'c_{\mathsf{a}}} \qquad\qquad\qquad\qquad c_{\mathsf{a}} \text{ is new}$$

$$(\mathsf{a} \neq \mathsf{b}) \quad \frac{}{c_{\mathsf{a}}1'c_{\mathsf{b}}} \qquad\qquad\qquad\qquad \text{for all } \mathsf{a} \neq \mathsf{b}$$

where $c_{\mathsf{a}} \in \mathbb{OC}^0_{\mathsf{RL}_{\mathcal{M},\varphi}}$ is *new* whenever it appears in a conclusion of the rule and does not appear in its premise;

- for every $c \in \mathbb{OC}^1_{\mathsf{RL}_{\mathcal{M},\varphi}}$ and for every $\mathsf{a} \in U$ such that $m(c) \neq \mathsf{a}$ we add the rules of the following form:

$$(\mathsf{ca}) \quad \frac{}{c1'c_{\mathsf{a}}}$$

- a set of formulas is assumed to be an axiomatic set whenever it includes either of the following subsets:
  - $\{c1'c_{\mathsf{a}}\}$, for every $c \in \mathbb{OC}^1_{\mathsf{RL}_{\mathcal{M},\varphi}}$ and for every $\mathsf{a} \in U$ such that $m(c) = \mathsf{a}$;
  - $\{c_{\mathsf{a}}Sc_{\mathsf{b}}\}$, for every atomic subterm S of R and for all $\mathsf{a}, \mathsf{b} \in U$ such that $(\mathsf{a}, \mathsf{b}) \in m(S)$;
  - $\{c_{\mathsf{a}}-Sc_{\mathsf{b}}\}$, for every atomic subterm S of R and for all $\mathsf{a}, \mathsf{b} \in U$ such that $(\mathsf{a}, \mathsf{b}) \notin m(S)$.

The correctness of all new rules and the validity of all new axiomatic sets follow directly from the definition of $\mathsf{RL}_{\mathcal{M},\varphi}$-semantics. For example, the correctness of the rule $(-S)$ follows from the following property of $n(S)$: $(v(x), v(y)) \in n(S)$ iff for all $\mathsf{a}, \mathsf{b} \in U$, either $(n(c_{\mathsf{a}}), n(c_{\mathsf{b}})) \notin n(S)$ or $v(x) \neq c_{\mathsf{a}}$ or $v(y) \neq c_{\mathsf{b}}$. Note that for every $x \in \mathbb{OS}_{\mathsf{RL}_{\mathcal{M},\varphi}}$ and for every valuation v in $\mathcal{N}$, there exists $c_{\mathsf{a}} \in \mathbb{OC}^0_{\mathsf{RL}_{\mathcal{M},\varphi}}$ such that the model $\mathcal{N}$ satisfies $v(x) = n(c_{\mathsf{a}})$, hence the rule $(1')$ is correct. The correctness of the rule $(\mathsf{a} \neq \mathsf{b})$ follows form the following property of $\mathcal{N}$-models: for all $\mathsf{a}, \mathsf{b} \in U$, if $\mathsf{a} \neq \mathsf{b}$, then $n(c_{\mathsf{a}}) \neq n(c_{\mathsf{b}})$.

The completion conditions are those of $\mathsf{RL}(1, 1')$-system adapted to the language of $\mathsf{RL}_{\mathcal{M},R}$ and additionally for every atomic subterm S of R we add the following conditions:

$\mathsf{Cpl}(-S)$ If $x-Sy \in b$, then for some $c_{\mathsf{a}}, c_{\mathsf{b}} \in \mathbb{OC}^0_{\mathsf{RL}_{\mathcal{M},\varphi}}$ all of the following conditions are satisfied: $x-1'c_{\mathsf{a}} \in b$, $y-1'c_{\mathsf{b}} \in b$ and $c_{\mathsf{a}}-Sc_{\mathsf{b}} \in b$.

$\mathsf{Cpl}(1')$ For every $x \in \mathbb{OV}_{\mathsf{RL}_{\mathcal{M},\varphi}}$ there exists $c_{\mathsf{a}} \in \mathbb{OC}^0_{\mathsf{RL}_{\mathcal{M},\varphi}}$ such that $x-1'c_{\mathsf{a}} \in b$.

$\mathsf{Cpl}(\mathsf{a} \neq \mathsf{b})$ For all $\mathsf{a}, \mathsf{b} \in U$ such that $\mathsf{a} \neq \mathsf{b}$, $c_{\mathsf{a}}1'c_{\mathsf{b}} \in b$.

$\mathsf{Cpl}(\mathsf{ca})$ For every $c \in \mathbb{OC}^1_{\mathsf{RL}_{\mathcal{M},\varphi}}$ and for every $\mathsf{a} \in U$ such that $n(c) \neq \mathsf{a}$, $c1'c_{\mathsf{a}} \in b$.

A branch model is a structure $\mathcal{N}^b = (W^b, n^b)$ satisfying the following conditions:

- $W^b = \mathbb{OS}_{\mathsf{RL}_{\mathcal{M},\varphi}}$;
- $n^b(c) = c$, for every $c \in \mathbb{OC}_{\mathsf{RL}_{\mathcal{M},\varphi}}$;

- $n^b(S) = \{(x,y) \in W^b \times W^b : xSy \notin b\}$, for $S \in \{1, 1'\}$;
- $n^b(S) = \{(x,y) \in W^b \times W^b :$ there exists $\mathsf{a}, \mathsf{b} \in U$ such that $\gamma(\mathsf{a}, \mathsf{b}, x, y)\}$, where $\gamma(\mathsf{a}, \mathsf{b}, x, y)$ is $[(\mathsf{a}, \mathsf{b}) \in m(S) \wedge (x, c_\mathsf{a}) \in n^b(1') \wedge (y, c_\mathsf{b}) \in n^b(1')]$;
- n^b extends to all compound terms as in $\mathsf{RL}(1, 1')$-models.

As in $\mathsf{RL}(1, 1')$-logic it is easy to prove that $n^b(1')$ and $n^b(1)$ are an equivalence relation and a universal relation, respectively.

Let $v^b : \mathbb{OS}_{\mathsf{RL}_{\mathcal{M}, \varphi}} \to W^b$ be a valuation in $\mathcal{N}^b$ such that $v^b(x) = x$ for every $x \in \mathbb{OS}_{\mathsf{RL}_{\mathcal{M}, \varphi}}$. Then the following holds:

Proposition 29
For every open branch b of an $\mathsf{RL}_{\mathcal{M}, \varphi}$-proof tree, and for every $\mathsf{RL}_{\mathcal{M}, \varphi}$-formula xRy:

$$(*) \quad \text{if} \quad \mathcal{N}^b, v^b \models xRy, \quad \text{then} \quad xRy \notin b.$$

Proof
The proof is similar to the proof of analogous proposition for $\mathsf{RL}(1, 1')$-logic. That is we need to show that $(*)$ holds for every atomic subterm S of R and its complement.

Let $\varphi = xSy$ for some atomic subterm S of R. Assume $\mathcal{N}^b, v^b \models xSy$. By the definition of $n^b(S)$ there exist $\mathsf{a}, \mathsf{b} \in U$ such that $(\mathsf{a}, \mathsf{b}) \in m(S)$, $x1'c_\mathsf{a} \notin b$ and $y1'c_\mathsf{b} \notin b$. Since $(\mathsf{a}, \mathsf{b}) \in m(S)$, $c_\mathsf{a} S c_{c_\mathsf{b}} \notin b$, otherwise b would be closed. Therefore the following holds: $c_\mathsf{a} S c_\mathsf{b} \notin b$, $x1'c_\mathsf{a} \notin b$ and $y1'c_\mathsf{b} \notin b$. Suppose $xSy \in b$. By the completion conditions for the rules $(1'1)$ and $(1'2)$, for all $c_\mathsf{a}, c_\mathsf{b} \in \mathbb{OC}^0_{\mathsf{RL}_{\mathcal{M}, \varphi}}$, at least one the following holds: $x1'c_\mathsf{a} \in b$ or $y1'c_\mathsf{b} \in b$ or $c_\mathsf{a} S c_{c_\mathsf{b}} \in b$, a contradiction.

Let $\varphi = x-Sy$, for some atomic subterm S of R. Assume $\mathcal{N}^b, v^b \models x-Sy$. Then for all $\mathsf{a}, \mathsf{b} \in U$, $(\mathsf{a}, \mathsf{b}) \notin m(S)$ or $x1'c_\mathsf{a} \in b$ or $y1'c_\mathsf{b} \in b$. Since $(\mathsf{a}, \mathsf{b}) \notin m(S)$, $c_\mathsf{a} - S c_{c_\mathsf{b}} \notin b$, otherwise b would be closed. Therefore for all $\mathsf{a}, \mathsf{b} \in U$, the following holds: if $c_\mathsf{a} - S c_\mathsf{b} \in b$, then $x1'c_\mathsf{a} \in b$ or $y1'c_\mathsf{b} \in b$. Suppose $x-Sy \in b$. By the completion condition for the rule $(-S)$, for some $c_\mathsf{a}, c_\mathsf{b} \in \mathbb{OC}^0_{\mathsf{RL}_{\mathcal{M}, \varphi}}$, the following holds: $x-1'c_\mathsf{a} \in b$ and $y-1'c_\mathsf{b} \in b$ and $c_\mathsf{a} - S c_{c_\mathsf{b}} \in b$, a contradiction. $\square$

Since $n^b(1')$ is an equivalence relation on W^b, we may define the quotient model $\mathcal{N}^b_q = (W^b_q, n^b_q)$ as follows:

- $W^b_q = \{\|x\| : x \in W^b\}$, where $\|x\|$ is the equivalence class of $n^b(1')$ generated by x;
- $n^b_q(c) = \|n^b(c)\|$, for every $c \in \mathbb{OC}_{\mathsf{RL}_{\mathcal{M}, \varphi}}$;
- $n^b_q(S) = \{(\|x\|, \|y\|) \in W^b_q \times W^b_q : (x, y) \in n^b(S)\}$, for every atomic S;
- n^b_q extends as in $\mathsf{RL}(1, 1')$-models.

Proposition 30
The quotient model $\mathcal{N}^b_q = (W^b_q, n^b_q)$ satisfies the following conditions:

1. $card(W^b_q) = card(W)$;
2. $c \in \|c_\mathsf{a}\|$ *iff* $n(c) = \mathsf{a}$
3. $n^b_q(S) = \{(\|c_\mathsf{a}\|, \|c_\mathsf{b}\|) \in W^b_q \times W^b_q : (n(c_\mathsf{a}), n(c_\mathsf{b})) \in n(S)\}$.

Proof

Proof of 1. For all $\mathsf{a}, \mathsf{b} \in U$, if $\mathsf{a} \neq \mathsf{b}$, then $c_{\mathsf{a}} 1' c_{\mathsf{b}} \in h$. Therefore for all $\mathsf{a}, \mathsf{b} \in U$ such that $\mathsf{a} \neq \mathsf{b}$, $(c_{\mathsf{a}}, c_{\mathsf{b}}) \notin n^b(1')$, hence $card(W_q^b) \geq card(W)$. By the completion condition for $(1')$, for every $x \in W^b$ there is $c_{\mathsf{a}} \in W^b$ such that $x - 1' c_{\mathsf{a}} \in b$. Therefore for every element x of W^b, $x \in \|c_{\mathsf{a}}\|$ for some $a \in U$. Hence $card(W_q^b) \leq card(W)$.

Proof of 2. For $c \in \mathbb{OC}^0_{\mathsf{RL}_{\mathcal{M};\varphi}}$ the proof is obvious. Let $c \in \mathbb{OC}^1_{\mathsf{RL}_{\mathcal{M};\varphi}}$. If $n(c) = \mathsf{a}$, then $c 1' c_{\mathsf{a}} \notin b$, since otherwise b would be closed. Therefore $c \in \|c_{\mathsf{a}}\|$. If $n(c) \neq \mathsf{a}$, then by the completion condition for (ca), $c 1' c_{\mathsf{a}} \in b$, hence $c \notin \|c_{\mathsf{a}}\|$.

Proof of 3. This follows directly from the definition of $n^b(S)$. $\qquad\qquad \square$

The above proposition implies that the function $f \colon W_q^b \to W$ defined as $f(\|c_{\mathsf{a}}\|) = a$ is an isomorphism between $\mathcal{N}_q^b$ and $\mathcal{N}$. Therefore $\mathcal{N}_q^b$ and $\mathcal{N}$ satisfy exactly the same formulas. Now the completeness $\mathsf{RL}_{\mathcal{M},\varphi}$ can be proved similarly as in $\mathsf{RL}(1, 1')$-logic.

Theorem 9 (Soundness and completeness of $\mathsf{RL}_{\mathcal{M},\varphi}$)
For every $\mathsf{RL}_{\mathcal{M},R}$-formula xRy the following conditions are equivalent:

- xRy is $\mathsf{RL}_{\mathcal{M},\varphi}$-provable.
- xRy is $\mathsf{RL}_{\mathcal{M},\varphi}$-valid.

Due to the above theorem and Proposition 28 we obtain the following:

Theorem 10
The following statements are equivalent:

- $\mathcal{M} \models xRy$,
- xRy is $\mathsf{RL}_{\mathcal{M},\varphi}$-provable.

The method presented above can be also used in the case of non-classical logics for which the problem of model checking is in question. By way of example consider the modal logic K. Let $\mathcal{M} = (U, m)$ be a K-model such that $U = \{\mathsf{a}, \mathsf{b}\}$, $m(p) = \{\mathsf{a}\}$ and the accessibility relation is defined as $m(R) = \{(\mathsf{a}, \mathsf{a}), (\mathsf{b}, \mathsf{a})\}$. Let φ be the formula of the form $\langle R \rangle p$. Let us consider the problem: 'is φ true in $\mathcal{M}$?' The translation of the formula φ is $\tau(\varphi) = (R; (P; 1))$, where $\tau'(p) = P$. Using the construction from the proof of Proposition 26 it is easy to prove that there exist an $\mathsf{RL}_\mathsf{K}(1, 1')$-model $\mathcal{M}'$ such that the following holds:

$$\mathcal{M} \models \varphi \quad \text{iff} \quad \mathcal{M}' \models x\tau(\varphi)y.$$

The $\mathsf{RL}_\mathsf{K}(1, 1')$-model $\mathcal{M}' = (U', m')$ is defined as follows:

- $U' = m'(1) = \{\mathsf{a}, \mathsf{b}\}$;
- $m'(P) = \{(\mathsf{a}, \mathsf{a}), (\mathsf{a}, \mathsf{b})\}$;
- $m'(R) = \{(\mathsf{a}, \mathsf{a}), (\mathsf{b}, \mathsf{a})\}$;

- $m'(1') = \{(\mathsf{a},\mathsf{a}),(\mathsf{b},\mathsf{b})\}$;
- m' extends to all compound terms as in $\mathsf{RL}(1,1')$-models.

Therefore the model checking problem 'is φ true in $\mathcal{M}$?' is equivalent to the problem 'is a formula $x\tau(\varphi)y$ true in $\mathcal{M}'$?'. For the latter we apply the method already presented above. The vocabulary of $\mathsf{RL}_{\mathcal{M}',x\tau(\varphi)y}$-language adequate for testing whether $\mathcal{M}' \models x\tau(\varphi)y$ consists of the following sets of symbols:

- $\mathbb{OV}_{\mathsf{RL}_{\mathcal{M}',x\tau(\varphi)y}}$ a countable infinite set of object variables;
- $\mathbb{OC}_{\mathsf{RL}_{\mathcal{M}',x\tau(\varphi)y}} = \{c_{\mathsf{a}}, c_{\mathsf{b}}\}$;
- $\mathbb{RC}_{\mathsf{RL}_{\mathcal{M}',x\tau(\varphi)y}} = \{R, P, 1, 1'\}$;
- $\mathbb{OP}_{\mathsf{RL}_{\mathcal{M}',x\tau(\varphi)y}} = \{-, \cup, \cap, ;, {}^{-1}\}$;
- a set of parentheses $\{(,)\}$.

An $\mathsf{RL}_{\mathcal{M}',x\tau(\varphi)y}$-model is the structure $\mathcal{N} = (W, n)$ defined as $\mathsf{RL}_\mathsf{K}(1,1')$-models with the following additional condition $n(c_{\mathsf{a}}) = \mathsf{a}$, $n(c_{\mathsf{b}}) = \mathsf{b}$.

The additional rules of $\mathsf{RL}_{\mathcal{M}',x\tau(\varphi)y}$-system are: $(-R)$, $(-P)$, $(\mathsf{a} \neq \mathsf{b})$ and $(1')$. Additional $\mathsf{RL}_{\mathcal{M}',x\tau(\varphi)y}$-axiomatic sets are those which include one of the following sets: $\{c_{\mathsf{a}}Rc_{\mathsf{a}}\}$, $\{c_{\mathsf{b}}Rc_{\mathsf{a}}\}$, $\{c_{\mathsf{b}}-Rc_{\mathsf{b}}\}$, $\{c_{\mathsf{a}}-Rc_{\mathsf{b}}\}$, $\{c_{\mathsf{a}}Pc_{\mathsf{a}}\}$, $\{c_{\mathsf{a}}Pc_{\mathsf{b}}\}$, $\{c_{\mathsf{b}}-Pc_{\mathsf{b}}\}$ or $\{c_{\mathsf{b}}-Pc_{\mathsf{a}}\}$

By Theorem 10, truth of φ in $\mathcal{M}$ is equivalent to $\mathsf{RL}_{\mathcal{M}',x\tau(\varphi)y}$-provability of φ. Figure 3 presents a closed $\mathsf{RL}_{\mathcal{M}',x\tau(\varphi)y}$-proof tree for $x\tau(\varphi)y$.

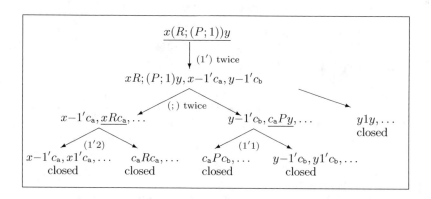

Fig. 3.

13 Applications to Verification of Satisfaction in Non-classical Logics

The logic $\mathsf{RL}_{df}(C)$ is used in the formalisms of relational logics whose problem of verification of satisfaction in a model is in question. Let $\mathcal{M} = (U, m)$ be a fixed $\mathsf{RL}^*(1,1')$-model, let a, b be elements of U and let $\varphi = xRy$ be an $\mathsf{RL}(1,1')$-formula, where R is a relational term and x, y are any object symbols. In order to obtain the relational formalism for the problem '$(a, b) \in m(R)$?', we consider an instance $\mathrm{RL}_{\mathcal{M},\varphi,\mathsf{a},\mathsf{b}}$ of the logic $\mathrm{RL}_{df}(C)$. The language, the models, and the system of $\mathrm{RL}_{\mathcal{M},\varphi,\mathsf{a},\mathsf{b}}$-logic are constructed in a similar way as in the model check-ing problem. The vocabulary of the logic $\mathrm{RL}_{\mathcal{M},\varphi,\mathsf{a},\mathsf{b}}$ is defined as in $\mathrm{RL}_{\mathcal{M},\varphi}$-logic with additional set of relational constants:

$$C = \{R_c : c \in \mathbb{OC}^0_{\mathrm{RL}_{\mathcal{M},\varphi,\mathsf{a},\mathsf{b}}}\}.$$

An $\mathrm{RL}_{\mathcal{M},\varphi,\mathsf{a},\mathsf{b}}$-models are defined as $\mathrm{RL}_{\mathcal{M},\varphi}$-models with the following addi-tional condition:

$$n(R_c) = \{n(c)\} \times W, \text{ for every } c \in \mathbb{OC}^0_{\mathrm{RL}_{\mathcal{M},\varphi,\mathsf{a},\mathsf{b}}}.$$

As in the case of $\mathrm{RL}_{\mathcal{M},\varphi}$-models, the class of $\mathrm{RL}_{\mathcal{M},\varphi,\mathsf{a},\mathsf{b}}$-models has exactly one element up to isomorphism. Therefore, $\mathrm{RL}_{\mathcal{M},\varphi,\mathsf{a},\mathsf{b}}$-validity is equivalent to the truth in a single $\mathrm{RL}_{\mathcal{M},\varphi,\mathsf{a},\mathsf{b}}$-model $\mathcal{N}$.

Proposition 31
The following statements are equivalent:

- $(\mathsf{a}, \mathsf{b}) \in m(R)$;
- $x[-(R_{c_a}; R_{c_b}^{-1}) \cup R]y$ *is* $\mathrm{RL}_{\mathcal{M},\varphi,\mathsf{a},\mathsf{b}}$-*valid.*

Proof
Note that $\mathrm{RL}_{\mathcal{M},\varphi,\mathsf{a},\mathsf{b}}$-validity of $x[-(R_{c_a}; R_{c_b}^{-1}) \cup R]y$ is equivalent to the following property: for every $x, y \in W$, if $(x, y) \in n(R_{c_a}; R_{c_b}^{-1})$, then $(x, y) \in n(R)$.

$(\rightarrow)$ Let $(\mathsf{a}, \mathsf{b}) \in m(R)$. Assume $x, y \in W$ and $(x, y) \in n(R_{c_a}; R_{c_b}^{-1})$. Then by the semantics of R_{c_a} and R_{c_b}, $x = \mathsf{a}$ and $y = \mathsf{b}$. Since $n(R) = m(R)$, $(x, y) \in n(R)$.

$(\leftarrow)$ Assume $(\mathsf{a}, \mathsf{b}) \notin m(R)$. We need to show that there exist $x, y \in W$ such that $(x, y) \in n(R_{c_a}; R_{c_b}^{-1})$ but $(x, y) \notin n(R)$. Let $x = \mathsf{a}$ and $y = \mathsf{b}$. Then $(x, y) \in n(R_{c_a}; R_{c_a}^{-1})$. Since $n(R) = m(R)$, $(x, y) \notin n(R)$. □

$\mathrm{RL}_{\mathcal{M},\varphi,\mathsf{a},\mathsf{b}}$-proof system consists of the rules and axiomatic sets of the systems of $\mathrm{RL}_{\mathcal{M},\varphi}$-logic and $\mathrm{RL}_{df}(C)$-logic adjusted to $\mathrm{RL}_{\mathcal{M},\varphi,\mathsf{a},\mathsf{b}}$-language. The complete-ness of $\mathrm{RL}_{\mathcal{M},\varphi,\mathsf{a},\mathsf{b}}$-system can be proved in a similar way as in in the case of $\mathrm{RL}_{\mathcal{M},\varphi}$-system.

Theorem 11 (Soundness and completeness of $\mathrm{RL}_{\mathcal{M},\varphi,\mathsf{a},\mathsf{b}}$)
For every $\mathrm{RL}_{\mathcal{M},\varphi,\mathsf{a},\mathsf{b}}$-formula φ the following conditions are equivalent:

- *φ is $\mathrm{RL}_{\mathcal{M},\varphi,\mathsf{a},\mathsf{b}}$-provable.*
- *φ is $\mathrm{RL}_{\mathcal{M},\varphi,\mathsf{a},\mathsf{b}}$-valid.*

Due to the above theorem and Proposition 31 we obtain the following:

Theorem 12
The following statements are equivalent:

- $(\mathsf{a}, \mathsf{b}) \in m(R)$;
- $x[-(R_{c_a}; R_{c_b}^{-1}) \cup R]y$ *is* $\mathrm{RL}_{\mathcal{M},\varphi,\mathsf{a},\mathsf{b}}$-*provable.*

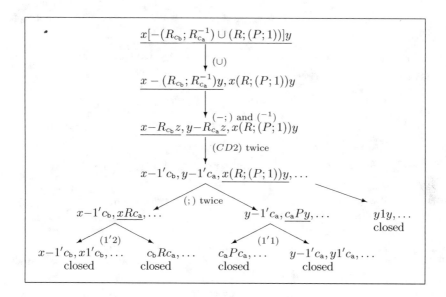

Fig. 4.

As an example of an application of the method presented above, consider the modal logic K. Let $\mathcal{M} = (U, m)$ be a K-model such that $U = \{a, b\}$, $m(p) = \{a\}$ and the accessibility relation is defined as $m(R) = \{(b, a)\}$. Let φ be the formula of the form $\langle R \rangle p$. Let us consider the problem: 'is φ satisfied in $\mathcal{M}$ by a state b?' The translation of the formula φ is $\tau(\varphi) = (R; (P; 1))$, where $\tau'(p) = P$. From the proof of Proposition 26 it follows that there exist an $\mathsf{RL_K}(1, 1')$-model $\mathcal{M}'$ and a valuation v_b such that the following holds:

$$(\star) \quad \mathcal{M}, b \models \varphi \text{ iff } \mathcal{M}', v_b \models x\tau(\varphi)y.$$

The $\mathsf{RL_K}(1, 1')$-model $\mathcal{M}' = (U', m')$ is defined as follows:

- $U' = m'(1) = \{a, b\}$;
- $m'(P) = \{(a, a), (a, b)\}$;
- $m'(R) = \{(b, a)\}$;
- $m'(1') = \{(a, a), (b, b)\}$;
- m' extends to all compound terms as in $\mathsf{RL}(1, 1')$-models.

Let v_b be a valuation such that $v_b(x) = b$ and $v_b(y) = a$. Then $\mathcal{M}'$ and v_b satisfy the condition $(\star)$.

Therefore the satisfiability problem 'is φ satisfied in $\mathcal{M}$ by a state b?' is equivalent to the problem 'is a formula $x\tau(\varphi)y$ satisfied in $\mathcal{M}'$ by v_b?'. By Theorem 12 this is equivalent to $\mathsf{RL}_{\mathcal{M}', x\tau(\varphi)y, b, a}$-provability of $x[-(R_{c_b}; R_{c_a}^{-1}) \cup \tau(\varphi)]y$.

$\mathsf{RL}_{\mathcal{M}', x\tau(\varphi)y, b, a}$-proof system contains the rules and axiomatic sets of $\mathsf{RL}_{df}(C)$-proof system adjusted to $\mathsf{RL}_{\mathcal{M}', x\tau(\varphi)y, b, a}$-language and additionally it contains:

– the rules $(-R)$, $(-P)$, $(1')$ and $(a \neq b)$ of $\mathsf{RL}_{\mathcal{M}', x\tau(\varphi)y}$-system adjusted to $\mathsf{RL}_{\mathcal{M}', x\tau(\varphi)y,b,a}$ language;

– axiomatic sets that include either of the following subsets: $\{c_b R c_a\}$, $\{c_a P c_a\}$, $\{c_a P c_b\}$, $\{c_a - R c_a\}$, $\{c_b - R c_b\}$, $\{c_a - R c_b\}$, $\{c_b - P c_b\}$, and $\{c_b - P c_a\}$.

Figure 4 presents a closed $\mathsf{RL}_{\mathcal{M}', x\tau(\varphi)y,b,a}$-proof tree for $x[-(R_{c_b}; R_{c_a}^{-1}) \cup \tau(\varphi)]y$. We recall that the rule $(CD2)$ used in that proof is presented in Section 9.

14   RRA Algebras, FRA Algebras and Relational Logics

RRA is a class of algebras isomorphic to an algebra $(\mathcal{P}(1), -, \cup, \cap, ^{-1}, ;, 1, 1')$, where 1 is an equivalence relation, $1'$ is an identity on the field of 1, $-, \cup$ and $\cap$ are Boolean operations, $^{-1}$ and ; are converse and composition of binary relations, respectively. FRA is a class of algebras isomorphic to an algebra $(\mathcal{P}(U \times U), -, \cup, \cap, ^{-1}, ;, U \times U, 1')$, where U is a non-empty set, $1'$ is an identity on U and $-, \cup, \cap, ^{-1}, ;$ are as above.

The theorem below states the connection between RRA-validity and $\mathsf{RLN}^*(1, 1')$-validity:

Theorem 13
Let $R \in \mathbb{RT}_{\mathsf{RL}(1,1')}$ and $x, y \in \mathbb{OV}_{\mathsf{RL}(1,1')}$. Then xRy is $\mathsf{RLN}^*(1, 1')$-valid iff $R = 1$ is RRA-valid.

Proof

Proof of $(\rightarrow)$ Assume xRy is $\mathsf{RLN}^*(1, 1')$-valid, that is for every $\mathsf{RLN}(1, 1')$-model $\mathcal{M} = (U, m)$, $m(1) \subseteq m(R)$. Suppose $R = 1$ is not RRA-valid. Then there exist RRA-algebra $\mathfrak{A}$ and an assignment a in $\mathfrak{A}$ such that $1^{\mathfrak{A}} \not\subseteq R^{\mathfrak{A}}(a)$, where $1^{\mathfrak{A}}$ is an equivalence relation. Consider a model $\mathcal{M}_{\mathfrak{A}} = (\text{field of } 1^{\mathfrak{A}}, m_{\mathfrak{A}})$ such that:

– $m_{\mathfrak{A}}(P) = P^{\mathfrak{A}}(a)$ for every relational variable P;
– $m_{\mathfrak{A}}(1) = 1^{\mathfrak{A}}$;
– $m_{\mathfrak{A}}(1') = 1'^{\mathfrak{A}}$;
– $m_{\mathfrak{A}}$ extends homomorphically to all compound terms as in the definition of an RL-model.

Since $\mathfrak{A}$ is an RRA algebra, so $1'^{\mathfrak{A}}$ is an equivalence relation on the field of $1^{\mathfrak{A}}$. Therefore $\mathcal{M}_{\mathfrak{A}}$ is an $\mathsf{RLN}(1, 1')$-model. Since xRy is $\mathsf{RLN}(1, 1')$-valid, hence $m_{\mathfrak{A}}(1) \subseteq m_{\mathfrak{A}}(R)$, that is $1^{\mathfrak{A}} \subseteq R^{\mathfrak{A}}(a)$, a contradiction.

Proof of $(\leftarrow)$ Assume $R = 1$ is RRA-valid. Suppose xRy is not $\mathsf{RLN}^*(1, 1')$-valid. Then there exists an $\mathsf{RLN}^*(1, 1')$-model $\mathcal{M} = (U, m)$ such that $m(1) \not\subseteq m(R)$. Consider an algebra $\mathfrak{A}_{\mathcal{M}} = (\mathcal{P}(m(1)), \cup, \cap, -, ;, ^{-1}, m(1'), m(1))$. It is easy to see that $\mathfrak{A}$ is an RRA-algebra. Let a be an assignment in $\mathfrak{A}_{\mathcal{M}}$ such that $P^{\mathfrak{A}_{\mathcal{M}}}(a) = m(P) \cap m(1)$ for every relational variable P. Since $R = 1$ is true in $\mathfrak{A}_{\mathcal{M}}$, so $R^{\mathfrak{A}_{\mathcal{M}}}(a) = 1^{\mathfrak{A}_{\mathcal{M}}} = m(1)$. Therefore $m(R) \cap m(1) = m(1)$, hence $m(1) \subseteq m(R)$, a contradiction.   □

Due to the above theorem and Theorem 5 we obtain the following:

Theorem 14

Let xRy be an RL$(1,1')$*-formula. Then* xRy *is* RL$(1,1')$*-provable iff* $R = 1$ *is* RRA*-valid.*

A non-trivial example of RLN$(1,1')$-valid equation is presented in the Appendix.

Similarly we can prove the following theorem which states the connection between FRA-validity and RL$(1,1')$-validity.

Theorem 15

Let $R \in \mathbb{RT}_{\mathsf{RL}(1,1')}$ *and* $x, y \in \mathbb{OV}_{\mathsf{RL}(1,1')}$*. Then* xRy *is* RL$^*(1,1')$*-valid iff* $R = 1$ *is* FRA*-valid.*

Due to Theorem 5 the above theorems imply the following well known result:

Theorem 16

The set of equations valid in RRA *and the set of equations valid in* FRA *are equal.*

15   Conclusion and Future Work

We presented a survey of relational logics, in particular, we discussed the logics which are the counterparts to the classes RRA and FRA and the logics which enable us reasoning both about relations and about individual elements of a domain on which the relations are defined. We extensively discussed the applications of those logics to the major logical tasks: verification of validity, verification of entailment, model checking and verification of satisfaction in a model. We explained how we can perform these tasks for non-classical logics after translating them into the appropriate relational logics.

An important open problem is to modify the proof systems presented in the paper for the relational logics RL$_\mathsf{L}$, where L is a modal logic, so that they become decision procedures. Another interesting problem is to establish bounds on the number of variables needed in the proofs of formulas of the relational logics presented in the paper.

References

1. A. Burrieza, M. Ojeda-Aciego, and E. Orłowska, *Relational approach to order of magnitude reasoning*, this volume, 2006.
2. J. Dallien and W. MacCaull, *RelDT: A relational dual tableaux automated theorem prover*, Preprint, 2005.
3. S. Demri, E. Orłowska, and I. Rewitzky, *Towards reasoning about Hoare relations*, Annals of Mathematics and Artificial Intelligence 12, 1994, 265-289.
4. S.Demri and E. Orłowska, *Logical analysis of demonic nondeterministic programs*, Theoretical Computer Science 166, 1996, 173-202.

5. A. Formisano, E. Omodeo, and E. Orłowska, *A PROLOG tool for relational translation of modal logics: A front-end for relational proof systems*, in: B. Beckert (ed) TABLEAUX 2005 Position Papers and Tutorial Descriptions, Universitt Koblenz-Landau, Fachberichte Informatik No 12, 2005, 1-10.
6. A. Formisano and M. Nicolosi Asmundo, *An efficient relational deductive system for propositional non-classical logics*, Journal of Applied Non-Classical Logics (2006), to appear.
7. A. Formisano, E. Omodeo, and E. Orłowska, *An environment for specifying properties of dyadic relations and reasoning about them. II: Relational presentation of non-classical logics*, this volume, 2006.
8. M. Frias and E. Orłowska, *A proof system for fork algebras and its applications to reasoning in logics based on intuitionism*, Logique et Analyse 150-151-152, 1995, 239-284.
9. J. Golińska-Pilarek and E. Orłowska, *Tableaux and dual Tableaux: Transformation of proofs*, Studia Logica (2006), to appear.
10. B. Konikowska, Ch. Morgan, and E. Orłowska, *A relational formalisation of arbitrary finite-valued logics*, Logic Journal of IGPL 6 No 5, 1998, 755-774.
11. W. MacCaull, *Relational proof theory for linear and other substructural logics*, Logic Journal of IGPL 5, 1997, 673-697.
12. W. MacCaull, *Relational tableaux for tree models, language models and information networks*, in: E. Orłowska (ed) Logic at Work. Essays dedicated to the memory of Helena Rasiowa, Springer-Physica Verlag, Heidelberg, 1998a.
13. W. MacCaull, *Relational semantics and a relational proof system for full Lambek Calculus*, Journal of Symbolic Logic 63, 2, 1998b, 623-637.
14. W. MacCaull and E. Orłowska, *Correspondence results for relational proof systems with applications to the Lambek calculus*, Studia Logica 71, 2002, 279-304.
15. E. Orłowska, *Relational interpretation of modal logics*, in: Andreka, H., Monk, D., and Nemeti, I. (eds) Algebraic Logic, Colloquia Mathematica Societatis Janos Bolyai 54, North Holland, Amsterdam, 1988, 443-471.
16. E. Orłowska, *Relational proof systems for relevant logics*, Journal of Symbolic Logic 57, 1992, 1425-1440.
17. E. Orłowska, *Dynamic logic with program specifications and its relational proof system*, Journal of Applied Non-Classical Logic 3, 1993, 147-171.
18. E. Orłowska, *Relational semantics for non-classical logics: Formulas are relations*, in: Woleński, J. (ed) Philosophical Logic in Poland, Kluwer, 1994, 167-186.
19. E. Orłowska, *Temporal logics in a relational framework*, in: Bolc, L. and Szałas, A. (eds) Time and Logic-a Computational Approach, University College London Press, 1995, 249-277.
20. E. Orłowska, *Relational proof systems for modal logics*, in: Wansing, H. (ed) Proof Theory of Modal Logics, Kluwer, 1996, 55-77.
21. E. Orłowska, *Relational formalisation of non-classical logics*, in: Brink, C. Kahl, W., and Schmidt, G. (eds) Relational Methods in Computer Science, Springer, Wien/New York, 1997, 90-105.
22. H. Rasiowa and R. Sikorski, *The Mathematics of Metamathematics*, Polish Scientific Publishers, Warsaw, 1963.
23. G. Schmidt and T. Ströhlein, *Relations and graphs*, EATCS Monographs on Theoretical Computer Science, Springer, Heidelberg.
24. A. Tarski, *On the calculus of relations*, The Journal of Symbolic Logic 6 (1941), 73-89.
25. A. Tarski and S. R. Givant, *A Formalization of Set Theory without Variables*, Colloquium Publications, vol. 41, American Mathematical Society, 1987.

Appendix

We present a construction of a closed $\mathsf{RL}(1,1')$ proof tree of an equation which is not valid in RA, while it is valid in RRA. It has the following form:

$$\tau = 1$$

where $\tau := (1; \rho; 1)$ and $\rho := (A \cup B \cup C \cup D \cup E)$ for:

- $A = -(1; R; 1)$;
- $B = [R \cap -[(N; N) \cap (R; N)]]$;
- $C = (N; N \cup R; R) \cap N$;
- $D = [(R \cup R^{-1} \cup 1') \cap N]$;
- $E = -(R \cup R^{-1} \cup 1' \cup N)$.

To prove validity of $\tau = 1$ we need to prove validity of the formula $u\tau w$, for $u, w \in \mathbb{OV}_{\mathsf{RL}(1,1')}$, $u \neq w$.

It is easy to show that in $\mathsf{RL}(1,1')$-proof tree for $u\tau w$, if a formula $u\tau w$ occurs in a node of this tree, then it is possible to build a subtree of $\mathsf{RL}(1,1')$-proof tree with this formula at the root which ends with exactly one non-axiomatic node containing at least one of the following formulas: zAv, zBv, zCv, zDv and zEv, for any variables z, v. Therefore, in such cases instead of building long subtrees we will use the following abbreviations which have a form of the rules:

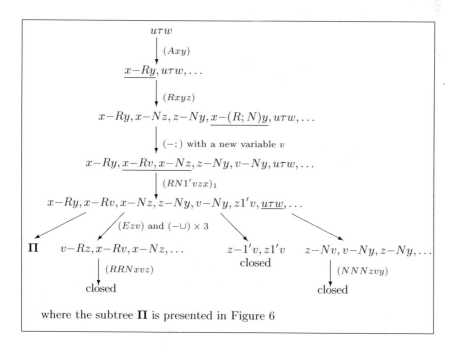

Fig. 5. $\mathsf{RL}(1,1')$-proof tree for $u\tau w$

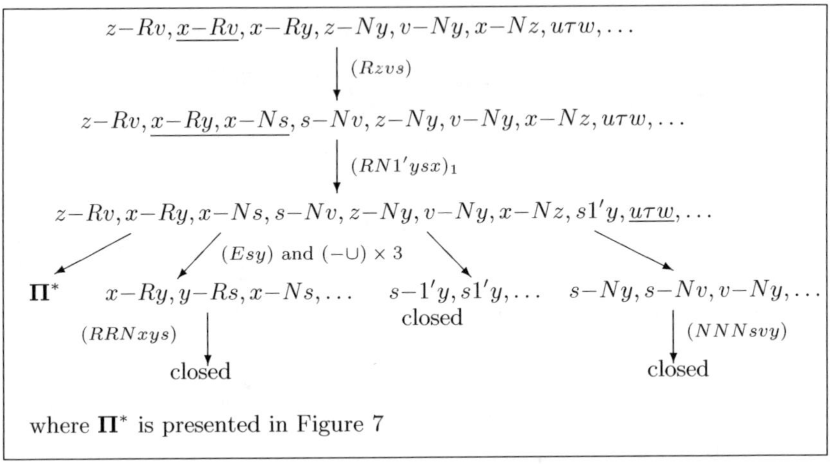

$$z-Rv, \underline{x-Rv}, x-Ry, z-Ny, v-Ny, x-Nz, u\tau w, \ldots$$

$$\downarrow (Rzvs)$$

$$z-Rv, \underline{x-Ry, x-Ns}, s-Nv, z-Ny, v-Ny, x-Nz, u\tau w, \ldots$$

$$\downarrow (RN1'ysx)_1$$

$$z-Rv, x-Ry, x-Ns, s-Nv, z-Ny, v-Ny, x-Nz, s1'y, \underline{u\tau w}, \ldots$$

(Esy) and $(-\cup) \times 3$

$\mathbf{\Pi^*}$ $x-Ry, y-Rs, x-Ns, \ldots$ $s-1'y, s1'y, \ldots$ $s-Ny, s-Nv, v-Ny, \ldots$

   $(RRNxys)$ closed $(NNNsvy)$

       closed   closed

where $\mathbf{\Pi^*}$ is presented in Figure 7

Fig. 6. The subtree $\mathbf{\Pi}$

$$(Azv) \ \frac{u\tau w}{zAv, u\tau w} \qquad (Bzv) \ \frac{u\tau w}{zBv, u\tau w} \qquad (Czv) \ \frac{u\tau w}{zCv, u\tau w}$$

$$(Dzv) \ \frac{u\tau w}{zDv, u\tau w} \qquad (Ezv) \ \frac{u\tau w}{zEv, u\tau w}$$

Similarly, we can admit the following derived rules:

$$(1'^*) \ \frac{x1'y}{y1'x} \qquad (Rxyz) \ \frac{x-Ry, u\tau w}{x-Ry, x-Nz, z-Ny, x-(R;N)y, u\tau w}$$

where z is a new variable,

$$(RN1'xyz)_1 \ \frac{z-Rx, z-Ny, u\tau w}{z-Rx, z-Ny, x1'y, u\tau w}, \ \frac{z-Rx, z-Ny, u\tau w}{z-Rx, z-Ny, y1'x, u\tau w}$$

$$(RN1'xyz)_2 \ \frac{x-Rz, y-Nz, u\tau w}{x-Rz, y-Nz, x1'y, u\tau w}, \ \frac{x-Rz, y-Nz, u\tau w}{x-Rz, y-Nz, y1'x, u\tau w}$$

$$(RRNxyz) \ \frac{x-Ry, y-Rz, x-Nz, u\tau w}{closed}$$

$$(NNNxyz) \ \frac{x-Ny, y-Nz, x-Nz, u\tau w}{closed}$$

By way of example, in Figures 8 and 9 we show how to obtain the derived rules $(Rxyz)$ and $(RN1'xyz)_1$, respectively. Similarly we may obtain the remaining derived rules. It is easy to check that the derived rule (Cxy) is needed to get $(RRNxyz)$ and $(NNNxyz)$, while (Dxy) is needed in the proofs of $(RN1'xyz)_1$ and $(RN1'xyz)_2$.

Figure 5 presents a closed RL$(1, 1')$-proof tree for $u\tau w$.

$$s-Ry, \underline{z-Rv, s-Nv}, z-Ny, v-Ny, x-Nz, u\tau w, \dots$$

$$\Big\downarrow (RN1'zsv)_2$$

$$s-Ry, z-Rv, s-Nv, s1'z, z-Ny, v-Ny, x-Nz, \underline{u\tau w}, \dots$$

(Esz) and $(-\cup) \times 3$

$s-Rz, z-Rv,$	$z-Rs, s-Ry$	$s-1'z, s1'z, \dots$	$x-Ns, s-Nz,$
$s-Nv \dots$	$z-Ny \dots$	closed	$x-Nz, \dots$

$\Big\downarrow (RRNszv)$ $\Big\downarrow (RRNzsy)$ $\Big\downarrow (NNNxsz)$

closed   closed   closed

Fig. 7. The subtree $\mathbf{\Pi}^*$

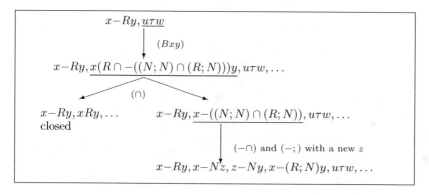

Fig. 8. A derivation of the rule $(Rxyz)$

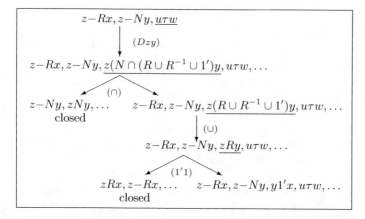

Fig. 9. A derivation of the rule $(RN1'xyz)_1$

Fuzzy Information Relations and Operators: An Algebraic Approach Based on Residuated Lattices*

Anna Maria Radzikowska[1] and Etienne E. Kerre[2]

[1] Faculty of Mathematics and Information Science, Warsaw University of Technology
Plac Politechniki 1, 00–661 Warsaw, Poland
`annrad@mini.pw.edu.pl`
[2] Department of Applied Mathematics and Computer Science, Ghent University
Krijgslaan 281 (S9), B-9000 Gent, Belgium
`Etienne.Kerre@Ugent.be`

Abstract. We discuss fuzzy generalisations of information relations taking two classes of residuated lattices as basic algebraic structures. More precisely, we consider commutative and integral residuated lattices and *extended residuated lattices* defined by enriching the signature of residuated lattices by an antitone involution corresponding to the De Morgan negation. We show that some inadequacies in representation occur when residuated lattices are taken as a basis. These inadequacies, in turn, are avoided when an extended residuated lattice constitutes the basic structure. We also define several fuzzy information operators and show characterizations of some binary fuzzy relations using these operators.

Keywords: Information relations, Information operators, Residuated lattices, Fuzzy sets, Fuzzy logical connectives.

1 Introduction

In real–life problems we usually deal with incomplete information. Generally speaking, there are two reasons for incompleteness of information. Firstly, we often have only partial data about a domain under considerations. Secondly, the acquired information, if available, is often imprecise (e.g. when expressed by means of linguistic terms like "quite good" or "rather tall"). Formal methods for representing and analyzing incomplete information have been extensively developed within the theory of rough sets ([26]). In these approaches an information relation is any relation defined on a set of objects of an information system and determined by the properties of these objects. Since properties of an object can be represented by a set of values of its attributes (properties), any information relation is formally a binary relation between two subsets of a domain in discourse. Examples of some information relations (in information systems) and

---

* This work was carried out in the framework of COST Action 274/TARSKI on *Theory and Applications of Relational Structures as Knowledge Instruments* (www.tarski.org).

H. de Swart et al. (Eds.): TARSKI II, LNAI 4342, pp. 162–184, 2006.

their theories can be found, for example, in [6], [10], [25], and [26]. A comprehensive exposition of logical and algebraic theories of information relations and their applications can be found in [7].

When imprecise information is admitted, it is clear that it cannot be adequately represented by means of standard methods based on classical two–valued structures. A natural solution seems to be fuzzy generalisations of the respective methods. Multi–valued generalisations of information relations based on residuated lattices were developed in [27].

In the present paper we continue our studies of fuzzy information relations and information operators. In [30], [31], and [33] we discussed fuzzy generalisations of information relations taking the unit interval $[0, 1]$ and traditional fuzzy logical connectives as the basis. In this framework the relationships between objects are real numbers from $[0, 1]$, so they are always comparable. In real–life problems, however, such relationships need not have this property. For instance, a child is usually similar to both parents, but it is often hard to say to which of his/her parents the child is more (or less) similar. Therefore, some lattice–based approaches seems to be more adequate.

We present some fuzzy generalisations of information relations and information operators taking two classes of residuated lattices ([4],[8],[15],[16],[22],[42]) as basic algebraic structures. Our approach is motivated by the role these algebras play in fuzzy set theory ([18],[19],[20],[23],[43]) and by the rough set–style data analysis ([26]). In a residuated lattice a product operator and its residuum are abstract counterparts of a triangular norm ([41]) and a fuzzy residual implication ([23]), respectively. However, traditional residuated lattices do not provide sufficiently general counterparts of other fuzzy logical connectives, in particular triangular conorms, fuzzy negations, and fuzzy S–implications. Consequently, in generalisations of information relations some inadequacies occur. From this reason, *double residuated lattices* were introduced ([28],[29]) and some fuzzy information relations and operators were investigated. In the signature of these algebras there are two independent operations corresponding to a triangular norm and a triangular conorm. Yet these structures do not give us the algebraic counterpart of the De Morgan negation. Therefore, while some inadequacies are avoided, other drawbacks in representation still remain. To cope with these problems, we propose yet another class of residuated lattices, called *extended residuated lattices* ([12]), which are an extension of residuated lattices by an antitone involution. This operation, together with the operations of residuated lattices, allows us to define algebraic counterparts of the main classes of fuzzy logical connectives. Basing on these algebras, we extend the results obtained in [36] and discuss another generalisation of information relations. We show how these representations allow us to avoid inadequacies occurring when residuated lattices are taken as a basis.

It is well–known that binary relations determine modal–like operators which, in turn, are the abstract counterparts of the information operators derived from information systems ([7]). Generally speaking, an information operator is any mapping defined on binary relations on a non–empty universe and subsets of this universe. A general theory of the classical abstract information operators was developed in [7], [10], and [11]. A fuzzy generalisation of some information operators, based on the interval $[0,1]$, were presented in [32], [33], and [38]. In [34], [35], [36], and [37] fuzzy approximation operators based on residuated lattices were discussed.

In this paper we propose a generalisation of information operators determined by information relations based on extended residuated lattices. This approach might be a basis for developing multi–valued logics and algebras. On the other hand, this is a generalisation of approximation operators, which are the main tools in rough set–style data analysis. We show that properties of main classes of fuzzy information relations can be expressed by means of these operators.

The paper is organized as follows. In Section 2 we provide some algebraic foundations to our discussion. In particular, the notions of residuated lattices and extended residuated lattices will be presented. Also, the notion of fuzzy sets and fuzzy relations will be recalled. In Section 3 we define several fuzzy information relations taking a commutative and integral residuated lattice as a basic structure. Main properties of these relations will be presented. We will point out some drawbacks of this representation and propose another generalisation of some information relations, where extended residuated lattices are taken as basic structures. Next, in Section 4, we discuss some fuzzy information operators. It will be shown that these operators are useful for characterisations of fuzzy binary relations. Some concluding remarks will complete the paper.

2   Algebraic Foundations

2.1   Residuated Lattices

A *monoid* is a system $(M, \circ, \varepsilon)$, where M is a non–empty set, $\circ$ is an associative operation in M, and $\varepsilon \in M$ is such that $\varepsilon \circ a = a \circ \varepsilon = a$ for every $a \in M$. A monoid $(M, \circ, \varepsilon)$ is called *commutative* iff $\circ$ is commutative.

Typical examples of monoid operations are *triangular norms* (t–norms) and *triangular conorms* (t–conorms). Recall ([41]) that a triangular norm t (resp. a triangular conorm s) is a $[0,1]^2 - [0,1]$ mapping, non-decreasing in both arguments, associative, commutative, and satisfying for every $a \in [0,1]$ the boundary condition $t(a, 1) = a$ (resp. $s(0, a) = a$). The well–known t–norms and t–conorms, t_Z and s_Z (the Zadeh's t–norm and the t–conorm), t_P and s_P (the algebraic product and the bounded sum), and t_L and s_L (the Łukasiewicz t–norm and the Łukasiewicz t–conorm), are given in Table 1.

Table 1. Well–known t–norms and t–conorms

$t_Z(a,b) = \min(a,b)$	$s_Z(a,b) = \max(a,b)$
$t_P(a,b) = a \cdot b$	$s_P(a,b) = a+b-a \cdot b$
$t_L(a,b) = \max(0, a+b-1)$	$s_L(a,b) = \min(1, a+b)$

Let $(L, \leqslant)$ be a poset and let $\circ$ be a binary operation in L. Define two binary operations in L, $\rightarrow_r$, $\rightarrow_l$, satisfying the *residuation conditions* for all $a, b, c \in L$,

$$a \circ b \leqslant c \quad iff \quad a \leqslant b \rightarrow_l c \tag{1}$$
$$a \circ b \leqslant c \quad iff \quad b \leqslant a \rightarrow_r c \tag{2}$$

The operations (1) and (2) are called the *left residuum of* $\circ$ and the *right residuum of* $\circ$, respectively. It can be easily shown that if the respective residua exist, then

$$a \rightarrow_l b = \sup\{c \in L : c \circ a \leqslant b\}$$
$$a \rightarrow_r b = \sup\{c \in L : a \circ c \leqslant b\}.$$

Clearly, if $\circ$ is commutative, then $\rightarrow_l = \rightarrow_r$.

Residua of left–continuous[1] t–norms are called *fuzzy residual implications* ([23]). Three well–known residual implications, $\rightarrow_Z$, $\rightarrow_P$, and $\rightarrow_L$, determined by t_Z, t_P and t_L, respectively, are given in Table 2.

Table 2. Well–known residual implications

Gödel implication	$i_Z(a, b) = 1$ iff $a \leqslant b$ and $i_Z(a, b) = b$ otherwise
Gaines implication	$i_P(a, b) = 1$ iff $a \leqslant b$ and $i_P(a, b) = \frac{b}{a}$ otherwise
Łukasiewicz implication	$i_L(a, b) = \min(1, 1-a+b)$

Definition 1. *A **residuated lattice** is an algebra* $(L, \wedge, \vee, \otimes, \rightarrow_l, \rightarrow_r, 0, 1, 1')$ *such that*

(i) $(L, \wedge, \vee, 0, 1)$ *is a bounded lattice with the least element 0 and the greatest element 1,*

(ii) $(L, \otimes, 1')$ *is a monoid, and*

(iii) $\rightarrow_l$ *and* $\rightarrow_r$ *are the left and the right residuum of* $\otimes$, *respectively.*

The operation $\otimes$ *of a residuated lattice L is called its product.* □

We say that a residuated lattice $(L, \wedge, \vee, \otimes, \rightarrow_l, \rightarrow_r, 0, 1, 1')$ is

- *integral* iff $1' = 1$,
- *commutative* iff $\otimes$ is commutative,
- *complete* iff the underlying lattice $(L, \wedge, \vee, 0, 1)$ is complete.

Remark 1. Some researchers (in particular, fuzzy logicians) assume that residuated lattices are commutative by definition (e.g. [3],[19]). Others, however, consider these structures in a more general framework and assume that the product operation of residuated lattices need not be commutative (see, for example, [4] and [22]). □

Throughout this paper we consider only integral and commutative residuated lattices, which will be referred to as *R–lattices* and written simply $(L, \wedge, \vee, \otimes, \rightarrow, 0, 1)$.

---

[1] A t–norm is *called left–continuous* iff it has left–continuous partial mappings.

Given an R–lattice $(L, \wedge, \vee, \otimes, \to, 0, 1)$, we define the following *precomplement* operation for every $a \subset L$:

$$\neg a = a \to 0. \tag{3}$$

Note that this operation is a generalisation of the pseudo–complement in a lattice ([39]). If $\wedge = \otimes$, then $\to$ is the relative pseudo–complement, $\neg$ is the pseudo–complement and $(L, \wedge, \vee, \to, \neg, 0, 1)$ is a Heyting algebra.

Example 1. Let t be a left–continuous t–norm and let i_t be the fuzzy residual implication based on t. Put $\mathcal{L} = [0, 1]$. Then the algebra $(\mathcal{L}, \min, \max, t, i_t, 0, 1)$ is an R–lattice. □

The following lemma will be useful later.

Lemma 1. *Let* $(L, \wedge, \vee, \otimes, \to, 0, 1)$ *be an R–lattice such that its product* $\otimes$ *satisfies the following condition: for all* $a, b \in L$,

$$a \neq 0 \ \& \ b \neq 0 \implies a \otimes b \neq 0. \tag{4}$$

Then for every $a \in L$, $\neg a = 0$ *iff* $a \neq 0$ *and* $\neg a = 1$ *iff* $a = 0$.

Proof. Analogous to the proof presented in [5]. ∎

Following the terminology from fuzzy set theory, we say that the product $\otimes$ satisfying (4) *has no zero divisors*. Notice that among t–norms given in Table 1, the Zadeh's t–norm t_Z and the algebraic product t_P have this property, while the Łukasiewicz t–norm t_L does not. The family of all R–lattices, which product satisfy (4), will be denoted by RL^+.

For the recent results on residuated lattices we refer to [2], [4], [21], and [22].

Given an R–lattice $(L, \wedge, \vee, \otimes, \to, 0, 1)$, its product $\otimes$ is an algebraic counterpart of a left–continuous t–norm, the residuum $\to$ of $\otimes$ corresponds to a fuzzy residual implication determined by $\otimes$, and the precomplement $\neg$ corresponds to a fuzzy negation.[2] However, in general $\neg$ is not involutive. Moreover, the signature of R–lattices do not give algebraic counterparts of t–conorms. From this reason *double residuated lattices* were proposed (see [28],[29]).

First, let us recall the following notions. Given a poset $(L, \leqslant)$, and a binary operation $\circ$ in L, the following binary operations in L, $\leftarrow_l$ and $\leftarrow_r$, respectively called the *dual left residuum of* $\circ$ and the *dual right residuum of* $\circ$, are defined as follows: for all $a, b \in L$,

$$c \leqslant a \circ b \quad \text{iff} \quad c \leftarrow_l b \leqslant a \tag{5}$$
$$c \leqslant a \circ b \quad \text{iff} \quad c \leftarrow_r a \leqslant b. \tag{6}$$

If the respective dual residua of $\circ$ exist, then

$$a \leftarrow_l b = \inf\{c \in L : a \leqslant c \circ b\}$$
$$a \leftarrow_r b = \inf\{c \in L : a \leqslant b \circ c\}.$$

---

[2] A fuzzy negation ([23]) is a non–increasing mapping $n : [0, 1] \to [0, 1]$ satisfying $n(0) = 1$ and $n(1) = 0$.

Dual residua of a lattice join were studied by Rauszer ([40]) in the context of Heyting–Brouwer logic. In [1] dual residua of a monoid operator are discussed. The dual residua of the most famous t–conorms are presented in Table 3.

Table 3. The dual residua of well–known t–conorms

$a \leftarrow_Z b = 0$ iff $b \leqslant a$ and $a \leftarrow_Z b = b$ otherwise
$a \leftarrow_P b = 0$ iff $b \leqslant a$ and $a \leftarrow_P b = \frac{b-a}{1-a}$ otherwise
$a \leftarrow_L b = \max(0, b-a)$

Definition 2. ([28],[29]) *A **double residuated lattice** is an algebra* $(L, \wedge, \vee, \otimes,$ $\oplus, \rightarrow_l, \rightarrow_r, \leftarrow_l, \leftarrow_r, 0, 1, 0', 1')$ *such that* $(L, \wedge, \vee, \otimes, \rightarrow_l, \rightarrow_r, 0, 1, 1')$ *is a residuated lattice,* $(L, \oplus, 0')$ *is a monoid, and* $\leftarrow_l$ *and* $\leftarrow_r$ *are respectively the dual left and the dual right residuum of* $\oplus$. $\square$

A double residuated lattice is called *commutative* (resp. *integral*) iff $\otimes$ and $\oplus$ are commutative (resp. $1' = 1$ and $0' = 0$). Commutative and integral double residuated lattices will be written $(L. \wedge, \vee, \otimes, \oplus, \rightarrow, \leftarrow, 0, 1)$.

Given a commutative and integral double residuated lattice, define the *dual precomplement* operation as

$$\ulcorner a = 1 \leftarrow a \quad \text{for every } a \in L. \tag{7}$$

This operation is a generalisation of a dual pseudo–complement ([39]). The dual pseudo–complement is one of the operations in double Stone algebras. However, it is a primitive operation there, residuation operations are not in the signature of Stone algebras.

Let $\mathcal{L} = (L, \wedge, \vee, 0, 1)$ be a bounded lattice with its ordering $\leqslant$. We write $\mathcal{L}^{-1}$ to denote the lattice obtained from $\mathcal{L}$ by reversing its ordering, i.e. the lattice with the ordering $\leqslant^{-1} = \geqslant$. Then the join $\vee^{-1}$ (resp. the meet $\wedge^{-1}$) of $\mathcal{L}^{-1}$ is the meet $\wedge$ (resp. the join $\vee$) of $\mathcal{L}$ and the greatest (resp. the least) element of $\mathcal{L}^{-1}$ is the least (resp. the greatest) element of $\mathcal{L}$. In other words, $\mathcal{L}^{-1} = (L, \vee, \wedge, 1, 0)$.

Proposition 1. [29] *Let* $(L, \wedge, \vee, \otimes, \oplus, \rightarrow_l, \rightarrow_r, \leftarrow_l, \leftarrow_r, 0, 1, 0', 1')$ *be a double residuated lattice. Then the algebras* $(L, \wedge, \vee, \otimes, \rightarrow_l, \rightarrow_r, 0, 1, 1')$ *and* $(L, \vee, \wedge, \oplus,$ $\leftarrow_l, \leftarrow_r, 1, 0, 0')$ *are residuated lattices.* ∎

In view of the above proposition it is easily observed that in double residuated lattices the analogon of Lemma 1 holds. Namely, if $\oplus$ satisfies the condition

$$a \neq 1 \ \& \ b \neq 1 \Longrightarrow a \oplus b \neq 1 \quad \text{for all } a, b \in L,$$

then $\ulcorner a = 1$ iff $a \neq 1$ and $\ulcorner a = 0$ iff $a = 1$. This means that $\ulcorner$ can be reduced to the binary case.

Observe that the signature of a commutative and integral double residuated lattice $(L, \wedge, \vee, \otimes, \oplus, \rightarrow, \leftarrow, 0, 1)$ gives two independent algebraic counterparts of a t–norm ($\otimes$) and a t–conorm ($\oplus$). Also, the residuum $\to$ of $\otimes$ corresponds to a residual implication (determined by $\otimes$). However, we do not have algebraic counterpart of the second main class of fuzzy implications called *S–implications*. Recall ([23]) that an S–implication, determined by a t–conorm s and a fuzzy negation n, is defined by: $i_{s,n}(a,b) = s(n(a),b)$ for all $a,b \in [0,1]$ (the most famous S–implications, based respectively on s_Z and n, s_P and n, and s_L and n, where n is the standard fuzzy negation $n(a)=1-a$ for $a \in [0,1]$, are given in Table 4). Moreover, neither the precomplement $\neg$ nor the dual precomplement $\llcorner$ are sufficiently general counterparts of fuzzy negations, since (under some conditions) can be reduced to the binary case. Therefore, in general case we cannot obtain the counterpart of the De Morgan negation. Having this on mind, the *extended residuated lattices* were defined ([12],[34],[36]).

Table 4. Well–known S–implications

Kleene–Dienes implication	$i_{s_Z,n}(a,b) = \max(1-a,n)$
Reichenbach implication	$i_{s_P,n}(a,b) = 1-a+a \cdot b$
Łukasiewicz implication	$i_{s_L,n}(a,b) = \min(1,1-a+b)$

Definition 3. *By an **extended residuated lattice** we mean a system $(L, \wedge, \vee, \otimes, \to_l, \to_r, \sim, 0, 1, 1')$ such that*

    (i) *$(L, \wedge, \vee, \otimes, \to_l, \to_r, 0, 1, 1')$ is a residuated lattice*
    (ii) *$\sim$ is an antitone involution satisfying $\sim0=1$ and $\sim1=0$.* □

Analogously, an extended residuated lattice is integral (resp. commutative) iff the underlying residuated lattice is integral (resp. commutative). Any integral and commutative extended residuated lattice will be referred to as an *ER–lattice* and written $(L, \wedge, \vee, \otimes, \to, \sim, 0, 1)$.

Let $(L, \wedge, \vee, \otimes, \to, \sim, 0, 1)$ be an ER–lattice. Let us define the following operations in L: for all $a,b \in L$,

$$a \oplus b = \sim(\sim a \otimes \sim b) \tag{8}$$
$$a \Rightarrow b = \sim a \oplus b \tag{9}$$
$$a \leftarrow b = \sim(\sim a \to \sim b) \tag{10}$$
$$a \Leftarrow b = \sim(\sim a \Rightarrow \sim b). \tag{11}$$

Remark 2. Assume that $\sim$ and $\to$ are respectively the classical negation and implication. From the definition (10) it follows that $a \leftarrow b = \sim(b \to a)$, so $a \leftarrow b$ is a generalisation of the classical conjunction $b \wedge \sim a$. The operation (11) has the similar interpretation. □

By straightforward verification one can easily check the following

Proposition 2. *Let* $(L, \wedge, \vee, \otimes, \rightarrow, \sim, 0, 1)$ *be an ER–lattice. Then*

(i) $(L, \oplus, 0)$ *is a commutative monoid*
(ii) $\leftarrow$ *is the dual residuum of* $\oplus$
(iii) $(L, \wedge, \vee, \otimes, \rightarrow, 0, 1)$ *and* $(L, \vee, \wedge, \oplus, \leftarrow, 1, 0)$ *are R–lattices*
(iv) $(L, \wedge, \vee, \otimes, \oplus, \rightarrow, \leftarrow, 0, 1)$ *is a commutative and integral double residuated lattice.* ∎

Given an ER–lattice $(L, \wedge, \vee, \otimes, \rightarrow, \sim, 0, 1)$, its product $\otimes$ and its sum $\oplus$ are algebraic counterparts of a triangular norm and a triangular conorm. Also, $\rightarrow$ and $\Rightarrow$ correspond to a fuzzy residual implication and an S–implication, respectively. Finally, $\sim$ corresponds to the De Morgan negation. Therefore, ER–lattices allow us to get algebraic counterparts of all main classes of fuzzy logical connectives.

Main properties of ER–lattices are given in the following two lemmas.

Lemma 2. *For every ER–lattice* $(L, \wedge, \vee, \otimes, \rightarrow, \sim, 0, 1)$ *and for all* $a, b, c \in L$, *the following properties hold:*

(i) $a \leqslant b$ *implies*
$$a \otimes c \leqslant b \otimes c$$
$$b \rightarrow c \leqslant a \rightarrow c$$
$$c \rightarrow a \leqslant c \rightarrow b$$
$$b \Rightarrow c \leqslant a \Rightarrow c$$
$$c \Rightarrow a \leqslant c \Rightarrow b$$
$$\neg b \leqslant \neg a$$

(ii) $a \otimes b \leqslant a$
(iii) $a \otimes b \leqslant a \wedge b$
(iv) $a \otimes 0 = 0$
(v) $a \leqslant b$ *iff* $a \rightarrow b = 1$
(vi) $1 \rightarrow a = 1 \Rightarrow a = a$
(vii) $a \otimes (a \rightarrow b) \leqslant b$
(viii) $a \otimes (b \rightarrow c) \leqslant b \rightarrow (a \otimes c)$
(ix) $(a \rightarrow b) \otimes (b \rightarrow c) \leqslant (a \rightarrow c)$
(x) $(a \Rightarrow c) \leqslant (a \Rightarrow b) \oplus (b \Rightarrow c)$
(xi) $(a \rightarrow b) \leqslant (c \rightarrow a) \rightarrow (c \rightarrow b)$
(xii) $(a \rightarrow b) \leqslant (a \otimes c) \rightarrow (b \otimes c)$
(xiii) $b \leqslant a \rightarrow (a \otimes b)$
(xiv) $a \rightarrow (b \rightarrow c) = (a \otimes b) \rightarrow c$
(xv) $a \Rightarrow (b \Rightarrow c) = (a \otimes b) \Rightarrow c$
(xvi) $a \rightarrow \neg b = \neg(a \otimes b)$
(xvii) $a \Rightarrow \sim b = \sim(a \otimes b)$
(xviii) $a \rightarrow b \leqslant \neg b \rightarrow \neg a$
(xix) $a \Rightarrow b = \sim b \Rightarrow \sim a$
(xx) $a \otimes \neg b \leqslant \neg(a \rightarrow b)$
(xxi) $a \leqslant \neg\neg a$

(i') $a \leqslant b$ *implies*
$$a \oplus c \leqslant b \oplus c$$
$$b \leftarrow c \leqslant a \rightarrow c$$
$$c \leftarrow a \leqslant c \rightarrow b$$
$$b \Leftarrow c \leqslant a \Rightarrow c$$
$$c \Leftarrow a \leqslant c \Rightarrow b$$
$$\ulcorner b \leqslant \ulcorner a$$

(ii') $a \leqslant a \oplus b$
(iii') $a \vee b \leqslant a \oplus b$
(iv') $a \oplus 1 = 1$
(v') $b \leftarrow a = 0$ *iff* $a \leqslant b$
(vi') $0 \leftarrow a = 0 \Leftarrow a = a$
(vii') $b \leqslant a \oplus (a \leftarrow b)$
(viii') $b \leftarrow (a \oplus c) \leqslant a \oplus (b \leftarrow c)$
(ix') $(a \leftarrow c) \leqslant (a \leftarrow b) \oplus (b \leftarrow c)$
(x') $(a \Leftarrow b) \otimes (b \Leftarrow c) \leqslant (a \Leftarrow c)$
(xi') $(c \leftarrow a) \leftarrow (c \leftarrow b) \leqslant (a \leftarrow b)$
(xii') $(a \oplus c) \leftarrow (b \oplus c) \leqslant (a \leftarrow b)$
(xiii') $a \leftarrow (a \oplus b) \leqslant b$
(xiv') $a \leftarrow (b \leftarrow c) = (a \oplus b) \leftarrow c$
(xv') $a \Leftarrow (b \Leftarrow c) = (a \oplus b) \Leftarrow c$
(xvi') $a \leftarrow \ulcorner b = \ulcorner(a \oplus b)$
(xvii') $a \Leftarrow \sim b = \sim (a \oplus b)$
(xviii') $\ulcorner b \leftarrow \ulcorner a \leqslant a \leftarrow b$
(xix') $a \Leftarrow b = \sim b \Leftarrow \sim a$
(xx') $a \otimes \sim b \leqslant \sim(a \Rightarrow b)$
(xxi') $\ulcorner\ulcorner a \leqslant a$.

Proof. Note that the properties in the right column can be easily obtained from the properties in the left column by the definitions (7)–(11). Moreover, all properties, where the operations $\otimes$, $\rightarrow$, and $\neg$ occur, are well–known properties of residuated lattices (see, e.g., [19],[22],[42]). Then it remains to show **(x)**, **(xv)**, **(xvii)**, and **(xix)**. By way of example we show **(x)** and **(xv)**.

(x) By **(ii')**, for all $a, b, c \in L$, $\sim a \oplus b \geqslant \sim a$ and $\sim b \oplus c \geqslant c$. Then, by the definition (9) and the property **(i')**, $(a \Rightarrow b) \oplus (b \Rightarrow c) \geqslant (\sim a \oplus c) = (a \Rightarrow c)$.

(xv) For all $a, b, c \in L$, it holds:

$$
\begin{aligned}
a \Rightarrow (b \Rightarrow c) \\
= \sim a \oplus (\sim b \oplus c) && \text{by the definition (9)} \\
= (\sim a \oplus \sim b) \oplus c && \text{by associativity of } \oplus \\
= \sim (a \otimes b) \oplus c && \text{by the definition (8)} \\
= (a \otimes b) \Rightarrow c. && \blacksquare
\end{aligned}
$$

Lemma 3. *For every ER–lattice* $(L, \wedge, \vee, \otimes, \rightarrow, \sim, 0, 1)$, *for every* $a \in L$, *and for all families* $(b_i)_{i \in I}$ *and* $(c_i)_{i \in I}$ *of elements of* L, *if the respective infima and suprema exist, then the following properties hold:*

(i) $a \otimes \sup_{i \in I} c_i = \sup_{i \in I}(a \otimes c_i)$ **(i')** $a \oplus (\inf_{i \in I} c_i) = \inf_{i \in I}(a \oplus c_i)$

(ii) $a \rightarrow \inf_{i \in I} c_i = \inf_{i \in I}(a \rightarrow c_i)$ **(ii')** $a \leftarrow (\sup_{i \in I} c_i) = \sup_{i \in I}(a \leftarrow c_i)$

(iii) $a \Rightarrow (\inf_{i \in I} c_i) = \inf_{i \in I}(a \Rightarrow c_i)$ **(iii')** $a \Leftarrow (\sup_{i \in I} c_i) = \sup_{i \in I}(a \Leftarrow c_i)$

(iv) $(\sup_{i \in I} c_i) \rightarrow a = \inf_{i \in I}(c_i \rightarrow a)$ **(iv')** $(\inf_{i \in I} c_i) \leftarrow a = \sup_{i \in I}(c_i \leftarrow a)$

(v) $(\sup_{i \in I} c_i) \Rightarrow a = \inf_{i \in I}(c_i \Rightarrow a)$ **(v')** $(\inf_{i \in I} c_i) \Leftarrow a = \sup_{i \in I}(c_i \Leftarrow a)$

(vi) $\sup_{i \in I} c_i \leqslant \neg \inf_{i \in I} \neg c_i$ **(vi')** $\sup_{i \in I} c_i = \sim \inf_{i \in I} \sim c_i$

(vii) $(\inf_{i \in I} b_i) \otimes (\inf_{i \in I} c_i)$ **(vii')** $\sup_{i \in I}(b_i \oplus c_i)$
$\leqslant \inf_{i \in I}(b_i \otimes c_i)$ $\leqslant (\sup_{i \in I} b_i) \oplus (\sup_{i \in I} c_i)$.

(viii) $\inf_{i \in I} \neg c_i = \neg \sup_{i \in I} c_i$

(ix) $\sup_{i \in I} \neg b_i \leqslant \neg (\inf_{i \in I} b_i)$.

Proof. As in Lemma 2, the properties in the right column are easily obtained from the respective properties in the left column using the definitions (7)–(11). Notice that all properties except from **(iii)** and **(v)** are known properties of residuated lattices.

(iii) By the definition of $\Rightarrow$ and **(i')**, we easily get $a \Rightarrow (\inf_{i \in I} c_i) = \sim a \oplus \inf_{i \in I} c_i = \inf_{i \in I}(\sim a \oplus c_i) = \inf_{i \in I}(a \Rightarrow c_i)$.

(v) can be proved in the analogous way. $\blacksquare$

Example 2. Let $\mathcal{L} = [0, 1]$ and let $(\mathcal{L}, \min, \max, t, i_t, 0, 1)$ be the R–lattice as in Example 1. Also, let n be the standard fuzzy negation $n(a) = 1 - a$ for every $a \in [0, 1]$. Then $(\mathcal{L}, \min, \max, t, i_t, n, 0, 1)$ is an ER–lattice. $\square$

Remark 3. Note that properties **(xviii)** and **(xix)** of Lemma 2 correspond to the contraposition law. In general, however, we do not have analogous links between $a \rightarrow b$ and $\sim a \rightarrow \sim b$. For example, consider the ER–lattice as in Example 2 and let $\rightarrow$ and $\sim$ be the Gödel implication (see Table 2) and the standard fuzzy negation.

Then for $a = 0.8$ and $b = 0.4$ we have: $a \to b = 0.4$ and $\sim b \to \sim a = 0.2$. Hence $a \to b > \sim b \to \sim a$. Taking $c = 0.1$, we easily get: $b \to c = 0.1$ and $\sim c \to \sim b = 0.6$, so $b \to c < \sim c \to \sim b$. $\square$

2.2 L–fuzzy Sets and L–fuzzy Relations

Fuzzy sets. Let L be a residuated lattice (in particular, R–lattice or ER–lattice) and let X be a non–empty domain. By an L–*fuzzy set in* X we mean any mapping $F : X \to L$. For every $x \in X$, $F(x)$ is the degree of membership of x to F. Two specific L–fuzzy sets in X, $\emptyset$ and X, are respectively defined by: $\emptyset(x) = 0$ and $X(x) = 1$ for every $x \in X$. The family of all L–fuzzy sets in X will be denoted by $\mathcal{F}_L(X)$.

Recall the basic operations on L–fuzzy sets. First, let L be an R–lattice. For all $A, B \in \mathcal{F}_L(X)$ and for every $x \in X$,

$$(A \sqcup_L B)(x) = A(x) \vee B(x)$$
$$(A \sqcap_L B)(x) = A(x) \wedge B(x)$$
$$(A \cap_L B)(x) = A(x) \otimes B(x)$$
$$(\neg_L A)(x) = \neg A(x).$$

If L is an ER–lattice, we additionally define:

$$(A \cup_L B)(x) = A(x) \oplus B(x)$$
$$(\sim_L A)(x) = \sim A(x)$$
$$({}_{\overline{L}}A)(x) = {}^\frown A(x).$$

For $A \in \mathcal{F}_L(X)$, we will write $A \gg \emptyset$ iff $A(x) \neq 0$ for every $x \in X$. Also, for two L–fuzzy sets $A, B \in \mathcal{F}_L(X)$, we will write $A \subseteq_L B$ iff $A(x) \leqslant B(x)$ for every $x \in X$ (Zadeh's inclusion). If L is complete, then for any indexed family $(A_i)_{i \in I}$ of L–fuzzy sets in X, $\bigcup_{i \in I} A_i$ and $\bigcap_{i \in I} A_i$ are L–fuzzy sets in X defined as: for every $x \in X$, $(\bigcup_{i \in I} A_i)(x) = \sup_{i \in I} A_i(x)$ and $(\bigcap_{i \in I} A_i)(x) = \inf_{i \in I} A_i(x)$.

Fuzzy relations. An L–*fuzzy relation on* X is a mapping $R : X \times X \to L$. The family of all L–fuzzy relations on X will be denoted by $\mathcal{R}_L(X)$.

An L–fuzzy relation $R \in \mathcal{R}_L(X)$ is called

- *reflexive* iff $R(x, x) = 1$ for every $x \in X$
- *irreflexive* iff $R(x, x) = 0$ for every $x \in X$
- *symmetric* iff $R(x, y) = R(y, x)$ for all $x, y \in X$
- $\otimes$–*transitive* iff $R(x, y) \otimes R(y, z) \leqslant R(x, z)$ for all $x, y, z \in X$
- $\oplus$–*cotransitive* iff $R(x, y) \oplus R(y, z) \geqslant R(x, z)$ for all $x, y, z \in X$
- $\otimes$–*quasi ordering* iff it is reflexive and $\otimes$–transitive
- $\otimes$–*equivalence* iff it is reflexive, symmetric, and $\otimes$–transitive
- *crisp* iff $R(x, y) \in \{0, 1\}$ for all $x, y \in X$.

Note that if R is crisp and $\oplus = \vee$, then cotransitivity of R means that the complement of R is transitive. Hence, $\oplus$–cotransitivity is a fuzzy generalisation of this property.

3   Fuzzy Information Relations

In this section we define several fuzzy information relations measuring degrees of relationship between two fuzzy sets. We take two classes of residuated lattices as an algebraic basis: complete R–lattices and complete ER–lattices.

Let L be a complete R–lattice or an ER–lattice and let $X \neq \emptyset$. By an *L–fuzzy information relation* we mean any L–fuzzy relation on $\mathcal{F}_L(X)$.

3.1   Fuzzy Information Relations Based on R–Lattices

Let us define several L–information relations.

Definition 4. *Let $(L, \wedge, \vee, \otimes, \rightarrow, 0, 1)$ be a complete R–lattice and let $X \neq \emptyset$. Define the following L–fuzzy information relations: for all $A, B \in \mathcal{F}_L(X)$,*

- **(i)** *L–fuzzy inclusion:*
$$inc_L(A, B) = \inf_{x \in X}(A(x) \rightarrow B(x))$$

- **(ii)** *L–fuzzy noninclusion:*
$$ninc_L(A, B) = \sup_{x \in X}(A(x) \otimes \neg B(x))$$

- **(iii)** *L–fuzzy compatibility:*
$$com_L(A, B) = \sup_{x \in X}(A(x) \otimes B(x))$$

- **(iv)** *L–fuzzy orthogonality:*
$$ort_L(A, B) = inc_L(A, \neg_L B)$$

- **(v)** *L–fuzzy exhaustiveness:*
$$exh_L(A, B) = \inf_{x \in X}(A(x) \vee B(x))$$

- **(vi)** *L–fuzzy nonexhaustiveness:*
$$nexh_L(A, B) = com_L(\neg_L A, \neg_L B)$$

- **(vii)** *L–fuzzy indiscernibility:*
$$ind_L(A, B) = inc_L(A, B) \otimes inc_L(B, A)$$

- **(viii)** *L–fuzzy diversity:*
$$div_L(A, B) = ninc_L(A, B) \vee ninc_L(\neg_L A, B). \qquad \square$$

For two L–fuzzy sets $A, B \in \mathcal{F}_L(X)$, $inc_L(A, B)$ (resp. $ninc_L(A, B)$) is the degree, to which A is included (resp. not included) in B. Note that the formula for $ninc_L$ is the straightforward generalisation of the classical equivalence: $A \not\subseteq B \Leftrightarrow (\exists x \in X)(x \in A \And x \notin B)$. Next, $com_L(A, B)$ (resp. $ort_L(A, B)$) represents the degree, to which A and B overlap (resp. are disjoint). The formulation for ort_L results from the generalisation of the classical equivalence: $A \cap B = \emptyset \Leftrightarrow A \subseteq -B$, where $-B = X \setminus B$. Furthermore, $exh_L(A, B)$ (resp. $nexh_L(A, B)$) is the degree, · to which A and B cover (resp. do not cover) the whole domain X. Note that in the classical case, $A \cup B \neq X \Leftrightarrow (-A \cap -B \neq \emptyset)$. This equivalence underlies the formulation for $nexh_L$. Finally, $ind_L(A, B)$ (resp. $div_L(A, B)$) is the degree, to which A is equal to B (resp. A differs from B). The formulation for div_L is again a generalisation of the classical equivalence: $A \neq B \Leftrightarrow (A \cap -B \neq \emptyset) \vee (-A \cap B \neq \emptyset)$.

The following proposition provides main properties of these relations.

Proposition 3. *Let $(L, \wedge, \vee, \otimes, \rightarrow 0, 1)$ be a complete R–lattice. Then*

(i) *inc_L is an L–quasi ordering*

(ii.1) *$ninc_L$ is irreflexive*

(ii.2) *if $L \in RL^+$, then for any $A \in \mathcal{F}_L(X)$ and for any $B \in \mathcal{F}_L(X)$ satisfying $B \gg \emptyset$, $ninc_L(A, B) = 0$*

(iii) *com_L and exh_L are symmetric*

(iv.1) *ort_L is symmetric*

(iv.2) *if $L \in RL^+$, then ort_L is crisp*

(v.1) *$nexh_L$ is symmetric;*

(v.2) *if $L \in RL^+$, then for all $A, B \in \mathcal{F}_L(X)$ such that $A(x) \neq 0$ or $B(x) \neq 0$ for any $x \in X$, $nexh_L(A, B) = 0$*

(vi) *ind_L is an L–fuzzy equivalence*

(vii.1) *div_L is irreflexive and symmetric*

(vii.2) *if $L \in RL^+$, then for all $A, B \in \mathcal{F}_L(X)$ such that $A \gg \emptyset$ and $B \gg \emptyset$, it holds $div_L(A, B) = 0$.*

Proof.

(i) See [3].

(ii.1) For every $A \in \mathcal{F}_L(X)$, we have: $ninc_L(A, A) = \sup_{x \in X}(A(x) \otimes \neg A(x)) = \sup_{x \in X}(A(x) \rightarrow (A(x) \rightarrow 0)) = 0$ by Lemma 2(vii).

(ii.2) Assume that $L \in RL^+$ (i.e. $\otimes$ has no zero divisors) and take an arbitrary $A \in \mathcal{F}_L(X)$ and $B \in \mathcal{F}_L(X)$ such that $B \gg \emptyset$, i.e. $B(x) \neq 0$ for every $x \in X$. Then by Lemma 1, $\neg_L B = \emptyset$, so we have: $ninc_L(A, B) = \sup_{x \in X}(A(x) \otimes \neg B(x)) = \sup_{x \in X}(A(x) \otimes 0) = 0$ by Lemma 2(iv).

(iii) Symmetry of com_L (resp. exh_L) directly follows from commutativity of $\otimes$ (resp. $\vee$).

(iv.1) By Lemma 2(xvi), for all $a, b \in L$, $a \rightarrow \neg b = \neg(a \otimes b) = \neg(b \otimes a) = b \rightarrow \neg a$. Then for every $A, B \in \mathcal{F}_L(X)$, $ort_L(A, B) = \inf_{x \in X}(A(x) \rightarrow \neg B(x)) = \inf_{x \in X}(B(x) \rightarrow \neg A(x)) = ort_L(B, A)$.

(iv.2) Assume that $L \in RL^+$. Let $A, B \in \mathcal{F}_L(X)$ and take an arbitrary $x \in X$. If $B(x) \neq 0$, then by Lemma 1, $\neg B(x) = 0$, so $A(x) \rightarrow \neg B(x) = \neg A(x) \in \{0, 1\}$. If $B(x) = 0$, then $A(x) \rightarrow \neg B(x) = A(x) \rightarrow 1 = 1$ by Lemma 2(v). Then $ort_L(A, B) = \inf_{x \in X}(A(x) \rightarrow \neg B(x)) \in \{0, 1\}$.

(v.1) Follows directly from symmetry of com_L.

(v.2) Assume that $L \in RL^+$ and consider $A, B \in \mathcal{F}_L(X)$ such that for every $x \in X$, $A(x) \neq 0$ or $B(x) \neq 0$. By Lemma 1, it implies that for every $x \in X$, $\neg A(x) = 0$ or $\neg B(x) = 0$, so using Lemma 2(iv), $\neg A(x) \otimes \neg B(x) = 0$ for every $x \in X$. Hence $nexh_L(A, B) = 0$.

(vi) Reflexivity and $\otimes$–transitivity of ind_L follows directly from (i), symmetry of ind_L results from commutativity of $\otimes$.

(vii.1) Let $A \in \mathcal{F}_L(X)$. For every $x \in X$, $A(x) \otimes \neg A(x) = A(x) \otimes (A(x) \rightarrow 0) = 0$ by Lemma 2(vii), so $com_L(A, \neg_L A) = 0$. By symmetry of com_L, $com_L(\neg_L A, A) = 0$. Hence $div_L(A, A) = 0$. Symmetry of div_L follows from symmetry of com_L.

(vii.2) Assume that $L \in RL^+$ and consider $A, B \in \mathcal{F}_L(X)$ such that $A(x) \neq 0$ and $B(x) \neq 0$ for every $x \in X$. By Lemma 1, $\neg A(x) = \neg B(x) = 0$ for every $x \in X$. Then we have $A(x) \otimes \neg B(x) = \neg A(x) \otimes B(x) = 0$ for every $x \in X$, which implies $\sup_{x \in X}(A(x) \otimes \neg B(x)) = \sup_{x \in X}(\neg A(x) \otimes B(x)) = 0$, so $div_L(A, B) = 0$. ∎

In the crisp case the relation of set inclusion (resp. compatibility, exhaustiveness, indiscernibility) is complementary to noninclusion (resp. orthogonality, nonexhaustiveness, diversity). While generalising these relations on the basis of R–lattices only the weaker form of complementarity holds, as the following proposition states.

Proposition 4. *For every complete R–lattice* $(L, \wedge, \vee, \otimes, \rightarrow, 0, 1)$,

  (i) $nincl_L \subseteq_L \neg_L incl_L$
  (ii) $ort_L = \neg_L com_L$ *and* $com_L \subseteq_L \neg_L ort_L$
  (iii) $exh_L \subseteq_L \neg_L nexh_L$
  (iv) $div_L \subseteq_L \neg_L ind_L$.

Proof.

(i) For every $A, B \in \mathcal{F}_L(X)$,

$$\begin{aligned}
\neg incl_L(A, B) \\
= \neg(\inf_{x \in X}(A(x) \rightarrow B(x))) \\
\geqslant \sup_{x \in X} \neg(A(x) \rightarrow B(x)) && \text{by Lemma 3(ix)} \\
\geqslant \sup_{x \in X} \neg(\neg B(x) \rightarrow \neg A(x)) && \text{by Lemma 2(xviii)} \\
= \sup_{x \in X} \neg\neg(\neg B(x) \otimes A(x)) && \text{by Lemma 2(xvi)} \\
\geqslant \sup_{x \in X}(\neg B(x) \otimes A(x)) && \text{by Lemma 2(xxi)} \\
= nincl_L(A, B).
\end{aligned}$$

(ii) For every $A, B \in \mathcal{F}_L(X)$,

$$\begin{aligned}
\neg com_L(A, B) \\
= \neg \sup_{x \in X}(A(x) \otimes B(x)) \\
= \inf_{x \in X} \neg(A(x) \otimes B(x)) && \text{by Lemma 3(viii)} \\
= \inf_{x \in X}(A(x) \rightarrow \neg B(x)) && \text{by Lemma 2(xvi)} \\
= ort_L(A, B).
\end{aligned}$$

Since $ort_L(A, B) = \neg com_L(A, B)$, from Lemma 2(xxi) we immediately obtain $\neg ort_L(A, B) = \neg\neg com_L(A, B) \geqslant com_L(A, B)$.

(iii) For all $A, B \in \mathcal{F}_L(X)$,

$$\begin{aligned}
\neg nexh_L(A, B) \\
= \neg \sup_{x \in X}(\neg A(x) \otimes \neg B(x)) \\
= \inf_{x \in X} \neg(\neg A(x) \otimes \neg B(x)) && \text{by Lemma 3(viii)} \\
\geqslant \inf_{x \in X} \neg(\neg A(x) \wedge \neg B(x)) && \text{by Lemma 2(iii)} \\
\geqslant \inf_{x \in X}(\neg\neg A(x) \vee \neg\neg B(x)) && \text{by Lemma 3(ix)} \\
\geqslant \inf_{x \in X}(A(x) \vee B(x)) && \text{by Lemma 2(xxi)}.
\end{aligned}$$

(iv) For every $A, B \in \mathcal{F}_L(X)$,

$\neg ind_L(A, B)$

$\quad = \neg((\inf_{x \in X}(A(x) \to B(x))) \otimes (\inf_{x \in X}(B(x) \to A(x))))$

$\quad \geqslant \neg(\inf_{x \in X}(A(x) \to B(x)) \wedge (\inf_{x \in X}(B(x) \to A(x))))$ by Lemma 2**(iii)**

$\quad \geqslant \neg(\inf_{x \in X}(A(x) \to B(x))) \vee \neg(\inf_{x \in X}(B(x) \to A(x)))$ by Lemma 3**(ix)**

$\quad \geqslant \sup_{x \in X} \neg(A(x) \to B(x)) \vee \sup_{x \in X} \neg(B(x) \to A(x))$ by Lemma 3**(ix)**

$\quad \geqslant \sup_{x \in X}(A(x) \otimes \neg B(x)) \vee \sup_{x \in X}(B(x) \otimes \neg A(x))$ by Lemma 2**(xx)**

$\quad = ninc_L(A, B) \vee ninc_L(B, A)$

$\quad = div_L(A, B).$ ∎

Proposition 3 shows that most properties of the L–fuzzy information relations discussed here coincide with their properties in the crisp case. Unfortunately, some properties are counterintuitive. First, fuzzy orthogonality should not reduce to the binary case. Moreover, the properties **(iv.2)**, **(v.2)**, and **(vii.2)** also do not coincide with what is expected, as the following example shows.

Example 3. Let $(\mathcal{L}, \min, \max, t, i_t, 0, 1)$ be the R–lattice as in Example 1, where $\mathcal{L} = [0, 1]$ and t is a left–continuous t–norm without zero divisors (e.g., t_Z or t_P). Consider an $\mathcal{L}$–fuzzy set A in $X \neq \emptyset$ given by: $A(x) = 0.001$ for every $x \in X$. The intuition dictates that X is not included in A to a very high degree. However, by Proposition 3**(ii.2)**, $ninc_\mathcal{L}(X, A) = 0$, which means that in fact X is totally included in A. Also, it is clear that $\emptyset$ and A do not cover the universe X to a high degree, but $nexh_\mathcal{L}(A, \emptyset) = 0$. Finally, A and X are totally different, yet $div_\mathcal{L}(A, X) = 0$.   □

In order to overcome these inadequacies, we take another class of residuated lattices, namely ER–lattices.

3.2   Fuzzy Information Relations Based on ER–Lattices

In this part we discuss another fuzzy generalisation of some information relations taking any complete ER–lattice $(L, \wedge, \vee, \otimes, \to, \sim, 0, 1)$ as a basic algebraic structure.

Definition 5. *For a complete ER–lattice* $(L, \wedge, \vee, \otimes, \to, \sim, 0, 1)$, *define the following* L–*fuzzy information relations: for all* $A, B \in \mathcal{F}_L(X)$,

    (i) L–*fuzzy noninclusion:*
$$Ninc_L(A, B) = \sup_{x \in X}(B(x) \leftarrow A(x))$$

    (ii) L–*fuzzy orthogonality:*
$$Ort_L(A, B) = \inf_{x \in X}(A(x) \Rightarrow \sim B(x))$$

    (iii) L–*fuzzy exhaustiveness:*
$$Exh_L(A, B) = \inf_{x \in X}(A(x) \oplus B(x))$$

    (iv) L–*fuzzy nonexhaustiveness:*
$$Nexh_L(A, B) = com_L(\sim_L A, \sim_L B)$$

    (v) L–*fuzzy diversity:*
$$Div_L(A, B) = Ninc_L(A, B) \otimes Ninc_L(B, A).$$ □

The definition of $Ninc_L$, Exh_L, and Div_L were presented in [29], where L was any complete double residuated lattice.

In view of Remark 2, $B(x) \leftarrow A(x)$ is a generalisation of the classical implication $A(x) \wedge \neg B(x)$, so $Ninc_L(A, B)$ is the fuzzy counterpart of the classical formula $(\exists x \in X)\,(x \in A\ \&\ x \notin B)$ and indeed represents the degree, to which A is not included in B. In the definition of Ort_L, Exh_L, $Nexh_L$, and Div_L we substitute the operations $\rightarrow$, $\neg$, and $\vee$ by $\Rightarrow$, $\sim$, and $\oplus$, respectively.

Proposition 5. *For every complete ER–lattice* $(L, \wedge, \vee, \otimes, \rightarrow, \sim, 0, 1)$,

  (i) $Ninc_L$ *is irreflexive and* $\oplus$*–cotransitive.*
  (ii) Ort_L, Exh_L, *and* $Nexh_L$ *are symmetric*
  (iii) Div_L *is irreflexive and symmetric.*

Proof.

(i) Irreflexivity of $Ninc_L$ results from Lemma 2(**v'**). To show that it is also $\oplus$–cotransitive, let us take $A, B, C \in \mathcal{F}_L(X)$. Then

$$Ninc_L(A, B) \oplus Ninc_L(B, C)$$
$$= \sup_{x \in X}(B(x) \leftarrow A(x)) \oplus \sup_{x \in X}(C(x) \leftarrow B(x))$$
$$\geqslant \sup_{x \in X}((B(x) \leftarrow A(x)) \oplus (C(x) \leftarrow B(x))) \qquad \text{by Lemma 3(\textbf{vii'})}$$
$$= \sup_{x \in X}((C(x) \leftarrow B(x)) \oplus (B(x) \leftarrow A(x))) \qquad \text{by commutativity of } \oplus$$
$$\geqslant \sup_{x \in X}(C(x) \leftarrow A(x)) \qquad \text{by Lemma 2(\textbf{ix'})}$$
$$= Ninc_L(A, C).$$

(ii) Symmetry of Ort_L follows from Lemma 2(**xvii**) and commutativity of $\otimes$, symmetry of Exh_L immediately follows from commutativity of $\oplus$, and symmetry of $Nexh_L$ results from symmetry of com_L.

(iii) Irreflexivity of Div_L follows from irreflexivity of $Ninc_L$, while symmetry of Div_L results from commutativity of $\oplus$. ∎

Example 4. Put $\mathcal{L} = [0, 1]$ and consider the lattice $(\mathcal{L}, \min, \max, t, i_t, n, 0, 1)$ as in Example 2 (recall that t is a left–continuous t–norm, i_t is the residual implication determined by t, and n is the standard fuzzy negation). Let $X \neq \emptyset$ be an arbitrary domain and let $A \in \mathcal{F}_{\mathcal{L}}(X)$ be defined as in Example 3, i.e. $A(x) = 0.001$ for every $x \in X$. By simple calculations we get $Ninc_{\mathcal{L}}(X, A) = 1$ for $\leftarrow\, \in \{\leftarrow_Z, \leftarrow_P\}$. Of course, this result coincides with our intuition. Let $B \in \mathcal{F}_{\mathcal{L}}(X)$ be such that $B(x) = 0.999$ for every $x \in X$. Then $Ort_{\mathcal{L}}(A, B) = t(n(0.001), n(0.999)) = t(0.999, 0.001) \notin \{0, 1\}$ for any t without zero divisors. Hence $Ort_{\mathcal{L}}$ does not reduce to a crisp relation. Moreover, for any t–norm t, $Nexh_{\mathcal{L}}(A, \emptyset) = 0.999$. Clearly, this is again the expected result: A and $\emptyset$ do not cover the universe X up to the very high degree. Finally, $Ninc_{\mathcal{L}}(A, X) = 0$ for the dual residuum of any (right–continuous) t–conorm s. Also, note that $Ninc_{\mathcal{L}}(X, A) = (0.001 \leftarrow 1) = 1$ for $\leftarrow\, \in \{\leftarrow_Z, \leftarrow_P\}$. Then $Div_{\mathcal{L}}(X, A) = 0 \oplus 1 = 1$. So, as expected, A differs from X to the very high degree. □

In view of the above example, it is now clear that the inadequacies in representation, which occur when R–lattices of the class RL^+ were taken as basic structures, are avoided.

Note also:

Proposition 6. *For every complete ER–lattice* $(L, \wedge, \vee, \otimes, \to, \sim, 0, 1)$,

(i) $Ninc_L(A, B) = \sim inc_L(\sim_L B, \sim_L A)$ *and* $Div_L(A, B) = \sim ind_L(\sim_L A, \sim_L B)$
for every $A, B \in \mathcal{F}_L(X)$,

(ii) $Ort_L = \sim_L com_L$ *and* $Exh_L = \sim_L Nexh_L$.

Proof.

(i) For every $A, B \in \mathcal{F}_L(X)$,

$Ninc_L(A, B)$
$$= \sup_{x \in X} (B(x) \leftarrow A(x))$$
$$= \sup_{x \in X} \sim (\sim B(x) \to \sim A(x)) \qquad \text{by (10)}$$
$$= \sim \inf_{x \in X} (\sim B(x) \to \sim A(x)) \qquad \text{by Lemma 3(vi')}$$
$$= \sim inc_L(\sim_L B, \sim_L A).$$

Similarly, using (10), Lemma 3(**vi'**), and (8), we get for every $A, B \in \mathcal{F}_L(X)$,

$Div_L(A, B)$
$$= Ninc_L(A, B) \oplus Ninc_L(B, A)$$
$$= \sup_{x \in X} (B(x) \leftarrow A(x)) \oplus \sup_{x \in X} (A(x) \leftarrow B(x))$$
$$= \sup_{x \in X} \sim (\sim B(x) \to \sim A(x)) \oplus \sup_{x \in X} \sim (\sim A(x) \to \sim B(x))$$
$$= \sim \inf_{x \in X} (\sim B(x) \to \sim A(x)) \oplus \sim \inf_{x \in X} (\sim A(x) \to \sim B(x))$$
$$= \sim (\inf_{x \in X} (\sim B(x) \to \sim A(x)) \otimes \inf_{x \in X} (\sim A(x) \to \sim B(x)))$$
$$= \sim ind_L(\sim_L A, \sim_L B).$$

The proof of (**ii**) is similar. ∎

Note that the properties stated in the above proposition coincide with the respective properties of these relations in the crisp case. Clearly, for every crisp subsets $A, B \subseteq X$, $A = B \Leftrightarrow -A = -B$. Yet in general $ind_L(A, B) \neq ind_L(\sim_L A, \sim_L B)$. Similarly, $inc_L(A, B) \neq inc_L(\sim_L B, \sim_L A)$. It follows from the fact that in an arbitrary ER–lattice L, we do not have any relationship between $a \to b$ and $\sim b \to \sim a$, as observed in Remark 2.

4 Fuzzy Information Operators

Let $(L, \wedge, \vee, \otimes, \to, \sim, 0, 1)$ be a complete ER–lattice. By an *L–fuzzy information operator* we mean any mapping $\Omega_L : \mathcal{R}_L(X) \times \mathcal{F}_L(X) \to \mathcal{F}_L(X)$. Below we define several *L*–information operators.

Definition 6. *For every complete ER–lattice* $(L, \wedge, \vee, \otimes, \to, \sim, 0, 1)$, *for every* $R \in \mathcal{R}_L(X)$, *for every* $A \in \mathcal{F}_L(X)$, *and for every* $x \in X$,

(**O.1**) $[R]_{\to} A(x) = \inf_{y \in X} (R(x, y) \to A(y))$
(**O.2**) $[R]_{\Rightarrow} A(x) = \inf_{y \in X} (R(x, y) \Rightarrow A(y))$
(**O.3**) $[R]_{\leftarrow} A(x) = \sup_{y \in X} (R(x, y) \leftarrow A(y))$
(**O.4**) $[R]_{\Leftarrow} A(x) = \sup_{y \in X} (R(x, y) \Leftarrow A(y))$
(**O.5**) $\langle R \rangle_{\otimes} A(x) = \sup_{y \in X} (R(x, y) \otimes A(y))$
(**O.6**) $\langle R \rangle_{\oplus} A(x) = \inf_{y \in X} (R(x, y) \oplus A(y))$. □

It is worth noting that $[]_\rightarrow$ (resp. $[]_\Rightarrow$) and $\langle\rangle_\otimes$ correspond to fuzzy modalities ([13],[14],[17]), i.e. $[R]_\rightarrow$ and $[R]_\Rightarrow$ are fuzzy generalisations of the necessity operator, while $\langle R\rangle_\otimes$ is the counterpart of the possibility operator. Also, these operators are fuzzy approximation operators well–known in the theory of fuzzy rough sets (see, e.g., [32], [34]), as well as fuzzy morphological operators which are basic tools in mathematical morphology ([24]).

Let $R \in \mathcal{R}_L(X)$. For any $x \in X$ we write xR to denote the L–fuzzy set in X defined as: $(xR)(y) = R(x, y)$ for every $y \in X$. Note that for every $A \in \mathcal{F}_L(X)$ and for every $x \in X$,

$$[R]_\rightarrow A(x) = inc_L(xR, A) \qquad [R]_\leftarrow A(x) = Ninc_L(xR, A)$$
$$\langle R\rangle_\otimes A(x) = com_L(xR, A) \qquad \langle R\rangle_\oplus A(x) = Exh_L(xR, A).$$

Definition 7. *Let $\Omega_1, \Omega_2 : \mathcal{R}_L(X) \times \mathcal{F}_L(X) \rightarrow \mathcal{F}_L(X)$ be two L–fuzzy information operators, let $\circ$ be a unary operation in L, and let $\circ_L : \mathcal{F}_L(X) \rightarrow \mathcal{F}_L(X)$ be such that $(\circ_L A)(x) = \circ A(x)$ for every $x \in X$. We say that Ω_1 and Ω_2 are*

- $\circ_L$*-dual iff* $\Omega_1(R, A) = \circ_L \Omega_2(R, \circ_L A)$ *for every* $R \in \mathcal{R}_L(X)$ *and for every* $A \in \mathcal{F}_L(X)$
- *weakly* $\circ_L$*-dual iff* $\Omega_1(R, A) \subseteq_L \circ_L \Omega_2(R, \circ_L A)$ *for every* $R \in \mathcal{R}_L(X)$ *and for every* $A \in \mathcal{F}_L(X)$
- $\circ_L$*-codual iff* $\Omega_1(R, A) = \circ_L \Omega_2(\circ_L R, \circ_L A)$ *for every* $R \in \mathcal{R}_L(X)$ *and for every* $A \in \mathcal{F}_L(X)$. □

Basic properties of the operators **(O.1)–(O.6)** are given in the following proposition.

Proposition 7. *For every complete ER–lattice L and for every $R \in \mathcal{F}_L(X)$,*

(i) $[R]_\rightarrow X = [R]_\Rightarrow X = \langle R\rangle_\oplus X = X, \quad [R]_\leftarrow \emptyset = [R]_\Leftarrow \emptyset = \langle R\rangle_\otimes \emptyset = \emptyset$

(ii) *for every $A, B \in \mathcal{F}_L(X)$ and for every $\Omega \in \{[]_\rightarrow, []_\Rightarrow, []_\leftarrow, []_\Leftarrow, \langle\rangle_\otimes, \langle\rangle_\oplus\}$, $A \subseteq_L B$ implies $\Omega(A) \subseteq_L \Omega(B)$*

(iii) *for every $A \in \mathcal{F}_L(X)$,*

$$[R]_\rightarrow A \subseteq_L \neg\langle R\rangle_\otimes \neg A \qquad\qquad [R]_\Rightarrow A = \sim\langle R\rangle_\otimes \sim A$$
$$\langle R\rangle_\otimes A \subseteq_L \neg[R]_\rightarrow \neg A \qquad\qquad \langle R\rangle_\otimes A = \sim[R]_\rightarrow \sim A$$
$$\ulcorner[R]_\leftarrow \ulcorner A \subseteq_L \langle R\rangle_\oplus A \qquad\qquad \langle R\rangle_\oplus A = \sim[R]_\Leftarrow \sim A$$
$$\ulcorner\langle R\rangle_\oplus \ulcorner A \subseteq_L [R]_\leftarrow A \qquad\qquad [R]_\Leftarrow A = \sim\langle R\rangle_\oplus \sim A$$

(iv) *for every $A \in \mathcal{F}_L(X)$,*

$$[R]_\rightarrow A = \sim[\sim R]_\leftarrow \sim A \qquad\qquad [R]_\Rightarrow A = \sim[\sim R]_\Leftarrow \sim A$$
$$[R]_\leftarrow A = \sim[\sim R]_\rightarrow \sim A \qquad\qquad [R]_\Leftarrow A = \sim[\sim R]_\Rightarrow \sim A$$

(v) *for every indexed family $(A_i)_{i \in I}$ of L–fuzzy sets in X,*

$$[R]_\rightarrow(\textstyle\bigcap_{i \in I} A_i) = \textstyle\bigcap_{i \in I}[R]_\rightarrow A_i \qquad [R]_\Rightarrow(\textstyle\bigcap_{i \in I} A_i) = \textstyle\bigcap_{i \in I}[R]_\Rightarrow A_i$$
$$[R]_\rightarrow(\textstyle\bigcup_{i \in I} A_i) {}_L\!\supseteq \textstyle\bigcup_{i \in I}[R]_\rightarrow A_i \qquad [R]_\Rightarrow(\textstyle\bigcup_{i \in I} A_i) {}_L\!\supseteq \textstyle\bigcup_{i \in I}[R]_\Rightarrow A_i$$
$$[R]_\leftarrow(\textstyle\bigcap_{i \in I} A_i) \subseteq_L \textstyle\bigcap_{i \in I}[R]_\leftarrow A_i \qquad [R]_\Leftarrow(\textstyle\bigcap_{i \in I} A_i) \subseteq_L \textstyle\bigcap_{i \in I}[R]_\Leftarrow A_i$$
$$[R]_\leftarrow(\textstyle\bigcup_{i \in I} A_i) = \textstyle\bigcup_{i \in I}[R]_\leftarrow A_i \qquad [R]_\Leftarrow(\textstyle\bigcup_{i \in I} A_i) = \textstyle\bigcup_{i \in I}[R]_\Leftarrow A_i$$
$$\langle R\rangle_\otimes(\textstyle\bigcap_{i \in I} A_i) \subseteq_L \textstyle\bigcap_{i \in I}\langle R\rangle_\otimes A_i \qquad \langle R\rangle_\oplus(\textstyle\bigcap_{i \in I} A_i) = \textstyle\bigcap_{i \in I}\langle R\rangle_\oplus A_i$$
$$\langle R\rangle_\otimes(\textstyle\bigcup_{i \in I} A_i) = \textstyle\bigcup_{i \in I}\langle R\rangle_\otimes A_i \qquad \langle R\rangle_\oplus(\textstyle\bigcup_{i \in I} A_i) {}_L\!\supseteq \textstyle\bigcup_{i \in I}\langle R\rangle_\oplus A_i.$$

Proof. Straightforward verification.   ∎

The property **(ii)** of the above proposition states the monotonicity of L–fuzzy information operators w.r.t. Zadeh's inclusion. Also, **(iii)** states the $\sim$–duality and weak $\neg$–duality between these operators, and **(iv)** establishes $\sim$–coduality between L–fuzzy information operators.

Corollary 1. *For every complete ER–lattice L,*

(i) $[\,]_{\Rightarrow}$ *and* $\langle\,\rangle_{\otimes}$, *as well as* $[\,]_{\Leftarrow}$ *and* $\langle\,\rangle_{\oplus}$, *are* $\sim$*–dual,*

(ii) $[\,]_{\rightarrow}$ *and* $\langle\,\rangle_{\otimes}$ *are weakly* $\neg$*–dual*

(iii) $[\,]_{\rightarrow}$ *and* $[\,]_{\leftarrow}$, *as well as* $[\,]_{\Rightarrow}$ *and* $[\,]_{\Leftarrow}$, *are* $\sim$*–codual.*   ∎

It is well-known that traditional information operators are useful for characterizing particular classes of (binary) relations. This is also the case for fuzzy information operators. The following theorem presents complete characterizations of some basic classes of fuzzy relations.

Theorem 1. *For every complete ER–lattice $(L, \wedge, \vee, \otimes, \rightarrow, \sim, 0, 1)$, for every $R \in \mathcal{R}_L(X)$, and for every $A \in \mathcal{F}_L(X)$ the following statements hold:*

(i) R *is reflexive*
  iff $[R]_{\rightarrow}A \subseteq_L A$
  iff $[R]_{\Rightarrow}A \subseteq_L A$
  iff $A \subseteq_L \langle R \rangle_{\otimes}A$

(ii) R *is irreflexive*
  iff $A \subseteq_L [R]_{\leftarrow}A$
  iff $A \subseteq_L [R]_{\Leftarrow}A$
  iff $\langle R \rangle_{\oplus}A \subseteq_L A$

(iii) R *is symmetric*
  iff $\langle R \rangle_{\otimes}[R]_{\rightarrow}A \subseteq_L A$
  iff $[R]_{\leftarrow}\langle R \rangle_{\oplus}A \subseteq_L A$
  iff $A \subseteq_L [R]_{\rightarrow}\langle R \rangle_{\otimes}A$
  iff $A \subseteq_L \langle R \rangle_{\oplus}[R]_{\leftarrow}A$

(iv) R *is* $\otimes$*–transitive*
  iff $[R]_{\rightarrow}A \subseteq_L [R]_{\rightarrow}[R]_{\rightarrow}A$
  iff $[R]_{\Rightarrow}A \subseteq_L [R]_{\Rightarrow}[R]_{\Rightarrow}A$
  iff $\langle R \rangle_{\otimes}\langle R \rangle_{\otimes}A \subseteq_L \langle R \rangle_{\otimes}A$

(v) R *is* $\oplus$*–cotransitive*
  iff $[R]_{\leftarrow}[R]_{\leftarrow}A \subseteq_L [R]_{\leftarrow}A$
  iff $[R]_{\Leftarrow}[R]_{\Leftarrow}A \subseteq_L [R]_{\Leftarrow}A$
  iff $\langle R \rangle_{\oplus}A \subseteq_L \langle R \rangle_{\oplus}\langle R \rangle_{\oplus}A.$

Proof. By way of example we prove **(ii)** and **(iv)**.

(ii) First, consider the inclusion $A \subseteq_L [R]_{\leftarrow}A$.

($\subseteq$) Assume that R is irreflexive. Then for every $A \in \mathcal{F}_L(X)$ and for every $x \in X$,

$$[R]_{\leftarrow}A(x) = \sup_{y \in X}(R(x,y) \leftarrow A(y)) \geqslant R(x,x) \leftarrow A(x) = 0 \leftarrow A(x) = A(x).$$

by Lemma 2**(vi')**.

($\supseteq$) Assume that R is not irreflexive. Then $R(x_0, x_0) \neq 0$ for some $x_0 \in X$. Put $A = x_0 R$. Then we have:

$$[R]_\leftarrow A(x_0) = \sup_{y \in X}(R(x_0, y) \leftarrow R(x_0, y)) = 0$$

by Lemma 2(**v'**). Hence $A(x_0) \rightarrow [R]_\leftarrow A(x_0) = R(x_0, x_0) \rightarrow 0 \neq 1$ by Lemma 2(**v**), so $A \not\subseteq_L [R]_\leftarrow A$.

Consider the second equivalence.
($\subseteq$) For every $A \in \mathcal{F}_L(X)$ and for every $x \in X$,

$$\begin{aligned}[R]_\Leftarrow A(x) &= \sup_{y \in X}(R(x, y) \Leftarrow A(y)) \\ &\geqslant R(x, x) \Leftarrow A(x) = 0 \Leftarrow A(x) = A(x).\end{aligned}$$

by Lemma 2(**vi'**).
($\supseteq$) As before, assume that R is not irreflexive, i.e. $R(x_0, x_0) \neq 0$ for some $x_0 \in X$. For $A = \{x_0\}$ we have:

$$\begin{aligned}[R]_\Leftarrow A(x_0) \\ = \sup_{y \in X}(R(x_0, y) \Leftarrow A(y)) \\ = \sup_{y \in X} \sim(\sim R(x_0, y) \Rightarrow \sim A(y)) &\qquad \text{by (11)} \\ = \sim \inf_{y \in X}(\sim R(x_0, y) \Rightarrow \sim A(y)) &\qquad \text{by Lemma 3(\textbf{vi'})} \\ = \sim \inf_{y \in X}(R(x_0, y) \oplus \sim A(y)) &\qquad \text{by (9)} \\ = \sim R(x_0, x_0).\end{aligned}$$

Since $R(x_0, x_0) \neq 0$, we have $\sim R(x_0, x_0) \neq 1$, so $[R]_\Leftarrow A(x_0) \neq 1$. But $A(x_0) = 1$. Therefore, $A \not\subseteq_L [R]_\Leftarrow A$.

Now, consider the third equivalence.
($\subseteq$) For any $A \in \mathcal{F}_L(X)$ and for any $x \in X$,

$$\langle R \rangle_\oplus A(x) = \inf_{y \in X}(R(x, y) \oplus A(y)) \leqslant R(x, x) \oplus A(x) = 0 \oplus A(x) = A(x).$$

($\supseteq$) Assume that R is not irreflexive, i.e. $R(x_0, x_0) \neq 0$ for some $x_0 \in X$. For $A = X \setminus \{x_0\}$ we have:

$$\langle R \rangle_\oplus A(x_0) = \inf_{y \in X}(R(x_0, y) \oplus A(y)) = R(x_0, x_0) \oplus 0 = R(x_0, x_0).$$

Since $A(x_0) = 0$, we get $\langle R \rangle_\oplus A(x_0) \not\leqslant A(x_0)$, which implies $\langle R \rangle_\oplus A \not\subseteq_L A$.

(**iv**) We show the first equivalence.
($\subseteq$) For every $A \in \mathcal{F}_L(X)$ and for every $x \in X$,

$$\begin{aligned}[R]_\rightarrow [R]_\rightarrow A(x) \\ = \inf_{y \in X}(R(x, y) \rightarrow (\inf_{z \in X}(R(y, z) \rightarrow A(z)))) \\ = \inf_{z \in X} \inf_{y \in X}(R(x, y) \rightarrow (R(y, z) \rightarrow A(z))) &\quad \text{by Lemma 3(\textbf{ii})} \\ = \inf_{z \in X} \inf_{y \in X}(R(x, y) \otimes R(y, z) \rightarrow A(z)) &\quad \text{by Lemma 2(\textbf{xiv})} \\ \geqslant \inf_{z \in X} \inf_{y \in X}(R(x, z) \rightarrow A(z)) &\quad \text{by assumption, Lemma 2(\textbf{i})} \\ = \inf_{z \in X}(R(x, z) \rightarrow A(z)) \\ = [R]_\rightarrow A(x).\end{aligned}$$

($\supseteq$) Assume now that R is not $\otimes$–transitive, i.e. $R(x_0, y_0) \otimes R(y_0, z_0) \not\leqslant R(x_0, z_0)$ for some $x_0, y_0, z_0 \in X$. By Lemma 2(**v**), this means that

(**iv.1**) $(R(x_0, y_0) \otimes R(y_0, z_0)) \rightarrow R(x_0, z_0) \neq 1$.

Consider $A = x_0 R$. Using again Lemma 2(**v**) we get

(**iv.2**) $[R]_\rightarrow A(x_0) = \inf_{y \in X}(R(x_0, y) \rightarrow R(x_0, y)) = 1$.

Next,

$$[R]_\to [R]_\to A(x_0)$$
$$= \inf_{y \in X}(R(x_0, y) \to (\inf_{z \in X}(R(y, z) \to R(x_0, z))))$$
$$= \inf_{z \in X} \inf_{y \in X}(R(x_0, y) \to (R(y, z) \to R(x_0, z))) \qquad \text{by Lemma 3(ii)}$$
$$= \inf_{z \in X} \inf_{y \in X}((R(x_0, y) \otimes R(y, z)) \to R(x_0, z)) \qquad \text{by Lemma 2(xiv)}$$
$$\leqslant (R(x_0, y_0) \otimes R(y_0, z_0)) \to R(x_0, z_0)$$
$$\neq 1 \qquad \text{by (iv.1).}$$

Therefore, we obtain

$$[R]_\to A(x_0) \to [R]_\to [R]_\to A(x_0)$$
$$= 1 \to [R]_\to [R]_\to A(x_0) \qquad \text{by (iv.2)}$$
$$= [R]_\to [R]_\to A(x_0) \qquad \text{by Lemma 2(vi)}$$
$$\neq 1.$$

Then, by Lemma 2(v), $[R]_\to A(x_0) \not\leqslant [R]_\to [R]_\to A(x_0)$, so $[R]_\to A \not\subseteq_L [R]_\to [R]_\to A$.

Now, we show the second equivalence.

($\subseteq$) For every $A \in \mathcal{F}_L(X)$ and for every $x \in X$,

$$[R]_\Rightarrow [R]_\Rightarrow A(x)$$
$$= \inf_{y \in X}(R(x, y) \Rightarrow (\inf_{z \in X}(R(y, z) \Rightarrow A(z))))$$
$$= \inf_{z \in X} \inf_{y \in Z}(R(x, y) \Rightarrow (R(y, z) \Rightarrow A(z))) \qquad \text{by Lemma 3(iii)}$$
$$= \inf_{z \in X} \inf_{y \in X}(R(x, y) \otimes R(y, z) \Rightarrow A(z)) \qquad \text{by Lemma 2(xv)}$$
$$\geqslant \inf_{z \in X} \inf_{y \in X}(R(x, z) \Rightarrow A(z)) \qquad \text{by assumption, Lemma 2(i)}$$
$$= [R]_\Rightarrow A(x).$$

($\supseteq$) Assume that R is not $\otimes$–transitive, i.e. there exist $x_0, y_0, z_0 \in X$ such that $R(x_0, y_0) \otimes R(y_0, z_0) \not\leqslant R(x_0, z_0)$. Then $\sim R(x_0, z_0) \not\leqslant \sim(R(x_0, y_0) \otimes R(y_0, z_0))$, which by Lemma 2(v) gives

(iv.3) $\sim R(x_0, z_0) \to \sim(R(x_0, y_0) \otimes R(y_0, z_0)) \neq 1.$

Take $A = X \backslash \{z_0\}$. Since for every $a \in L$, $a \oplus 1 = 1$, we easily get for every $y \in X$,

(iv.4) $[R]_\Rightarrow A(y) = \inf_{z \in X}(\sim R(y, z) \oplus A(z)) = \sim R(y, z_0).$

Furthermore,

$$[R]_\Rightarrow [R]_\Rightarrow A(x_0)$$
$$= \inf_{y \in X}(R(x_0, y) \Rightarrow [R]_\Rightarrow A(y))$$
$$= \inf_{y \in X}(R(x_0, y) \Rightarrow \sim R(y, z_0)) \qquad \text{by (iv.4)}$$
$$= \inf_{y \in X}(\sim R(x_0, y) \oplus \sim R(y, z_0)) \qquad \text{by (9)}$$
$$= \inf_{y \in X} \sim(R(x_0, y)) \otimes R(y, z_0)) \qquad \text{by (8)}$$
$$= \sim \sup_{y \in X}(R(x_0, y)) \otimes R(y, z_0)) \qquad \text{by Lemma 3(vi').}$$

Then we get

$$[R]_\Rightarrow A(x_0) \to [R]_\Rightarrow [R]_\Rightarrow A(x_0)$$
$$= \sim R(x_0, z_0) \to \sim \sup_{y \in X}(R(x_0, y) \otimes R(y, z_0)) \qquad \text{by (iv.4)}$$
$$\leqslant \sim R(x_0, z_0) \to \sim(R(x_0, y_0) \otimes R(y_0, z_0)) \qquad \text{by Lemma 2(i)}$$
$$\neq 1 \qquad \text{by (iv.3).}$$

By Lemma 2(**v**), this implies $[R]_{\Rightarrow}A(x_0) \not\leq [R]_{\Rightarrow}[R]_{\Rightarrow}A(x_0)$. Therefore, we get $[R]_{\Rightarrow}A \not\subseteq_L [R]_{\Rightarrow}[R]_{\Rightarrow}A$.

In the similar way the third equivalence can be proved.

5   Conclusions

In this paper we have presented fuzzy generalisations of several information relations and operators. Two classes of residuated lattices have been taken as basic algebraic structures: traditional residuated lattices (commutative and integral) and so–called extended residuated lattices (ER–lattices). It has been shown that ER–lattices allow us to define abstract counterparts of the main classes of fuzzy logical connectives. We have indicated that some inadequacies in representation occur when residuated lattices constitute the basic structures and that these drawbacks can be avoided on the basis of ER–lattices. Some fuzzy information operators have been presented. We have shown that these operators are useful for characterizations of main classes of fuzzy relations.

Acknowledgements

The idea of the extension of residuated lattices by antitone involutions was suggested by Prof. Dr Ivo Düntsch during the stay of Anna M. Radzikowska in St. Catharines in 2003 in the framework of the short–term scientific mission of the COST Action 274/TARSKI project. The authors would like to thank Prof. Duntsch for this suggestion, which inspired them to develop generalisations of information relations in the more adequate way.

References

1. Allwein, G. and Dunn, M. (1993). Kripke models for linear logic, Journal of Symbolic Logic 58(2), 514–545.
2. Bahls, P., Cole, J., Galatos, N., Jipsen, P., Tsinakis, C. (2003). Cancellative residuated lattices, Algebra Universalis 50(1), 83–106.
3. Belohlavek, R. (2002). *Fuzzy Relational Systems: Foundations and Principles*, Kluwer Academic Publishers.
4. Blount, K. and Tsinakis, C. (2003). The structure of residuated lattices. Int. J. of Algebra Comput. 13(4), 437–461.
5. De Baets, B. and Mesiar, R. (1997). Pseudo-metrics and T–equivalences, Journal of Fuzzy Mathematics 5(2), 471–481.
6. Demri, S., Orłowska, E., and Vakarelov, D. (1999). Indiscernibility and complementarity relations in information systems, JFAK, Essays Dedicated to Johan van Benthem on the Occasion of his 50th Birthday, Gerbrandy, J., Marx, M., de Rijke, M., and Venema, Y. (eds), Amsterdam University Press.
7. Demri, S. and Orłowska, E. (2002). *Incomplete Information: Structure, Inference, Complexity*, EATCS Monographs in Theoretical Computer Science, Springer–Verlag.

8. Dilworth, R. P. and Ward, N. (1939). Residuated lattices, Transactions of the American Mathematical Society 45, 335–354.

9. Dilworth, R. P. (1939). Non–commutative residuated lattices, Transactions of the American Mathematical Society 46, 426–444.

10. Düntsch, I. and Orłowska, E. (2000). Logics of complementarity in information systems, Mathematical Logic Quarterly 46, 267–288.

11. Düntsch, I. and Orłowska, E. (2000). Beyond modalities: sufficiency and mixed algebras, in: *Relational Methods for Computer Science Applications*, Orłowska, E. and Szałas, A. (eds), Physica Verlag, Heidelberg, 263–285.

12. Düntsch, I. (2003). Private communication.

13. Fitting, M. (1991). Many–valued modal logics, Fundamenta Informaticae 15, 235–254.

14. Fitting, M. (1992). Many–valued modal logics (II), Fundamenta Informaticae 17, 55–73.

15. Flondor, P. and Sularia, M. (2003). On a class of residuated semilattice monoids, Fuzzy Sets and Systems 138(1), 149–176.

16. Georgescu, G. and Iorgulescu, A. (2001). Pseudo–MV algebras, Multiple Valued Logic 6, 95–135.

17. Godo, L. and Rodriguez, R. O. (1999). A Fuzzy Modal Logic for Similarity Reasoning, in: *Fuzzy Logic and Soft Computing*, Chen G., Ying M., and Cai K.-Y. (eds), Kluwer, 33–48.

18. Gougen, J. A. (1967). L–fuzzy sets, Journal of Mathematical Analysis and Applications 18, 145–174.

19. Hajek, P. (1998). *Metamathematics of Fuzzy Logic*, Kluwer, Dordrecht.

20. Hajek, P. (2003). Fuzzy Logics with Noncommutative Conjunctions, Journal of Logic and Computation 13(4), 469–479.

21. Hart, J. B., Rafter, L., and Tsinakis, C. (2002). The structure of commutative residuated lattices, Int. J. Algebra Comput. 12(4), 509–524.

22. Jipsen, P. and Tsinaksis, C. (2002). A Survey of Residuated Lattices, in: *Ordered Algebraic Structures*, Martinez, J. (ed), Kluwer Academic Publishers, Dordrecht, 19–56.

23. Klir, G. and Yuan, B. (1995). *Fuzzy Sets and Fuzzy Logic: Theory and Applications*, Prentice–Hall, Englewood Cliffs, NJ.

24. Nachtegael, M. and Kerre, E. E. (eds) (2000). *Fuzzy Techniques in Image Processing*, Springer–Verlag, Heidelberg.

25. Orłowska, E. (1988). Kripke models with relative accessibility and their application to inferences from incomplete information, in: *Mathematical Problems in Computation Theory*, Mirkowska, G. and Rasiowa, H. (eds), Banach Center Publications 21, 329–339.

26. Orłowska, E. (ed) (1998). *Incomplete Information – Rough Set Analysis*, Studies in Fuzziness and Soft Computing, Springer-Verlag.

27. Orłowska, E. (1999). Multi–valuedness and uncertainty, Multiple–Valued Logics 4, 207–227.

28. Orłowska, E. and Radzikowska, A. M. (2001). Information relations and operators based on double residuated lattices, in: Proceedings of *6th International Workshop on Relational Methods in Computer Science RelMiCS–2001*, Oisterwijk, The Netherlands, 185–199.

29. Orłowska, E. and Radzikowska, A. M. (2002). Double residuated lattices and their applications, in: *Relational Methods in Computer Science*, de Swart, H. C. M. (ed), Lecture Notes in Computer Science 2561, Springer–Verlag, Heidelberg, 171–189.

30. Radzikowska, A. M. and Kerre, E. E. (2001). On Some Classes of Fuzzy Information Relations, in: Proceedings of *31st IEEE International Symposium of Multiple–Valued Logics ISMVL–2001*, Warsaw, 75–80.
31. Radzikowska, A. M. and Kerre, E. E. (2001). Towards studying of fuzzy information relations, in: Proceedings of *International Conference in Fuzzy Logics and Technology EUSFLAT–2001*, Leicester, UK, 365–369.
32. Radzikowska, A. M. and Kerre, E. E. (2002). A Comparative Study of Fuzy Rough Sets, Fuzzy Sets and Systems 126, 137–155.
33. Radzikowska, A. M. and Kerre, E. E. (2002). A fuzzy generalisation of information relations, in: *Beyond Two: Theory and Applications of Multiple–Valued Logics*, Orłowska, E. and Fitting, M. (eds), Springer–Velag, 264–290.
34. Radzikowska, A. M. and Kerre, E. E. (2004). On L–valued fuzzy rough sets, Lecture Notes in Computer Science 3070, 526–531.
35. Radzikowska, A. M. and Kerre, E. E. (2004). An algebraic characterisation of fuzzy rough sets, in: Proceedings of *IEEE International Conference on Fuzzy Systems FUZZ–IEEE 2004*, Budapest, Hungary, vol.1, 115–120.
36. Radzikowska, A. M. and Kerre, E. E. (2004). Lattice–based fuzzy information relations and operators, in: *Current Issues in Data and Knowledge Engineering*, De Baets, B., De Caluve, R., Kacprzyk, J., De Tré, G., and Zadrożny, S. (eds), EXIT, Warsaw, 433–443.
37. Radzikowska, A. M. and Kerre, E. E. (2004). Fuzzy Rough Sets based on Residuated Lattices, in: *Transactions on Rough Sets II: Rough Sets and Fuzzy Sets*, Peters, J. F., Skowron, A., Dubois. D., Grzymała–Busse, J. W., Inuiguchi, M., and Polkowski, L. (eds), Lecture Notes in Computer Science 3135, Springer–Verlag, 278–296.
38. Radzikowska, A. M. and Kerre, E. E. (2005). Characterisations of main classes of fuzzy relations using fuzzy modal operators, Fuzzy Sets and Systems 152(2), 223–247.
39. Rasiowa, H. and Sikorski, R. (1970). *The Mathematics of Metamathematics*, Warsaw.
40. Rauszer, C. (1974). Semi-Boolean algebras and their applications to intuitionistic logic with dual operations, Fundamenta Mathematicae 83, 219-249.
41. Schweizer, B. and Sklar, A. (1983). *Probabilistic Metric Spaces*, North Holland, Amsterdam.
42. Turunen, E. (1999). *Mathematics Behind Fuzzy Logic*, Springer–Verlag.
43. Zadeh, L. A. (1965). Fuzzy Sets, Information and Control 8, 338–358.

Aggregation of Fuzzy Relations and Preservation of Transitivity

Susanne Saminger[1], Ulrich Bodenhofer[2], Erich Peter Klement[1],
and Radko Mesiar[3,4]

[1] Department of Knowledge-Based Mathematical Systems,
Johannes Kepler University, A-4040 Linz, Austria
{susanne.saminger,ep.klement}@jku.at
[2] Institute of Bioinformatics,
Johannes Kepler University, A-4040 Linz, Austria
bodenhofer@bioinf.jku.at
[3] Department of Mathematics and Descriptive Geometry,
Faculty of Civil Engineering, Slovak University of Technology,
SK-813 68 Bratislava, Slovakia
mesiar@math.sk
[4] Institute for Research and Applications of Fuzzy Modeling
University of Ostrava, CZ-701 03 Ostrava 1, Czech Republic

Abstract. This contribution provides a comprehensive overview on the
theoretical framework of aggregating fuzzy relations under the premise
of preserving underlying transitivity conditions. As such it discusses the
related property of dominance of aggregation operators. After a thorough
introduction of all necessary and basic properties of aggregation operators, in particular dominance, the close relationship between aggregating
fuzzy relations and dominance is shown. Further, principles of building
dominating aggregation operators as well as classes of aggregation operators dominating one of the basic t-norms are addressed. In the paper
by Bodenhofer, Küng and Saminger, also in this volume, the interested
reader finds an elaborated (real world) example, i.e., an application of
the herein contained theoretical framework.

1   Introduction

Flexible (fuzzy) querying systems are designed not just to give results that match
a query exactly, but to give a list of possible answers ranked by their closeness to
the query—which is particularly beneficial if no record in the database matches
the query in an exact way (see [11, 12, 28, 29] for overviews and [7, 8, 9, 10] for
particular related examples). The closeness of a single value of a record to the
respective value in the query is usually measured by a fuzzy equivalence relation,
that is, a reflexive, symmetric, and T-transitive fuzzy relation. Recently, a generalization has been proposed [7, 8, 9] which also allows flexible interpretation of
ordinal queries (such as "at least" and "at most") by using fuzzy orderings [5].
In any case, if a query consists of at least two expressions that are to be interpreted vaguely, it is necessary to combine the degrees of matching with respect

H. de Swart et al. (Eds.): TARSKI II, LNAI 4342, pp. 185–206, 2006.
© Springer-Verlag Berlin Heidelberg 2006

to the different fields in order to obtain an overall degree of matching — a typical example of an aggregation task. More precisely, assume that we have a query $(q_1, \ldots, q_n)$, where each $q_i \in X_i$ is a value referring to the i-th field of the query. Given a data record $(x_1, \ldots, x_n)$ such that $x_i \in X_i$ for all $i = 1, \ldots, n$, the overall degree of matching is computed as

$$\tilde{R}((q_1, \ldots, q_n), (x_1, \ldots, x_n)) = \mathbf{A}(R_1(q_1, x_1), \ldots, R_n(q_n, x_n)),$$

where every R_i is a T-transitive binary fuzzy relation on X_i which measures the degree to which the value x_i matches the query value q_i.

It is natural to require that $\tilde{R}$ is fuzzy relation on the Cartesian product of all X_i and, therefore, that the range of the operation $\mathbf{A}$ should be the unit interval, i.e., $\mathbf{A} : [0,1]^n \to [0,1]$. Furthermore, it is desirable that if a data record matches one of the criteria of the query better than a second one, then the overall degree of matching for the first should be higher or at least the same as the overall degree of matching for second one. Clearly, if some data record matches all criteria, i.e., all $R_i(x_i, q_i) = 1$, then the overall degree of matching should also be 1. On the other hand, if a data record fulfills none of the criteria to any level, i.e., all $R_i(x_i, q_i) = 0$, then the overall degree should vanish to 0. Aggregation operators are exactly such functions which guarantee all these properties [13, 14, 15, 21].

In addition, it would be desirable that, if all relations R_i on X_i are T-transitive, also $\tilde{R}$ is still T-transitive in order to have a clear interpretation of the aggregated fuzzy relation $\tilde{R}$. It is, therefore, necessary to investigate which aggregation operators are particularly able to guarantee that $\tilde{R}$ maintains T-transitivity.

This contribution provides an overview on results on the aggregation of fuzzy relations and the related property of dominance of aggregation operators which have been achieved by collaboration among different research groups within the EU COST Action TARSKI. The present part focusses on the theoretical background, as such provides a comprehensive overview of the theory of aggregation operators dominating triangular norms as well as depends on results already published in [27, 30, 32]. In addition, in [10], the interested reader finds an elaborated (real world) example, i.e., an application of the herein contained theoretical framework. Next, we provide a thorough introduction of all necessary and basic properties of aggregation operators, in particular dominance. Then we turn to the close relationship between the aggregation of fuzzy relations and dominance. In Section IV, we discuss principles of building dominating aggregation operators and focus in Section V on the class of aggregation operators dominating one of the basic t-norms.

2   Basic Definitions and Preliminaries

In order to be self-contained and to provide a compact overview we provide basic definitions and results about aggregation operators and dominance. For more details on aggregation operators as well as t-norms we refer the interested reader to [2, 14, 21].

2.1   Aggregation Operators

Definition 1. [14] An *aggregation operator* is a function $\mathbf{A} : \bigcup_{n\in\mathbb{N}}[0,1]^n \to [0,1]$ which fulfills the following properties:

(AO1) $\mathbf{A}(x_1,\ldots,x_n) \leq \mathbf{A}(y_1,\ldots,y_n)$ whenever $x_i \leq y_i$ for all $i \in \{1,\ldots,n\}$,
(AO2) $\mathbf{A}(x) = x$ for all $x \in [0,1]$,
(AO3) $\mathbf{A}(0,\ldots,0) = 0$ and $\mathbf{A}(1,\ldots,1) = 1$.

Each aggregation operator $\mathbf{A}$ can be represented by a family $(\mathbf{A}_{(n)})_{n\in\mathbb{N}}$ of n-ary operations, i.e., functions $\mathbf{A}_{(n)} : [0,1]^n \to [0,1]$ given by

$$\mathbf{A}_{(n)}(x_1,\ldots,x_n) = \mathbf{A}(x_1,\ldots,x_n)$$

being non-decreasing and fulfilling $\mathbf{A}_{(n)}(0,\ldots,0) = 0$ and $\mathbf{A}_{(n)}(1,\ldots,1) = 1$. Such operations $\mathbf{A}_{(n)}$ are referred to as *n-ary aggregation operators*. Note also that in such a case $\mathbf{A}_{(1)} = \mathrm{id}_{[0,1]}$. Usually, the aggregation operator $\mathbf{A}$ and the corresponding family $(\mathbf{A}_{(n)})_{n\in\mathbb{N}}$ of n-ary operations are identified with each other.

Unless explicitly mentioned otherwise, we will restrict to aggregation operators acting on the unit interval (according to Definition 1). With only simple and obvious modifications, aggregation operators can be defined to act on any closed interval $I = [a,b] \subseteq [-\infty,\infty]$. Consequently, we will speak of an *aggregation operator acting on I*.

Particularly, such operators can be constructed by rescaling the input and output data, and as such creating isomorphic aggregation operators.

Consider an aggregation operator $\mathbf{A} : \bigcup_{n\in\mathbb{N}} [a,b]^n \to [a,b]$ on $[a,b]$ and a monotone bijection $\varphi : [c,d] \to [a,b]$. The operator $\mathbf{A}_\varphi : \bigcup_{n\in\mathbb{N}} [c,d]^n \to [c,d]$ defined by

$$\mathbf{A}_\varphi(x_1,\ldots,x_n) = \varphi^{-1}\big(\mathbf{A}(\varphi(x_1),\ldots,\varphi(x_n))\big)$$

is an aggregation operator on $[c,d]$, which is isomorphic to $\mathbf{A}$.

A particularly important transformation is duality induced by $\varphi_d : [0,1] \to [0,1]$, $\varphi_d(x) = 1 - x$. Applying this transformation to an aggregation operator $\mathbf{A}$ on the unit interval leads to the so-called *dual aggregation operator* $\mathbf{A}^d$.

Couples of dual aggregation operators are, e.g., the minimum and the maximum. The arithmetic mean is dual to itself. Such aggregation operators, i.e., $\mathbf{A} = \mathbf{A}^d$, are called *self-dual* (compare also [38] where these operators are called symmetric sums).

Let us now briefly summarize further properties of aggregation operators.

Definition 2. Consider some aggregation operator $\mathbf{A} : \bigcup_{n\in\mathbb{N}} [0,1]^n \to [0,1]$.

(i) $\mathbf{A}$ is called *symmetric*, if for all $n \in \mathbb{N}$ and for all $x_1,\ldots,x_n \in [0,1]$:

$$\mathbf{A}(x_1,\ldots,x_n) = \mathbf{A}(x_{\alpha(1)},\ldots,x_{\alpha(n)})$$

for all permutations $\alpha = (\alpha(1),\ldots,\alpha(n))$ of $\{1,\ldots,n\}$.

(ii) **A** is called *associative* if for all $n, m \in \mathbb{N}$ and for all $x_1, \ldots, x_n, y_1, \ldots, y_m \in [0, 1]$:

$$\mathbf{A}(x_1, \ldots, x_n, y_1, \ldots, y_m) = \mathbf{A}(\mathbf{A}(x_1, \ldots, x_n), \mathbf{A}(y_1, \ldots, y_m)).$$

(iii) An element $e \in [0, 1]$ is called a *neutral element* of **A** if for all $n \in \mathbb{N}$ and for all $x_1, \ldots, x_n \in [0, 1]$:

$$\mathbf{A}(x_1, \ldots, x_n) = \mathbf{A}(x_1, \ldots, x_{i-1}, x_{i+1}, \ldots, x_n)$$

whenever $x_i = e$ for some $i \in \{1, \ldots, n\}$.

(iv) **A** is *subadditive* on $[0, 1]$, if the following inequality holds for all $x_i, y_i \in [0, 1]$ with $x_i + y_i \in [0, 1]$:

$$\mathbf{A}(x_1 + y_1, \ldots, x_n + y_n) \leq \mathbf{A}(x_1, \ldots, x_n) + \mathbf{A}(y_1, \ldots, y_n).$$

Observe that, for a given aggregation operator **A**, the operators $\mathbf{A}_{(n)}$ and $\mathbf{A}_{(m)}$ need not be related in general, if $n \neq m$. However, if **A** is an associative aggregation operator, all n-ary operators $\mathbf{A}_{(n)}$, $n \geq 3$, can be identified with recursive extensions of the binary operator $\mathbf{A}_{(2)}$. Therefore, in case of associative aggregation operators, the distinction between $\mathbf{A}_{(2)}$ and **A** itself is often omitted.

Example 1. A typical example of a symmetric, but non-associative aggregation operator without neutral element is the *arithmetic mean* $\mathbf{M} \colon \bigcup_{n \in \mathbb{N}} [a, b]^n \to [a, b]$ defined for any interval $[a, b] \subseteq [-\infty, \infty]$ by

$$\mathbf{M}(x_1, \ldots, x_n) = \frac{1}{n} \sum_{i=1}^{n} x_i.$$

If for some practical purposes some of the properties of the arithmetic mean do not fit the demands of the aggregation process the arithmetic mean is usually modified with respect to the violated property but by preserving as many as possible other properties of the original aggregation operator. Three different approaches can be mentioned — introduction of weights, ordering of the inputs and transformation of the aggregation operator.

We briefly summarize the formal definitions of weighted means, (weighted) quasi-arithmetic means and OWA operators (see also, e.g., [14, 40]). Recall that for a fixed $n \in \mathbb{N}$, weighting vectors $\overrightarrow{w} = (w_1, \ldots, w_n)$ are characterized by fulfilling $\overrightarrow{w} \in [0, 1]^n$ and $\sum_{i=1}^{n} w_i = 1$.

Definition 3. For a continuous strictly monotone function $f : [a, b] \to [-\infty, \infty]$, the *quasi-arithmetic mean* $\mathbf{M}_f \colon \bigcup_{n \in \mathbb{N}} [a, b]^n \to [a, b]$ is given by

$$\mathbf{M}_f(x_1, \ldots, x_n) = f^{-1}\left(\frac{1}{n} \sum_{i=1}^{n} f(x_i)\right).$$

Consider for arbitrary $n \in \mathbb{N}$, a weighting vector $\overrightarrow{w}$. Then the *weighted mean* $\mathbf{W} \colon [a, b]^n \to [a, b]$ is given by

$$\mathbf{W}(x_1, \ldots, x_n) = \sum_{i=1}^{n} w_i x_i$$

and the *weighted quasi-arithmetic mean* $\mathbf{W}_f \colon \bigcup_{n \in \mathbb{N}} [a,b]^n \to [a,b]$ by

$$\mathbf{W}_f(x_1, \ldots, x_n) = f^{-1}\left(\sum_{i=1}^{n} w_i f(x_i)\right).$$

with $f \colon [a,b] \to [-\infty, \infty]$ again some continuous strictly monotone function. An *OWA operator* $\mathbf{W}' \colon \bigcup_{n \in \mathbb{N}} [a,b]^n \to [a,b]$ is characterized by

$$\mathbf{W}'(x_1, \ldots, x_n) = \sum_{i=1}^{n} w_i x_i'$$

where x_i' denotes the i-th order statistics from the sample $(x_1, \ldots, x_n)$ and w_i the corresponding weights.

2.2   Triangular Norms

Triangular norms can be interpreted as a particular class of aggregation operators which were originally introduced in the context of probabilistic metric spaces [25, 35, 36]. We just briefly state the formal definitions and introduce the four basic t-norms. For further details and properties about t-norms we refer to [22, 23, 24] or to the monographs [2, 21].

Definition 4. A *triangular norm* (t-norm for short) is a binary operation T on the unit interval which is commutative, associative, non-decreasing in each component, and has 1 as a neutral element.

Example 2. The following are the four basic t-norms:

$$
\begin{aligned}
&\text{Minimum:} && T_{\mathbf{M}}(x,y) = \min(x,y), \\
&\text{Product:} && T_{\mathbf{P}}(x,y) = x \cdot y, \\
&\text{Łukasiewicz t-norm:} && T_{\mathbf{L}}(x,y) = \max(x+y-1, 0), \\
&\text{Drastic product:} && T_{\mathbf{D}}(x,y) = \begin{cases} 0 & \text{if } (x,y) \in [0,1[^2, \\ \min(x,y) & \text{otherwise.} \end{cases}
\end{aligned}
$$

Several construction principles are known for t-norms. Here we just mention the concept of ordinal sums which allow to define t-norms by a particular behaviour on subdomains and, moreover, gave rise for a construction principle for aggregation operators.

Definition 5. Let $(T_i)_{i \in I}$ be a family of t-norms and let $(]a_i, e_i[)_{i \in I}$ be a family of non-empty, pairwise disjoint open subintervals of $[0,1]$. Then the following function $T \colon [0,1]^2 \to [0,1]$ is a t-norm [21]:

$$T(x,y) = \begin{cases} T_i^*(x,y) = a_i + (e_i - a_i) \cdot T(\frac{x-a_i}{e_i-a_i}, \frac{y-a_i}{e_i-a_i}), & \text{if } (x,y) \in [a_i, e_i]^2, \\ \min(x,y), & \text{otherwise.} \end{cases}$$

The t-norm T is called the *ordinal sum* of the *summands* $\langle a_i, e_i, T_i \rangle, i \in I$, and we shall write $T = (\langle a_i, e_i, T_i \rangle)_{i \in I}$.

Corresponding to t-norms, aggregation operators can also be constructed from several aggregation operators acting on non-overlapping domains. We will use the *lower ordinal sum* of aggregation operators [14,26]. Observe that this ordinal sum was originally proposed only for finitely many summands, however, we generalize this concept to an arbitrary (countable) number of summands.

Definition 6. Consider a family of aggregation operators

$$\left(\mathbf{A}_i : \bigcup_{n\in\mathbb{N}} [a_i, e_i]^n \to [a_i, e_i]\right)_{i\in\{1,\ldots,k\}}$$

acting on non-overlapping domains $[a_i, e_i]$ with $i \in \{1, \ldots, k\}$ and

$$0 \leq a_1 < e_1 \leq a_2 < e_2 \leq \ldots \leq e_k \leq 1.$$

The aggregation operator $\mathbf{A}^{(w)}$ defined by [14]

$$\mathbf{A}^{(w)}(x_1, \ldots, x_n) = \begin{cases} 0, & \text{if } u < a_1, \\ \mathbf{A}_i\big(\min(x_1, e_i), \ldots, \min(x_n, e_i)\big), & \text{if } a_i \leq u < a_{i+1}, \\ 1, & \text{if } u = 1. \end{cases}$$

with $u = \min(x_1, \ldots, x_n)$ is called the *lower ordinal sum* (of aggregation operators $\mathbf{A}_i$) and it is the weakest aggregation operator (with respect to the standard ordering of n-ary functions) that coincides with $\mathbf{A}_i$ at inputs from $[a_i, e_i]$.

If $(\mathbf{A}_i)_{i\in I}$ is a family of aggregation operators on $[0, 1]$ and $(]a_i, e_i[)_{i\in I}$ a (countable) family of non-empty, pairwise disjoint open subintervals of $[0, 1]$, then the lower ordinal sum of this family $\mathbf{A}^{(w)} = (\langle a_i, e_i, \mathbf{A}_i\rangle)_{i\in I}$ can be constructed in the following way:

$$\mathbf{A}^{(w)}(x_1, \ldots, x_n) = \begin{cases} \sup_{i\in I}\{\mathbf{A}_i^*\big(\min(x_1, e_i), \ldots, \min(x_n, e_i)\big) \mid a_i \leq u\}, \\ \qquad\qquad\qquad\qquad\qquad\qquad \text{if } u < 1, \\ 1, \qquad\qquad\qquad\qquad\qquad\qquad \text{otherwise,} \end{cases}$$

with $\sup \emptyset = 0$ and $u = \min(x_1, \ldots, x_n)$. $\mathbf{A}_i^*$ denotes the aggregation operator $\mathbf{A}_i$, scaled for acting on $[a_i, e_i]$ by

$$\mathbf{A}_i^*(x_1, \ldots, x_n) = a_i + (e_i - a_i) \cdot \mathbf{A}_i\left(\tfrac{x_1-a_i}{e_i-a_i}, \ldots, \tfrac{x_n-a_i}{e_i-a_i}\right).$$

2.3   Transitivity and Preservation of Transitivity

We have already mentioned that binary fuzzy relations R_i on the subspaces X_i can be used for the comparison of two objects on the subspaces' level. For details on fuzzy relations, especially fuzzy equivalence relations we recommend [3, 16, 17, 19, 42] and for fuzzy orderings [4, 5, 6, 20, 42]. We only recall the definition of T-transitivity, since we are interested in its preservation during the aggregation process.

Definition 7. Consider a binary fuzzy relation R on some universe X and an arbitrary t-norm T. R is called T-*transitive* if and only if, for all $x, y, z \in X$ the following property holds

$$T\big(R(x, y), R(y, z)\big) \leq R(x, z).$$

Definition 8. An aggregation operator $\mathbf{A}$ *preserves* T-*transitivity* if, for all $n \in \mathbb{N}$ and for all binary T-transitive fuzzy relations R_i on X_i with $i \in \{1, \ldots, n\}$, the aggregated relation $\tilde{R} = \mathbf{A}(R_1, \ldots, R_n)$ on the Cartesian product of all X_i, i.e.,

$$\tilde{R}(A, B) = \tilde{R}((a_1, \ldots, a_n), (b_1, \ldots, b_n)) = \mathbf{A}\big(R_1(a_1, b_1), \ldots, R_n(a_n, b_n)\big),$$

is also T-transitive, that means, for all $A, B, C \in \prod_{i=1}^{n} X_i$,

$$T\big(\tilde{R}(A, B), \tilde{R}(B, C)\big) \leq \tilde{R}(A, C).$$

Without loss of generality, we will restrict our considerations to fuzzy relations on the same universe $X_i = X$.

2.4   Dominance — Basic Notions and Properties

Similar to t-norms, the concept of dominance has been introduced in the framework of probabilistic metric spaces [37,39] when constructing the Cartesian products of such spaces. In the framework of t-norms, dominance is also needed when constructing T-equivalence relations and fuzzy orderings [4, 6, 16, 17] on some Cartesian product.

Definition 9. Consider two t-norms T_1 and T_2. We say that T_1 *dominates* T_2 if for all $x, y, u, v \in [0, 1]$ the following inequality holds

$$T_2(T_1(x, y), T_1(u, v)) \leq T_1(T_2(x, u), T_2(y, v)).$$

It can be easily verified (see also, e.g., [21]) that for any t-norm T, it holds that T itself and $T_{\mathbf{M}}$ dominate T. Furthermore, for any two t-norms T_1, T_2, $T_1 \gg T_2$ implies $T_1 \geq T_2$ and, therefore, we know that $T_{\mathbf{D}} \gg T$ if and only if $T = T_{\mathbf{D}}$ and $T \gg T_{\mathbf{M}}$ if and only if $T = T_{\mathbf{M}}$, since $T_{\mathbf{D}}$ is the weakest and $T_{\mathbf{M}}$ the strongest t-norm.

We have already mentioned before that t-norms can be interpreted as particular aggregation operators. Therefore, we extend the concept of dominance to the framework of aggregation operators [32].

Definition 10. Consider an n-ary aggregation operator $\mathbf{A}_{(n)}$ and an m-ary aggregation operator $\mathbf{B}_{(m)}$. We say that $\mathbf{A}_{(n)}$ *dominates* $\mathbf{B}_{(m)}$, $\mathbf{A}_{(n)} \gg \mathbf{B}_{(m)}$, if, for all $x_{i,j} \in [0, 1]$ with $i \in \{1, \ldots, m\}$ and $j \in \{1, \ldots, n\}$, the following property holds

$$\mathbf{B}_{(m)}\big(\mathbf{A}_{(n)}(x_{1,1}, \ldots, x_{1,n}), \ldots, \mathbf{A}_{(n)}(x_{m,1}, \ldots, x_{m,n})\big)$$
$$\leq \mathbf{A}_{(n)}\big(\mathbf{B}_{(m)}(x_{1,1}, \ldots, x_{m,1}), \ldots, \mathbf{B}_{(m)}(x_{1,n}, \ldots, x_{m,n})\big). \quad (1)$$

Note that if either n or m or both are equal to 1, because of the boundary condition **(AO2)**, $\mathbf{A}_{(n)} \gg \mathbf{B}_{(m)}$ is trivially fulfilled for any two aggregation operators $\mathbf{A}, \mathbf{B}$.

Definition 11. Let $\mathbf{A}$ and $\mathbf{B}$ be aggregation operators. We say that $\mathbf{A}$ *dominates* $\mathbf{B}$, $\mathbf{A} \gg \mathbf{B}$, if $\mathbf{A}_{(n)}$ dominates $\mathbf{B}_{(m)}$ for all $n, m \in \mathbb{N}$.

Note that, if two aggregation operators $\mathbf{A}$ and $\mathbf{B}$ are both acting on some closed interval $I = [a, b] \subseteq [-\infty, \infty]$, then the property of dominance can be easily adapted by requiring that (1) must hold for all arguments $x_{i,j} \in I$ and for all $n, m \in \mathbb{N}$. Further note that the concept of dominance relates to the fact that aggregation operators are operators on posets. Therefore, dominance can and has been introduced for arbitrary operations on posets (see, e.g., [37]).

Due to the monotonicity of aggregation operators, the minimum $T_{\mathbf{M}}$ dominates not only all t-norms, but also any aggregation operator $\mathbf{A}$,

$$\mathbf{A}(\min(x_1, y_1), \ldots, \min(x_n, y_n)) \leq \min(\mathbf{A}(x_1, \ldots, x_n), \mathbf{A}(y_1, \ldots, y_n)).$$

however, as will be shown later, not all aggregation operators dominate $T_{\mathbf{D}}$. Similarly, not all aggregation operators dominate the weakest aggregation operator

$$\mathbf{A}_w(x_1, \ldots, x_n) = \begin{cases} 1, & \text{if } x_1 = \ldots = x_n = 1, \\ 0. & \text{otherwise.} \end{cases}$$

Further on, we will denote the class of all aggregation operators $\mathbf{A}$ which dominate an aggregation operator $\mathbf{B}$ by

$$\mathcal{D}_{\mathbf{B}} = \{\mathbf{A} \mid \mathbf{A} \gg \mathbf{B}\}.$$

Since t-norms are special kinds of associative aggregation operators, the following proposition will be helpful for considering the dominance of an aggregation operator over a t-norm T.

Proposition 1. [32] *Let $\mathbf{A}, \mathbf{B}$ be two aggregation operators. Then the following holds:*

(i) If $\mathbf{B}$ is associative and $\mathbf{A}_{(n)} \gg \mathbf{B}_{(2)}$ for all $n \in \mathbb{N}$, then $\mathbf{A} \gg \mathbf{B}$.
(ii) If $\mathbf{A}$ is associative and $\mathbf{A}_{(2)} \gg \mathbf{B}_{(m)}$ for all $m \in \mathbb{N}$, then $\mathbf{A} \gg \mathbf{B}$.

Consequently, if two aggregation operators $\mathbf{A}$ and $\mathbf{B}$ are both associative, as it would be in the case of two t-norms, it is sufficient to show that $\mathbf{A}_{(2)} \gg \mathbf{B}_{(2)}$ for proving that $\mathbf{A} \gg \mathbf{B}$.

In case of a common neutral element, the property of dominance induces the order of the involved aggregation operators.

Lemma 1. [30] *Consider two aggregation operators $\mathbf{A}, \mathbf{B}$ with a common neutral element $e \in [0, 1]$. If $\mathbf{A}$ dominates $\mathbf{B}$, i.e., $\mathbf{A} \gg \mathbf{B}$, then $\mathbf{A} \geq \mathbf{B}$.*

As a consequence, it is clear that dominance is a reflexive and antisymmetric relation on the set of all t-norms, but it is not transitive as could be shown in [34] (for a counter example see also [33]). Note that transitivity of dominance in the framework of aggregation operators does not hold in general, since, e.g., $\mathbf{A}_w \gg T_{\mathbf{M}}$ and $T_{\mathbf{M}} \gg \mathbf{M}$ but $\mathbf{A}_w$ does not dominate $\mathbf{M}$ (see also [30]).

Further note, that the property of selfdominance of an aggregation operator, i.e., $\mathbf{A} \gg \mathbf{A}$, is nothing else than the property of bisymmetry in the sense of Aczél [1], i.e., for all $n, m \in \mathbb{N}$ and all $x_{i,j} \in [0,1]$ with $i \in \{1, \ldots, m\}$ and $j \in \{1, \ldots, n\}$

$$\mathbf{A}_{(m)}\big(\mathbf{A}_{(n)}(x_{1,1}, \ldots, x_{1,n}), \ldots, \mathbf{A}_{(n)}(x_{m,1}, \ldots, x_{m,n})\big)$$
$$= \mathbf{A}_{(n)}\big(\mathbf{A}_{(m)}(x_{1,1}, \ldots, x_{m,1}), \ldots, \mathbf{A}_{(m)}(x_{1,n}, \ldots, x_{m,n})\big).$$

Another interesting aspect is the invariance of dominance with respect to transformations.

Proposition 2. [32] *Consider two aggregation operators* $\mathbf{A}$ *and* $\mathbf{B}$ *on* $[a, b]$.

(i) $\mathbf{A} \gg \mathbf{B}$ *if and only if* $\mathbf{A}_\varphi \gg \mathbf{B}_\varphi$ *for all strictly increasing bijections* $\varphi : [c, d] \to [a, b]$.
(ii) $\mathbf{A} \gg \mathbf{B}$ *if and only if* $\mathbf{B}_\varphi \gg \mathbf{A}_\varphi$ *for all strictly decreasing bijections* $\varphi : [c, d] \to [a, b]$.

3 T-Transitivity and Dominance

Standard aggregation of fuzzy equivalence relations and fuzzy orderings preserving the T-transitivity has been done either by means of T itself or $T_{\mathbf{M}}$, but in fact, any t-norm $\tilde{T}$ dominating T can be applied, i.e., if R_1, R_2 are two T-transitive binary relations on a universe X and $\tilde{T} \gg T$, then also $\tilde{T}(R_1, R_2)$ is T-transitive [4, 6, 16].

As already mentioned above, in several applications, other types of aggregation processes preserving T-transitivity are required [8, 10] Especially the introduction of different weights (degrees of importance) for input fuzzy equivalences and orderings cannot be properly done by aggregation with t-norms, because of the commutativity. Therefore, we investigated aggregation operators preserving the T-transitivity of the aggregated fuzzy relations. The following theorem generalizes the result known for triangular norms [16].

Theorem 1. [32] *Let* $|X| \geq 3$ *and let* T *be an arbitrary t-norm. An aggregation operator* $\mathbf{A}$ *preserves the* T-*transitivity of fuzzy relations on* X *if and only if* $\mathbf{A} \in \mathcal{D}_T$.

4 Construction of Dominating Aggregation Operators

Since we have shown the close relationship between the preservation of T-transitivity and the dominance of the involved aggregation operator $\mathbf{A}$ over T, we are interested in the characterization of $\mathcal{D}_T$ for some t-norm T. Particularly, we are interested in the introduction of weights, respectively determining operations by its behaviour on subdomains.

4.1   Generated and Weighted T-Norms

Before turning to aggregation operators dominating a continuous, Archimedean t-norm T, recall that they are characterized by having a continuous *additive generator*, i.e., a continuous, strictly decreasing function $t : [0,1] \to [0,\infty]$ which fulfils $t(1) = 0$, and for all $x, y \in [0,1]$:

$$T(x,y) = t^{-1}\big(\min(t(0), t(x) + t(y))\big).$$

Then we also have that $T(x_1, \dots, x_n) = t^{-1}\big(\min(t(0), \sum_{i=1}^n t(x_i))\big)$.

Theorem 2. [32] *Consider some continuous, Archimedean t-norm T with an additive generator $t : [0,1] \to [0,c]$, with $t(0) = c$ and $c \in \,]0,\infty]$. Furthermore, let $\mathbf{A} : \bigcup_{n\in\mathbb{N}}[0,1]^n \to [0,1]$ be an aggregation operator. Then $\mathbf{A} \in \mathcal{D}_T$ if and only if the aggregation operator $\mathbf{H} : \bigcup_{n\in\mathbb{N}} [0,c]^n \to [0,c]$ defined by*

$$\mathbf{H}(z_1, \dots, z_n) = t(\mathbf{A}(t^{-1}(z_1), \dots, t^{-1}(z_n))) \tag{2}$$

for all $n \in \mathbb{N}$ and all $z_i \in [0,c]$ with $i \in \{1, \dots, n\}$ is subadditive on $[0,c]$.

One of the main purposes for investigating aggregation operators dominating t-norms was the request for introducing weights into the aggregation process. Hence, considering continuous Archimedean t-norms, we have to find subadditive aggregation operators, which provide this possibility.

Example 3. Consider some some weights $w_1, \dots, w_n \in [0,\infty]$, $n \geq 2$, and some $c \in \,]0,\infty]$, then $\mathbf{H}_{(n)} : [0,c]^n \to [0,c]$ given by

$$\mathbf{H}_{(n)}(x_1, \dots, x_n) = \min(c, \sum_{i=1}^n w_i x_i)$$

is an n-ary, subadditive aggregation operator on $[0,c]$, fulfilling $\mathbf{H}_{(n)}(c, \dots, c) = c$, whenever $c \leq c \cdot \sum_{i=1}^n p_i$. This means, with convention $0 \cdot \infty = 0$, if $c = \infty$, the sum must fulfill $\sum_{i=1}^n w_i > 0$ and if $c < \infty$, then also $\sum_{i=1}^n w_i \geq 1$.

If we combine such an aggregation operator with an additive generator of a continuous Archimedean t-norm by applying the construction method as proposed in Theorem 2 we can introduce weights into the aggregation process without losing T-transitivity.

Corollary 1. *Consider a continuous Archimedean t-norm T with additive generator t, $t(0) = c$, and a weighting vector $\vec{w} = (w_1, \dots, w_n)$, $n \geq 2$, with weights $w_i \in [0,\infty]$ fulfilling $c \leq c \cdot \sum_{i=1}^n w_i$. Further, let $\mathbf{A}_{(n)} : [0,1]^n \to [0,1]$ be an n-ary aggregation operator defined by Eq. (2) from the aggregation operator $\mathbf{H}_{(n)}$ introduced in Example 3. Then the n-ary aggregation operator can be rewritten by*

$$\mathbf{A}_{(n)}(x_1, \dots, x_n) = t^{-1}\big(\min(t(0), \sum_{i=1}^n w_i \cdot t(x_i))\big) \tag{3}$$

and it dominates the t-norm T, i.e., $\mathbf{A}_{(n)} \gg T$.

Remark 1. Note that the n-ary aggregation operator defined by Equation (3) is also called weighted t-norm $T_{\vec{w}}$ ([15, 21]). Further, for any strict t-norm T, it holds, that not only $T_{\vec{w}} \gg T$, but also $T \gg T_{\vec{w}}$. In case of some nilpotent t-norm T it is clear, that $T_{\vec{w}} \gg T$, but $T \gg T_{\vec{w}}$ only if all weights $w_i \notin \,]0,1[$. In case that $\sum_{i=1}^{n} w_i = 1$ we can apply Corollary 1 independently of $t(0)$. Thus for a continuous Archimedean t-norm T with additive generator t, any weighted quasi-arithmetic mean $\mathbf{W}_t$ dominates T. Especially, any weighted arithmetic mean $\mathbf{W}$ dominates $T_{\mathbf{L}}$ and any weighted geometric mean dominates $T_{\mathbf{P}}$.

Example 4. The strongest subadditive aggregation operator acting on $[0, c]$ is given by $\mathbf{H} : \bigcup_{n \in \mathbb{N}} [0, c]^n \to [0, c]$ with

$$\mathbf{H}(u_1, \ldots, u_n) = \begin{cases} 0, & \text{if } u_1 = \ldots = u_n = 0, \\ c, & \text{otherwise.} \end{cases}$$

Then, for any additive generator $t : [0, 1] \to [0, \infty]$ with $t(0) = c$, we have

$$t(\mathbf{A}(x_1, \ldots, x_n)) = \mathbf{H}(t(x_1), \ldots, t(x_n)),$$

for all $x_i \in [0, 1]$ with $i \in \{1, \ldots, n\}$ and some $n \in \mathbb{N}$, if and only if

$$\mathbf{A}(x_1, \ldots, x_n) = \begin{cases} 1, & \text{if } x_1 = \ldots = x_n = 1, \\ 0, & \text{otherwise,} \end{cases}$$

i.e., $\mathbf{A} = \mathbf{A}_w$ is the weakest aggregation. Observe that $\mathbf{A}_w$ dominates all t-norms, but not all aggregation operators, e.g., $\mathbf{A}_w$ does not dominate the arithmetic mean.

4.2   Ordinal Sums

Proposition 3. [32] *Let $(T_i)_{i \in I}$ be a family of t-norms, $(\mathbf{A}_i)_{i \in I}$ a family of aggregation operators, and $(]a_i, e_i[)_{i \in I}$ a family of non-empty, pairwise disjoint open subintervals of $[0, 1]$. If for all $i \in I : \mathbf{A}_i \in \mathcal{D}_{T_i}$, then the lower ordinal sum $A^{(w)} = (\langle a_i, e_i, \mathbf{A}_i \rangle)_{i \in I}$ dominates the ordinal sum $T = (\langle a_i, e_i, T_i \rangle)_{i \in I}$, i.e., $A^{(w)} \in \mathcal{D}_T$.*

Note that not all dominating aggregation operators are lower ordinal sums of dominating aggregation operators, e.g., the aggregation operator $\mathbf{A}_w$ introduced in Example 4 dominates all t-norms T, but is not a lower ordinal sum constructed by means of some index set I (in fact it is the empty lower ordinal sum). On the other hand, in case of summand t-norms the lower ordinal sum $\mathbf{A}_w = (\langle a_i, e_i, T_i \rangle)_{i \in I}$ coincides with the standard ordinal sum of t-norms $T = (\langle a_i, e_i, T_i \rangle)_{i \in I}$. Moreover, as shown in [31], the condition of Proposition 3 is not only sufficient but also necessary.

The following example also shows that weighted t-norms as proposed by Calvo and Mesiar [15] dominate the original t-norm but are no lower ordinal sums as proposed here. As a consequence we can conclude that $(\langle a_i, e_i, \mathcal{D}_{T_i'}\rangle)_{i \in I} \subset \mathcal{D}_T$, whenever $T = (\langle a_i, e_i, T_i\rangle_{i \in I}$.

Let $(]a_i, e_i[)_{i \in I}$ be a family of non-empty, pairwise disjoint open subintervals of $[0, 1]$ and let $t_i : [a_i, e_i] \to [0, \infty]$ be continuous, strictly decreasing mappings fulfilling $t_i(e_i) = 0$. Then (and only then) the following function $T : [0, 1]^2 \to [0, 1]$ is a continuous t-norm [15]:

$$T(x, y) = \begin{cases} t_i^{-1}\big(\min(t_i(0), t_i(x) + t_i(y))\big), & \text{if } (x, y) \in [a_i, e_i], \\ \min(x, x), & \text{otherwise.} \end{cases}$$

The corresponding weighted t-norm $T_{\vec{w}}$ in the sense of Calvo and Mesiar [15] is defined by

$$T_{\vec{w}}(x_1, \ldots, x_n) = \begin{cases} t_i^{-1}(\min(t_i(a_i), \sum_{i=1}^n w_i \cdot t_i(\min(x_i, e_i)))), & \text{if } u \in [a_i, e_i[, \\ \min(x_i \mid w_i > 0), & \text{otherwise,} \end{cases}$$

with $u = \min(x_i \mid w_i > 0)$ and some weighting vector $\vec{w} = (w_1, \ldots, w_n) \neq (0, \ldots, 0)$ such that, if $a_i = 0$ for some $i \in I$ and the corresponding $t_i(a_i)$ is finite, then $\sum_{i=1}^n w_i \geq 1$.

Example 5. Consider the t-norm $T = (\langle 0, \frac{1}{2}, T_{\mathbf{P}}\rangle)$, i.e.,

$$T(x, y) = \begin{cases} 2xy, & \text{if } (x, y) \in [0, \frac{1}{2}]^2, \\ \min(x, y), & \text{otherwise.} \end{cases}$$

We know that the geometric mean $G(x, y) = \sqrt{x \cdot y} = T_{\mathbf{P}(\frac{1}{2}, \frac{1}{2})}$ dominates $T_{\mathbf{P}}$. Therefore we can construct

- the lower ordinal sum $\mathbf{A}^{(w)} = (\langle 0, \frac{1}{2}, G\rangle)$ with

$$\mathbf{A}^{(w)}(x, y) = \begin{cases} 1, & \text{if}(x, y) = (1, 1), \\ \sqrt{\min(x, \frac{1}{2}) \cdot \min(y, \frac{1}{2})}, & \text{otherwise} \end{cases}$$

- and the weighted t-norm $T_{\vec{w}} = T_{(\frac{1}{2}, \frac{1}{2})}$ by

$$T_{(\frac{1}{2}, \frac{1}{2})}(x, y) = \begin{cases} \min(x, y), & \text{if } (x, y) \in]\frac{1}{2}, 1]^2, \\ \sqrt{\min(x, \frac{1}{2}) \cdot \min(y, \frac{1}{2})}, & \text{otherwise.} \end{cases}$$

Both aggregation operators — $\mathbf{A}^{(w)}$ as well as $T_{\vec{w}}$ — dominate the t-norm T and they coincide in any values except for arguments $(x, y) \in]\frac{1}{2}, 1]^2 \setminus \{(1, 1)\}$. Observe that this example also demonstrates that not all aggregation operators dominating an ordinal sum t-norm T are necessarily lower ordinal sums of dominating aggregation operators as given in Proposition 3.

5   Dominance of Basic T-Norms

Finally we will discuss the classes of aggregation operators dominating one of the basic t-norms as introduced in Example 2.

5.1   Dominance of the Minimum

As already observed, T_M dominates any t-norm T and any aggregation operator $\mathbf{A}$, but no t-norm T, except T_M itself, dominates T_M. The class of all aggregation operators dominating T_M is described in the following proposition.

Proposition 4. [32] *For any $n \in \mathbb{N}$, the class of all n-ary aggregation operators $\mathbf{A}_{(n)}$ dominating the strongest t-norm T_M is given by*

$$\mathcal{D}^{(n)}_{\min} = \{\min_{\mathcal{F}} \mid \mathcal{F} = (f_1, \dots, f_n),$$
$$f_i : [0,1] \to [0,1], \; non\text{-}decreasing, \; with$$
$$f_i(1) = 1 \; for \; all \; i \in \{1, \dots, n\},$$
$$f_i(0) = 0 \; for \; at \; least \; one \; i \in \{1, \dots, n\}\},$$

where $\min_{\mathcal{F}}(x_1, \dots, x_n) = \min(f_1(x_1), \dots, f_n(x_n))$.

Evidently, $\mathbf{A}_{(n)} \in \mathcal{D}^{(n)}_{\min}$ is symmetric if and only if

$$\mathbf{A}_{(n)}(x_1, \dots, x_n) = f\big(\min(x_1, \dots, x_n)\big)$$

for some non-decreasing function $f : [0,1] \to [0,1]$ with $f(0) = 0$ and $f(1) = 1$.

Example 6. As already observed in Example 4, the weakest aggregation operator $\mathbf{A}_w$ dominates all t-norms T. Since this aggregation operator is symmetric, it can be described by $\mathbf{A}_w(x_1, \dots, x_n) = f\big(\min(x_1, \dots, x_n)\big)$ with $f : [0,1] \to [0,1]$ given by

$$f(x) = \begin{cases} 1, & \text{if } x = 1, \\ 0, & \text{otherwise.} \end{cases}$$

Remark 2. Any aggregation operator $\mathbf{A}$ dominating T_M is also dominated by T_M, i.e., for arbitrary $n, m \in \mathbb{N}$ and for all $x_{i,j} \in [0,1]$ with $i \in \{1, \dots, n\}$ and $j \in \{1, \dots, m\}$ the following equality holds

$$\mathbf{A}\big(\min(x_{1,1}, \dots, x_{1,n}), \dots, \min(x_{m,1}, \dots, x_{m,n})\big)$$
$$= \min\big(\mathbf{A}(x_{1,1}, \dots, x_{m,1}), \dots, \mathbf{A}(x_{1,n}, \dots, x_{m,n})\big).$$

5.2   Dominance of the Drastic Product

Oppositely to the case of T_M, the weakest t-norm $T_D : [0,1]^2 \to [0,1]$ is dominated by any t-norm T. This can also be seen from the characterization of all aggregation operators dominating T_D as given in the next proposition.

Proposition 5. [32] *Consider an arbitrary $n \in \mathbb{N}$ and an n-ary aggregation operator $\mathbf{A}_{(n)} \cdot [0,1]^n \to [0,1]$. Then $\mathbf{A}_{(n)} \gg T_{\mathbf{D}}$ if and only if there exists a non-empty subset $I = \{k_1, \ldots, k_m\} \subseteq \{1, \ldots, n\}, k_1 < \ldots < k_m$, and a non-decreasing mapping $B : [0,1]^m \to [0,1]$ satisfying the following conditions*

> *(i) $B(0, \ldots, 0) = 0$,*
> *(ii) $B(u_1, \ldots, u_m) = 1$ if and only if $u_1 = \ldots = u_m = 1$,*

such that $\mathbf{A}(x_1, \ldots, x_n) = B(x_{k_1}, \ldots, x_{k_m})$.

Observe that the mapping B in the above proposition is an m-ary aggregation operator whenever $m \geq 2$. However, if $m = 1$, i.e., $I = \{k\}$, then $B : [0,1] \to [0,1]$ is a non-decreasing mapping with strict maximum $B(1) = 1$ and $B(0) = 0$ as well as $A(x_1, \ldots, x_n) = B(x_k)$ and is therefore a distortion of the k-th projection.

Concerning t-norms, for any t-norm T, we have $T(x_1, \ldots, x_n) = 1$ if and only if $x_i = 1$ for all $i \in \{1, \ldots, n\}$ and thus $I = \{1, \ldots, n\}$. Therefore $B = T$ and $T \in \mathcal{D}_{T_{\mathbf{D}}}$.

5.3   Dominance of the Łukasiewicz T-Norm

Summarizing the results from Section 4.1 we can characterize aggregation operators dominating the Łukasiewicz t-norm $T_{\mathbf{L}}$ by means of the subadditivity of the corresponding dual operator.

Theorem 3. [27] *An aggregation operator $\mathbf{A} \colon \bigcup_{n \in \mathbb{N}} [0,1]^n \to [0,1]$ dominates $T_{\mathbf{L}}$ if and only if its dual aggregation operator $\mathbf{A}^d \colon \bigcup_{n \in \mathbb{N}} [0,1]^n \to [0,1]$ is subadditive.*

Note that as a consequence of Proposition 2 an aggregation operator is dominated by $T_{\mathbf{L}}$ if and only if its dual aggregation operator $\mathbf{A}^d$ is superadditive.

As already mentioned in Remark 1, any weighted arithmetic mean $\mathbf{W}$ dominates $T_{\mathbf{L}}$. Moreover, due to Corollary 1, for any constant $c \in [1, \infty[$ we have also that $\mathbf{B} \colon \bigcup_{n \in \mathbb{N}} [0,1]^n \to [0,1]$, defined by

$$\mathbf{B}(x_1, \ldots, x_n) = \max(0, c \cdot \mathbf{W}(x_1, \ldots, x_n) + 1 - x)$$

dominates $T_{\mathbf{L}}$.

Based on Theorem 3 several other aggregation operators dominating $T_{\mathbf{L}}$ can be introduced. For example, the function $H \colon \bigcup_{n \in \mathbb{N}} [0, \infty]^n \to [0, \infty]$ given by

$$H(x_1, \ldots, x_n) = (\sum_{i=1}^{n} x_i^\lambda)^{\frac{1}{\lambda}}$$

is subadditive for any $\lambda \geq 1$. Therefore, also the Yager t-conorm $S_\lambda^{\mathbf{Y}} = \min(H, 1)$ is subadditive such that the Yager t-norm $T_\lambda^{\mathbf{Y}}$ dominates $T_{\mathbf{L}}$ for all $\lambda \in [1, \infty[$.

Similarly any root-power operator [18] $\mathbf{A}_\lambda \colon \bigcup_{n \in \mathbb{N}} [0,1]^n \to [0,1]$ given by

$$\mathbf{A}_\lambda(x_1, \ldots, x_n) = (\frac{1}{n} \sum_{i=1}^{n} x_i^\lambda)^{\frac{1}{\lambda}}$$

is subadditive for any $\lambda \geq 1$. As a consequence its dual aggregation operator $\mathbf{A}_\lambda^d : \bigcup_{n \in \mathbb{N}} [0,1]^n \to [0,1]$

$$\mathbf{A}_\lambda^d(x_1, \ldots, x_n) = 1 - \left(\tfrac{1}{n} \sum_{i=1}^n (1 - x_i)^\lambda \right)^{\frac{1}{\lambda}}$$

dominates $T_{\mathbf{L}}$.

For the aggregation of fuzzy relations, the introduction of weights in the aggregation process has been of importance. Therefore, the dominance of OWA operators over $T_{\mathbf{L}}$ is an interesting problem.

Proposition 6. [27] *Consider an n-ary OWA operator $\mathbf{W}'_{(n)}$, $n \in \mathbb{N}$, with weights $w_1, \ldots, w_n$. Then $\mathbf{W}'_{(n)}$ dominates $T_{\mathbf{L}}$ if and only if $w_1 \geq w_2 \geq \ldots \geq w_n$.*

If we consider an OWA operator $\mathbf{W}' : \bigcup_{n \in \mathbb{N}} [0,1]^n \to [0,1]$, it is clear that $\mathbf{W}' \gg T_{\mathbf{L}}$ if and only if $\mathbf{W}'_{(n)} \gg T_{\mathbf{L}}$ for all $n \in \mathbb{N}$.

It has been proposed in [41] to derive the weights for an OWA operator from some quantifier function $q : [0,1] \to [0,1]$, which is a monotone real function such that $\{0,1\} \subseteq \operatorname{Ran} q$. As a consequence, q can either be non-decreasing with $q(0) = 0$ and $q(1) = 1$ or can be non-increasing with $q(0) = 1$ and $q(1) = 0$.

Since we are looking for aggregation operators dominating $T_{\mathbf{L}}$, the corresponding weights for each n-ary operator must be non-increasing. Therefore we are looking for additional properties for the quantifier function, such that the non-increasingness of the weights is guaranteed. It will turn out, that non-increasingness of the weights is closely related to the concavity, resp. the convexity of the involved quantifier.

Definition 12. A function f on some convex domain A is *convex*, if the following inequality

$$f(\lambda x + (1 - \lambda)y) \leq \lambda f(x) + (1 - \lambda)f(y)$$

holds for all $\lambda \in [0,1]$ and $x, y \in A$. The function is said to be *concave*, if the inequality

$$f(\lambda x + (1 - \lambda)y) \geq \lambda f(x) + (1 - \lambda)f(y)$$

holds for all $\lambda \in [0,1]$ and $x, y \in A$.

First, we will restrict our considerations to non-decreasing quantifiers. Some examples for such functions are shown in Fig. 1. The weights derived from such a quantifier can be computed by

$$w_{in} = q(\tfrac{i}{n}) - q(\tfrac{i-1}{n}).$$

Lemma 2. *If $q : [0,1] \to [0,1]$ is a non-decreasing quantifier for some OWA operator and the generated weights fulfill $w_{1,n} \geq \ldots \geq w_{n,n}$ for all $n \in \mathbb{N}$ and $i \in \{1, \ldots, n\}$, then q is continuous on $]0,1[$.*

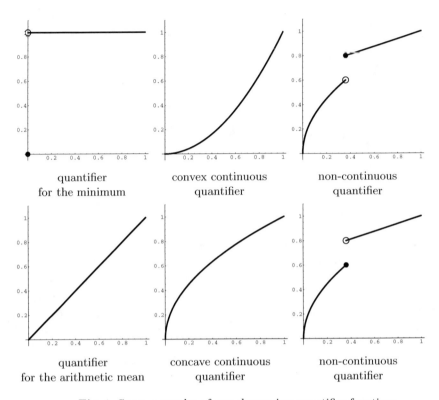

Fig. 1. Some examples of non-decreasing quantifier functions

Proposition 7. [27] *Consider some OWA operator with non-decreasing quantifier* $q : [0,1] \rightarrow [0,1]$ *and generated weights* $w_{1,n}, \ldots, w_{n,n}$ *for all* $n \in \mathbb{N}$. *Then these weights fulfill* $w_{1,n} \geq \ldots \geq w_{n,n}$ *for all* $n \in \mathbb{N}$ *if and only if* q *is concave on* $]0,1]$, *i.e.,* $\forall x, y \in [0,1], \forall \lambda \in [0,1]$

$$q(\lambda x + (1-\lambda)y) \geq \lambda q(x) + (1-\lambda)q(y).$$

Example 7. A typical example of an OWA operator $\mathbf{W}'$ dominating $T_{\mathbf{L}}$ is generated by the quantifier function $q(x) = 2x - x^2$. Observe that for any $n \in \mathbb{N}$ the corresponding weights are given by

$$w_{in} = \frac{2(n-i)+1}{n^2}, \quad i \in \{1, \ldots, n\}.$$

If a quantifier function is non-increasing then the weights can be computed by

$$w_{in} = q(\tfrac{i-1}{n}) - q(\tfrac{i}{n}).$$

For a few examples of non-increasing quantifiers see Fig. 2. The following properties can be shown analogously to the case of non-decreas-ing quantifiers.

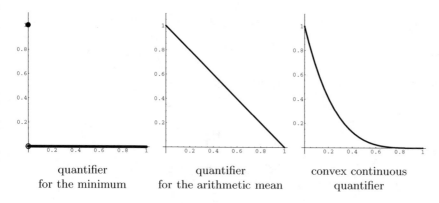

| quantifier | quantifier | convex continuous |
| for the minimum | for the arithmetic mean | quantifier |

Fig. 2. Some examples of non-increasing quantifier functions

Corollary 2. *If $q : [0,1] \to [0,1]$ is a non-increasing quantifier for some OWA operator and the generated weights fulfill $w_{1n} \geq \ldots \geq w_{nn}$ for all $n \in \mathbb{N}$ and $i \in \{1, \ldots, n\}$, then q is continuous on $]0,1]$.*

Corollary 3. *Consider some OWA operator with non-increasing quantifier $q : [0,1] \to [0,1]$. Then the generated weights fulfill $w_{1n} \geq \ldots \geq w_{nn}$ for all $n \in \mathbb{N}$ if and only if q is convex on $]0,1]$, i.e., $\forall x, y \in [0,1], \forall \lambda \in [0,1]$*

$$q(\lambda x + (1-\lambda)y) \leq \lambda q(x) + (1-\lambda)q(y).$$

Remark 3. Any nilpotent t-norm T is isomorphic to the Łukasiewicz t-norm $T_{\mathbf{L}}$, i.e., $T = (T_{\mathbf{L}})_\varphi$ with $\varphi : [0,1] \to [0,1]$ a strictly increasing bijection. According to Proposition 2, we know that if $T_{\mathbf{L}}$ is dominated by an OWA operator $\mathbf{W}'$ then an isomorphic t-norm $T = (T_{\mathbf{L}})_\varphi$ is dominated by the aggregation operator $\mathbf{W}'_\varphi$. In fact $\mathbf{W}'_\varphi$ is nothing else than an ordered weighted quasi-arithmetic mean (OWQA) with respect to the strictly increasing bijection $\varphi : [0,1] \to [0,1]$ with corresponding weights $w_{1n} \geq w_{2n} \geq \ldots \geq w_{nn}$ for all $n \in \mathbb{N}$, i.e.,

$$\mathbf{W}'_\varphi(x_1, \ldots, x_n) = \varphi^{-1}(\mathbf{W}'(\varphi(x_1), \ldots, \varphi(x_n)))$$

$$= \varphi^{-1}(\tfrac{1}{n} \sum_{i=1}^{n} w_{in} \varphi(x_i)') = \varphi^{-1}(\tfrac{1}{n} \sum_{i=1}^{n} w_{in} \varphi(x_i')).$$

5.4   Dominance of the Product

Concerning dominance over the product $T_{\mathbf{P}}$, Theorem 2 transforms as follows.

Theorem 4. *[27] An aggregation operator $\mathbf{A}: \bigcup_{n \in \mathbb{N}}[0,1]^n \to [0,1]$ dominates $T_{\mathbf{P}}$ if and only if the function $f_n: [0,\infty]^n \to [0,\infty]$ given by*

$$f_n(x_1, \ldots, x_n) = -\log(\mathbf{A}(e^{-x_1}, \ldots, e^{-x_n}))$$

is subadditive for each $n \in \mathbb{N}$.

Again an aggregation operator $\mathbf{A}$ is dominated by $T_{\mathbf{P}}$ if and only if each f_n as given by Theorem 4 is superadditive.

As already mentioned any weighted geometric mean dominates $T_{\mathbf{P}}$. Moreover, for any $n \geq 2$ and any $\vec{w} = (w_1, \ldots, w_n)$ with $\sum_{i=1}^{n} w_i > 0$ and $w_i \in [0, \infty]$, the function $\mathbf{H} : [0, \infty]^n \to [0, \infty]$ defined by

$$\mathbf{H}(x_1, \ldots, x_n) = \sum_{i=1}^{n} w_i \cdot x_i$$

is an n-ary, subadditive aggregation operator acting on $[0, \infty]$. Therefore, any n-ary aggregation operator

$$\mathbf{A}_{\vec{w}}(x_1, \ldots, x_n) = \prod_{i=1}^{n} x_i^{w_i}$$

dominates the product $T_{\mathbf{P}}$.

However, observing that for all $\lambda \geq 1$, the function

$$\mathbf{H}_\lambda : [0, \infty]^2 \to [0, \infty], \mathbf{H}_\lambda(x, y) = (x^\lambda + y^\lambda)^{\frac{1}{\lambda}},$$

is also a binary, subadditive aggregation operator acting on $[0, \infty]$, also any member of the Aczél-Alsina family of t-norms $(T_\lambda^{\mathbf{AA}})_{\lambda \in [1, \infty]}$, is contained in $\mathcal{D}_{T_{\mathbf{P}}}$ because of Theorem 2.

Similar as in the case of the Łukasiewicz t-norm $T_{\mathbf{L}}$, we can show the next result.

Proposition 8. [27] *For a fixed $n \in \mathbb{N}$ and some weighting vector $\vec{w} = (w_1, \ldots, w_n)$, let $\mathbf{A} : [0, 1]^n \to [0, 1]$ be an ordered weighted geometric mean, i.e., $\mathbf{A}(x_1, \ldots, x_n) = \prod_{i=1}^{n} (x_i')^{w_i}$ where x_i' is again the i-th order statistic of $(x_1, \ldots, x_n)$. Then $\mathbf{A} \gg T_{\mathbf{P}}$ if and only if $w_1 \geq w_2 \geq \ldots \geq w_n$.*

Due to the isomorphism of any strict t-norm to the product $T_{\mathbf{P}}$, similar considerations are valid for any strict t-norm.

5.5   Final Remarks Related to Continuous Archimedean T-Norms

In Section 5.3 we have shown how the dominance of $T_{\mathbf{L}}$ by an OWA operator $\mathbf{W}'$ restricts the possible choices for weights. When considering some similar constraints reflecting $\mathbf{W}' \gg T$ for some other continuous Archimedean t-norm T, we cannot exploit the isomorphism of $T_{\mathbf{L}}$ and nilpotent t-norms (then also $\mathbf{W}'$ should be isomorphically transformed). Thus as a separate problem let us consider a continuous Archimedean t-norm T with additive generator t and an OWA operator $\mathbf{W}' : \bigcup_{n \in \mathbb{N}} [0, 1]^n \to [0, 1]$ which is supposed to dominate T, i.e., for all $n \in \mathbb{N}$ and for all $x_i, y_i \in [0, 1], i \in \{1, \ldots, n\}$

$$\mathbf{W}'(T(x_1, y_1), \ldots, T(x_n, y_n)) \geq T(\mathbf{W}'(x_1, \ldots, x_n), \mathbf{W}'(y_1, \ldots, y_n)).$$

If we concentrate on the binary case and choose $x_1 = 0$, $y_1 = 1$, $x_2 = 1$, $y_2 > 0$ then we see that necessarily

$$\mathbf{W}'(0, y_2) = w_2 y_2 \geq T(w_2, w_1 y_2 + w_2)$$
$$= t^{-1}(\min(t(0), t(w_2) + t(w_1 y_2 + w_2))),$$

i.e., for all $y_2 \in [0, 1[$

$$t(w_2 y_2) \leq t(w_2) + t(w_1 y_2 + w_2).$$

Evidently if $t(0) = +\infty$ then we get that $w_2 = 0$ because of the continuity of t. Similarly we can show in the general case with $n \in \mathbb{N}$ that $w_i = 0$ for $i > 1$. It follows that for any strict t-norm T only one OWA dominates T, namely the minimum.

In the case of nilpotent t-norms, equation (5.5) gives a necessary condition for $\mathbf{W}' \gg T$.

For $y_2 \to 0^+$ we get that for normed additive generators $1 \leq 2t(w_2)$, i.e., $w_2 \leq t^{-1}(\frac{1}{2})$ holds. This fact can be exploited in determination of OWA operators dominating a specific t-norm. For example, it can be conjectured that an OWA operator with weights $(w_1, \ldots, w_n)$ dominates

- Yager's t-norm T_p^Y [21] with parameter $p \in]0, \infty[$ and normed additive generator $t_p(x) = (1 - x)^p$ if and only if

$$w_i \geq \frac{1}{2^{1/p} - 1} w_{i+1}, i = 1, \ldots, n - 1,$$

- Schweizer-Sklar's t-norm T_λ^{SS} [21] with parameter $\lambda \in]0, \infty[$ and normed additive generator $t_\lambda(x) = 1 - x^\lambda$ if and only if

$$w_i \geq (2^{1/\lambda} - 1) w_{i+1}.$$

Observe that the arithmetic mean $\mathbf{M} \gg T_p^Y$ if and only if $p \leq 1$ and $\mathbf{M} \gg T_\lambda^{SS}$ if and only if $\lambda \geq 1$. Recall that $T_{\mathbf{L}} = T_1^Y = T_1^{SS}$.

6   Conclusion

We have discussed the aggregation of fuzzy relations and the preservation of their transitivity. In particular, the aggregation operator $\mathbf{A}$ preserves the T-transitivity of fuzzy relations if and only if it dominates the corresponding t-norm T ($\mathbf{A} \in \mathcal{D}_T$). Several methods for constructing aggregation operators within a certain class $\mathcal{D}_T$ have been mentioned with a particular emphasis on the introduction of weights. Further, a characterization of $\mathcal{D}_T$ for the four basic t-norms has been provided.

Acknowledgements

The authors gratefully acknowledge support by COST Action 274 "TARSKI". When this paper was mainly written, Ulrich Bodenhofer was affiliated with Software Competence Center Hagenberg, A-4232 Hagenberg, Austria. Therefore, he thanks for support by the Austrian Government, the State of Upper Austria, and the Johannes Kepler University Linz in the framework of the K*plus* Competence Center Program. Radko Mesiar also gratefully acknowledges the support of VEGA 1/3006/06 and VZ MSM 6198898701.

References

1. J. Aczél. *Lectures on Functional Equations and their Applications*. Academic Press, New York, 1966.
2. C. Alsina, M. Frank, and B. Schweizer. *Associative Functions: Triangular Norms and Copulas*. World Scientific Publishing Company, Singapore, 2006.
3. J. C. Bezdek and J. D. Harris. Fuzzy partitions and relations: An axiomatic basis for clustering. *Fuzzy Sets and Systems*, 1:111–127, 1978.
4. U. Bodenhofer. *A Similarity-Based Generalization of Fuzzy Orderings*, volume C 26 of *Schriftenreihe der Johannes-Kepler-Universität Linz*. Universitätsverlag Rudolf Trauner, 1999.
5. U. Bodenhofer. A similarity-based generalization of fuzzy orderings preserving the classical axioms. *Internat. J. Uncertain. Fuzziness Knowledge-Based Systems*, 8(5):593–610, 2000.
6. U. Bodenhofer. Representations and constructions of similarity-based fuzzy orderings. *Fuzzy Sets and Systems*, 137(1):113–136, 2003.
7. U. Bodenhofer, P. Bogdanowicz, G. Lanzerstorfer, and J. Küng. Distance-based fuzzy relations in flexible query answering systems: Overview and experiences. In I. Düntsch and M. Winter, editors, *Proc. 8th Int. Conf. on Relational Methods in Computer Science*, pages 15–22, St. Catharines, ON, February 2005. Brock University.
8. U. Bodenhofer and J. Küng. Enriching vague queries by fuzzy orderings. In *Proc. 2nd Int. Conf. in Fuzzy Logic and Technology (EUSFLAT 2001)*, pages 360–364, Leicester, UK, September 2001.
9. U. Bodenhofer and J. Küng. Fuzzy orderings in flexible query answering systems. *Soft Computing*, 8(7):512–522, 2004.
10. U. Bodenhofer, J. Küng, and S. Saminger. Flexible query answering using distance-based fuzzy relations. In H. de Swart, E. Orlowska, M. Roubens, and G. Schmidt, editors, *Theory and Applications of Relations Structures as Knowledge Instruments II*, Lecture Notes in Artificial Intelligence. Springer, 2006.
11. P. Bosc, B. Buckles, F. Petry, and O. Pivert. Fuzzy databases: Theory and models. In J. Bezdek, D. Dubois, and H. Prade, editors, *Fuzzy Sets in Approximate Reasoning and Information Systems*, pages 403–468. Kluwer Academic Publishers, Boston, 1999.
12. P. Bosc, L. Duval, and O. Pivert. Value-based and representation-based querying of possibilistic databases. In G. Bordogna and G. Pasi, editors, *Recent Issues on Fuzzy Databases*, pages 3–27. Physica-Verlag, Heidelberg, 2000.
13. B. Bouchon-Meunier, editor. *Aggregation and Fusion of Imperfect Information*. Physica-Verlag, Heidelberg, 1998.

14. T. Calvo, A. Kolesárová, M. Komorníková, and R. Mesiar. Aggregation operators: properties, classes and construction methods. In T. Calvo, G. Mayor, and R. Mesiar, editors. *Aggregation Operators. New Trends and Applications*, pages 3–104. Physica-Verlag, Heidelberg, 2002.

15. T. Calvo and R. Mesiar. Weighted means based on triangular conorms. *Internat. J. Uncertain. Fuzziness Knowledge-Based Systems*, 9(2):183–196, 2001.

16. B. De Baets and R. Mesiar. Pseudo-metrics and T-equivalences. *J. Fuzzy Math.*, 5(2):471–481, 1997.

17. B. De Baets and R. Mesiar. T-partitions. *Fuzzy Sets and Systems*, 97:211–223, 1998.

18. J.J. Dujmovic. Weighted conjunctive and disjunctive means and their application in system evaluation. *Univ. Beograd Publ. Electrotech. Fak*, pages 147–158, 1975.

19. U. Höhle. Fuzzy equalities and indistinguishability. In *Proc. 1st European Congress on Fuzzy and Intelligent Technologies*, volume 1, pages 358–363, Aachen, 1993.

20. U. Höhle and N. Blanchard. Partial ordering in L-underdeterminate sets. *Inform. Sci.*, 35:133–144, 1985.

21. E. P. Klement, R. Mesiar, and E. Pap. *Triangular Norms*. Kluwer Academic Publishers, Dordrecht, 2000.

22. E. P. Klement, R. Mesiar, and E. Pap. Triangular norms. Position paper I: basic analytical and algebraic properties. *Fuzzy Sets and Systems*, 143:5–26, 2004.

23. E. P. Klement, R. Mesiar, and E. Pap. Triangular norms. Position paper II: general constructions and parameterized families. *Fuzzy Sets and Systems*, 145:411–438, 2004.

24. E. P. Klement, R. Mesiar, and E. Pap. Triangular norms. Position paper III: continuous t-norms. *Fuzzy Sets and Systems*, 145:439–454, 2004.

25. K. Menger. Statistical metrics. *Proc. Nat. Acad. Sci. U.S.A.*, 8:535–537, 1942.

26. R. Mesiar and B. De Baets. New construction methods for aggregation operators. In *Proc. 8th Int. Conf. on Information Processing and Management of Uncertainty in Knowledge-Based Systems*, volume 2, pages 701–706, Madrid, 2000.

27. R. Mesiar and S. Saminger. Domination of ordered weighted averaging operators over t-norms. *Soft Computing*, 8:562–570, 2004.

28. F. E. Petry and P. Bosc. *Fuzzy Databases: Principles and Applications*. International Series in Intelligent Technologies. Kluwer Academic Publishers, Boston, 1996.

29. A. Rosado, J. Kacprzyk, R. A. Ribeiro, and S. Zadrozny. Fuzzy querying in crisp and fuzzy relational databases: An overview. In *Proc. 9th Int. Conf. on Information Processing and Management of Uncertainty in Knowledge-Based Systems*, volume 3, pages 1705–1712, Annecy, July 2002.

30. S. Saminger. *Aggregation in Evaluation of Computer-Assisted Assessment*, volume C 44 of *Schriftenreihe der Johannes-Kepler-Universität Linz*. Universitätsverlag Rudolf Trauner, 2005.

31. S. Saminger, B. De Baets, and H. De Meyer. On the dominance relation between ordinal sums of conjunctors. *Kybernetika*, 42(3):337–350, 2006.

32. S. Saminger, R. Mesiar, and U. Bodenhofer. Domination of aggregation operators and preservation of transitivity. *Internat. J. Uncertain. Fuzziness Knowledge-Based Systems*, 10(Suppl.):11–35, 2002.

33. S. Saminger, P. Sarkoci, and B. De Baets. The dominance relation on the class of continuous t-norms from an ordinal sum point of view. In H. de Swart, E. Orlowska, M. Roubens, and G. Schmidt, editors, *Theory and Applications of Relations Structures as Knowledge Instruments II*, Lecture Notes in Artificial Intelligence. Springer, 2006.

34. P. Sarkoci. Dominance is not transitive on continuous triangular norms. *Aequationes Mathematicae*, 2006. submitted.

35. B. Schweizer and A. Sklar. Statistical metric spaces. *Pacific J. Math*, 10:313–334, 1960.

36. B. Schweizer and A. Sklar. Associative functions and statistical triangle inequalities. *Publ. Math. Debrecen*, 8:169–186, 1961.

37. B. Schweizer and A. Sklar. *Probabilistic Metric Spaces*. North-Holland, New York, 1983.

38. W. Silvert. Symmetric summation: A class of operations on fuzzy sets. *IEEE Trans. Systems Man Cybernet.*, 9:657–659, 1979.

39. R. M. Tardiff. Topologies for probabilistic metric spaces. *Pacific J. Math.*, 65:233–251, 1976.

40. R. R. Yager. On ordered weighted averaging aggregation operators in multicriteria decisionmaking. *IEEE Trans. Systems Man Cybernet.*, 18:183–190, 1988.

41. R. R. Yager and D. P. Filev. *Essentials of Fuzzy Modelling and Control*. J. Wiley & Sons, New York, 1994.

42. L. A. Zadeh. Similarity relations and fuzzy orderings. *Inform. Sci.*, 3:177–200, 1971.

Flexible Query Answering Using Distance-Based Fuzzy Relations

Ulrich Bodenhofer[1], Josef Küng[2], and Susanne Saminger[3]

[1] Institute of Bioinformatics,
Johannes Kepler University, A-4040 Linz, Austria
bodenhofer@bioinf.jku.at
[2] Institute for Applied Knowledge Processing
Johannes Kepler University, A-4040 Linz, Austria
jkueng@faw.uni-linz.ac.at
[3] Dept. of Knowledge-Based Mathematical Systems
Johannes Kepler University, A-4040 Linz, Austria
susanne.saminger@jku.at

Abstract. This paper addresses the added value that is provided by using distance-based fuzzy relations in flexible query answering. To use distances and/or concepts of gradual similarity in that domain is not new. Within the last ten years, however, results in the theory of fuzzy relations have emerged that permit a smooth and pragmatic, yet expressive and effective, integration of ordinal concepts too. So this paper primarily highlights the benefits of integrating fuzzy orderings in flexible query answering systems, where the smooth interplay of fuzzy equivalence relations and fuzzy orderings allows to use simple distances as a common basis for defining both types of relations. As one case study, we discuss a pragmatic variant of a flexible query answering system—the so-called Vague Query System (VQS). The integration of fuzzy orderings into that system is provided in full detail along with the necessary methodological background and demonstrative examples.

1 Introduction

From a naive viewpoint, databases are nothing else but means to store and retrieve data in an appropriately structured way. Conventional database systems available on the market offer powerful mechanisms to retrieve data according to complex criteria. A large majority of systems supports the *Structured Query Language (SQL)* that has become a widely accepted standard.

No matter how complex a query might be, SQL is based on logical expressions that a given record either fulfills or not. The use of classical binary logic for data retrieval poses severe limitations. Firstly, real-world data, in particular, numeric data, are often perturbed by noise or measurement errors. This may result in unstable behavior in the sense that minimal variations of the data can change the result of a query dramatically. Secondly, no structural information is available about how close a rejected record was to the fulfillment of the query. This loss

H. de Swart et al. (Eds.): TARSKI II, LNAI 4342, pp. 207–228, 2006.
© Springer-Verlag Berlin Heidelberg 2006

of information is particularly harmful if the user would still be interested in potentially close records if the query gives an empty result. Thirdly, constructs that are closer to natural language, like vague and qualitative expressions, would mean a strong enrichment of a query interface in terms of usability and flexibility. SQL, however, does not support such kinds of elements.

These fundamental needs have created an own discipline at the interface between database and fuzzy logic research. On the one hand, researchers in fuzzy logic have soon been interested in the question how to cope with imprecise and/or qualitative data and relations in database systems. The concepts developed in this direction are nowadays often subsumed under the term *"fuzzy databases"* [4,9,28,32]. A second branch of research, on the other hand, has been concerned with the problem how query interfaces to conventional databases with crisp data can be extended such that a flexible interpretation of queries is possible [6, 7, 8, 16, 19, 21, 22, 26, 29, 37, 38]—in particular, with the motivation to suggest alternatives that are close to matching the criteria in case that a query gives an empty result. This area is often referred to as *"flexible querying"*. As recent overviews demonstrate [4, 5, 34], significant progress has been made in both directions.

Fuzzy relations have a long tradition in flexible querying. In particular, fuzzy equivalence relations[1] have often been used for modeling the similarity between two records in a gradual way [7, 8, 10, 11]—particularly with the motivation to have a degree of closeness to which a record matches a query. Fuzzy orderings [17,30,44], on the other hand, have not contributed to applications in flexible querying so far. As elaborated in detail in [1], the limited applicability of fuzzy orderings is not a coincidence, but a systematic consequence of not relating fuzzy orderings to a proper concept of fuzzy similarity (see also [18]). This paper addresses the question how a generalized notion of fuzzy orderings that overcomes these limitations [1, 18] can be integrated fruitfully in flexible query answering systems. We will see that the strong connection between fuzzy equivalence relations and fuzzy orderings allows to use simple distances as a common basis for defining both types of relations. Distances or gradual similarity are common in flexible query answering, but this paper demonstrates that the formulation in the frame of distance-based fuzzy relations and the integration of fuzzy orderings provide higher expressiveness and interpretability while maintaining simplicity.

This paper does not only give theoretical and methodological background. Instead it is built around an existing system from practice—the so-called *Vague Query System (VQS)* [26,27] which is a pragmatic approach to handling queries with crisp data in a flexible way by incorporating a certain tolerance for imprecision. Consequently, this paper is organized as follows. In Section 2, we highlight VQS as the basis for further investigations. Section 3 gives the necessary background from the theory of fuzzy relations that is necessary to re-formulate VQS in the frame of fuzzy relations and to integrate ordering-based queries.

---

[1] Other names for this fundamental class of fuzzy relations are—some of them assuming a specific t-norm—fuzzy equality [23,25], indistinguishability operators [39,40], similarity relations [44], likeness relation [14], and proximity relation [15].

VQLExpression := "SELECT FROM" DataSource "WHERE" Conditions
                "INTO" destinationTableName;
DataSource   := ([ownerName"."]rootTableName) |
                ([ownerName"."]rootViewName) |
                "("sqlSelectStatement")";
Conditions   := columnName "IS" ValueExpression
                {"AND" columnName "IS" ValueExpression};
ValueExpression := ("`"alphaNumericValue"´" | numericValue)
                ["WEIGHTED BY" numericValue];

Fig. 1. The syntax of VQL

Section 4 addresses the fundamental issue of aggregating multiple queries. Then Section 5 summarizes all findings and puts them in context with VQS again; a first demonstrative example is presented. Section 6 discusses the choice of an important degree of freedom—the underlying t-norm. Finally, Section 7 provides a non-trivial real-world example. Note that this paper is a state-of-the-art summary that integrates two previously published papers [2,3].

2   The Vague Query System (VQS)

VQS is an add-on to conventional relational databases which acts as a proxy between the user and the database [26,27]. Since VQS communicates with the underlying database only on the basis of standard SQL, no adaptations to the database system or the data model have to be made, which allows easy integration into existing applications.

Flexible interpretation of queries requires semantic information about the attributes. In case of numeric attributes, considering Euclidean distances is most often sufficient. For non-numeric attributes, most other systems [19,29] use similarity tables, which often implies serious limitations in terms of storage and computational effort. VQS avoids these problems by using a so-called NCR table (numeric coordinate representation), i.e. an assignment of (possibly multidimensional) numeric values to all possible instances of a non-numeric attribute (e.g. assignment of RGB color values to natural language names of colors or assignment of GPS coordinates to city names). This approach is only applicable in case that the number of possible instances of an attribute is finite and under the assumption that a meaningful numeric representation is available. In practice, these requirements can most often be met (e.g. in tourist information systems, where this approach has been applied already [31]).

Figure 1 shows the syntax of the *Vague Query Language (VQL)* used by VQS (in [27], an extension to vague joins has been proposed. As joins are not in the main focus of this paper, we restrict to the simpler variant from [26]).

The question arises how VQS implements the "IS" operator (which should be understood in the sense of "is similar to"). Provided that there is one single "IS" condition in the query, VQS retrieves all records from the data source and ranks them according to the distance from the query value. In case that the

Table 1. Hotel data set (artificial toy data set)

#	Location	EUR/night	*
1	Salzburg Center (S)	120.00	5
2	Salzburg Center (S)	85.00	4
3	Salzburg Liefering (SL)	70.00	4
4	Anif (A)	80.00	4
5	Mattsee (M)	60.00	4
6	Salzburg Aigen (SA)	70.00	3
7	Salzburg Aigen (SA)	77.00	4
8	Linz (L)	70.00	4
9	Salzburg Maxglan (SM)	60.00	3

column contains numeric values, the distance between two values x, y can easily be computed as the absolute value of the difference $|x - y|$ (Euclidean norm for the one-dimensional real space $\mathbb{R}$). If the column under consideration is non-numeric, the distance is computed as the distance of the associated values in the corresponding NCR table. VQS works with normalized distances, i.e. any raw distance value is divided by the maximum of possible distances of values in the column under consideration. Every condition, therefore, is assigned a distance value normalized to the unit interval $[0, 1]$. This value then corresponds to the closeness of the record to the query. In case that two or more "IS" conditions are combined with "AND", a weighted average of the distances in the different columns is used to rank the results (equal weights are used by default, which can be overridden using the optional "WEIGHTED BY" expression).

Example 1. We consider a toy data set describing hotels. For each hotel, the location, the price per night, and the category (in no. of stars) is stored in a table. The data set is shown in Table 1. Consider the following query:

```
SELECT FROM HotelTable
    WHERE Location IS 'Salzburg Center'
        AND Price    IS 70
        AND Category IS 4
INTO ResultSet
```

Assume that the distances between locations are given as in Table 2 (as the result of computing a distance measure for corresponding values in an NCR table).

To compute the result set for this query, all distances are first normalized, which means that the distance of any two locations is divided by 147.8, each discrepancy in the price is divided by 60, and each discrepancy in the category is divided by 2 (as stressed above, distances are divided by the maximal possible distance in the data set). Using equal weights, i.e. the overall degree of matching is computed by means of the arithmetic mean, we obtain the result set shown in Table 2 sorted by the closeness to the query (d_1, d_2, and d_3 denote the normalized distances with respect to location, price, and category, respectively; d is the aggregated distance).

Table 2. Distance matrix for locations in the hotel data set

[km]	S	SL	A	M	SA	L	SM
S	0.0	4.4	9.0	21.3	4.2	133.2	3.8
SL	4.4	0.0	17.3	22.1	7.0	133.8	3.8
A	9.0	17.3	0.0	29.9	8.6	147.8	13.7
M	21.3	22.1	29.9	0.0	23.4	138.6	25.1
SA	4.2	7.0	8.6	23.4	0.0	135.1	5.0
L	133.2	133.8	147.8	138.6	135.1	0.0	137.4
SM	3.8	3.8	13.7	25.1	5.0	137.4	0.0

Table 3. VQS result set

#	Location	d_1	d_2	d_3	d
3	Salzburg Liefering	0.0298	0.0000	0.0000	**0.0099**
7	Salzburg Aigen	0.0284	0.1167	0.0000	**0.0484**
4	Anif	0.0609	0.1667	0.0000	**0.0759**
2	Salzburg Center	0.0000	0.2500	0.0000	**0.0833**
5	Mattsee	0.1441	0.1667	0.0000	**0.1036**
6	Salzburg Aigen	0.0284	0.0000	0.5000	**0.1761**
9	Salzburg Maxglan	0.0257	0.1667	0.5000	**0.2308**
8	Linz	0.9012	0.0000	0.0000	**0.3004**
1	Salzburg Center	0.0000	0.8333	0.5000	**0.4444**

Example 1 clearly demonstrates two severe shortcomings of VQS:

1. VQS is restricted to "IS" queries that are interpreted with a certain tolerance for imprecision. For the location column, this is not a serious restriction. For the price column, however, this is a painful limitation, as the user is not necessarily interested in a price that is as close to EUR 70 as possible, but more likely in a price that exceeds EUR 70 as little as possible. This once more underlines the need for integrating ordinal information into VQS.
2. The normalization of distances is done for all columns independently solely on the basis of the largest distance between two values in the column. The result is that the normalization is biased to the actual data in the column and, consequently, that two distance values for different columns may be difficultly comparable. In the above example, prices differ within a range of EUR 20, while the maximal distance of locations is 147.8 km. This means that a considerable distance of 7.39 km corresponds to an almost negligible difference in price of EUR 1. This undesired bias can only be compensated by adjusting the weights by trial and error—an effort that the user is most probably not willing to spend.

The main focus, of course, is to tackle the first problem. However, it will turn out that, by reformulating VQS in the framework of fuzzy relations, also the second shortcoming can be overcome.

In order to enrich VQS with ordinal constructs, we first define the language. The following sections are then devoted to a detailed step-by-step description of

VQLExpression	:= "SELECT FROM" DataSource "WHERE" Conditions "INTO" destinationTableName;
DataSource	:= ([ownerName "."]rootTableName) | ([ownerName "."]rootViewName) | "("sqlSelectStatement")";
Conditions	:= Condition {"AND" Condition};
Condition	:= NonNumericCond ParameterExpression | NumericCond ParameterExpression;
NonNumericCond	:= columnName "IS" alphaNumericValue;
NumericCond	:= columnName "IS" numericValue | columnName "IS AT LEAST" numericValue | columnName "IS AT MOST" numericValue | columnName "IS WITHIN" (" numericValue "," numericValue ")";
ParameterExpression	:= ["TOLERATE UP TO" numericValue] ["WEIGHTED BY" numericValue];

Fig. 2. The syntax of oVQL

the corresponding semantics. Figure 2 shows the syntax of the *Ordering-Enriched Vague Query Language (oVQL)*. It is obvious that oVQL differs from VQL in the respect that there is an explicit distinction between numeric and non-numeric attributes. For non-numeric ones, only the "IS" condition is defined like in VQL. For numeric ones, three new types of conditions

"IS AT LEAST", "IS AT MOST", and "IS WITHIN"[2]

are added. An ordering might also be defined for a non-numeric attribute. However, we leave this aspect aside for this paper, as the main motivation for ordinal structures are numeric attributes (for which an ordering is most often given in a straightforward way). Another major difference is the "TOLERATE UP TO" expression. Its role will become clearer later. For the moment, let us just mention that it corresponds to an upper bound of tolerance the user can set to specify his/her radius of interest around a query value.

3   Conditions Based on Fuzzy Relations

We first approach the question how the semantics of the single conditions "IS", "IS AT LEAST", "IS AT MOST", and "IS WITHIN" are modeled. For the "IS" condition, the original VQS already provides a reasonable definition using distances. From an intuitive point of view, a query like, e.g.,

```
Price IS AT MOST 70
```

should be perfectly fulfilled by a price x if $x \leq 70$. If $x > 70$, some flexible interpretation taking the distance between x and 70 into account should take

---

[2] Specifically meaning "is within the range of ...".

place. This ad-hoc approach is not only intuitive, but has a sound theoretical foundation in the general framework of fuzzy orderings [1,18]. The present section provides the necessary basics to demonstrate this fact.

Given a non-empty domain X, a mapping $R : X^2 \to [0,1]$ that assigns a degree of relationship $R(x,y) \in [0,1]$ to each pair $(x,y) \in X^2$ is called (binary) fuzzy relation on X. Two types of fuzzy relations will be central objects in this paper—fuzzy equivalence relations and fuzzy orderings. We use triangular norms (t-norms) as generalized models of conjunction [24]. A triangular norm is a binary operation on the unit interval (i.e. a $[0,1]^2 \to [0,1]$ mapping) which is associative, commutative, non-decreasing, and has 1 as neutral element.

Definition 1. A binary fuzzy relation $E : X^2 \to [0,1]$ is called *fuzzy equivalence relation* with respect to a t-norm T, for brevity *T-equivalence*, if and only if the following three axioms are fulfilled for all $x, y, z \in X$:

(i) Reflexivity: $E(x,x) = 1$
(ii) Symmetry: $E(x,y) = E(y,x)$
(iii) T-transitivity: $T\big(E(x,y), E(y,z)\big) \leq E(x,z)$

Definition 2. A fuzzy relation $L : X^2 \to [0,1]$ is called *fuzzy ordering* with respect to a t-norm T and a T-equivalence E, for brevity *T-E-ordering*, if and only if it is T-transitive and fulfills the following two axioms for all $x, y \in X$:

(i) E-Reflexivity: $E(x,y) \leq L(x,y)$
(ii) T-E-antisymmetry: $T\big(L(x,y), L(y,x)\big) \leq E(x,y)$

Moreover, we call a T-E-ordering L *strongly complete* if $\max\big(L(x,y), L(y,x)\big) = 1$ for all $x, y \in X$.

We now briefly mention a crucial result that helps to clarify the connection between the two types of fuzzy relations and the semantics of the four oVQL conditions.

Definition 3. A crisp ordering $\preceq$ on a domain X and a T-equivalence $E : X^2 \to [0,1]$ are called *compatible* if and only if the following holds for all $x, y, z \in X$:

$$x \preceq y \preceq z \;\Rightarrow\; E(x,z) \;\leq\; \min\big(E(x,y), E(y,z)\big)$$

Compatibility between a crisp ordering $\preceq$ and a fuzzy equivalence relation E can be interpreted as follows: the two outer elements of any three-element chain are at most as similar as any two inner elements.

Theorem 1. [1] *Consider a fuzzy relation L on a domain X and a T-equivalence E. Then the following two statements are equivalent:*

(i) L is a strongly complete T-E-ordering.
(ii) There exists a linear ordering $\preceq$ the relation E is compatible with such that L can be represented as follows:

$$L(x,y) = \begin{cases} 1 & \text{if } x \preceq y \\ E(x,y) & \text{otherwise} \end{cases}$$

Theorem 1 particularly implies that the "combination" of a crisp linear ordering $\preceq$ and a fuzzy equivalence relation compatible with $\preceq$ has a clear theoretical interpretation as a vague concept of ordering (a "linear ordering with imprecision"). This is exactly what we need to define the semantics of the four oVQL conditions in a theoretically sound way: if we manage to transfer the distance-based interpretation of the "IS" condition into one using a fuzzy equivalence relation that is compatible with the underlying linear ordering, then Theorem 1 provides the perfect justification of the above ad-hoc idea to "combine" a crisp linear ordering with a flexible interpretation based on distances.

The question arises how to transform distances into a fuzzy equivalence relation in a meaningful way such that the closer a value is to a query, the higher the degree of fulfillment is. For this purpose, a well-established result is available if the t-norm T under consideration is continuous Archimedean [24, 36] (simplistically, this means that T is a continuous mapping fulfilling $T(x, x) < x$ for all $x \in]0, 1[$). Such a t-norm can always be represented by means of a so-called *additive generator*, i.e. a continuous and strictly decreasing bijection $f : [0, 1] \rightarrow [0, \infty]$, such that the following representation holds:

$$T(x, y) = f^{-1}\big(\min(f(x) + f(y), f(0))\big)$$

Note that there is a major difference between t-norms with a generator for which $f(0) = \infty$ holds and those with a generator that fulfills $f(0) < \infty$. In the former case, we speak of a *strict* t-norm with the product $T_P(x, y) = x \cdot y$ being the most important representative. In the latter case, T belongs to the class of *nilpotent* t-norms, out of which the so-called Łukasiewicz t-norm $T_L(x, y) = \max(x+y-1, 0)$ is the most prominent representative. Note that the additive generator f is determined up to a positive multiplicative constant. Therefore, we can assume without any loss of generality that $f(0) = 1$ in the nilpotent case $f(0) < \infty$.

Theorem 2. [14] *Consider a continuous Archimedean t-norm T with additive generator f, a pseudo-metric $d : X^2 \rightarrow]0, \infty[$, and a real constant $C > 0$. Then the following mapping is a T-equivalence:*

$$E_{d,C}(x, y) = f^{-1}\big(\min(\tfrac{1}{C} \cdot d(x, y), f(0))\big) \tag{1}$$

Theorem 2 states that we can transform a (pseudo-)metric into a T-equivalence if we consider a continuous Archimedean t-norm T. The following example shows how this can be done for the two basic t-norms T_L and T_P.

Example 2. Let us consider a metric $d : X^2 \rightarrow [0, \infty[$. It is clear that $d'(x, y) = \tfrac{1}{C} \cdot d(x, y)$ (for $C > 0$) is a metric as well. Since $f_L(x) = 1 - x$ is a self-inverse additive generator of the Łukasiewicz t-norm T_L fulfilling $f(0) = 1$, Theorem 2 implies that

$$E_{d,C}(x, y) = \max\big(1 - \tfrac{1}{C} \cdot d(x, y), 0\big)$$

defines a T_L-equivalence [14, 25]. The value C is obviously the maximal distance of two objects x and y up to which $E(x, y) > 0$ can hold.

The function $f_{\mathbf{P}}(x) = -\ln x$ is an additive generator of product t-norm $T_{\mathbf{P}}$, where $f_{\mathbf{P}}^{-1}(x) = \exp(-x)$. Then Theorem 2 yields that

$$E'_{d,C}(x,y) = \exp\left(-\tfrac{1}{C} \cdot d(x,y)\right)$$

is a $T_{\mathbf{P}}$-equivalence. The value C does not have an as intuitive interpretation in this example as above. Anyway, C has the same quantitative influence: the larger C is chosen, the slower $E(x,y)$ decreases with increasing distance $d(x,y)$.

The only question remaining open is how a fuzzy equivalence relation can be constructed from a (pseudo-)metric such that compatibility with a given crisp ordering is fulfilled. The following proposition provides the vehicle to answer this question.

Proposition 1. *Let T be a continuous Archimedean t-norm with an additive generator f and let $\preceq$ be an ordering of the domain X. If a pseudo-metric $d : X^2 \to [0,\infty[$ fulfills*

$$x \preceq y \preceq z \;\Rightarrow\; d(x,z) \geq \max\left(d(x,y), d(y,z)\right) \tag{2}$$

for all $x,y,z \in X$, then its induced fuzzy equivalence relation E_d, defined as in (1), is compatible with $\preceq$.

Note that scaling a given pseudo-metric d with a factor $\tfrac{1}{C}$ (with $C > 0$) does not change the compatibility with an ordering in the sense of (2), i.e. a pseudo-metric d fulfills (2) if and only if $d'(x,y) = \tfrac{1}{C} \cdot d(x,y)$ does so.

Example 3. Consider the real numbers $\mathbb{R}$ (or any subset of them) and the usual linear ordering of real numbers. For a given sequence $x \leq y \leq z$, we trivially have that

$$|x - y| = |y - x| = y - x$$
$$|x - z| = |z - x| = z - x$$
$$|y - z| = |z - y| = z - y$$

and we obtain (with $d(a,b) = |a - b|$) $d(x,z) = z - x \geq y - x = d(x,y)$ and $d(x,z) = z - x \geq z - y = d(y,z)$, which proves that the absolute distance of real numbers is compatible with the linear ordering $\leq$. As a consequence, we obtain that

$$E_C(x,y) = \max\left(1 - \tfrac{1}{C} \cdot |x - y|, 0\right)$$

is a $T_{\mathbf{L}}$-equivalence on $\mathbb{R}$ that is compatible with $\leq$ and that

$$E'_C(x,y) = \exp\left(-\tfrac{1}{C} \cdot |x - y|\right)$$

is a $T_{\mathbf{P}}$-equivalence on $\mathbb{R}$ that is compatible with $\leq$. Then Theorem 1 implies that

$$L_C(x,y) = \begin{cases} 1 & \text{if } x \leq y \\ \max\left(1 - \tfrac{1}{C} \cdot |x - y|, 0\right) & \text{otherwise} \end{cases}$$

is a $T_{\mathbf{L}}$-$E_{d,C}$-ordering on $\mathbb{R}$ and that

$$L'_C(x, y) = \begin{cases} 1 & \text{if } x \leq y \\ \exp\left(-\frac{1}{C} \cdot |x - y|\right) & \text{otherwise} \end{cases}$$

is a $T_{\mathbf{P}}$-$E'_{d,C}$-ordering on $\mathbb{R}$. Of course, these constructions can be carried out analogously for any other continuous Archimedean t-norm T.

The conclusion of Example 3 is that we now have a meaningful and theoretically sound way to construct fuzzy orderings on the real numbers by "combining" ordering and distance. This enables us to define the semantics of the four types of oVQL conditions. In the following, assume that we are given a continuous Archimedean t-norm T with additive generator f.

For a given non-numeric column, a condition "x IS q" is evaluated in the following way: for a concrete value x_0, the degree to which x_0 fulfills the condition is computed as[3]

$$t(\text{"}x \text{ IS } q\text{"} \mid x_0) = E_{d,C}(x_0, q) \tag{3}$$

where

$$E_{d,C}(x, y) = f^{-1}\left(\min(\tfrac{1}{C} \cdot d(x, y), f(0))\right) \tag{4}$$

and d is a metric for the column under investigation which is constructed using an NCR.

For a numeric attribute, we are able to define the degrees to which the four conditions are fulfilled by

$$t(\text{"}x \text{ IS } q\text{"} \mid x_0) = E_C(x_0, q) \tag{5}$$
$$t(\text{"}x \text{ IS AT LEAST } q\text{"} \mid x_0) = L_C(q, x_0) \tag{6}$$
$$t(\text{"}x \text{ IS AT MOST } q\text{"} \mid x_0) = L_C(x_0, q) \tag{7}$$
$$t(\text{"}x \text{ IS WITHIN } (a, b)\text{"} \mid x_0) = \min\left(L_C(\min(a, b), x_0), L_C(x_0, \max(a, b))\right) \tag{8}$$

with

$$E_C(x, y) = f^{-1}\left(\min(\tfrac{1}{C} \cdot |x - y|, f(0))\right)$$

and

$$L_C(x, y) = \begin{cases} 1 & \text{if } x \leq y, \\ f^{-1}\left(\min(\tfrac{1}{C} \cdot |x - y|, f(0))\right) & \text{otherwise.} \end{cases}$$

Note that this formulation of the semantics of the four oVQL conditions uses distances that need not be normalized. Instead of a normalization that is only determined by the data, the user has influence on how the raw distances are converted into matching degrees. If C is chosen independently of the data set, no bias to specific records in the data set can occur. If $T = T_{\mathbf{L}}$, C even has a clear interpretation (see Example 2). Therefore, it makes sense to leave the choice of C to the user (at least as an option). If he/she does not choose a particular value, a reasonable default value can be chosen. In the context of our hotel example, this means that either the user specifies the maximum acceptable distance C or a common sense default value (e.g. $C = 20$km) is chosen by the system.

---

[3] The mapping $t(Q \mid x_0)$ is a dummy function that evaluates the degree to which a concrete value x_0 fulfills the query Q.

Analogously for price and category: the user may decide which deviation from the query value q is tolerable. The optional choice of the value C, no matter whether we consider a numeric attribute or a non-numeric one, is possible through the "TOLERATE UP TO" expression.

4   The Aggregation Issue

The semantics of single oVQL conditions have been defined in the previous section. It remains to clarify how to proceed if the query involves two or more such conditions. In analogy to the original VQS, it would be desirable to aggregate the degrees to which a given record fulfills the individual conditions into one overall matching degree.

In the original VQS, this aggregation is done by means of a weighted average of distances, which is a reasonable choice, since a weighted average of metrics is again a metric. In the framework of fuzzy equivalence relations and fuzzy orderings, it is not clear whether this way of aggregating degrees of fulfillment is appropriate, as we are dealing with truth values instead of distances. Weighted averages still seem appropriate from a purely intuitive point of view. This viewpoint, however, needs further justification.

Suppose that we have a VQS query consisting of n conditions. Let us denote the degrees to which a given record fulfills these conditions with $t_1, \ldots, t_n$. We want to aggregate these degrees into one global degree to which the record fulfills the entire query. If we model the aggregation by means of a mapping $\mathbf{A} : [0,1]^n \to [0,1]$, it appears natural to require at least the following properties:

1. If all conditions are perfectly fulfilled, i.e. all degrees $t_1, \ldots, t_n$ are 1, then the global degree should of course be 1. In mathematical terms:

$$\mathbf{A}(1, \ldots, 1) = 1 \tag{9}$$

2. If none of the single queries is fulfilled at all, i.e. all degrees $t_1, \ldots, t_n$ are 0, then the global degree of fulfillment should be 0, too:

$$\mathbf{A}(0, \ldots, 0) = 0$$

3. If one degree t_i is increased while the others are kept constant, the overall degree must not decrease, i.e. $\mathbf{A}$ should be non-increasing in each component.
4. VQS allowed to introduce weights that allow the user to assign degrees of relative importance to each condition. As these weights (optional expression "WEIGHTED BY") are also defined in oVQL, $\mathbf{A}$ must offer the possibility to integrate weights.

The first three requirements are exactly the conditions that $\mathbf{A}$ is an n-ary aggregation operator [12]. Aggregation operators are a very general class of functions. The question arises which other properties to demand from $\mathbf{A}$ in order to have meaningful and interpretable results. The use of weighted averages for aggregating distances in VQS is justified by the fact that weighted averages/sums are metric-preserving aggregation operators [33,35]. In the investigations of this paper, we consider single conditions whose semantics are modeled by means of fuzzy equivalence relations and fuzzy orderings. Both classes of fuzzy relations are reflexive and T-transitive (for some fixed continuous Archimedean t-norm T). Analogously to the distance-based setting of VQS, it would be desirable that both properties are preserved if the fuzzy relations are aggregated by $\mathbf{A}$. That would guarantee that the global degree of matching fulfills the same properties as the individual fuzzy relations that are used to evaluate single conditions.

The preservation of reflexivity is guaranteed by (9). The preservation of T-transitivity is a much more complex matter. As a recent investigation has shown [35], this preservation is equivalent to the dominance of the t-norm T by the aggregation operator $\mathbf{A}$.

Definition 4. Consider an n-ary aggregation operator $\mathbf{A}$ and a t-norm T. We say that $\mathbf{A}$ *dominates* T if and only if, for all sequences $(x_1, \ldots, x_n) \in [0,1]^n$ and $(y_1, \ldots, y_n) \in [0,1]^n$, the following property holds:

$$T\big(\mathbf{A}(x_1, \ldots, x_n), \mathbf{A}(y_1, \ldots, y_n)\big) \leq \mathbf{A}\big(T(x_1, y_1), \ldots, T(x_n, y_n)\big)$$

Dominance is a highly non-trivial matter in the theory of triangular norms and aggregation operators. For this paper, we suffice with a basic result that enables us to define weighted aggregation operators which dominate a given continuous Archimedean t-norm in a straightforward way.

Theorem 3. [35] *Consider a continuous Archimedean t-norm T with additive generator f and a weighting vector $\boldsymbol{w} = (w_1, \ldots, w_n)$ (where $w_i \in [0, \infty[$). If $f(0) = \infty$, we require that $\sum_{i=1}^{n} w_i > 0$, and if $f(0) < \infty$, we require that $\sum_{i=1}^{n} w_i \geq 1$. Then the following function is an aggregation operator that dominates T:*

$$\mathbf{A}_{\boldsymbol{w}}(x_1, \ldots, x_n) = f^{-1}\big(\min(f(0), \sum_{i=1}^{n} w_i \cdot f(x_i)))\big) \tag{10}$$

It is obvious that, for $w_1 = \cdots = w_n = 1$, $\mathbf{A}_{\boldsymbol{w}}$ coincides with the n-ary extension of the t-norm T. If, more generally, $\sum_{i=1}^{n} w_i = n$, a kind of weighted n-ary variant of the t-norm T is obtained. In case $\sum_{i=1}^{n} w_i = 1$, $\mathbf{A}_{\boldsymbol{w}}$ can be considered as the *weighted quasi-arithmetic mean* induced by the generator f [35].

Example 4. Let us apply this construction to the Łukasiewicz t-norm $T_{\mathbf{L}}$:

$$\mathbf{A}_{\boldsymbol{w}}(x_1, \ldots, x_n) = 1 - \big(\min(1, \sum_{i=1}^{n} w_i \cdot (1 - x_i)))\big)$$

$$= \max\big(0, \sum_{i=1}^{n} w_i \cdot x_i - \sum_{i=1}^{n} w_i + 1)\big)$$

It is easy to see that in the case $\sum_{i=1}^{n} w_i = 1$ nothing else than the ordinary weighted arithmetic mean is obtained. Analogously, the same construction can be applied for the product t-norm $T_{\mathbf{P}}$:

$$\mathbf{A}'_{\boldsymbol{w}}(x_1, \ldots, x_n) = \exp\Big(\sum_{i=1}^{n} w_i \cdot \ln x_i\Big) = \prod_{i=1}^{n} x_i^{w_i}$$

This operator coincides with the weighted geometric mean if $\sum_{i=1}^{n} w_i = 1$.

Example 4 shows that the ad-hoc idea of using weighted averages like in VQS is justified in the fuzzy relation-based framework of this paper too, but only for $T = T_{\mathbf{L}}$. If $T \neq T_{\mathbf{L}}$, the corresponding weighted quasi-arithmetic mean with respect to T must be chosen to preserve T-transitivity. In case $T_{\mathbf{P}}$, the weighted geometric mean is obtained.

Remark 1. The main benefit of a system like VQS is that a ranking of records according to the closeness to the query is obtained. Practically, the matching degrees themselves are of minor importance—what matters is the ranking. In case that the chosen t-norm T is strict, the ranking result is invariant with respect to the sum of the weights (of course, as long as this sum is positive; cf. Theorem 3). If T is nilpotent, the sum of the weights has more influence. If $\sum_{i=1}^{n} w_i = 1$, the term $\sum_{i=1}^{n} w_i \cdot f(x_i)$ in (10) cannot exceed $f(0)$. If $\sum_{i=1}^{n} w_i > 1$, it may happen that $\sum_{i=1}^{n} w_i \cdot f(x_i) > f(0)$, which means a loss of information. Therefore, we suggest to normalize all weights in the query such that their sum is 1, which is simple and results in a minimal loss of information. We suggest the following strategy:

- If the user specifies weights for all conditions (i.e. a sequence of weights $\tilde{w}_1, \ldots, \tilde{w}_n$ that does not necessarily sum up to 1), the weights for aggregation are chosen as $(i = 1, \ldots, n)$

$$w_i = \frac{\tilde{w}_i}{\sum_{i=1}^{n} \tilde{w}_i}.$$

- If the user specifies no weights or only for some conditions, the remaining "raw weights" are filled up with 1's. Then the interpretation is transparent for the user: With a weight $\tilde{w}_i > 1$, he/she can strengthen the importance of a condition. With a weight $\tilde{w}_i < 1$, he/she correspondingly weakens the importance of a condition.

Note that the aggregation methods described in this section are tailored to the specific needs of a framework that makes use of T-transitive fuzzy relations. In fuzzy querying, also other means of aggregation are commonly used, such as, Ordered Weighted Average (OWA) operators [12, 20, 41, 43] or linguistic quantifiers [22, 42, 45].

5   Summary and Demonstrative Examples

Sections 3 and 4 together provide the mechanisms that are necessary to define the semantics of oVQL. The following things have to be fixed in order to make the ordering-enriched VQS work:

Table 4. Result sets for the unweighted query for $T = T_{\mathbf{L}}$ (above) and $T = T_{\mathbf{P}}$ (below)

#	Location	t_1	t_2	t_3	t
3	Salzburg Liefering	0.7800	1.0000	1.0000	**0.9267**
7	Salzburg Aigen	0.7900	0.5333	1.0000	**0.7744**
9	Salzburg Maxglan	0.8100	1.0000	0.5000	**0.7700**
6	Salzburg Aigen	0.7900	1.0000	0.5000	**0.7633**
1	Salzburg Center	1.0000	0.0000	1.0000	**0.6667**
2	Salzburg Center	1.0000	0.0000	1.0000	**0.6667**
5	Mattsee	0.0000	1.0000	1.0000	**0.6667**
8	Linz	0.0000	1.0000	1.0000	**0.6667**
4	Anif	0.5500	0.3333	1.0000	**0.6278**

#	Location	t_1	t_2	t_3	t
3	Salzburg Liefering	0.8025	1.0000	1.0000	**0.9293**
7	Salzburg Aigen	0.8106	0.6271	1.0000	**0.7981**
9	Salzburg Maxglan	0.8270	1.0000	0.6065	**0.7945**
6	Salzburg Aigen	0.8106	1.0000	0.6065	**0.7893**
2	Salzburg Center	1.0000	0.3679	1.0000	**0.7165**
5	Mattsee	0.3447	1.0000	1.0000	**0.7011**
4	Anif	0.6376	0.5134	1.0000	**0.6892**
1	Salzburg Center	1.0000	0.0357	1.0000	**0.3292**
8	Linz	0.0013	1.0000	1.0000	**0.1086**

1. NCR tables for the non-numeric attributes and corresponding distance measures (corresponding to d in (4))
2. A continous Archimedean t-norm T with additive generator f
3. For all attributes, a default value for the tolerance radius C

Then, for a given query and a given record, the degrees to which the individual conditions are fulfilled can be computed as specified in (3) and (5)–(8). The final aggregation of these degrees into the overall degree to which the record fulfills the query is done with the aggregation operator specified in (10). This degree of fulfillment is computed for each record in the table under consideration. Finally, these degrees of fulfillment are sorted and the sorted list of records is presented to the user in descending order (better fitting records first).

Example 5. We reconsider the hotel data set of Example 1 presented in Table 1. The matrix of distances between locations is again given as in Table 2. Table 4 shows the results that are obtained for the following query:

```
SELECT FROM HotelTable
    WHERE Location IS 'Salzburg Center'
            TOLERATE UP TO 20
        AND Price IS WITHIN (60,70)
            TOLERATE UP TO 10
        AND StarCategory IS AT LEAST 4
            TOLERATE UP TO 2
    INTO ResultSet
```

Table 5. Result sets for the weighted query for $T = T_{\mathbf{L}}$ (above) and $T = T_{\mathbf{P}}$ (below)

#	Location	t_1	t_2	t_3	t
3	Salzburg Liefering	0.7800	1.0000	1.0000	**0.8533**
1	Salzburg Center	1.0000	0.0000	1.0000	**0.8333**
2	Salzburg Center	1.0000	0.0000	1.0000	**0.8333**
9	Salzburg Maxglan	0.8100	1.0000	0.5000	**0.7900**
7	Salzburg Aigen	0.7900	0.5333	1.0000	**0.7822**
6	Salzburg Aigen	0.7900	1.0000	0.5000	**0.7767**
4	Anif	0.5500	0.3333	1.0000	**0.5889**
5	Mattsee	0.0000	1.0000	1.0000	**0.3333**
8	Linz	0.0000	1.0000	1.0000	**0.3333**

#	Location	t_1	t_2	t_3	t
3	Salzburg Liefering	0.8025	1.0000	1.0000	**0.8636**
2	Salzburg Center	1.0000	0.3679	1.0000	**0.8465**
9	Salzburg Maxglan	0.8270	1.0000	0.6065	**0.8106**
7	Salzburg Aigen	0.8106	0.6271	1.0000	**0.8043**
6	Salzburg Aigen	0.8106	1.0000	0.6065	**0.7998**
4	Anif	0.6376	0.5134	1.0000	**0.6629**
1	Salzburg Center	1.0000	0.0357	1.0000	**0.5738**
5	Mattsee	0.3447	1.0000	1.0000	**0.4916**
8	Linz	0.0013	1.0000	1.0000	**0.0118**

Now consider the following query:

```
SELECT FROM HotelTable
    WHERE Location IS 'Salzburg Center'
                TOLERATE UP TO 20 WEIGHTED BY 4
        AND Price IS WITHIN (60,70)
                TOLERATE UP TO 10
        AND StarCategory IS AT LEAST 4
                TOLERATE UP TO 2
    INTO ResultSet
```

The weight for the condition referring to the location means that the closeness of the location receives a four times as large importance as price and category. The weights used for aggregation (according to Remark 1) are, therefore, $(w_1, w_2, w_3) = (\frac{2}{3}, \frac{1}{6}, \frac{1}{6})$. Table 5 shows the results obtained for $T = T_{\mathbf{L}}$ and $T = T_{\mathbf{P}}$. In all four tables, the columns labeled t_1, t_2, and t_3 contain the degrees of fulfillment of the three single conditions in the query (referring to location, price, and category). The rightmost column displays the overall degree of matching.

6   The Choice of the Underlying t-Norm

In Example 5, the following rankings are obtained for the query with equal weights (we denote the degree of matching for the i-th record with t^j):

$$t^3 > t^7 > t^9 > t^6 > t^1 = t^2 = t^5 = t^8 > t^4 \qquad \text{for } T = T_{\mathbf{L}}$$
$$t^3 > t^7 > t^9 > t^6 > t^2 > t^5 > t^4 > t^1 > t^8 \qquad \text{for } T = T_{\mathbf{P}}$$

The following rankings are obtained for the second query with weights:

$$t^3 > t^1 = t^2 > t^9 > t^7 > t^6 > t^4 > t^5 = t^8 \qquad \text{for } T = T_{\mathbf{L}}$$
$$t^3 > t^2 > t^9 > t^7 > t^6 > t^4 > t^1 > t^5 > t^8 \qquad \text{for } T = T_{\mathbf{P}}$$

We clearly see, at least for this specific example, that the results depend on the choice of the underlying t-norm. However, we also see that, for instance, the best four matches for the unweighted query are equally ranked for $T_{\mathbf{L}}$ and $T_{\mathbf{P}}$. This is not a coincidence, but has a clear mathematical explanation. To explain that, we consider a query consisting of n conditions "x_i IS q_i". We restrict to this case for simplicity just to investigate the role of the underlying t-norm. With additional effort, analogous arguments can be constructed involving also the three other types of conditions. Then, for each attribute x_i, a metric d_i is defined (using an NCR for a non-numeric attribute and the absolute distance for a numeric attribute). Given a record $(\tilde{x}_1, \ldots, \tilde{x}_n)$ and corresponding query values $(\tilde{q}_1, \ldots, \tilde{q}_n)$, the degrees of fulfillment are computed as

$$t_i = t(\text{"}x_i \text{ IS } q_i\text{"} \mid \tilde{x}_i) = E_{d_i, C_i}(\tilde{x}_i, q_i) = f^{-1}\big(\min(\tfrac{1}{C_i} \cdot d_i(\tilde{x}_i, q_i), f(0))\big).$$

Then the overall degree of fulfillment is given as

$$\mathbf{A_w}(t_1, \ldots, t_n) = f^{-1}\big(\min(f(0), \sum_{i=1}^{n} w_i \cdot f(t_i)))$$
$$= f^{-1}\big(\min(f(0), \sum_{i=1}^{n} w_i \cdot \min(\tfrac{1}{C_i} \cdot d_i(\tilde{x}_i, q_i), f(0)))) = (*)$$

Let us first consider the strict case, i.e. $f(0) = \infty$. Then the overall degree of matching simplifies to

$$(*) = f^{-1}\Big(\sum_{i=1}^{n} \frac{w_i}{C_i} \cdot d_i(\tilde{x}_i, q_i)\Big). \qquad (11)$$

This implies that the ranking of a set of records only depends on the ranking of the weighted sum $\sum_{i=1}^{n} \frac{w_i}{C_i} \cdot d_i(\tilde{x}_i, q_i)$. Firstly, this means that all strict t-norms give exactly the same ranking of records (as f is a strictly decreasing continuous bijection $[0, \infty] \to [0, 1]$). It is therefore, absolutely sufficient to pick out one convenient representative of this class of t-norms, for which $T_{\mathbf{P}}$ is the canonical choice. Moreover, if each value C_i is chosen as the maximal distance of the values for attribute x_i, then even exactly the same ranking as in the original VQS is obtained. This fact might indicate at first glance that the results in this paper—at least for strict t-norms—do not provide any added value compared with the original VQS. This may be true under the restrictive assumptions of this example. However, we (1) have gained fundamental theoretical insight and a sound justification from the theory of fuzzy relations, (2) there is now the possibility to formulate also the other three types of conditions, and (3) the values C_i provide additional influence on the rating of distances for the user.

In the nilpotent case, i.e. $f(0) = 1$, the situation is slightly more complicated. The overall degree of fulfillment is given as

$$(*) = f^{-1}\Big(\min(1, \sum_{i=1}^{n} w_i \cdot \min(\tfrac{1}{C_i} \cdot d_i(\tilde{x}_i, q_i), 1))\Big).$$

In case that, for all $i = 1, \ldots, n$, $\frac{1}{C_i} \cdot d_i(\tilde{x}_i, q_i) \leq 1$ and that $\sum_{i=1}^{n} \frac{w_i}{C_i} \cdot d_i(\tilde{x}_i, q_i) \leq 1$, the overall degree of fulfillment is the same as in (11). That also explains why the rankings of the four best-matching records $t^3 > t^7 > t^9 > t^6$ are the same for $T_\mathbf{L}$ and $T_\mathbf{P}$ for the unweighted query in Example 5.

Let us briefly summarize our findings about the choice of the underlying t-norm T. There are basically two choices for T. Any strict t-norm leads to the same results, therefore, there is no point in choosing any other than the simplest representative, $T_\mathbf{P}$. If a nilpotent t-norm is chosen, the particular choice does have influence on the result. From a practical perspective, however, $T_\mathbf{L}$ is a pragmatic and justifiable choice. Choosing a nilpotent t-norm has the advantage that, for a given condition, the tolerance radius C has a clear and unambiguous interpretation. However, any information outside this radius is lost, which is not the case for strict t-norms. Whether this is desired or not has to be decided according to the requirements of the concrete application.

7   A Real-World Example

The concepts introduced in this paper have been evaluated with a prototype implemented by two students of the first author [2]. The goal was to develop

Table 6. Intermediate query result before flexible interpretation; the rightmost column provides the distance from Linz (zip 4020)

#	Location	HP	Year	Mileage (km)	Price (EUR)	Distance (km)
1	4364 St. Thomas	90	1994	164000	3750	35
2	4232 Münzbach	116	2000	120000	13950	31
3	4871 Zipf	101	2000	17500	18500	64
4	4651 Stadl-Paura	90	1991	187900	2800	39
5	4064 Oftering	107	1991	109000	2900	13
6	5350 Strobl	101	1997	137000	8750	88
7	5222 Munderfing	90	1996	156000	5900	86
8	4905 Thomasroith	90	1994	214500	4590	54
9	4840 Vöcklabruck	110	1998	n.a.	5700	56
10	4656 Kirchham	116	1991	200000	1600	46
11	4141 Pfarrkirchen	90	1995	189000	3950	42

Table 7. Result sets for $T_{\mathbf{L}}$ (left) and $T_{\mathbf{P}}$ (right; numbers rounded to three digits)

#	t_1	t_2	l_3	t_4	t_5	**t**
1	0.00	0.00	0.00	0.00	1.00	**0.20**
2	0.00	0.40	1.00	0.00	0.00	**0.28**
3	0.00	1.00	1.00	1.00	0.00	**0.60**
4	0.00	0.00	0.00	0.00	1.00	**0.20**
5	0.35	1.00	0.00	0.00	1.00	**0.47**
6	0.00	1.00	0.50	0.00	1.00	**0.50**
7	0.00	0.00	0.00	0.00	1.00	**0.20**
8	0.00	0.00	0.00	0.00	1.00	**0.20**
9	0.00	1.00	1.00	$n.a.$	1.00	**0.75**
10	0.00	0.40	0.00	0.00	1.00	**0.28**
11	0.00	0.00	0.00	0.00	1.00	**0.20**

#	t_1	t_2	t_3	t_4	t_5	**t**
1	0.174	0.368	0.135	0.004	1.000	**0.126**
2	0.212	0.549	1.000	0.069	0.019	**0.173**
3	0.041	1.000	1.000	1.000	>0.000	**0.096**
4	0.142	0.368	0.030	0.001	1.000	**0.065**
5	0.522	1.000	0.030	0.145	1.000	**0.296**
6	0.012	1.000	0.607	0.022	1.000	**0.176**
7	0.014	0.368	0.368	0.006	1.000	**0.103**
8	0.067	0.368	0.135	>0.000	1.000	**0.053**
9	0.061	1.000	1.000	$n.a.$	1.000	**0.497**
10	0.100	0.549	0.030	>0.000	1.000	**0.056**
11	0.122	0.368	0.223	0.001	1.000	**0.093**

a flexible query answering interface to a relational database containing cars for sale.

The most important table in the database is the list of available cars. This table has 53 columns and a total of approx. 65000 rows/records. Technical data, features, age, mileage, and the zip code where it is available can be stored for each car. Roughly half of the columns are categorical and half are numerical. The different models and brands are stored in separate auxiliary tables in a normalized way. For the zip code, two more tables are available, one that maps a zip code to a town name and one table that assigns a distance (in km) to each pair of zip codes.

The prototype in its current version mainly complies with the principles presented in Section 2, but does not make use of NCRs. For the zip code, a complete distance table is available anyway, so there is no particular need for an NCR. All other categorical attributes are treated in a crisp way without any flexible interpretation. It is possible to choose between two t-norms, $T_{\mathbf{L}}$ and $T_{\mathbf{P}}$.

Let us consider the following query:

```
SELECT FROM CarTable
  WHERE Model IS 'Volkswagen Passat'
    AND Layout IS 'Wagon'
    AND Location IS 'Linz' TOLERATE UP TO 20
    AND HorsePower IS WITHIN (100,110) TOLERATE UP TO 10
    AND YearBuilt IS AT LEAST 1998 TOLERATE UP TO 2
    AND Mileage IS AT MOST 80000 TOLERATE UP TO 15000
    AND Price IS AT MOST 10000 TOLERATE UP TO 1000
  INTO ResultSet
```

The first two conditions are referring to categorical attributes that are not interpreted in a flexible way. Hence, we only need to consider records fulfilling those two conditions. Table 6 shows a list of 11 cars to be considered.

Then Table 7 shows the results obtained for $T_{\mathbf{L}}$ and $T_{\mathbf{P}}$. In these tables, the columns labeled $t_1, \ldots, t_5$ contain the degrees to which records fulfill the five conditions (in this order: distance, horsepower, year, mileage, price) that are interpreted as described previously (cf. Section 5). The final matching degree is shown in the last columns labeled **t**.

Note that the table contains relatively many missing values. In [2], we proceeded in the following way: if a value was missing in the record under consideration, the respective condition was supposed to be fulfilled with a degree of 1. The rationale behind this strategy was that there is no evidence that the record would not fulfill this condition (which corresponds to a kind of optimistic approach). We have figured out, however, that this ad-hoc approach could lead to significantly distorted matching degrees. That is why we propose to leave such values uninterpreted and to consequently leave them out during aggregation (corresponding to a kind of neutral approach).

The following rankings are obtained for the query (we denote the degree of matching for the j-th record/car with t^j):

$$t^9 > t^3 > t^6 > t^5 > t^2 = t^{10} > t^1 = t^4 = t^7 = t^8 = t^{11} \qquad \text{for } T = T_{\mathbf{L}}$$
$$t^9 > t^5 > t^6 > t^2 > t^1 > t^7 > t^3 > t^{11} > t^4 > t^{10} > t^8 \qquad \text{for } T = T_{\mathbf{P}}$$

Obviously, the rankings significantly differ for the two basic t-norms which is obvious from the discussion in Section 6.

Extensive experiments were carried out with the prototype. The goal was to evaluate the general concept of the ordering-enriched VQS and its possible advantages over classical querying. The following points are worth mentioning:

1. The language of VQS is easy to use and easy to interpret for humans. Even non-skilled persons were easily able to interpret the queries and the result lists.
2. VQS is computationally efficient, mainly because of its pragmatic approach, i.e. the use of Euclidean distances.
3. At least for numeric attributes, the degrees to which records fulfill queries depend on the query values in a continuous way. Therefore, the approach is robust with respect to noisy data and the choice of a particular query value.

Finally, note that the framework of VQS permits much more sophisticated concepts to deal with categorical attributes (instead of dealing with them crisply as in this prototypical case study). For some attributes, an NCR would be straightforward. For categorical attributes with a small number of possible instances, distance/similarity tables seem feasible. In this example, none of the two ways is feasible for the car model column, as the database currently contains around 1000 models from 90 manufacturers. An idea in this direction would be to derive the similarities from the data describing the individual cars by PCA, clustering, or machine learning [13].

8   Concluding Remarks

We have proposed an approach how to support queries involving ordering conditions in the vague query system VQS. This has been accomplished by utilizing the correspondence between (pseudo-)metrics and fuzzy equivalence relations and applying results from the theory of fuzzy orderings. It is worth to mention that VQS has just been considered as a case study. In fact, the applicability of fuzzy orderings to realizing ordering-based flexible queries is not restricted to VQS, but can be carried out analogously for any distance-based flexible query answering system.

Acknowledgements

The authors gratefully acknowledge support by COST Action 274 "TARSKI". When this paper was mainly written, Ulrich Bodenhofer was affiliated with Software Competence Center Hagenberg, A-4232 Hagenberg, Austria. Therefore, he thanks for support by the Austrian Government, the State of Upper Austria, and the Johannes Kepler University Linz in the framework of the K*plus* Competence Center Program. The authors are further grateful to Peter Bogdanowicz and Gerhard Lanzerstorfer for implementing the prototype presented in Section 7.

References

1. U. Bodenhofer. A similarity-based generalization of fuzzy orderings preserving the classical axioms. *Internat. J. Uncertain. Fuzziness Knowledge-Based Systems*, 8(5):593–610, 2000.
2. U. Bodenhofer, P. Bogdanowicz, G. Lanzerstorfer, and J. Küng. Distance-based fuzzy relations in flexible query answering systems: Overview and experiences. In I. Düntsch and M. Winter, editors, *Proc. 8th Int. Conf. on Relational Methods in Computer Science*, pages 15–22, St. Catharines, ON, February 2005. Brock University.
3. U. Bodenhofer and J. Küng. Fuzzy orderings in flexible query answering systems. *Soft Computing*, 8(7):512–522, 2004.
4. P. Bosc, B. Buckles, F. Petry, and O. Pivert. Fuzzy databases: Theory and models. In J. Bezdek, D. Dubois, and H. Prade, editors, *Fuzzy Sets in Approximate Reasoning and Information Systems*, volume 5 of *The Handbooks of Fuzzy Sets*, pages 403–468. Kluwer Academic Publishers, Boston, 1999.
5. P. Bosc, L. Duval, and O. Pivert. Value-based and representation-based querying of possibilistic databases. In G. Bordogna and G. Pasi, editors, *Recent Issues on Fuzzy Databases*, volume 53 of *Studies in Fuzziness and Soft Computing*, pages 3–27. Physica-Verlag, Heidelberg, 2000.
6. P. Bosc, M. Galibourg, and G. Hamon. Fuzzy querying with SQL: Extension and implementation aspects. *Fuzzy Sets and Systems*, 28(3):333–349, 1988.
7. P. Bosc and O. Pivert. Fuzzy querying in conventional databases. In L. A. Zadeh and J. Kacprzyk, editors, *Fuzzy Logic for the Management of Uncertainty*, pages 645–671. John Wiley & Sons, New York, 1992.
8. P. Bosc and O. Pivert. SQLf: A relational database language for fuzzy querying. *IEEE Trans. Fuzzy Systems*, 3:1–17, 1995.
9. B. P. Buckles and F. E. Petry. Fuzzy databases and their applications. In M. M. Gupta and E. Sanchez, editors, *Fuzzy Information and Decision Processes*, pages 361–371. North-Holland, New York, 1982.

10. B. P. Buckles and F. E. Petry. Query languages for fuzzy databases. In J. Kacprzyk and R. R. Yager, editors, *Management Decision Support Systems Using Fuzzy Sets and Possibility Theory*, pages 241–252. Verlag TÜV Rheinland, Köln, 1985.
11. B. P. Buckles, F. E. Petry, and H. Sachar. Design of similarity-based relational databases. In C. V. Negotia and H. Prade, editors, *Fuzzy Logic in Knowledge Engineering*, pages 3–17. Verlag TÜV Rheinland, Cologne, 1986.
12. T. Calvo, G. Mayor, and R. Mesiar, editors. *Aggregation Operators*, volume 97 of *Studies in Fuzziness and Soft Computing*. Physica-Verlag, Heidelberg, 2002.
13. B. C. Csáji, J. Küng, J. Palkoska, and R. Wagner. On the automation of similarity information maintenance in flexible query answering systems. In F. Galindo, M. Takizawa, and R. Traunmüller, editors, *Proc. 15th Int. Conf. on Database and Expert Systems Applications*, volume 3180 of *Lecture Notes in Computer Science*, pages 130–140. Springer, Berlin, 2004.
14. B. De Baets and R. Mesiar. Pseudo-metrics and T-equivalences. *J. Fuzzy Math.*, 5(2):471–481, 1997.
15. D. Dubois and H. Prade. Similarity-based approximate reasoning. In J. M. Zurada, R. J. Marks, and C. J. Robinson, editors, *Computational Intelligence Imitating Life*, pages 69–80. IEEE Press, New York, 1994.
16. D. Dubois and H. Prade. Using fuzzy sets in flexible querying: Why and how? In *Proc. Workshop on Flexible Query-Answer Systems (FQAS'96)*, pages 89–103, Roskilde, May 1996.
17. J. Fodor and M. Roubens. *Fuzzy Preference Modelling and Multicriteria Decision Support*. Kluwer Academic Publishers, Dordrecht, 1994.
18. U. Höhle and N. Blanchard. Partial ordering in L-underdeterminate sets. *Inform. Sci.*, 35:133–144, 1985.
19. T. Ichikawa and M. Hirakawa. ARES: A relational database with the capability of performing flexible interpretation of queries. *IEEE Trans. Software Eng.*, 12(5):624–634, 1986.
20. J. Kacprzyk and S. Zadrozny. Implementation of OWA operators in fuzzy querying for Microsoft Access. In R. R. Yager and J. Kacprzyk, editors, *The Ordered Weighted Averaging Operators: Theory and Applications*, pages 293–306. Kluwer Academic Publishers, Boston, 1997.
21. J. Kacprzyk, S. Zadrozny, and A. Ziolkowski. FQUERY III+: a "human-consistent" database querying system based on fuzzy logic with linguistic quantifiers. *Inform. Sci.*, 14(6):443–453, 1989.
22. J. Kacprzyk and A. Ziolkowski. Database queries with fuzzy linguistic quantifiers. *IEEE Trans. Syst. Man Cybern.*, 16:474–479, 1986.
23. F. Klawonn. Fuzzy sets and vague environments. *Fuzzy Sets and Systems*, 66:207–221, 1994.
24. E. P. Klement, R. Mesiar, and E. Pap. *Triangular Norms*, volume 8 of *Trends in Logic*. Kluwer Academic Publishers, Dordrecht, 2000.
25. R. Kruse, J. Gebhardt, and F. Klawonn. *Foundations of Fuzzy Systems*. John Wiley & Sons, New York, 1994.
26. J. Küng and J. Palkoska. VQS—a vague query system prototype. In *Proc. 8th Int. Workshop on Database and Expert Systems Applications*, pages 614–618. IEEE Computer Society Press, Los Alamitos, CA, 1997.
27. J. Küng and J. Palkoska. Vague joins—an extension of the vague query system VQS. In *Proc. 9th Int. Workshop on Database and Expert Systems Applications*, pages 997–1001. IEEE Computer Society Press, Los Alamitos, CA, 1998.
28. J. M. Medina, O. Pons, and M. A. Vila. GEFRED: a generalized model of fuzzy relational databases. *Inform. Sci.*, 76(1–2):87–109, 1994.

29. A. Motro. VAGUE: A user interface to relational databases that permits vague queries. *ACM Trans. Off. Inf. Syst.*, 6(3):187–214, 1988.

30. S. V. Ovchinnikov. Similarity relations, fuzzy partitions, and fuzzy orderings. *Fuzzy Sets and Systems*, 40(1):107–126, 1991.

31. J. Palkoska, A. Dunzendorfer, and J. Küng. Vague queries in tourist information systems. In *Information and Communication Technologies in Tourism (ENTER 2000)*, pages 61–70. Springer, Vienna, 2000.

32. F. E. Petry and P. Bosc. *Fuzzy Databases: Principles and Applications.* International Series in Intelligent Technologies. Kluwer Academic Publishers, Boston, 1996.

33. A. Pradera and E. Trillas. A note of pseudometrics aggregation. *Int. J. General Systems*, 31(1):41–51, 2002.

34. A. Rosado, J. Kacprzyk, R. A. Ribeiro, and S. Zadrozny. Fuzzy querying in crisp and fuzzy relational databases: An overview. In *Proc. 9th Int. Conf. on Information Processing and Management of Uncertainty in Knowledge-Based Systems*, volume 3, pages 1705–1712, Annecy, July 2002.

35. S. Saminger, R. Mesiar, and U. Bodenhofer. Domination of aggregation operators and preservation of transitivity. *Internat. J. Uncertain. Fuzziness Knowledge-Based Systems*, 10(Suppl.):11–35, 2002.

36. B. Schweizer and A. Sklar. *Probabilistic Metric Spaces.* North-Holland, Amsterdam, 1983.

37. V. Tahani. A conceptual framework for fuzzy query processing: A step towards very intelligent database systems. *Inf. Proc. and Manag.*, 13:289–303, 1977.

38. Y. Takahashi. Fuzzy database query languages and their relational completeness theorem. *IEEE Trans. Knowl. Data Eng.*, 5(1):122–125, 1993.

39. E. Trillas and L. Valverde. An inquiry into indistinguishability operators. In H. J. Skala, S. Termini, and E. Trillas, editors, *Aspects of Vagueness*, pages 231–256. Reidel, Dordrecht, 1984.

40. L. Valverde. On the structure of F-indistinguishability operators. *Fuzzy Sets and Systems*, 17(3):313–328, 1985.

41. R. R. Yager. On ordered weighted averaging aggregation operators in multicriteria decisionmaking. *IEEE Trans. Syst. Man Cybern. B*, 18(1):183–190, 1988.

42. R. R. Yager. Interpreting linguistically quantified propositions. *Int. J. Intell. Syst.*, 9:149–184, 1994.

43. R. R. Yager and J. Kacprzyk, editors. *The Ordered Weighted Averaging Operators: Theory and Applications.* Kluwer Academic Publishers, Boston, 1997.

44. L. A. Zadeh. Similarity relations and fuzzy orderings. *Inform. Sci.*, 3:177–200, 1971.

45. L. A. Zadeh. A computational approach to fuzzy quantifiers in natural languages. *Comput. Math. Appl.*, 9:149–184, 1983.

General Representation Theorems for Fuzzy Weak Orders

Ulrich Bodenhofer[1], Bernard De Baets[2], and János Fodor[3]

[1] Institute of Bioinformatics,
Johannes Kepler University, A-4040 Linz, Austria
bodenhofer@bioinf.jku.at
[2] Dept. of Applied Mathematics, Biometrics, and Process Control,
Ghent University, Coupure links 653, B-9000 Gent, Belgium
Bernard.DeBaets@ugent.be
[3] Institute of Intelligent Engineering Systems,
Budapest Tech, H-1034 Budapest, Hungary
fodor@bmf.hu

Abstract. The present paper gives a state-of-the-art overview of general representation results for fuzzy weak orders. We do not assume that the underlying domain of alternatives is finite. Instead, we concentrate on results that hold in the most general case that the underlying domain is possibly infinite. This paper presents three fundamental representation results: (i) score function-based representations, (ii) inclusion-based representations, (iii) representations by decomposition into crisp linear orders and fuzzy equivalence relations.

1 Introduction

Weak orders are among the most fundamental concepts in preference modeling. A binary relation $\lesssim$ on a given non-empty domain X is called a *weak order* if it has the following three properties for all $x, y, z \in X$:

$$x \lesssim x \qquad \text{(reflexivity)}$$
$$\text{if } x \lesssim y \text{ and } y \lesssim z \text{ then } x \lesssim z \quad \text{(transitivity)}$$
$$x \lesssim y \text{ or } y \lesssim x \qquad \text{(completeness)}$$

Obviously the only difference between weak orders and linear orders is that weak orders need not be antisymmetric, i.e., a weak order $\lesssim$ is a linear order if and only if the additional property

$$\text{if } x \lesssim y \text{ and } y \lesssim x \text{ then } x = y \text{ (antisymmetry)}$$

holds for all $x, y \in X$. It is easy to see that the ranking of linearly ordered properties of objects constitutes a weak order, e.g., ranking cars by their maximum speed, ranking persons by their height or weight, ranking products by their price, and so forth. This basic fact is not only a fundamental construction principle, but a fundamental representation of weak orders.

H. de Swart et al. (Eds.): TARSKI II, LNAI 4342, pp. 229–244, 2006.

Theorem 1. *A binary relation $\lesssim$ on a non-empty domain X is a weak order if and only if there exists a linearly ordered non-empty set Y and a mapping $f : X \to Y$ such that $\lesssim$ can be represented in the following way for all $x, y \in X$:*

$$x \lesssim y \quad \text{if and only if} \quad f(x) \leq f(y) \tag{1}$$

The proof that a relation defined as in Eq. (1) is a weak order is straightforward. To prove the existence of a set Y and a mapping f such that representation (1) holds for a given weak order $\lesssim$, one has to follow the following steps: (a) define an equivalence relation $\sim$ as the symmetric kernel of $\lesssim$, (b) define Y as the factor set $X_{/\sim}$, (c) define f as the projection $f(x) = \langle x \rangle_\sim$, (d) prove that the projection of $\lesssim$ onto $X_{/\sim}$ is a linear order on $X_{/\sim}$, (e) prove that representation (1) holds. From this perspective, we can view weak orders as linear orders of equivalence classes. In the context of Theorem 1, the equivalence classes contain exactly those elements that share the same property, i.e., those elements for which f yields the same value.

Note that there is an alternative construction of Y and f. Let us define the *foreset* of an element $x \in X$, denoted $C(x)$, as the set of elements smaller than or equivalent to x, i.e., $C(x) = \{y \in X \mid y \lesssim x\}$. Then define Y as the set of all foresets, i.e., $Y = \{C(x) \mid x \in X\}$. It is straightforward to prove that $x \lesssim y$ if and only $C(x) \subseteq C(y)$, and it follows directly from the completeness of $\lesssim$ that Y is linearly ordered with respect to ordinary set inclusion. Thus, we can also conclude that weak orders on X can be represented by embedding into linearly ordered subsets of the partially ordered set $(\mathcal{P}(X), \subseteq)$.

In the case that X is at most countable, Theorem 1 can be strengthened in the following way: it is always possible to choose $Y = [0, 1]$, i.e., for each weak order, we can find a mapping $f : X \to [0, 1]$ such that representation (1) holds. In other words, weak orders on countable domains can always be embedded into the linear order on the unit interval. This is a classic result that goes back to Cantor [7, 17, 21].

Weak orders are not only simple and fundamental concepts (as the above examples illustrate), they are the basis for representing other fundamental concepts in preference modeling and order theory: it is known that *preorders*, i.e., reflexive and transitive binary relations, are uniquely characterized as intersections of weak orders.

In analogy to the crisp case, fuzzy weak orders are fundamental concepts in fuzzy preference modeling [8, 11, 12, 19]. Given a non-empty set of alternatives X, a fuzzy relation $R : X^2 \to [0, 1]$ is a *fuzzy weak order* if it has the following three properties for all $x, y, z \in X$, where T denotes a left-continuous t-norm:

$$R(x, x) = 1 \qquad \text{(reflexivity)}$$
$$T(R(x, y), R(y, z)) \leq R(x, z) \quad (T\text{-transitivity})$$
$$R(x, y) = 1 \text{ or } R(y, x) = 1 \quad \text{(strong completeness)}$$

The goal of this paper is to provide an overview of representation results for fuzzy weak orders. We concentrate on those results that hold for all possible domains X. Results holding only for finite and/or countable domains will not be considered. Consequently, this paper is organized as follows. After providing some preliminaries in Section 2, we discuss score function-based representations in depth in Section 3 that will be complemented by inclusion-based representations in Section 4. Section 5 is devoted to decomposing fuzzy weak orders into crisp linear orders and fuzzy equivalence relations—in direct analogy to the factor set representation discussed above.

Note that this paper is a state-of-the-art review that mainly integrates results from previously published papers on similarity-based fuzzy orders [3, 4, 5]. This paper consistently views the results from the perspective of fuzzy weak orders.

2   Preliminaries

In this paper, we solely use values from the unit interval to express degrees of order/preference. This is not a serious restriction from a practical point of view, and it is also the standard setting widely used in fuzzy preference modeling. Correspondingly, we use left-continuous triangular norms as standard models for fuzzy conjunctions [16].

Definition 1. An associative, commutative, and non-decreasing binary operation on the unit interval (i.e. a $[0,1]^2 \to [0,1]$ mapping) which has 1 as neutral element is called *triangular norm*, short *t-norm*. A t-norm T is called *left-continuous* if the equality

$$T(\sup_{i \in I} x_i, y) = \sup_{i \in I} T(x_i, y)$$

holds for all families $(x_i)_{i \in I} \in [0,1]^I$ and all $y \in [0,1]$.

The three basic t-norms are denoted as $T_{\mathbf{M}}(x,y) = \min(x,y)$, $T_{\mathbf{P}}(x,y) = x \cdot y$, and $T_{\mathbf{L}}(x,y) = \max(x+y-1,0)$. Further assume that

$$\vec{T}(x,y) = \sup\{u \in [0,1] \mid T(x,u) \leq y\}$$

denotes the unique residual implication of T. For the sake of completeness, let us list the following fundamental properties (valid for all $x,y,z \in [0,1]$) [13, 15, 16]:

(I1) $x \leq y$ if and only if $\vec{T}(x,y) = 1$
(I2) $T(x,y) \leq z$ if and only if $x \leq \vec{T}(y,z)$
(I3) $T(\vec{T}(x,y), \vec{T}(y,z)) \leq \vec{T}(x,z)$
(I4) $\vec{T}(1,y) = y$
(I5) $T(x, \vec{T}(x,y)) \leq y$
(I6) $y \leq \vec{T}(x, T(x,y))$

Furthermore, $\vec{T}$ is non-increasing and left-continuous in the first argument and non-decreasing and right-continuous in the second argument.

If T is a continuous t-norm, then the following holds for all $x, y, z \in [0,1]$:

(I7) if $z \geq x$ then $\vec{T}(x,y) = \vec{T}(\vec{T}(z,x), \vec{T}(z,y))$

The biimplication of T is defined as $\overset{\leftrightarrow}{T}(x,y) = T(\vec{T}(x,y), \vec{T}(y,x))$ and fulfills the following assertions for all $x, y, z \in [0,1]$, see [13, 15]:

(B1) $\overset{\leftrightarrow}{T}(x,y) = 1$ if and only if $x = y$
(B2) $\overset{\leftrightarrow}{T}(x,y) = \overset{\leftrightarrow}{T}(y,x)$
(B3) $\overset{\leftrightarrow}{T}(x,y) = \min(\vec{T}(x,y), \vec{T}(y,x))$
(B4) $T(\overset{\leftrightarrow}{T}(x,y), \overset{\leftrightarrow}{T}(y,z)) \leq \overset{\leftrightarrow}{T}(x,z)$
(B5) $\overset{\leftrightarrow}{T}(x,y) = \vec{T}(\max(x,y), \min(x,y))$

In this paper, uppercase letters will be used synonymously for fuzzy sets/relations and their corresponding membership functions. The fuzzy power set of X will be denoted with $\mathcal{F}(X) = \{A \mid A : X \to [0,1]\}$.

A binary fuzzy relation $R : X^2 \to [0,1]$ is called

- *reflexive* if $R(x,x) = 1$ for all $x \in X$,
- *symmetric* if $R(x,y) = R(y,x)$ for all $x, y \in X$,
- *T-transitive* if $T(R(x,y), R(y,z)) \leq R(x,z)$ for all $x, y, z \in X$,
- *strongly complete* if $\max(R(x,y), R(y,x)) = 1$ for all $x, y \in X$.

Fuzzy relations that are reflexive and T-transitive are called *fuzzy preorders* with respect to T, short *T-preorders*. Symmetric T-preorders are called *fuzzy equivalence relations* with respect to T, short *T-equivalences*. As mentioned in Section 1 already, strongly complete T-preorders are called *fuzzy weak orders* with respect to T, short *weak T-orders*. Given a T-equivalence $E : X^2 \to [0,1]$, a binary fuzzy relation $L : X^2 \to [0,1]$ is called a *fuzzy order* with respect to T and E, short *T-E-order*, if it is T-transitive and additionally has the following two properties:

- *E-reflexivity:* $E(x,y) \leq L(x,y)$ for all $x, y \in X$
- *T-E-antisymmetry:* $T(L(x,y), L(y,x)) \leq E(x,y)$ for all $x, y \in X$

Given a binary fuzzy relation $R : X^2 \to [0,1]$ and an $x \in X$, analogously to the crisp case (cf. Section 1), the *foreset* of x is defined as the fuzzy set $C(x) \in \mathcal{F}(X)$ that expresses the degree to which a given value $y \in X$ is smaller than or equivalent to x, i.e., $C(x)(y) = R(y,x)$ [2].

3   Score Function-Based Representations

The starting point of this section is Theorem 1. It is natural to first ask the question whether there is a straightforward generalization of this theorem to the case of fuzzy weak orders.

Theorem 2. *A binary fuzzy relation $R : X^2 \to [0,1]$ is a weak T-order if and only if there exist a non-empty domain Y, a T-equivalence $E : Y^2 \to [0,1]$, a strongly complete T-E-order $L : Y^2 \to [0,1]$, and a mapping $f : X \to Y$ such that the following equality holds for all $x, y \in X$:*

$$R(x,y) = L(f(x), f(y)) \tag{2}$$

Theorem 2 can be viewed from two different angles. On the one hand, it is a nice straightforward generalization of Theorem 1 and demonstrates the smooth interplay between fuzzy weak orders and strongly complete fuzzy orders (analogously to the crisp case). On the other hand, fuzzy weak orders and strongly complete fuzzy orders are basically the same concepts. From this point of view, Theorem 2 does not provide us with a new construction method or any new insight. More insight would potentially be obtained if we could restrict the choice of Y or E to certain standard cases that could be utilized for constructions in an easier way.

One interesting question is, for instance, whether Y, L, and f can be chosen such that Theorem 2 holds for E being the crisp equality (i.e., with L being a so-called T-order [4, 12, 13, 14], which, in the case that $T = T_{\mathbf{M}}$, is nothing else but a fuzzy partial order in the sense of Zadeh [18, 19, 26]). The answer is quick and negative: as demonstrated in [4, Subsection 2.3], strongly complete fuzzy orders with respect to some t-norm T and the crisp equality can only be crisp orders. Thus, it is never possible to embed a non-crisp weak order into a strongly complete T-order, so it is impossible to strengthen Theorem 2 by fixing E as the crisp equality.

So the question remains whether there is any standard choice Y, E, L, f into which we can embed all, or at least a subclass of, weak T-orders. As shown by Ovchinnikov, it is possible to embed a weak T-order into a continuous weak T-order on the real numbers $\mathbb{R}$, but it is necessary to restrict to strict t-norms and finite domains X [20]. Since this is outside the scope of this paper, we turn our attention to a different investigation. The standard crisp case consists of the unit interval $[0, 1]$ equipped with its natural linear order. Given a left-continuous t-norm T, the canonical fuzzification of the natural linear order on $[0, 1]$ consists in the residual implication $\vec{T}$ [4, 13, 15]. The following proposition, therefore, provides us with a construction that can be considered a straightforward counterpart of (1).

Proposition 1. *Given a function $f : X \to [0, 1]$, the relation defined by*

$$R(x, y) = \vec{T}\big(f(x), f(y)\big) \tag{3}$$

is a weak T-order.

The function f in Proposition 1 can also be understood as a fuzzy set on X. In this section, we rather leave this aspect aside and adopt the classical interpretation as a *score function*.

Note that the simple construction of Proposition 1 is not a unique characterization, i.e., there are weak T-orders that cannot be represented by means of a single score function. In order to demonstrate that, let us consider a set X with at least three elements. We choose an arbitrary linear order of the elements of X (which always exists due to basic results from order theory [22, 23]) and define R as the crisp linear order itself:

$$R(x, y) = \begin{cases} 1 & \text{if } x \leq y \\ 0 & \text{otherwise} \end{cases}$$

Clearly, R is a fuzzy weak order with respect to every t-norm T. Now assume that there exists a score function $f : X \to [0, 1]$ such that representation (3) holds. Let us choose an arbitrary chain of three distinct elements $x < y < z$. Then it clearly follows that $R(z, x) = \vec{T}(f(z), f(x)) = 0$ and $R(z, y) = \vec{T}(f(z), f(y)) = 0$. Since the monotonicity of $\vec{T}$ and (14) imply $\vec{T}(x, y) \geq \vec{T}(1, y) = y$, it trivially follows that $\vec{T}(x, y) = 0$ can hold only if $y = 0$. Thus, we obtain that $f(x) = f(y) = 0$. This entails

$$R(y, x) = \vec{T}(f(y), f(x)) = \vec{T}(0, 0) = 1,$$

which is a contradiction. Hence, we obtain that the most basic fuzzy weak orders—crisp linear orders—are never representable as in Proposition 1, no matter which t-norm we choose. It is, therefore, justified to introduce the representability according to Proposition 1 as a distinct notion.

Definition 2. Consider a weak T-order $R : X^2 \to [0, 1]$. R is called *representable* if there exists a function $f : X \to [0, 1]$, called *generating (score) function*, such that Eq. (3) holds.

Example 1. Let us consider $X = [0, 5]$ and the following two score functions $f_1, f_2 : X \to [0, 1]$:

$$f_1(x) = \min \left(1, \max(0, x - 2)\right)$$

$$f_2(x) = \begin{cases} 0 & \text{if } x \in [0, 1[\\ 0.4 \cdot (x - 1) & \text{if } x \in]1, 2[\\ 0.7 + 0.3 \cdot (x - 2) & \text{if } x \in [2, 3[\\ 1 & \text{if } x \in [3, 5] \end{cases}$$

Figure 1 depicts six fuzzy weak orders defined according to Proposition 1:

$$R_1(x, y) = \vec{T}_\mathbf{M}(f_1(x), f_1(y)) \qquad R_2(x, y) = \vec{T}_\mathbf{M}(f_2(x), f_2(y))$$
$$R_3(x, y) = \vec{T}_\mathbf{P}(f_1(x), f_1(y)) \qquad R_4(x, y) = \vec{T}_\mathbf{P}(f_2(x), f_2(y))$$
$$R_5(x, y) = \vec{T}_\mathbf{L}(f_1(x), f_1(y)) \qquad R_6(x, y) = \vec{T}_\mathbf{L}(f_2(x), f_2(y))$$

The fuzzy relations plotted in Figure 1 have one common feature: the lower right edge always corresponds to the generating score function. More specifically, all fuzzy weak orders in the left column fulfill $R(5, y) = f_1(y)$, while $R(5, y) = f_2(y)$ holds for the fuzzy weak orders in the right column. Note that this is true independent of the t-norm chosen (at least for the three basic t-noVrms). The question arises whether this is a coincidence or whether there is a principle behind. The following theorem tells us that the latter is the case, but even more than that, we obtain a unique characterization of representable fuzzy weak orders (at least for continuous t-norms).

Theorem 3. *Assume that T is continuous. Then a weak T-order R is representable if and only if the following function is a generating function of R:*

$$\bar{f}(x) = \inf_{z \in X} R(z, x)$$

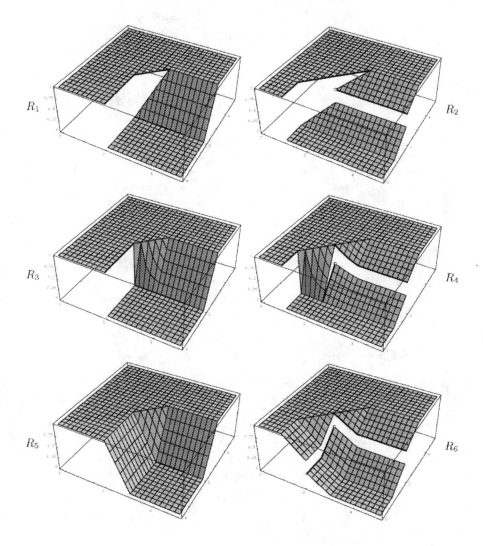

Fig. 1. Fuzzy weak orders constructed from the two score functions f_1 and f_2 by means of Proposition 1 using the three basic t-norms

Theorem 3 provides us with an easy-to-use tool for checking whether a fuzzy weak order is representable—we only have to check whether one specific function is a generating score function. Note, however, that the generating function need not be unique, i.e., it may happen that a fuzzy weak order R is generated by some score function f that does not coincide with $\bar{f}$ defined as in Theorem 3. Let us shortly consider this issue and ask ourselves under which condition $\bar{f}$ coincides with some generating score function f. So assume that R is representable as $R(x,y) = \vec{T}(f(x), f(y))$, then we obtain the following:

$$\bar{f}(x) = \inf_{z \in X} R(z, x) = \inf_{z \in X} \vec{T}(f(z), f(x)) = \vec{T}\left(\sup_{z \in X} f(z), f(x)\right)$$

Then $\sup_{z \in X} f(z) = 1$ is a sufficient criterion for f and $\bar{f}$ to coincide:

$$\bar{f}(x) = \vec{T}\left(\sup_{z \in X} f(z), f(x)\right) = \vec{T}(1, f(x)) \overset{(14)}{=} f(x)$$

It should be clear now that by far not all fuzzy weak orders are representable by single score functions—for all left-continuous t-norms, there exist non-representable fuzzy weak orders. What has not been answered so far is the question whether fuzzy weak orders can be represented by more than one score function. The following well-known theorem provides us with a starting point to this investigation.

Theorem 4. [24] *Consider a binary fuzzy relation $R : X^2 \to [0, 1]$. Then the following two statements are equivalent:*

(i) R is a T-preorder.
(ii) There exists a non-empty family of $X \to [0, 1]$ score functions $(f_i)_{i \in I}$ such that the following representation holds:

$$R(x, y) = \inf_{i \in I} \vec{T}(f_i(x), f_i(y)) \tag{4}$$

Theorem 4 is essential for two main reasons: (1) it shows that every T-preorder is an intersection of representable weak T-orders, (2) as weak T-orders are a special kind of T-preorders, we know for sure that, for each weak T-order R, there exists a family of score functions such that R can be represented as in Eq. (4). Be aware, however, that this is only a representation of theoretical nature. We do not know yet how to choose a family of score functions $(f_i)_{i \in I}$ such that fuzzy relation defined as in Eq. (4) is guaranteed to fulfill strong completeness. The following theorem provides us with a unique characterization of weak T-orders.

Theorem 5. *Consider a binary fuzzy relation $R : X^2 \to [0, 1]$. Then the following two statements are equivalent:*

(i) R is a weak T-order.
(ii) There exists a crisp weak order $\lesssim$ and a non-empty family of $X \to [0, 1]$ score functions $(f_i)_{i \in I}$ that are non-decreasing with respect to $\lesssim$ such that representation (4) holds.

If we want to use Theorem 5 to construct fuzzy weak orders on the real numbers (or a subset of them), one can start from the natural linear order of real numbers, since this order is a crisp weak order, of course. The question arises whether each fuzzy weak order can be represented by a family of score functions that are monotonic with respect to a linear order. The following theorem gives a positive answer and characterizes weak T-orders as intersections of representable weak T-orders that are generated by score functions that are monotonic at the same time with respect to the same crisp linear order.

Theorem 6. *Consider a binary fuzzy relation* $R : X^2 \to [0, 1]$. *Then the following two statements are equivalent:*

(i) R *is a weak* T-*order.*
(ii) *There exists a linear order* $\preceq$ *and a non-empty family of* $X \to [0, 1]$ *score functions* $(f_i)_{i \in I}$ *that are non-decreasing with respect to* $\preceq$ *such that representation (4) holds.*

Example 2. We consider $X = [0, 5]$ again and a family of five functions that are defined as follows:

$$g_1(x) = \min(1, x)$$
$$g_2(x) = \min\big(1, \max(0, x - 1)\big)$$
$$g_3(x) = \min\big(1, \max(0, x - 2)\big)$$
$$g_4(x) = \min\big(1, \max(0, x - 3)\big)$$
$$g_5(x) = \min\big(1, \max(0, x - 4)\big)$$

It is immediate that all five functions are non-decreasing with respect to the natural order of real numbers. Figure 2 depicts six fuzzy weak orders defined in accordance with Theorem 6:

$$R_7(x, y) = \min_{i \in \{1,3,5\}} \vec{T}_{\mathbf{M}}(g_i(x), g_i(y)) \qquad R_8(x, y) = \min_{i \in \{1,...,5\}} \vec{T}_{\mathbf{M}}(g_i(x), g_i(y))$$

$$R_9(x, y) = \min_{i \in \{1,3,5\}} \vec{T}_{\mathbf{P}}(g_i(x), g_i(y)) \qquad R_{10}(x, y) = \min_{i \in \{1,...,5\}} \vec{T}_{\mathbf{P}}(g_i(x), g_i(y))$$

$$R_{11}(x, y) = \min_{i \in \{1,3,5\}} \vec{T}_{\mathbf{L}}(g_i(x), g_i(y)) \qquad R_{12}(x, y) = \min_{i \in \{1,...,5\}} \vec{T}_{\mathbf{L}}(g_i(x), g_i(y))$$

Example 2 uses the natural linear order of real numbers and rather simple monotonic score functions. The next example constructs some more complicated weak $T_{\mathbf{L}}$-orders on the basis of a non-trivial order on the real numbers.

Example 3. Let us consider the following transformation function:

$$\varphi(x) = \begin{cases} 4 - x & \text{if } x \in [1, 3] \\ x & \text{otherwise} \end{cases}$$

It is immediate that φ is a bijective $\mathbb{R} \to \mathbb{R}$ mapping that equals the identity in $] - \infty, 1[\, \cup \,]3, \infty[$ and flips the values in $[1, 3]$. It is clear, therefore, that the binary relation

$$x \preceq y \text{ if and only if } \varphi(x) \leq \varphi(y)$$

is a linear order on the real numbers. Taking the score functions $g_1, \ldots, g_5$ from Example 2, we can define another family of score functions $h_1, \ldots, h_5$ as

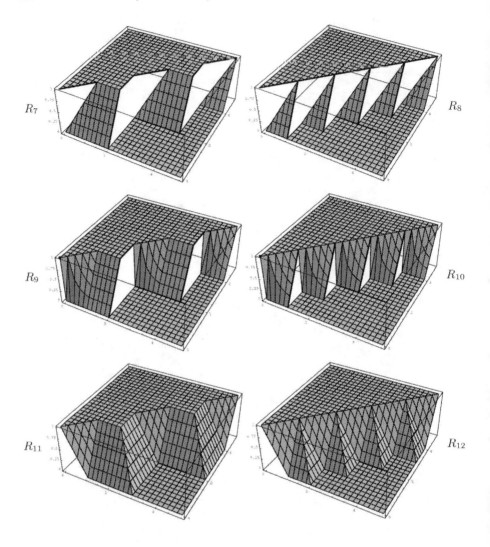

Fig. 2. Six fuzzy weak orders constructed by means of Theorem 6 using the three basic t-norms

$h_i(x) = g_i(\varphi(x))$ (for all $x \in [0,5]$). It is easy to see that all functions h_i are non-decreasing with respect to $\preceq$. Thus, we can use them to define fuzzy weak orders. Figure 3 shows three weak $T_\mathbf{L}$-orders defined as follows:

$$R_{13}(x,y) = \vec{T}_\mathbf{L}(h_3(x), h_3(y))$$
$$R_{14}(x,y) = \min_{i \in \{1,3,5\}} \vec{T}_\mathbf{L}(h_i(x), h_i(y))$$
$$R_{15}(x,y) = \min_{i \in \{1,\dots,5\}} \vec{T}_\mathbf{L}(h_i(x), h_i(y))$$

R_{13} R_{14} R_{15}

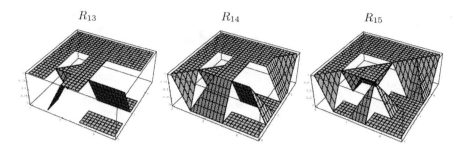

Fig. 3. Three weak T_{L}-orders with a non-trivial underlying crisp linear order

4   Inclusion-Based Representations

As mentioned in Section 1, Theorem 1 can also be proved by embedding the given weak order into the partially ordered set $(\mathcal{P}(X), \subseteq)$. Technically, this is done by mapping the elements $x \in X$ to their foresets $C(x)$. The question arises whether an analogous technique works for fuzzy weak orders as well. This section is devoted to this topic.

Consider the fuzzy power set $\mathcal{F}(X)$. Then the well-known crisp inclusion of fuzzy sets

$$A \subseteq B \text{ if and only if } A(x) \leq B(x) \text{ for all } x \in X$$

is a crisp partial order on $\mathcal{F}(X)$ [25]. Given a left-continuous t-norm T, we can define the following two binary fuzzy relations on $\mathcal{F}(X)$ [1, 4, 13]:

$$\mathrm{INCL}_T(A, B) = \inf_{x \in X} \vec{T}(A(x), B(x))$$

$$\mathrm{SIM}_T(A, B) = \inf_{x \in X} \overleftrightarrow{T}(A(x), B(x))$$

It was proved in [4] that SIM_T is a T-equivalence on $\mathcal{F}(X)$ and that INCL_T is a T-SIM_T-order on $\mathcal{F}(X)$. Moreover, it is easy to see from elementary properties of residual (bi)implications that $\mathrm{INCL}_T(A, B) = 1$ if and only if $A \subseteq B$ and that $\mathrm{SIM}_T(A, B) = 1$ if and only if $A = B$.

The following theorem provides us with a unique characterization of fuzzy weak orders that is based on an embedding of the given fuzzy weak order into the fuzzy power set.

Theorem 7. *Consider a binary fuzzy relation $R : X^2 \to [0, 1]$. Then the following two statements are equivalent:*

(i) R is a weak T-order.
(ii) There exists a non-empty family of fuzzy sets $S \subseteq \mathcal{F}(X)$ that are linearly ordered with respect to the inclusion relation $\subseteq$ and a mapping $\varphi : X \to S$ such that the following representation holds for all $x, y \in X$:

$$R(x, y) = \mathrm{INCL}_T(\varphi(x), \varphi(y)) \tag{5}$$

We can formulate an equivalent result that appears a bit more appealing than Theorem 7.

Corollary 1. *Consider a binary fuzzy relation $R : X^2 \to [0,1]$. Then the following two statements are equivalent:*

(i) R is a weak T-order.
(ii) There exists a mapping $\varphi : X \to \mathcal{F}(X)$ fulfilling $\varphi(x) \subseteq \varphi(y)$ or $\varphi(y) \subseteq \varphi(x)$ for all $x, y \in X$ such that representation (5) holds.

If we omit the linearity conditions in Theorem 7 and Corollary 1, a unique representation of T-preorders is obtained: a fuzzy relation R is a T-preorder if and only if there exists a mapping $\varphi : X \to \mathcal{F}(X)$ such that Eq. (5) holds [5]. In this sense, the T-preorder INCL_T on $\mathcal{F}(X)$ "contains" all T-preorders that can be defined on X. Weak T-orders are then the sub-class that is obtained by restricting to linearly ordered subsets of $\mathcal{F}(X)$.

The proof of Theorem 7 (and Corollary 1) is based on mapping each $x \in X$ to its foreset. However, there is no restriction to only use foresets in (5), as long as the range of the embedding mapping $\varphi(X)$ is linearly ordered. Thus, Theorem 7 and Corollary 1 give rise to potentially interesting constructions. For infinite domains, however, $\mathrm{INCL}_T(A, B)$ is mostly difficult to compute, as an infimum over an infinite set has to be determined. Only under very restrictive assumptions, for instance, that all membership functions of the fuzzy sets $\varphi(x)$ are piecewise linear or differentiable, practically feasible constructions are imaginable. One can overall conclude that Theorem 7 and Corollary 1 provide us with nice theoretical insight, but they do not have much practical value. That is why we do not provide an example in this section.

5   Decompositions into Crisp Linear Orders and T-Equivalences

The standard proof of Theorem 1 is based on the factorization with respect to the symmetric kernel of a given weak order (cf. Section 1). One can state, in other words, that a crisp weak order can always be decomposed into a crisp linear order and an equivalence relation. This section follows this idea and presents corresponding results for fuzzy weak orders. Before coming to the main result, let us shortly introduce an important prerequisite.

Definition 3. Let $\preceq$ be a crisp order on X and let $E : X^2 \to [0,1]$ be a fuzzy equivalence relation (regardless of the underlying t-norm T). E is called *compatible with* $\preceq$ if and only if the following inequality holds for all ascending three-element chains $x \preceq y \preceq z$ in X:

$$E(x, z) \leq \min(E(x, y), E(y, z))$$

Compatibility of a crisp order and a fuzzy equivalence relation can be understood as follows: the two outer elements of an ascending three-element chain are at most as similar as any two elements of this chain.

Theorem 8. *Consider a binary fuzzy relation $R : X^2 \to [0,1]$. Then the following two statements are equivalent:*

(i) R is a weak T-order.

(ii) There exists a crisp linear order $\preceq$ and a T-equivalence E that is compatible with $\preceq$ such that R can be represented as follows:

$$R(x, y) = \begin{cases} 1 & \text{if } x \preceq y \\ E(x, y) & \text{otherwise} \end{cases} \tag{6}$$

Representation (6) simply says the following: weak T-orders are characterized as unions of crisp linear orders and compatible T-equivalences. In other words, we can say that weak T-orders are a fuzzification of crisp linear orders, and the fuzzy component can solely be attributed to a T-equivalence.

To utilize Theorem 8 for constructing weak T-orders, we have to know more about how to construct T-equivalences that are compatible with a given crisp linear order. Let us start with a well-known result on T-equivalences.

Theorem 9. *[24] Consider a binary fuzzy relation $E : X^2 \to [0,1]$. Then the following two statements are equivalent:*

(i) E is a T-equivalence.

(ii) There exists a non-empty family of $X \to [0,1]$ functions $(f_i)_{i \in I}$ such that the following representation holds:

$$E(x, y) = \inf_{i \in I} \overset{\leftrightarrow}{T}(f_i(x), f_i(y)) \tag{7}$$

The following theorem finally provides a unique characterization of T-equivalences that are compatible with a given crisp linear order.

Theorem 10. *Consider a crisp linear order $\preceq$ on X and a binary fuzzy relation $E : X^2 \to [0,1]$. Then the following two statements are equivalent:*

(i) E is a T-equivalence that is compatible with $\preceq$.

(ii) There exists a non-empty family of $X \to [0,1]$ functions $(f_i)_{i \in I}$ that are non-decreasing with respect to $\preceq$ such that representation (7) holds.

Note that Theorem 10 remains valid if we replace "non-decreasing" in (ii) by "non-increasing".

Example 4. It is easy to see that $E_1(x, y) = \exp(-|x - y|)$ is a $T_{\mathbf{P}}$-equivalence on the real numbers $X = \mathbb{R}$ that is compatible with the natural order $\leq$ and that $E_2(x, y) = \max(1 - |x - y|, 0)$ is a $T_{\mathbf{L}}$-equivalence on the real numbers $X = \mathbb{R}$ that is also compatible with $\leq$ [6,9,10]. Hence, Theorem 8 entails that

$$R_{16}(x, y) = \begin{cases} 1 & \text{if } x \leq y \\ \exp(-|x - y|) & \text{otherwise} \end{cases}$$
$$= \min(1, \exp(y - x))$$

is a weak $T_{\mathbf{P}}$-order and

$$R_{17}(x, y) = \begin{cases} 1 & \text{if } x \le y \\ \max(1 - |x - y|, 0) & \text{otherwise} \end{cases}$$
$$= \min(1, \max(1 - x + y, 0))$$

is a weak $T_{\mathbf{L}}$-order. Figure 4 shows these two fuzzy weak orders (where the plots are restricted to $[0, 5]^2$).

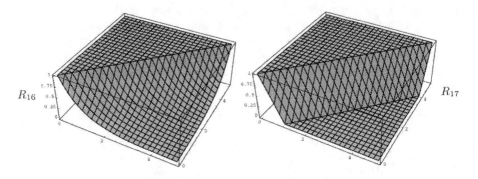

Fig. 4. Two fuzzy weak orders constructed from the absolute distance of real numbers

6   Concluding Remarks

In this contribution, we have highlighted various representations of fuzzy weak orders. Score function-based representations and the decomposition of fuzzy weak orders into crisp linear orders and fuzzy equivalence relations also provided us with practically feasible construction methods. Unlike most of the existing literature, we have not assumed that the underlying domain is finite.

Acknowledgements

The authors gratefully acknowledge support by COST Action 274 "TARSKI". When this paper was mainly written, Ulrich Bodenhofer was affiliated with Software Competence Center Hagenberg, A-4232 Hagenberg, Austria. Therefore, he thanks for support by the Austrian Government, the State of Upper Austria, and the Johannes Kepler University Linz in the framework of the K*plus* Competence Center Program. János Fodor was partly supported by the grant OTKA T046762.

References

1. W. Bandler and L. J. Kohout. Fuzzy power sets and fuzzy implication operators. *Fuzzy Sets and Systems*, 4:183–190, 1980.
2. W. Bandler and L. J. Kohout. Fuzzy relational products as a tool for analysis and synthesis of the behaviour of complex natural and artificial systems. In S. K. Wang and P. P. Chang, editors, *Fuzzy Sets: Theory and Application to Policy Analysis and Information Systems*, pages 341–367. Plenum Press, New York, 1980.
3. U. Bodenhofer. Representations and constructions of strongly linear fuzzy orderings. In *Proc. EUSFLAT-ESTYLF Joint Conference*, pages 215–218, Palma de Mallorca, September 1999.
4. U. Bodenhofer. A similarity-based generalization of fuzzy orderings preserving the classical axioms. *Internat. J. Uncertain. Fuzziness Knowledge-Based Systems*, 8(5):593–610, 2000.
5. U. Bodenhofer. Representations and constructions of similarity-based fuzzy orderings. *Fuzzy Sets and Systems*, 137(1):113–136, 2003.
6. U. Bodenhofer and J. Küng. Fuzzy orderings in flexible query answering systems. *Soft Computing*, 8(7):512–522, 2004.
7. G. Cantor. Beiträge zur Begründung der transfiniten Mengenlehre. *Math. Ann.*, 46:481–512, 1895.
8. B. De Baets, J. Fodor, and E. E. Kerre. Gödel representable fuzzy weak orders. *Internat. J. Uncertain. Fuzziness Knowledge-Based Systems*, 7(2):135–154, 1999.
9. B. De Baets and R. Mesiar. Pseudo-metrics and T-equivalences. *J. Fuzzy Math.*, 5(2):471–481, 1997.
10. B. De Baets and R. Mesiar. Metrics and T-equalities. *J. Math. Anal. Appl.*, 267:331–347, 2002.
11. J. Fodor and S. V. Ovchinnikov. On aggregation of T-transitive fuzzy binary relations. *Fuzzy Sets and Systems*, 72:135–145, 1995.
12. J. Fodor and M. Roubens. *Fuzzy Preference Modelling and Multicriteria Decision Support*. Kluwer Academic Publishers, Dordrecht, 1994.
13. S. Gottwald. *Fuzzy Sets and Fuzzy Logic*. Vieweg, Braunschweig, 1993.
14. S. Gottwald. *A Treatise on Many-Valued Logics*. Studies in Logic and Computation. Research Studies Press, Baldock, 2001.
15. P. Hájek. *Metamathematics of Fuzzy Logic*, volume 4 of *Trends in Logic*. Kluwer Academic Publishers, Dordrecht, 1998.
16. E. P. Klement, R. Mesiar, and E. Pap. *Triangular Norms*, volume 8 of *Trends in Logic*. Kluwer Academic Publishers, Dordrecht, 2000.
17. D. H. Krantz, R. D. Luce, P. Suppes, and A. Tversky. *Foundations of Measurement*. Academic Press, San Diego, CA, 1971.
18. S. V. Ovchinnikov. Similarity relations, fuzzy partitions, and fuzzy orderings. *Fuzzy Sets and Systems*, 40(1):107–126, 1991.
19. S. V. Ovchinnikov. An introduction to fuzzy relations. In D. Dubois and H. Prade, editors, *Fundamentals of Fuzzy Sets*, volume 7 of *The Handbooks of Fuzzy Sets*, pages 233–259. Kluwer Academic Publishers, Boston, 2000.
20. S. V. Ovchinnikov. Numerical representation of transitive fuzzy relations. *Fuzzy Sets and Systems*, 126:225–232, 2002.
21. F. S. Roberts. *Measurement Theory*. Addison-Wesley, Reading, MA, 1979.
22. J. G. Rosenstein. *Linear Orderings*, volume 98 of *Pure and Applied Mathematics*. Academic Press, New York, 1982.

23. E. Szpilrajn. Sur l'extension de l'ordre partiel. *Fund. Math.*, 16:386–389, 1930.

24. L. Valverde. On the structure of F-indistinguishability operators. *Fuzzy Sets and Systems*, 17(3):313–328, 1985.

25. L. A. Zadeh. Fuzzy sets. *Inf. Control*, 8:338–353, 1965.

26. L. A. Zadeh. Similarity relations and fuzzy orderings. *Inform. Sci.*, 3:177–200, 1971.

Relational Representation Theorems for Lattices with Negations: A Survey

Wojciech Dzik[1], Ewa Orłowska[2], and Clint van Alten[3]

[1] Institute of Mathematics, Silesian University
Bankowa 12, 40–007 Katowice, Poland
dzikw@silesia.top.pl
[2] National Institute of Telecommunications
Szachowa 1, 04–894 Warsaw, Poland
orlowska@itl.waw.pl
[3] School of Mathematics, University of the Witwatersrand
Private Bag 3, Wits 2050, Johannesburg, South Africa
cvalten@maths.wits.ac.za

Abstract. Relational representation theorems are presented in a unified framework for general (including non-distributive) lattices endowed with various negation operations.

1 Introduction

We present relational representation theorems in a unified framework for the lattice based algebras of logics with various negation operations, for both general (including non-distributive) lattices and for distributive lattices.

The negation operations include sufficiency or negative necessity as negation, Heyting negation, pseudo-complement, De Morgan negation and ortho-negation. Part of the results are carried out within the framework of Urquhart's representation theorem for lattices [17] and Allwein–Dunn developments on Kripke semantics for linear logic [1] which we jointly call Urquhart–Allwein–Dunn – framework, generalized to a duality between the algebras and abstract frames (relational systems). In order to have it in the same unified framework, we also include representations of distributive lattices with relative pseudo-complement, with relative pseudo-complement and minimal negation (of Johansson), with De Morgan negation, and Boolean algebras with sufficiency (negative necessity) operator. The distributive lattice cases contain known results, but we include them to present all results together in the unified framework.

Our framework, based on a generalization of the Urquhart–Allwein–Dunn representation, requires the following steps:

Step 1. A class of algebras is given. Its signature is that of lattices extended by a unary operation corresponding to negation.

Step 2. We define a class of relational structures (frames) that provide a Kripke-style semantics for the logic whose algebraic semantics is determined by the class of algebras in question.

H. de Swart et al. (Eds.): TARSKI II, LNAI 4342, pp. 245–266, 2006.
© Springer-Verlag Berlin Heidelberg 2006

Step 3. For any algebra W of the given class we define its canonical frame. The universe $X(W)$ of this frame consists of all pairs (x_1, x_2) such that x_1 is a filter and x_2 is an ideal of the lattice reduct of W and (x_1, x_2) is a maximal disjoint pair. Relations are defined on $X(W)$ which correspond in an appropriate way to the operations of the algebra.

Step 4. For any frame X we define its complex algebra. The universe of the complex algebra is a family $L(X)$ of special subsets of X referred to as ℓ-stable sets.

Step 5. We prove a representation theorem saying that every algebra W is embeddable into the complex algebra of its canonical frame, i.e., $L(X(W))$. The universe of the representation algebra consists of subrelations of $X(W)$.

Below we list several well known examples of classical representations giving, in particular, the algebras, frames, complex algebras and canonical frames.

The class of Boolean algebras has the class of sets as its class of frames which can be seen as relational systems with the empty family of relations. A canonical frame is the set of ultrafilters of a given algebra. A complex algebra is the powerset algebra of the set of ultrafilters. The Stone representation theorem says that a given Boolean algebra is embeddable into this powerset algebra.

The class of distributive lattices has the class of partial orders as its class of frames. A canonical frame is the set of prime filters of a given distributive lattice with set inclusion. A complex algebra of a frame is a family of $\leq$-increasing subsets with the set union and intersection. The representation theorem says that a given distributive lattice is embeddable into the complex algebra of the canonical frame.

The class of ortholattices has the class of orthogonality spaces (sets with orthogonality relations, i.e., irreflexive and symmetric relations $\perp$, first defined by Foulis and Randall) as its class of frames. A canonical frame is the set of proper filters of a given ortholattice with the set inclusion and ortho-negation defined by orthogonality relation $\perp$: for two proper filters x and y, $x \perp y$ iff there is an element a such that $-a \in x$ and $a \in y$. A complex algebra of a frame is a family of regular subsets of this frame defined as follows: first $a \perp Y$ iff for all $b \in Y$, $a \perp b$ and $Y^* = \{a : a \perp Y\}$; now Y is $\perp$-*regular* iff $Y = Y^{**}$. The representation theorem of Goldblatt [9] says that a given ortholattice is embeddable into the lattice of regular subsets of the orthogonality space.

The framework described above serves, on the one hand, as a tool for investigation of classes of lattices with negation operations and, on the other hand, as a means for developing Kripke-style semantics for the logics whose algebraic semantics is given. Then representation theorems play an essential role in proving completeness of the logics with respect to a Kripke-style semantics determined by a class of frames associated with a given class of algebras. In this paper we deal mainly with the algebraic aspects of lattices with negation. The framework presented above has been used in [13] and [7] in the context of lattice-based modal logics. It has been applied to lattice-based relation algebras in [6] and to double residuated lattices in [11] and [12]. In our relational representations we will provide definitions of abstract relational systems or frames such that particular properties of the relations in frames correspond to particular types of negations.

2   Negations

We follow J.M. Dunn's analysis of negations, also known as "Dunn's Kite of Negations". Dunn's study of negation in non-classical logics as a negative modal operator is an application of his gaggle theory, cf. [5], which is a generalization of the Jonsson-Tarski Theorem. In gaggle theory, negation $\neg$ is treated as a Galois connective on an underlying poset or bounded lattice. This treatment requires the Galois condition:

(Gal) $a \leq \neg b \iff b \leq \neg a$

Further analysis of negation on a bounded lattice leads to the following conditions for $\neg$ (we always assume that 0 is the least element and 1 the greatest):

(Suff1)	$\neg(a \vee b) = \neg a \wedge \neg b$	(Sufficiency 1)
(Suff2)	$\neg 0 = 1$	(Sufficiency 2)
(WCon)	$a \leq b \Rightarrow \neg b \leq \neg a$	(Weak Contrapositive, Preminimal)
(Weak$\neg\neg$)	$a \leq \neg\neg a$	(Weak Double Negation)
(Abs)	$a \wedge \neg a = 0$	(Absurdity, Intuitionistic)
(DeM)	$\neg\neg a \leq a$	(De Morgan, Strong Double Negation)

Lemma 2.1. *In any bounded lattice with an operation $\neg$ the following implications hold:*

(a)   (Suff1) $\Rightarrow$ (WCon)
(b)   (Gal) $\Rightarrow$ (Suff2)
(c)   (Gal) $\iff$ (Suff1) and (Weak$\neg\neg$)
(d)   (Gal) $\iff$ (WCon) and (Weak$\neg\neg$)

Proof. We show only the implication (Gal) $\Rightarrow$ (Suff1) of (c). By (Weak$\neg\neg$), $a \leq a \vee b \leq \neg\neg(a \vee b)$, hence by (Gal), $\neg(a \vee b) \leq \neg a$. Similarly, $\neg(a \vee b) \leq \neg b$, so we have $\neg(a \vee b) \leq \neg a \wedge \neg b$. By (Gal), $a \leq \neg(\neg a \wedge \neg b)$ and $b \leq \neg(\neg a \wedge \neg b)$, so $a \vee b \leq \neg(\neg a \wedge \neg b)$ hence $\neg a \wedge \neg b \leq \neg(a \vee b)$.

As noted in (b), one may derive $\neg 0 = 1$ from (Gal) or its equivalents. If one has, in addition, either (Abs) or (DeM) then one may also derive $\neg 1 = 0$. Also note that from (WCon) one may derive $\neg a \vee \neg b \leq \neg(a \wedge b)$. Lastly, by (c), note that the class of bounded lattices with negation satisfying the Galois condition (Gal) is a variety (i.e., an equational class) with equational axioms (Suff1) and (Weak$\neg\neg$).

We shall consider five types of negation on bounded (non-distributive) lattices. In each case, the negation satisfies (Suff1) and (Suff2); in the first and weakest case we consider just these two axioms. The next case is Heyting negation in which the negation satisfies (WCon), (Weak$\neg\neg$) (or, equivalently, just (Gal)) and (Abs); such algebras are also called 'weakly pseudo-complemented lattices'. Thereafter, we consider 'pseudo-complemented lattices' which satisfy, in addition, the following pseudo-complement quasi-identity:

(Pcq) $a \wedge b = 0 \quad \Rightarrow \quad a \leq \neg b.$

Its converse is derivable from the identity (Abs). In the case of De Morgan negation the identities (Gal) and (DeM) are assumed giving the class of 'De Morgan lattices'. Finally, ortho-negation is considered which satisfies (Gal) and both (Abs) and (DeM); these algebras are known as 'ortholattices'.

In the distributive lattice case, we consider 'relatively pseudo-complemented lattices', that is, where the 'residuum' (or relative pseudo-complement) of $\wedge$ exists, denoted $\rightarrow$. One may induce a negation by choosing any element ∂ in the lattice and defining $\neg x = x \rightarrow \partial$. The negation induced in this way is a minimal negation in the sense of Johansson and Rasiowa. This negation satisfies (Gal) (hence also (WCon) and (Weak¬¬)) but not necessarily (Abs) (unless the chosen element ∂ is the least element, in which case we have Heyting algebras). It does not necessarily satisfy (DeM) either, so we also consider distributive lattices in which (DeM) is added, namely, 'De Morgan algebras'. Adding both (Abs) and (DeM) to (Gal) results in the class of Boolean algebras. At the end we consider Boolean algebras with sufficiency (or negative necessity) operator.

Part I   Non-distributive Lattices

3   Preliminaries

We give here the necessary background on the relational representation of non-distributive lattices in the style of Urquhart [17] (see also [6] and [13]). The representations of non-distributive lattices with negations is built on top of this framework.

Let X be a non-empty set and let $\leq_1$ and $\leq_2$ be two quasi-orders on X. The structure $\langle X, \leq_1, \leq_2 \rangle$ is called a *doubly ordered set* if it satisfies:

$$(\forall x, y)((x \leq_1 y \text{ and } x \leq_2 y) \Rightarrow x = y). \tag{1}$$

For a doubly ordered set $\boldsymbol{X} = \langle X, \leq_1, \leq_2 \rangle$, $A \subseteq X$ is $\leq_1$–*increasing* (resp., $\leq_2$–*increasing*) if, for all $x, y \in X$, $x \in A$ and $x \leq_1 y$ (resp., $x \leq_2 y$) imply $y \in A$. We define two mappings $\ell, r : 2^X \rightarrow 2^X$ by

$$\ell(A) = \{x \in X : \forall y (x \leq_1 y \Rightarrow y \notin A)\} \tag{2}$$
$$r(A) = \{x \in X : \forall y (x \leq_2 y \Rightarrow y \notin A)\}. \tag{3}$$

Then $A \subseteq X$ is called ℓ–*stable* (resp., r–*stable*) if $\ell(r(A)) = A$ (resp., $r(\ell(A)) = A$). The set of all ℓ-stable subsets of X will be denoted by $L(\boldsymbol{X})$.

Lemma 3.1. [6],[13] *If $\langle X, \leq_1, \leq_2 \rangle$ is a doubly ordered set then, for all $A \subseteq X$,*

(a) $\ell(A)$ is $\leq_1$–increasing and $r(A)$ is $\leq_2$–increasing,
(b) if A is $\leq_1$–increasing, then $A \subseteq \ell(r(A))$,
(c) if A is $\leq_2$–increasing, then $A \subseteq r(\ell(A))$.

Lemma 3.2. [17] *Let $\langle X, \leqslant_1, \leqslant_2 \rangle$ be a doubly ordered set. Then the mappings ℓ and r form a Galois connection between the lattices of $\leqslant_1$-increasing and $\leqslant_2$-increasing subsets of X. In particular, for every $\leqslant_1$-increasing set A and $\leqslant_2$-increasing set B,*

$$A \subseteq \ell(B) \text{ iff } B \subseteq r(A).$$

Let $\boldsymbol{X} = \langle X, \leqslant_1, \leqslant_2 \rangle$ be a doubly ordered set. Define two binary operations $\wedge^C$ and $\vee^C$ on 2^X and two constants 0^C and 1^C as follows: for all $A, B \subseteq X$,

$$A \wedge^C B = A \cap B \tag{4}$$
$$A \vee^C B = \ell(r(A) \cap r(B)) \tag{5}$$
$$0^C = \emptyset \tag{6}$$
$$1^C = X. \tag{7}$$

Observe that the definition of $\vee^C$ in terms of $\wedge^C$ resembles a De Morgan law with two different negations. In [17], $\boldsymbol{L}(\boldsymbol{X}) = \langle L(\boldsymbol{X}), \wedge^C, \vee^C, 0^C, 1^C \rangle$ is shown to be a bounded lattice; it is called the **complex algebra** of $\boldsymbol{X}$.

Let $\boldsymbol{W} = \langle W, \wedge, \vee, 0, 1 \rangle$ be a bounded lattice. By a *filter-ideal pair* of $\boldsymbol{W}$ we mean a pair (x_1, x_2) such that x_1 is a filter of $\boldsymbol{W}$, x_2 is an ideal of $\boldsymbol{W}$ and $x_1 \cap x_2 = \emptyset$. The family of all filter-ideal pairs of $\boldsymbol{W}$ will be denoted by $FIP(\boldsymbol{W})$. Define the following three quasi-ordering relations: for any (x_1, x_2), $(y_1, y_2) \in FIP(\boldsymbol{W})$,

$$(x_1, x_2) \leqslant_1 (y_1, y_2) \text{ iff } x_1 \subseteq y_1$$
$$(x_1, x_2) \leqslant_2 (y_1, y_2) \text{ iff } x_2 \subseteq y_2$$
$$(x_1, x_2) \leqslant (y_1, y_2) \text{ iff } (x_1, x_2) \leqslant_1 (y_1, y_2) \text{ and } (x_1, x_2) \leqslant_2 (y_1, y_2).$$

We say that $(x_1, x_2) \in FIP(\boldsymbol{W})$ is *maximal* if it is maximal with respect to $\leqslant$. We denote by $X(\boldsymbol{W})$ the set of all maximal filter-ideal pairs of $\boldsymbol{W}$. Note that $X(\boldsymbol{W})$ is a binary relation on 2^W. In the sequel, if we write $x \in X(\boldsymbol{W})$, we shall assume that $x = (x_1, x_2)$ where x_1 denotes the filter and x_2 denotes the ideal. The same convention holds for y, z, etc. It was shown in [17] that for any $x \in FIP(\boldsymbol{W})$ there exists $y \in X(\boldsymbol{W})$ such that $x \leqslant y$; in this case, we say that x has been *extended* to y.

If $\boldsymbol{W} = \langle W, \wedge, \vee, 0, 1 \rangle$ is a bounded lattice then the **canonical frame** of $\boldsymbol{W}$ is defined as the relational structure $\boldsymbol{X}(\boldsymbol{W}) = \langle X(\boldsymbol{W}), \leqslant_1, \leqslant_2 \rangle$.

Consider the complex algebra $\boldsymbol{L}(\boldsymbol{X}(\boldsymbol{W}))$ of the canonical frame of a bounded lattice $\boldsymbol{W}$. Note that $\boldsymbol{L}(\boldsymbol{X}(\boldsymbol{W}))$ is an algebra of subrelations of $X(\boldsymbol{W})$. Define a mapping $h : W \to 2^{X(\boldsymbol{W})}$ by

$$h(a) = \{x \in X(\boldsymbol{W}) : a \in x_1\}.$$

Then h is a map from $\boldsymbol{W}$ to $L(\boldsymbol{X}(\boldsymbol{W}))$ and, moreover, we have the following result.

Proposition 3.1. [17] *For every bounded lattice $\boldsymbol{W}$, h is a lattice embedding of $\boldsymbol{W}$ into $\boldsymbol{L}(\boldsymbol{X}(\boldsymbol{W}))$.*

The following theorem is a weak version of Urquhart's result.

Theorem 3.1 (Representation theorem for lattices). *Every bounded lattice is embeddable into the complex algebra of its canonical frame.*

4   Lattices with Sufficiency (Negative Necessity) Operator

By a *lattice with a sufficiency operator* we mean an algebra $\boldsymbol{W} = \langle W, \wedge, \vee, \neg, 0, 1 \rangle$ which is a bounded lattice with a unary operation $\neg$, called a *sufficiency operator*, satisfying:

(Suff1) $\neg(a \vee b) = \neg a \wedge \neg b$
(Suff2) $\neg 0 = 1$.

Such operators are also called 'negative necessity'. (Note that such operators are antitone.) The name is due to its modal interpretation (cf. Orłowska, E., Vakarelov, D. [13]). The operator $[R]\neg$, which is the composition of the classical necessity operator $[R]$ with the classical negation, is a sufficiency operator. Recall that, given a Kripke frame $\langle X, R \rangle$, where R is a binary relation on X and $A \subseteq X$, the classical necessity is defined by

$$[R]A = \{x \in X : \forall y(xRy \Rightarrow y \in A)\}.$$

Let $\mathcal{LS}$ denote the variety of all lattices with a sufficiency operator. The following definitions and results are based on the treatment of sufficiency in [13].

Let $\mathcal{R}_{LS}$ denote the class of all *sufficiency frames*, i.e., relational structures of the type $\boldsymbol{X} = \langle X, \leqslant_1, \leqslant_2, R, S \rangle$, where $\langle X, \leqslant_1, \leqslant_2 \rangle$ is a doubly ordered set (i.e., $\leqslant_1$ and $\leqslant_2$ are quasi-orders satisfying (1)) and R and S are binary relations on X such that the following hold:

(Mono R) $(x' \leqslant_1 x$ and xRy and $y \leqslant_2 y') \Rightarrow x'Ry'$
(Mono S) $(x \leqslant_2 x'$ and xSy and $y' \leqslant_1 y) \Rightarrow x'Sy'$
(SC R_S) $xRy \Rightarrow (\exists x' \in X)(x \leqslant_1 x'$ and $x'Sy)$
(SC S_R) $xSy \Rightarrow (\exists y' \in X)(y \leqslant_1 y'$ and $xRy')$.

The conditions (Mono R) and (Mono S) are called sufficiency monotonicity conditions, and (SC R_S) and (SC S_R) are called sufficiency stability conditions.

Unary operators $[R]$ and $\langle S \rangle$ are defined on 2^X as follows. For all $A \subseteq X$,

$[R]A = \{x \in X : \forall y(xRy \Rightarrow y \in A)\}$,
$\langle S \rangle A = \{x \in X : \exists y(xSy$ and $y \in A)\}$.

For each $\boldsymbol{W} \in \mathcal{LS}$ we define the **canonical frame** of $\boldsymbol{W}$ as the relational structure $\boldsymbol{X}(\boldsymbol{W}) = \langle X(\boldsymbol{W}), \leqslant_1, \leqslant_2, R^c, S^c \rangle$, where $X(\boldsymbol{W})$ is the set of all maximal disjoint filter-ideal pairs of $\boldsymbol{W}$ and, for all $x = (x_1, x_2), y = (y_1, y_2) \in X(\boldsymbol{W})$,

$$x \leqslant_1 y \text{ iff } x_1 \subseteq y_1$$
$$x \leqslant_2 y \text{ iff } x_2 \subseteq y_2$$
$$xR^cy \text{ iff } \forall a(\neg a \in x_1 \Rightarrow a \in y_2)$$
$$xS^cy \text{ iff } \forall a(a \in y_1 \Rightarrow \neg a \in x_2).$$

Lemma 4.1. [13] *If* $W \in \mathcal{LS}$ *then* $X(W) \in \mathcal{R}_{LS}$.

Let $X = \langle X, \leqslant_1, \leqslant_2, R, S \rangle \in \mathcal{R}_{LS}$. Then $\langle X, \leqslant_1, \leqslant_2 \rangle$ is a doubly ordered set hence we may consider its complex algebra $\langle L(X), \wedge^C, \vee^C, 0^C, 1^C \rangle$, where $L(X)$ is the set of ℓ-stable sets (see definitions (2) and (3)) and the operations are defined as in (4–7). We extend this definition to define the ***complex algebra*** of X as $L(X) = \langle L(X), \wedge^C, \vee^C, \neg^C, 0^C, 1^C \rangle$ where, for all $A \subseteq X$,

$$\neg^C A = [R] r(A).$$

Lemma 4.2. [13] *If* $X \in \mathcal{R}_{LS}$ *then* $L(X) \in \mathcal{LS}$.

Let $W = \langle W, \wedge, \vee, \neg, 0, 1 \rangle \in \mathcal{LS}$. By the above lemmas, we have $L(X(W)) \in \mathcal{LS}$ as well. Recall that the function $h : W \to L(X(W))$ defined by

$$h(a) = \{ x \in X(W) : a \in x_1 \}$$

is an embedding of the lattice part of W into $L(X(W))$. Moreover, h also preserves negation, hence we have the following result.

Theorem 4.1. [13] *Each* $W \in \mathcal{LS}$ *is embeddable into* $L(X(W))$.

5   Lattices with Heyting Negation

A *weakly pseudo-complemented lattice* is an algebra $W = \langle W, \wedge, \vee, \neg, 0, 1 \rangle$ which is a bounded lattice with a unary operation $\neg$ satisfying:

(WCon) $a \leq b \Rightarrow \neg b \leq \neg a$
(Weak¬¬) $a \leq \neg\neg a$
(Abs) $a \wedge \neg a = 0$

We denote by $\mathcal{W}$ the variety of all weakly pseudo-complemented lattices. By Lemma 2.1, $\mathcal{W}$ also satisfies (Gal), (Suff1) and (Suff2), as well as $\neg 1 = 0$ and $\neg a \vee \neg b \leq \neg(a \wedge b)$.

We shall need the following lemma. We use $(X]$ to denote the downward closure of a subset X of a lattice and $[X)$ for the upward closure. Also, for any subset X of a a weakly pseudo-complemented lattice, we define

$$\neg X = \{ \neg b : b \in X \}.$$

Lemma 5.1. *Let F be a proper filter of $W \in \mathcal{W}$. Then the following hold.*

(a) $(\neg F]$ is an ideal.
(b) $F \cap (\neg F] = \emptyset$.
(c) For all $a \in W$, $\neg a \in F$ iff $a \in (\neg F]$.

Proof. (a) Note that $(\neg F]$ is downward closed. Suppose that $a, b \in (\neg F]$. Then $a \leq \neg c$ and $b \leq \neg d$ for some $c, d \in F$. Since F is a filter, $c \wedge d \in F$ so $\neg(c \wedge d) \in \neg F$. Since $a \vee b \leq \neg c \vee \neg d \leq \neg(c \wedge d)$, we have $a \vee b \in (\neg F]$. Thus, $(\neg F]$ is an ideal.

(b) Suppose there is some $a \in F \cap (\neg F]$. Then $a \leq \neg b$ for some $b \in F$, so $b \leq \neg a$. Thus, $\neg a \in F$ hence $0 = a \wedge \neg a \in F$, which is a contradiction.

(c) If $\neg a \in F$ then $\neg\neg a \in (\neg F]$ hence $a \in (\neg F]$ since $a \leq \neg\neg a$. If $a \in (\neg F]$ then $a \leq \neg b$ for some $b \in F$, so $b \leq \neg a$ hence $\neg a \in F$.

We will denote by $\mathcal{R}_W$ the class of all relational structures of type $\boldsymbol{X} = \langle X, \leqslant_1,$ $\leqslant_2, C \rangle$, where $\langle X, \leqslant_1, \leqslant_2 \rangle$ is a doubly ordered set and C is a binary relation on X such that the following hold:

(FC1) $(\forall x, y, z)((xCy \text{ and } z \leqslant_1 x) \Rightarrow zCy)$
(FC2) $(\forall x, y, z)((xCy \text{ and } y \leqslant_2 z) \Rightarrow xCz)$
(FC3) $(\forall x)(\exists y)(xCy \text{ and } x \leqslant_1 y)$
(FC4) $(\forall x, y)(xCy \Rightarrow \exists z(yCz \text{ and } x \leqslant_1 z))$
(FC5) $(\forall s, t, y)[(yCs \text{ and } s \leqslant_2 t) \Rightarrow \exists z(y \leqslant_1 z \text{ and } \forall u(z \leqslant_2 u \Rightarrow tCu))].$

For each $\boldsymbol{W} \in \mathcal{W}$ we define the **canonical frame** of $\boldsymbol{W}$ as the relational structure $\boldsymbol{X}(\boldsymbol{W}) = \langle X(\boldsymbol{W}), \leqslant_1, \leqslant_2, C \rangle$, where $X(\boldsymbol{W})$ is the set of all maximal disjoint filter-ideal pairs of $\boldsymbol{W}$ and, for all $x = (x_1, x_2)$, $y = (y_1, y_2) \in X(\boldsymbol{W})$,

$$x \leqslant_1 y \text{ iff } x_1 \subseteq y_1$$
$$x \leqslant_2 y \text{ iff } x_2 \subseteq y_2$$
$$xCy \text{ iff } \forall a(\neg a \in x_1 \Rightarrow a \in y_2).$$

Lemma 5.2. *If $\boldsymbol{W} \in \mathcal{W}$ then $\boldsymbol{X}(\boldsymbol{W}) \in \mathcal{R}_W$.*

Proof. We know that $\langle X(\boldsymbol{W}), \leqslant_1, \leqslant_2 \rangle$ is a doubly ordered set. Properties (FC1) and (FC2) are straightforward to prove. For (FC3), suppose $x \in X(\boldsymbol{W})$. By Lemma 5.1, $\langle x_1, (\neg x_1] \rangle$ is a disjoint filter-ideal pair, so we can extend it to a maximal one, say y. If $\neg a \in x_1$ then $a \in (\neg x_1]$ (by Lemma 5.1(c)) hence $a \in y_2$. Thus, xCy. Also, $x_1 \subseteq y_1$, i.e., $x \leqslant_1 y$, so we have found the required y.

For (FC4), suppose $x, y \in X(\boldsymbol{W})$ and xCy. By Lemma 5.1(a), $(\neg y_1]$ is an ideal. If $a \in x_1 \cap (\neg y_1]$ then $a \in x_1$ implies $\neg\neg a \in x_1$, which implies $\neg a \in y_2$. But $a \in (\neg y_1]$ implies $\neg a \in y_1$ (by Lemma 5.1(c)), which contradicts the fact that $y_1 \cap y_2 = \emptyset$. Thus, $x_1 \cap (\neg y_1] = \emptyset$. Thus, we can extend $\langle x_1, (\neg y_1] \rangle$ to a maximal disjoint filter-ideal pair, say z. If $\neg a \in y_1$ then $a \in (\neg y_1]$ hence $a \in z_2$, so yCz. Also, $x \leqslant_1 z$, so we have proved (FC4).

For (FC5), suppose that $s, t, y \in X(\boldsymbol{W})$ such that yCs and $s \leqslant_2 t$. First, we show that $y_1 \cap (\neg t_1] = \emptyset$. Suppose $a \in y_1 \cap (\neg t_1]$. Then, $\neg\neg a \in y_1$ hence $\neg a \in s_2$. Since $s \leqslant_2 t$ we have $\neg a \in t_2$. Also, $a \leq \neg b$ for some $b \in t_1$, so $\neg a \geq \neg\neg b \geq b$ hence $\neg a \in t_1$. This contradicts the fact that t_1 and t_2 are disjoint.

We therefore have that $\langle y_1, (\neg t_1] \rangle$ is a disjoint filter-ideal pair, so we may extend it to a maximal one, say z. Then, $y_1 \subseteq z_1$, i.e., $y \leqslant_1 z$. Suppose $z \leqslant_2 w$ and $\neg a \in t_1$. Then $\neg\neg a \in \neg t_1$ so $a \in (\neg t_1] \subseteq z_2 \subseteq w_2$ hence $a \in w_2$. Thus, we have proved (FC5).

Let $\boldsymbol{X} = \langle X, \leqslant_1, \leqslant_2, C \rangle \in \mathcal{R}_W$. Since $\langle X, \leqslant_1, \leqslant_2 \rangle$ is a doubly ordered set we may consider its complex algebra $\langle L(\boldsymbol{X}), \wedge^C, \vee^C, 0^C, 1^C \rangle$, where $L(\boldsymbol{X})$ is the set of ℓ-stable sets with operations defined as in (4–7). Extending this definition we define the **complex algebra** of $\boldsymbol{X}$ as $\boldsymbol{L}(\boldsymbol{X}) = \langle L(\boldsymbol{X}), \wedge^C, \vee^C, \neg^C, 0^C, 1^C \rangle$, where, for $A \in L(X)$,

$$\neg^C A = \{x \in X : \forall y(xCy \Rightarrow y \notin A)\}.$$

Lemma 5.3. *If A is ℓ-stable then so is $\neg^C A$.*

Proof. We have $\neg^C A = \{x : \forall y(xCy \Rightarrow y \notin A)\}$ and

$$\ell r(\neg^C A) = \{x : \forall s(x \leqslant_1 s \Rightarrow \exists t(s \leqslant_2 t \text{ and } \forall u(tCu \Rightarrow u \notin A)))\}.$$

Let $x \in \neg^C A$ and suppose that $x \leqslant_1 s$ for some s. We claim that $t = s$ satisfies the required properties. Clearly, $s \leqslant_2 s$. If sCu, then xCu since $x \leqslant_1 s$, by (FC1) hence $u \notin A$. Thus, $x \in \ell r(\neg^C A)$ so $\neg^C A \subseteq \ell r(\neg^C A)$.

For the reverse inclusion, note that, since A is ℓ-stable, we have

$$\neg^C A = \neg^C \ell r(A) = \{x : \forall y(xCy \Rightarrow \exists z(y \leqslant_1 z \text{ and } \forall u(z \leqslant_2 u \Rightarrow u \notin A)))\}.$$

Let $x \in \ell r(\neg^C A)$ and suppose that xCy for some y. By (FC4), there exists s such that

$$x \leqslant_1 s \text{ and } yCs.$$

Then, since $x \in \ell r(\neg^C A)$ and $x \leqslant_1 s$, there exists t such that

$$s \leqslant_2 t \text{ and } \forall u(tCu \Rightarrow u \notin A).$$

Since yCs and $s \leqslant_2 t$, by (FC5) there exists z such that

$$y \leqslant_1 z \text{ and } \forall u(z \leqslant_2 u \Rightarrow tCu).$$

Thus, $\forall u(z \leqslant_2 u \Rightarrow u \notin A)$, so we have found the required z, so $x \in \neg^C \ell r(A) = \neg^C A$.

Lemma 5.4. *If $X \in \mathcal{R}_W$ then $L(X) \in \mathcal{W}$.*

Proof. To see that (WCon) holds, suppose A, B are ℓ-stable sets and $A \subseteq B$. Let $x \in \neg^C B$. Then, for all y, xCy implies $y \notin B$ hence also $y \notin A$, so $x \in \neg^C A$.

To see that (Weak$\neg\neg$) holds, note that

$$\neg^C \neg^C A = \{x : \forall y(xCy \Rightarrow \exists z(yCz \text{ and } z \in A))\}.$$

Let $x \in A$ and suppose that xCy for some y. By (FC4), there exists z such that yCz and $x \leqslant_1 z$. Since A is $\leqslant_1$-increasing and $x \in A$, we have $z \in A$. Thus, the required z exists, showing that $x \in \neg^C \neg^C A$.

To see that (Abs) holds, let A be an ℓ-stable set and suppose there exists $x \in A \cap \neg^C A$. By (FC3), there exists a y such that xCy and $x \leqslant_1 y$. Since $x \in \neg^C A$ and xCy we have $y \notin A$. But $x \in A$ and A is ℓ-stable, hence $\leqslant_1$-increasing, so $x \leqslant_1 y$ implies $y \in A$, a contradiction.

The above lemmas show that if $W \in \mathcal{W}$ then so is $L(X(W))$. Recall that the function $h : W \to L(X(W))$ defined by

$$h(a) = \{x \in X(W) : a \in x_1\}$$

is an embedding of the lattice part of W into $L(X(W))$. We show that h also preserves negation.

Theorem 5.1. [8] *Each $W \in \mathcal{W}$ is embeddable into $L(X(W))$.*

Proof. We need only show that $h(\neg a) = \neg^C h(a)$ for all $a \in W$, where

$$h(\neg a) = \{x : \neg a \in x_1\}$$

and

$$\neg^C h(a) = \{x : \forall y(xCy \Rightarrow a \notin y_1)\}.$$

First, let $x \in h(\neg a)$ and suppose that xCy for some y. Then $\neg a \in x_1$ so $a \in y_2$ hence $a \notin y_1$, as required.

Next, let $x \in \neg^C h(a)$ and suppose that $\neg a \notin x_1$. Then $a \notin (\neg x_1]$ (by Lemma 5.1(c)) so $\langle [a), (\neg x_1] \rangle$ forms a disjoint filter-ideal pair which we can extend to a maximal one, say y. If $\neg c \in x_1$ then $c \in (\neg x_1]$ so xCy hence $a \notin y_1$, a contradiction since $[a) \subseteq y_1$.

6   Pseudo-complemented Lattices

A *pseudo-complemented lattice* is an algebra $W = \langle W, \wedge, \vee, \neg, 0, 1 \rangle$ which is a bounded lattice with a unary operation $\neg$ satisfying:

$$a \wedge b = 0 \quad \Leftrightarrow \quad a \leq \neg b.$$

The class of all pseudo-complemented lattices is denoted $\mathcal{P}$. Note that (Gal) is derivable by

$$a \leq \neg b \Leftrightarrow a \wedge b = 0 \Leftrightarrow b \wedge a = 0 \Leftrightarrow b \leq \neg a.$$

Thus, (Suff1), (Suff2), (WCon) and (Weak$\neg\neg$) are derivable and, from $a \leq \neg\neg a$, we get $a \wedge \neg a = 0$, so (Abs) is derivable hence also $\neg 1 = 0$. The class $\mathcal{W}$ of weakly pseudo-complemented lattices is easily seen to satisfy the quasi-identity

$$a \leq \neg b \quad \Rightarrow \quad a \wedge b = 0,$$

hence $\mathcal{P}$ is a subclass of $\mathcal{W}$ defined by the quasi-identity

(Pcq) $a \wedge b = 0 \quad \Rightarrow \quad a \leq \neg b.$

As an example that shows that $\mathcal{P}$ is a proper subclass of $\mathcal{W}$ consider the lattice with 6 elements $1, 0, a, b, c, d$, where 1 is the top, 0 is the bottom and a, b, c, d are incomparable. Let $\neg a = b$, $\neg b = a$, $\neg c = d$ and $\neg d = c$. This example is in $\mathcal{W}$ but not in $\mathcal{P}$ since $a \wedge c = 0$ but $a \not\leq \neg c$.

We will denote by $\mathcal{R}_P$ the class of all relational structures of type $X = \langle X, \leq_1, \leq_2, C \rangle$, where $\langle X, \leq_1, \leq_2 \rangle$ is a doubly ordered set and C is a binary relation on X such that (FC1–FC5) hold as well as

(FC6) $(\forall x, y)(xCy \Rightarrow \exists z(x \leq_1 z \text{ and } y \leq_1 z)).$

That is, $\mathcal{R}_P$ is the subclass of $\mathcal{R}_W$ defined by (FC6).

If $\boldsymbol{W} \in \mathcal{P}$ then $\boldsymbol{W} \in \mathcal{W}$ as well hence its **canonical frame** is the relational structure $\boldsymbol{X}(\boldsymbol{W}) = \langle X(\boldsymbol{W}), \leqslant_1, \leqslant_2, C \rangle$, where $X(\boldsymbol{W})$ is the set of all maximal disjoint filter-ideal pairs of $\boldsymbol{W}$ and, for all $x, y \in X(\boldsymbol{W})$,

$$x \leqslant_1 y \text{ iff } x_1 \subseteq y_1$$
$$x \leqslant_2 y \text{ iff } x_2 \subseteq y_2$$
$$xCy \text{ iff } \forall a(\neg a \in x_1 \Rightarrow a \in y_2).$$

Lemma 6.1. *If $\boldsymbol{W} \in \mathcal{P}$ then $\boldsymbol{X}(\boldsymbol{W}) \in \mathcal{R}_P$.*

Proof. We need only show that (FC6) holds. So, let $x, y \in X(\boldsymbol{W})$ such that xCy. Consider the filter generated by $x_1 \cup y_1$, denoted $Fi(x_1 \cup y_1)$. We claim that $0 \notin Fi(x_1 \cup y_1)$. If we suppose otherwise, then there exist $a_1, \ldots, a_n \in x_1$ and $b_1, \ldots, b_m \in y_1$ such that

$$(\textstyle\bigwedge_{i=1}^n a_i) \wedge (\bigwedge_{j=1}^m b_j) = 0.$$

If we set $a = \bigwedge_{i=1}^n a_i$ and $b = \bigwedge_{j=1}^m b_j$, then $a \in x_1$ and $b \in y_1$ such that $a \wedge b = 0$. But this implies that $a \leq \neg b$, by (Pcq), hence $\neg b \in x_1$. Finally, since xCy and $\neg b \in x_1$, we have $b \in y_2$. Thus, $b \in y_1 \cap y_2$, a contradiction.

This shows that $0 \notin Fi(x_1 \cup y_1)$ so $\langle Fi(x_1 \cup y_1), \{0\} \rangle$ is a disjoint filter-ideal pair. This can be extended to a maximal disjoint filter-ideal pair, say z. Then $x \leqslant_1 z$ and $y \leqslant_1 z$, as required.

Let $\boldsymbol{X} = \langle X, \leqslant_1, \leqslant_2, C \rangle \in \mathcal{R}_P$ (so $\boldsymbol{X}$ satisfies (FC1–FC6)). Then $\boldsymbol{X}$ is also in $\mathcal{R}_W$ hence we may consider its complex algebra $\boldsymbol{L} = \langle L(\boldsymbol{X}), \wedge^C, \vee^C, \neg^C, 0^C, 1^C \rangle$, where $L(\boldsymbol{X})$ is the set of ℓ-stable sets, the lattice operations are defined as in (4–7) and, for $A \in L(\boldsymbol{X})$,

$$\neg^C A = \{x \in X : \forall y(xCy \Rightarrow y \notin A)\}.$$

Lemma 6.2. *If $\boldsymbol{X} \in \mathcal{R}_P$ then $\boldsymbol{L}(\boldsymbol{X}) \in \mathcal{P}$.*

Proof. We need only show that $\boldsymbol{L}(\boldsymbol{X})$ satisfies the quasi-identity (Pcq), i.e., for $A, B \in L(\boldsymbol{X})$,

$$A \cap B = \emptyset \quad \Rightarrow \quad A \subseteq \neg^C B = \{x \in X : \forall y(xCy \Rightarrow y \notin B)\}.$$

Suppose that $A \cap B = \emptyset$ and let $x \in A$. Let $y \in X$ such that xCy. By (FC6), there exists $z \in X$ such that $x \leqslant_1 z$ and $y \leqslant_1 z$. Since $x \in A$ and A is $\leqslant_1$-increasing, we have $z \in A$ as well. If $y \in B$ then, since B is $\leqslant_1$-increasing, it would follow that $z \in B$ and hence that $z \in A \cap B$, contradicting our assumption that $A \cap B = \emptyset$. Thus, $y \notin B$ hence $x \in \neg^C B$, as required.

Thus, we have shown that if $\boldsymbol{W} \in \mathcal{P}$ then so is $\boldsymbol{L}(\boldsymbol{X}(\boldsymbol{W}))$. Moreover, from the previous section we know that h is an embedding of $\boldsymbol{W}$ into $\boldsymbol{L}(\boldsymbol{X}(\boldsymbol{W}))$, hence we have the following result.

Theorem 6.1. [8] *Each $\boldsymbol{W} \in \mathcal{P}$ is embeddable into $\boldsymbol{L}(\boldsymbol{X}(\boldsymbol{W}))$.*

7   Lattices with De Morgan Negation

By a *De Morgan lattice* we mean an algebra $\boldsymbol{W} = \langle W, \wedge, \vee, \neg, 0, 1 \rangle$ which is a bounded lattice with a unary operation $\neg$ satisfying:

(Gal) $a \leq \neg b \Rightarrow b \leq \neg a$

(DeM) $\neg \neg a \leq a$

Let $\mathcal{M}$ denote the variety of all De Morgan lattices. Recall that from (Gal) and (DeM) one may derive (Suff1), (Suff2), (WCon), (Weak¬¬) and $\neg 1 = 0$. The following are also derivable in $\mathcal{M}$:

$$\neg \neg a = a$$
$$\neg (a \wedge b) = \neg a \vee \neg b$$
$$\neg a = \neg b \Rightarrow a = b.$$

We will denote by $\mathcal{R}_M$ the class of all relational structures of type $\boldsymbol{X} = \langle X, \leqslant_1, \leqslant_2, N \rangle$, where $\langle X, \leqslant_1, \leqslant_2 \rangle$ is a doubly ordered set, $N : X \to X$ is a function and, for all $x, y \in X$,

(M1) $N(N(x)) = x$,

(M2) $x \leqslant_1 y \Rightarrow N(x) \leqslant_2 N(y)$,

(M3) $x \leqslant_2 y \Rightarrow N(x) \leqslant_1 N(y)$.

The representation in this section essentially comes from [1], where the function N is called a 'generalized Routley-Meyer star operator'. We give full details here and in the next section show how the method may be extended to ortholattices.

For each $\boldsymbol{W} \in \mathcal{M}$, define the ***canonical frame*** of $\boldsymbol{W}$ as the relational structure $\boldsymbol{X}(\boldsymbol{W}) = \langle X(\boldsymbol{W}), \leqslant_1, \leqslant_2, N \rangle$, where $X(\boldsymbol{W})$ is the set of all maximal disjoint filter-ideal pairs of $\boldsymbol{W}$ and, for $x, y \in X(\boldsymbol{W})$,

$$x \leqslant_1 y \quad \text{iff} \quad x_1 \subseteq y_1,$$
$$x \leqslant_2 y \quad \text{iff} \quad x_2 \subseteq y_2,$$
$$N(x) = (\neg x_2, \neg x_1), \text{ where } \neg A = \{\neg a : a \in A\} \text{ for any } A \subseteq W.$$

Lemma 7.1. *If $\boldsymbol{W} \in \mathcal{M}$ then $\boldsymbol{X}(\boldsymbol{W}) \in \mathcal{R}_M$.*

Proof. We have already observed that $\langle X(\boldsymbol{W}), \leqslant_1, \leqslant_2 \rangle$ is a doubly ordered set. Condition (M1) follows from (DeM) and conditions (M2) and (M3) are immediate. Thus, we need only show that N is a function from $\boldsymbol{X}(\boldsymbol{W})$ to $\boldsymbol{X}(\boldsymbol{W})$. That is, if $x \in X(\boldsymbol{W})$, we must show that $N(x)$ is a maximal disjoint filter-ideal pair. (This is also done by Allwein and Dunn.) Let $a_1, a_2 \in x_2$ hence $\neg a_1, \neg a_2 \in \neg x_2$. Then $\neg a_1 \wedge \neg a_2 = \neg (a_1 \vee a_2)$ and $a_1 \vee a_2 \in x_2$, hence $\neg x_2$ is closed under $\wedge$. If $\neg a_1 \leq b$ then $\neg b \leq \neg \neg a_1 = a_1$, so $\neg b \in x_2$. Then $b = \neg \neg b \in \neg x_2$, so $\neg x_2$ is upward closed. Thus, $\neg x_2$ is a filter. Similarly, $\neg x_1$ is an ideal. Also, $\neg x_1$ and $\neg x_2$ can be shown disjoint using the implication: $\neg b = \neg c \Rightarrow b = c$ and the fact that x_1 and x_2 are disjoint. To show maximality, suppose $y \in X(\boldsymbol{W})$ and $\neg x_1 \subseteq y_1$ and $\neg x_2 \subseteq y_2$. Then $\neg \neg x_1 \subseteq \neg y_1$, i.e., $x_1 \subseteq \neg y_1$ and also $x_2 \subseteq \neg y_2$. Since $(\neg y_2, \neg y_1)$ is a disjoint filter-ideal pair, the maximality of x implies $x_1 = \neg y_2$ and $x_2 = \neg y_1$. Thus, $\neg x_1 = y_2$ and $\neg x_2 = y_1$ so $N(x)$ is maximal.

If $\boldsymbol{X} = \langle X, \leqslant_1, \leqslant_2, N \rangle \in \mathcal{R}_M$, then $\langle X, \leqslant_1, \leqslant_2 \rangle$ is a doubly ordered set, so we may consider its complex algebra $\langle L(\boldsymbol{X}), \wedge^C, \vee^C, 0^C, 1^C \rangle$, where $L(\boldsymbol{X})$ is the set of ℓ-stable sets and the operations are as in (4–7). We extend this definition to define the **complex algebra** of $\boldsymbol{X}$ as $L(\boldsymbol{X}) = \langle L(\boldsymbol{X}), \wedge^C, \vee^C, \neg^C, 0^C, 1^C \rangle$ where, for $A \in L(\boldsymbol{X})$,

$$\neg^C A = \{x \in X : N(x) \in r(A)\}.$$

Lemma 7.2. *If $\boldsymbol{X} \in \mathcal{R}_M$ then $L(\boldsymbol{X}) \in \mathcal{M}$.*

Proof. We need to show that $\neg^C A$ is ℓ-stable, i.e., $\ell r(\neg^C A) = \neg^C A$, and that $L(\boldsymbol{X})$ satisfies (Gal) and (DeM). Since ℓ and r form a Galois connection, by Lemma 3.2, we have $\neg^C A \subseteq \ell r(\neg^C A)$ iff $r(\neg^C A) \subseteq r(\neg^C A)$. For the converse, suppose that for every y, if $x \leqslant_1 y$ then $y \notin r(\neg^C A)$ and assume, to the contrary, that $x \notin \neg^C A$. Then $N(x) \notin r(A)$ and there is z such that $N(x) \leqslant_2 z$ and $z \in A$. It follows by (M3) and (M1) that $x \leqslant_1 N(z)$ and hence, by the above assumption, $N(z) \notin r(\neg^C A)$. Thus, there is t such that $N(z) \leqslant_2 t$ and $t \in \neg^C A$. By application of N and (M3) and (M1), we have that $z \leqslant_1 N(t)$ and $N(t) \in r(A)$, in particular $N(t) \notin A$. But $z \in A$ and A is $\leqslant_1$–increasing, as $A = \ell r(A)$, hence $N(t) \in A$, a contradiction.

To prove (Gal), suppose that $A \subseteq \neg^C B$. Then, for every x, if $x \in A$ then $N(x) \in r(B)$. Suppose that $x \in B$ and, to the contrary, that $x \notin \neg^C A$, i.e., $N(x) \notin r(A)$, in which case $N(x) \leqslant_2 y$ and $y \in A$, for some y. By (M3) and (M1), $x \leqslant_1 N(y)$ hence $N(y) \in B$ since $B = \ell r(B)$ is $\leqslant_1$–increasing. But also $y \in \neg^C B$, by the assumption, and $N(y) \in r(B)$, a contradiction since $B \cap r(B) = \emptyset$.

To prove (DeM), let $x \in \neg^C \neg^C A$, hence $N(x) \in r(\neg^C A)$. We show that $x \in \ell(r(A))$ which equals A since A is ℓ-closed. Let $x \leqslant_1 w$. Then $N(x) \leqslant_2 N(w)$, by (M2), hence $N(w) \in r(\neg^C A)$ since $r(\neg^C A)$ is $\leqslant_2$–increasing. Thus, $N(w) \notin \neg^C A$, i.e., $w = N(N(w)) \notin r(A)$. Thus, $x \in \ell(r(A)) = A$.

The above lemmas imply that if $\boldsymbol{W} \in \mathcal{M}$, then $L(\boldsymbol{X}(\boldsymbol{W})) \in \mathcal{M}$ as well. Recall that the function $h : \boldsymbol{W} \to L(\boldsymbol{X}(\boldsymbol{W}))$ defined by

$$h(a) = \{x \in X(\boldsymbol{W}) : a \in x_1\}$$

is an embedding of the lattice part of $\boldsymbol{W}$ into $L(\boldsymbol{X}(\boldsymbol{W}))$. As in the case of Heyting negation, we shall show that h also preserves De Morgan negation.

Theorem 7.1. [8] *Each $\boldsymbol{W} \in \mathcal{M}$ is embeddable into $L(\boldsymbol{X}(\boldsymbol{W}))$.*

Proof. We need only show that $h(\neg a) = \neg^C h(a)$ for all $a \in W$, where

$$h(\neg a) = \{x \in X(\boldsymbol{W}) : \neg a \in x_1\}$$

and

$$\neg^C h(a) = \{x \in X(\boldsymbol{W}) : N(x) \in r(h(a))\}$$
$$= \{x \in X(\boldsymbol{W}) : (\forall y \in X(\boldsymbol{W}))(\neg x_1 \subseteq y_2 \Rightarrow a \notin y_1)\}.$$

First, let $x \in h(\neg a)$. Then $\neg a \in x_1$, hence $a = \neg\neg a \in \neg x_1$. Suppose that $\neg x_1 \subseteq y_2$. Then $a \notin y_1$, since y_1 and y_2 are disjoint.

Next, let $x \in \neg^C h(a)$. Suppose, to the contrary, that $\neg a \notin x_1$. Then $a \notin (\neg x_1]$ and so $\langle [a), (\neg x_1] \rangle$ is a disjoint filter-ideal pair, which can be extended to a maximal one, say y. Thus, $(\neg x_1] \subseteq y_1$, so $a \notin y_1$, but $[a) \subseteq y_1$, a contradiction.

8   Lattices with Ortho-negation (Ortholattices)

An *ortholattice* is an algebra $\boldsymbol{W} = \langle W, \wedge, \vee, \neg, 0, 1 \rangle$ which is a bounded lattice with a unary operation $\neg$ which satisfies (Gal), (DeM) and (Abs). That is, the negation in an ortholattice is both De Morgan and Intuitionistic. Let $\mathcal{O}$ denote the variety of all ortholattices. Since $\mathcal{O}$ is a subclass of both $\mathcal{W}$ and $\mathcal{M}$, it satisfies all the identities satisfied by either class. We extend the relational representation for De Morgan lattices to ortholattices.

We will denote by $\mathcal{R}_O$ the class of all relational structures of type $\boldsymbol{X} = \langle X, \leqslant_1, \leqslant_2, N \rangle$, where $\langle X, \leqslant_1, \leqslant_2 \rangle$ is a doubly ordered set and $N : X \to X$ is a function such that, for all $x, y \in X$,

(M1) $N(N(x)) = x$
(M2) $x \leqslant_1 y \Rightarrow N(x) \leqslant_2 N(y)$
(M3) $x \leqslant_2 y \Rightarrow N(x) \leqslant_1 N(y)$
(O) $(\forall x)(\exists y)(x \leqslant_1 y$ and $N(x) \leqslant_2 y)$

That is, $\mathcal{R}_O$ is the subclass of $\mathcal{R}_M$ defined by (O). If $\boldsymbol{W} \in \mathcal{O}$, then $\boldsymbol{W} \in \mathcal{M}$ hence its canonical frame is the relational structure $\boldsymbol{X}(\boldsymbol{W}) = \langle X(\boldsymbol{W}), \leqslant_1, \leqslant_2, N \rangle$, where $X(\boldsymbol{W})$ is the set of all maximal disjoint filter-ideal pairs of $\boldsymbol{W}$ and, for x, $y \in X(\boldsymbol{W})$,

$x \leqslant_1 y$ iff $x_1 \subseteq y_1$
$x \leqslant_2 y$ iff $x_2 \subseteq y_2$
$N(x) = (\neg x_2, \neg x_1)$, where $\neg A = \{\neg a : a \in A\}$ for $A \subseteq W$.

Lemma 8.1. *If* $\boldsymbol{W} \in \mathcal{O}$ *then* $\boldsymbol{X}(\boldsymbol{W}) \in \mathcal{R}_{(O)}$.

Proof. We need only show that $\boldsymbol{X}(\boldsymbol{W})$ satisfies (O). Let $x \in X(\boldsymbol{W})$. Observe that x_1 and $\neg x_1$ are disjoint, for if $a \in x_1 \cap (\neg x_1)$ then $a \in x_1$ and $a \in \neg x_1$, so $\neg a \in \neg\neg x_1 = x_1$, hence $a \wedge \neg a \in x_1$. But, by (Abs), $a \wedge \neg a = 0$, so $x_1 = W$, a contradiction. Thus, we may extend $(x_1, \neg x_1)$ to a maximal disjoint filter-ideal pair y. Then $x_1 \subseteq y_1$ and $\neg x_1 \subseteq y_2$, so we have found a y that satisfies the required conditions of (O).

If $\boldsymbol{X} = \langle X, \leqslant_1, \leqslant_2, N \rangle \in \mathcal{R}_O$, then $\boldsymbol{X} \in \mathcal{R}_M$ so it has a canonical algebra $\boldsymbol{L}(\boldsymbol{X}) = \langle L(X), \wedge^C, \vee^C, \neg^C, 0^C, 1^C \rangle$ defined as in the De Morgan negation case.

Lemma 8.2. *If* $\boldsymbol{X} \in \mathcal{R}_O$ *then* $\boldsymbol{L}(\boldsymbol{X}) \in \mathcal{O}$.

Proof. We need only show that $\boldsymbol{L}(\boldsymbol{X})$ satisfies $A \wedge^C (\neg^C A) = 0^C$. Suppose, to the contrary, that there exists $A \in L(\boldsymbol{X})$ such that $A \cap (\neg^C A) \neq \emptyset$, and let

$x \in A \cap (\neg^C A)$. By (O), there exists y such that $x \leqslant_1 y$ and $N(x) \leqslant_2 y$. Since A is $\leqslant_1$–increasing, $y \in A$. Since $x \in \neg^C A$, $N(x) \in r(A)$. But then $N(x) \leqslant_2 y$ implies $y \notin A$, a contradiction.

Thus, the above lemmas imply that if $\boldsymbol{W} \in \mathcal{O}$, then $\boldsymbol{L}(\boldsymbol{X}(\boldsymbol{W})) \in \mathcal{O}$ as well. Since the map h is an embedding of De Morgan lattices, we have the following result.

Theorem 8.1. [8] *Each $\boldsymbol{W} \in \mathcal{O}$ is embeddable into $\boldsymbol{L}(\boldsymbol{X}(\boldsymbol{W}))$.*

Part II   Distributive Lattices

9   Relatively Pseudo-complemented Lattices

A *relatively pseudo-complemented lattice* is an algebra $\boldsymbol{W} = \langle W, \wedge, \vee, \rightarrow \rangle$ where $\langle W, \wedge, \vee \rangle$ is a lattice and $\rightarrow$ is a binary operation on W satisfying:

$$a \wedge c \leq b \quad \Leftrightarrow \quad c \leq a \rightarrow b.$$

The operation $\rightarrow$ is the 'residuum' of $\wedge$. For properties of relatively pseudo-complemented lattices, see [15] or [2]). It is known that every relatively pseudo-complemented lattice is distributive and has a constant 1 definable by $1 = a \rightarrow a$, which is the greatest element of the lattice. We include 1 in the language so that $\boldsymbol{W} = \langle W, \wedge, \vee, \rightarrow, 1 \rangle$. It is known that all relatively pseudo-complemented lattices form a variety and we denote this variety by $\mathcal{RP}$. $\mathcal{RP}$ satisfies the following:

$a \rightarrow b = 1 \Leftrightarrow a \leq b$
$1 \rightarrow b = b, \quad a \rightarrow 1 = 1$
$a \rightarrow b = 1$ and $a = 1 \Rightarrow b = 1$
$a \rightarrow (b \rightarrow c) = b \rightarrow (a \rightarrow c)$
$a \wedge (a \rightarrow b) = a \wedge b$
$b \leq a \rightarrow b$
$a \leq b \Rightarrow c \rightarrow a \leq c \rightarrow b.$

In the case of distributive lattices such as $\mathcal{RP}$ the relational representation is built on the set of prime ideals of the lattice rather than the maximal disjoint filter-ideal pairs used in the non-distributive cases. The underlying relational structures are of the type $\langle X, \leq \rangle$, where X is a set and $\leq$ a quasi-order on X. The class of all such relational structures is denoted by $\mathcal{R}_{RP}$.

For each $\boldsymbol{W} \in \mathcal{RP}$ we define the **canonical frame** of $\boldsymbol{W}$ as the relational structure $\boldsymbol{X}(\boldsymbol{W}) = \langle X(\boldsymbol{W}), \leq^C \rangle$, where $X(\boldsymbol{W})$ is the set of all prime filters of $\boldsymbol{W}$ and $\leq^C = \subseteq$.

Lemma 9.1. *If $\boldsymbol{W} \in \mathcal{RP}$ then $\boldsymbol{X}(\boldsymbol{W}) \in \mathcal{R}_{RP}$.*

For each $\langle X, \leq \rangle \in \mathcal{RP}$, we define the operation $[\leq] : 2^X \rightarrow 2^X$ by

$$[\leq]A = \{x \in X : \forall y(x \leq y \Rightarrow y \in A)\}.$$

Observe that $[\leq]A$ is the largest upward closed subset of A. Note also that $[\leq]$ is monotonic and, for any $A \subseteq X$, $[\leq]A = A$ iff A is upward closed, and $[\leq][\leq]A = [\leq]A$.

If $\boldsymbol{X} = \langle X, \leq \rangle \in \mathcal{R}_{RP}$ we define the ***complex algebra*** of $\boldsymbol{X}$ as $\boldsymbol{L}(\boldsymbol{X}) = \langle L(\boldsymbol{X}), \wedge^C, \vee^C, \rightarrow^C, 1^C \rangle$ where $L(\boldsymbol{X}) = \{A \subseteq X : [\leq]A = A\}$ and, for all $A, B \in L(\boldsymbol{X})$,

$$A \wedge^C B = A \cap B,$$
$$A \vee^C B = A \cup B,$$
$$A \rightarrow^C B = [\leq](-A \cup B), \quad \text{where } -A \text{ is the set complement of } A \text{ in } X,$$
$$1^C = X.$$

Lemma 9.2. *If $\boldsymbol{X} \in \mathcal{R}_{RP}$ then $\boldsymbol{L}(\boldsymbol{X}) \in \mathcal{RP}$.*

Proof. It is clear that $L(\boldsymbol{X})$ is closed under $\wedge^C$ and $\vee^C$ and that these operations describe a distributive lattice with greatest element 1^C. We need only show that $\rightarrow^C$ is the residuum of $\cap$, i.e., for all $A, B, C \in L(\boldsymbol{X})$,

$$A \cap C \subseteq B \quad \text{iff} \quad C \subseteq A \rightarrow^C B = [\leq](-A \cup B).$$

Assume that $A \cap C \subseteq B$ and let $x \in C$. Take any $y \in X$ such that $x \leq y$. Then $y \in C$ since C is a filter. If $y \in A$ then $y \in A \cap C$ hence $y \in B$ so $y \in -A \cup B$. If $y \notin A$ then, trivially, $y \in -A \cup B$. Conversely, assume $C \subseteq [\leq](-A \cup B)$ and let $x \in A \cap C$. Then $x \in C$ hence $x \in [\leq](-A \cup B)$. Since $x \leq x$, we have $x \in -A \cup B$, but $x \in A$, so we must have $x \in B$, as required.

The above lemmas show that if $\boldsymbol{W} \in \mathcal{RP}$, then so is $\boldsymbol{L}(\boldsymbol{X}(\boldsymbol{W}))$. To show that $\boldsymbol{W}$ embeds into $\boldsymbol{L}(\boldsymbol{X}(\boldsymbol{W}))$ we define the map $f : W \rightarrow L(\boldsymbol{X}(\boldsymbol{W}))$ by

$$f(a) = \{F \in X(\boldsymbol{W}) : a \in F\}.$$

For the proof of next theorem we need the following observations. Let F be a (lattice) filter of a relatively pseudo-complemented lattice $\boldsymbol{W}$. Then the following hold for all $a, b \in W$:

$a \in F$ and $a \rightarrow b \in F \Rightarrow b \in F$;
if $b \notin F$, then there is a prime filter F' such that $F \subseteq F'$ and $b \notin F'$.

Theorem 9.1. *Each $\boldsymbol{W} \in \mathcal{RP}$ is embeddable into $\boldsymbol{L}(\boldsymbol{X}(\boldsymbol{W}))$.*

Proof. That the map f is a lattice embedding follows by standard arguments of M.H. Stone [16] (see also [2]). We need only show the preservation of relative pseudo-complement by f, i.e., that $f(a \rightarrow b) = f(a) \rightarrow^C f(b) = [\leq^C](-f(a) \cup f(b))$. Let $F \in f(a \rightarrow b)$, i.e., $a \rightarrow b \in F$. It follows that $a \notin F$ or $b \in F$, hence $F \notin f(a)$ or $F \in f(b)$, i.e., $F \in -f(a) \cup f(b)$, so $f(a \rightarrow b) \subseteq -f(a) \cup f(b)$. Since for every $a \in W$, $f(a) = [\leq^C]f(a)$ we have, by monotonicity of $[\leq^C]$, that $f(a \rightarrow b) = [\leq^C]f(a \rightarrow b) \subseteq [\leq^C](-f(a) \cup f(b))$. For the converse inclusion, suppose $F \in [\leq^C](-f(a) \cup f(b))$. Then, for all G,

$$F \subseteq G \Rightarrow a \notin G \text{ or } b \in G. \tag{8}$$

In particular, $a \notin F$ or $b \in F$. If $b \in F$ then, since $b \leq a \to b$, we have $a \to b \in F$. If $b \notin F$, then $a \notin F$. We show that also in this case $a \to b \in F$. Suppose, to the contrary, that $a \to b \notin F$. Set $H = \{c : a \to c \in F\}$. Since $a \to (c \wedge d) = (a \to c) \wedge (a \to d)$, it follows that H is closed under meets. Since $c \leq d$ implies $a \to c \leq a \to d$, H is upward closed. Thus, H is a filter of $\boldsymbol{W}$. Moreover, $F \subseteq H$, $a \in H$ and $b \notin H$. Thus, we may extend H to a prime filter H' such that $b \notin H'$, but $F \subseteq H'$ and $a \in H'$, contradicting (8).

10 Relatively Pseudo-complemented Lattices with Minimal Negation

Now we consider relatively pseudo-complemented lattices with minimal negation, also called minimal negation of Johansson [10], (cf. Dunn and Hardegree [5]) or contrapositional negation, (cf. Rasiowa [14]). This is a relatively pseudo-complemented lattice enriched with an operation corresponding to minimal negation, (i.e., minimal negation of Johansson, or contrapositional negation).

By a *relatively pseudo-complemented lattice with minimal negation* we mean an algebra $\boldsymbol{W} = \langle W, \wedge, \vee, \to, \neg, \partial, 1 \rangle$, where $\langle W, \wedge, \vee, \to, 1 \rangle$ is a relatively pseudo-complemented lattice, $\partial \in W$ (not necessarily the smallest element) and $\neg$ is a unary operator satisfying:

(RPM1) $a \to \neg b \leq b \to \neg a$,
(RPM2) $\neg 1 = \partial$.

Let $\mathcal{RPM}$ denote the variety of all relatively pseudo-complemented lattices with minimal negation. Note that (RPM1) is equivalent to $a \to \neg b = b \to \neg a$ and corresponds to the condition for quasi-minimal, or Galois, negation (Gal): $a \leq \neg b \Rightarrow b \leq \neg a$.

Lemma 10.1

(a) If $\boldsymbol{W} \in \mathcal{RPM}$, then $\neg a = a \to \partial$ for all $a \in W$.
(b) Let $\boldsymbol{W} \in \mathcal{RP}$ and let ∂ be any element of W. If we define a unary operation $\neg$ by $\neg a = a \to \partial$ for all $a \in W$, then $\neg$ is a minimal negation.

Proof. (a) For all $a \in W$ we have $\neg a = 1 \to \neg a = a \to \neg 1 = a \to \partial$. (b) For (RPM1), for all $a, b \in W$ we have $a \to \neg b = a \to (b \to \partial) = b \to (a \to \partial) = b \to \neg a$. For (RPM2), we have $\neg 1 = 1 \to \partial = \partial$.

We will denote by $\mathcal{R}_{RPM}$ the class of all relational structures of type $\boldsymbol{X} = \langle X, \leq, D \rangle$, where $\leq$ is a quasi-order on X and $D \subseteq X$.

For each $\boldsymbol{W} \in \mathcal{RPM}$ we define the **canonical frame** of $\boldsymbol{W}$ as $\boldsymbol{X}(\boldsymbol{W}) = \langle X(\boldsymbol{W}), \leq^C, D^C \rangle$, where $X(\boldsymbol{W})$ is the set of all prime filters of $\boldsymbol{W}$, $\leq^C = \subseteq$ and

$$D^C = \{F \in X(\boldsymbol{W}) : \partial \in F\}.$$

Lemma 10.2. *If $W \in \mathcal{RPM}$ then $X(W) \in \mathcal{R}_{RPM}$.*

If $X = \langle X, \leq, D \rangle \in \mathcal{R}_{RPM}$, then $\langle X, \leq \rangle \in \mathcal{R}_{RP}$ hence it has a complex algebra $\langle L(X), \wedge^C, \vee^C, \rightarrow^C, 1^C \rangle$ as defined in the previous section. The **complex algebra** of X, denoted $L(X)$, is the extension of this algebra by the constant ∂^C and the operation $\neg^C$ defined by

$$\partial^C = [\leq]D,$$
$$\neg^C A = A \rightarrow^C \partial^C \text{ for } A \in L(X).$$

Lemma 10.3. *If $X \in \mathcal{R}_{RPM}$ then $L(X) \in \mathcal{RPM}$.*

Proof. Since $[\leq][\leq]D = [\leq]D$, we have $\partial^C \in L(X)$ and hence $L(X)$ is also closed under $\neg^C$. Since $L(X)$ is a relatively pseudo-complemented lattice, (RPM1) follows from properties of $\rightarrow$. (RPM2) follows from $\neg^C 1 = [\leq](-X \cup [\leq]D) = [\leq][\leq]D = [\leq]D = \partial^C$.

Thus, if $W \in \mathcal{RPM}$ so is $L(X(W))$.

Theorem 10.1. *Each $W \in \mathcal{RPM}$ is embeddable into $L(X(W))$.*

Proof. From the previous section we know that the function $f : W \rightarrow L(X(W))$ defined by

$$f(a) = \{F \in X(W) : a \in F\}$$

is an embedding on the reduct $\langle W, \wedge, \vee, \rightarrow, 1 \rangle$. We have $f(\partial) = \{F \in X(W) : \partial \in F\} = D^C$ and $f(\partial)$ is an upward closed subset of $X(W)$ so $f(\partial) = [\leq]D^C = \partial^C$. Since $\rightarrow$ is preserved it follows that $\neg$ is too.

11   Distributive Lattices with De Morgan Negation

Now we consider distributive lattices with negation operation corresponding to De Morgan negation (i.e., satisfying (Gal) and (DeM)). We will see the difference in techniques of representation between the previous non-distributive case and the distributive case. The representation theorem below is a modification of the result of Białynicki-Birula and Rasiowa [3] to the unified framework.

By a *De Morgan algebra* (also called a *distributive lattice with involution*) we mean a De Morgan lattice $\langle W, \wedge, \vee, \neg, 0, 1 \rangle$ whose lattice reduct is distributive. Let $\mathcal{DM}$ denote the variety of all De Morgan algebras. Thus, $\mathcal{DM}$ satisfies (Gal) and (DeM), as well as (Suff1), (Suff2), (WCon), (Weak¬¬), $\neg 1 = 0$ and

$$\neg\neg a = a$$
$$\neg(a \wedge b) = \neg a \vee \neg b$$
$$\neg a = \neg b \Rightarrow a = b.$$

For $W \in \mathcal{DM}$ and $A \subseteq W$, let $\neg A = \{\neg a : a \in A\}$. Then the following hold:

(A1) $\neg A = \{a : \neg a \in A\}$
(A2) $\neg(W - A) = W - (\neg A)$
(A3) $\neg\neg A = A$
(A4) A is a prime filter iff $\neg A$ is a prime ideal.

We will denote by $\mathcal{R}_{DM}$ the class of all relational structures of type $\boldsymbol{X} = \langle X, \leq, N \rangle$, where $\leq$ is a quasi-order on X, $N : X \to X$ is a function and, for all $x, y \in X$,

(DM1) $x \leq y \Rightarrow N(y) \leq N(x)$,
(DM2) $N(N(x)) = x$.

Compare these with (M1–M3). If we let $N(A) = \{N(x) : x \in A\}$, for $A \subseteq X$, then the following hold:

(A5) $N(A) = \{x : N(x) \in A\}$
(A6) $N(X - A) = X - N(A)$
(A7) $N(A \cup B) = N(A) \cup N(B)$
(A8) $NN(A) = A$.

The only non-trivial property is (A6), but this follows since: $x \in N(X - A)$ iff $N(x) \in X - A$ iff $N(x) \notin A$ iff $x \notin N(A)$.

For each $\boldsymbol{W} \in \mathcal{DM}$ we define the **canonical frame** of $\boldsymbol{W}$ as the relational structure $\boldsymbol{X}(\boldsymbol{W}) = \langle X(\boldsymbol{W}), \leq^C, N^C \rangle$, where $X(\boldsymbol{W})$ is the set of all prime filters of $\boldsymbol{W}$, $\leq^C = \subseteq$ and, for $F \in X(\boldsymbol{W})$,

$$N^C(F) = W - (\neg F).$$

Lemma 11.1. *If $\boldsymbol{W} \in \mathcal{DM}$ then $\boldsymbol{X}(\boldsymbol{W}) \in \mathcal{R}_{DM}$.*

Proof. We first show that N is a function from $X(\boldsymbol{W})$ to $X(\boldsymbol{W})$. Let $F \in X(\boldsymbol{W})$, so F is a prime filter. It is routine to check that $N^C(F)$ is a filter. For primeness, suppose that $a \vee b \in N^C(F) = W - (\neg F)$. Then $a \vee b \notin \neg F$ so $\neg(a \vee b) = \neg a \wedge \neg b \notin F$. Thus, either $\neg a \notin F$ or $\neg b \notin F$, so $a \notin \neg F$ or $b \notin \neg F$, hence $a \in W - (\neg F)$ or $b \in W - (\neg F)$.

For (DM1), suppose $F, G \in X(\boldsymbol{W})$ and $F \subseteq G$. Now, by (A2), (A1) and definitions we have $a \in N^C(G)$ iff $a \in W - (\neg G)$ iff $a \notin \neg G$ iff $\neg a \notin G$ hence, by the assumption, $\neg a \notin F$ iff $a \notin \neg F$ iff $a \in W - (\neg F)$ iff $a \in N^C(F)$.

For (DM2), by (A6) and (A7) we have $N^C(N^C(F)) = W - (\neg N^C(F)) = W - (\neg(W - (\neg F))) = W - (W - \neg\neg F) = \neg\neg F = F$.

If $\boldsymbol{X} = \langle X, \leq \rangle \in \mathcal{R}_{DM}$ we define the **complex algebra** of $\boldsymbol{X}$ as $\boldsymbol{L}(\boldsymbol{X}) = \langle L(\boldsymbol{X}), \wedge^C, \vee^C, \neg^C, 0^C, 1^C \rangle$ where $L(\boldsymbol{X}) = \{A \subseteq X : [\leq]A = A\}$ and, for all $A, B \in L(\boldsymbol{X})$,

$A \wedge^C B = A \cap B,$
$A \vee^C B = A \cup B,$
$\neg^C A = X - N(A),$
$1^C = X,$
$0^C = \emptyset.$

Recall that, for $A \subseteq X$,

$$[\leq]A = \{x \in X : \forall y(x \leq y \Rightarrow y \in A)\}.$$

Lemma 11.2. *If* $X \in \mathcal{R}_{DM}$ *then* $L(X) \in \mathcal{DM}$.

Proof. We show that if $A \in L(X)$, then $\neg^C A \in L(X)$, that is $\neg^C A = [\leq]\neg^C A$. Let $x \in \neg^C A$, so $N(x) \notin A$. Suppose that $x \notin [\leq]\neg^C A$. Then there is y such that $x \leq y$ and $y \notin \neg^C A$, thus $N(y) \in A = [\leq]A$, that is $\forall z (N(y) \leq z \Rightarrow z \in A)$. Since $x \leq y$, we have $N(y) \leq N(x)$, and taking $z = N(x)$ we get $N(x) \in A$, a contradiction. For the converse, let $x \in [\leq]\neg^C A$. Then $\forall y (x \leq y \Rightarrow N(y) \notin A)$; suppose that $x \notin \neg^C A$, hence $N(x) \in A$. Taking $y = x$ we get a contradiction.

Now we show that $\neg^C \neg^C A = A$. Using (A6) and (A8) we have $X - N(\neg^C A) = X - N(X - N(A)) = X - (X - NN(A)) = NN(A) = A$. This proves (DeM) and (Weak¬¬) hence (Gal) follows by Lemma 2.1. Next we show (Suff1), i.e., that $\neg^C (A \cup B) = \neg^C A \cap \neg^C B$. By (A7) we have $x \in X - N(A \cup B)$ iff $x \notin N(A \cup B)$ iff $N(x) \notin A$ and $N(x) \notin B$ iff $x \in \neg^C A \cap \neg^C B$.

The above lemmas imply that if $W \in \mathcal{DM}$, then $L(X(W)) \in \mathcal{DM}$ as well. Recall that the function $f : W \to L(X(W))$ defined by

$$f(a) = \{F \in X(W) : a \in F\}$$

is an embedding of the lattice parts of W and $L(X(W))$. We show that it preserves negation as well.

Theorem 11.1. *Each* $W \in \mathcal{DM}$ *is embeddable into* $L(X(W))$.

Proof. We need only show the preservation of negation. We have, by definition,

$$\begin{aligned}
\neg^C f(a) &= X(W) - (N^C(f(a)) \\
&= X(W) - \{N^C(F) : F \in f(a)\} \\
&= X(W) - \{W - (\neg F) : a \in F\}
\end{aligned}$$

and

$$f(\neg a) = \{G : \neg a \in G\}.$$

Note that $a \in F$ iff $\neg a \in \neg F$ iff $\neg a \notin W - (\neg F)$. Thus, $\{W - (\neg F) : a \in F\}$ consists of all $G \in X(W)$ for which $\neg a \notin G$. Therefore $X(W) - \{W - (\neg F) : a \in F\}$ consists of all $G \in X(W)$ such that $\neg a \in G$, i.e., $\neg^C f(a) = f(\neg a)$.

12   Boolean Algebras with Sufficiency Operator

By a *Boolean algebra with sufficiency* (or *negative necessity*) *operator* we mean an algebra $W = \langle W', \neg \rangle$, where $W' = \langle W, \wedge, \vee, -, 0, 1 \rangle$ is a Boolean algebra, and $\neg$ a unary operation satisfying:

(Suff1) $\neg(a \vee b) = \neg a \wedge \neg b$
(Suff2) $\neg 0 = 1.$

Let $\mathcal{SUA}$ denote the variety of all Boolean algebras with sufficiency operator. We extend the relational representation to Boolean algebras with sufficiency operator.

A *frame* is a relational structure of type $\boldsymbol{X} = \langle X, R \rangle$, where $R \subseteq X \times X$. Let $\mathcal{R}$ be a class of all frames.

For each $\boldsymbol{W} \in \mathcal{SUA}$ we define the **canonical frame** of $\boldsymbol{W}$ as the relational structure $\boldsymbol{X}(\boldsymbol{W}) = \langle X(\boldsymbol{W}), R^C \rangle$, where $X(\boldsymbol{W})$ is the set of all prime filters of $\boldsymbol{W}$ and, for $F, G \in X(\boldsymbol{W})$,

$$FR^CG \text{ iff } \neg G \cap F \neq \emptyset$$

where $\neg A = \{a \in W : \neg a \in A\}$ for each $A \subseteq X$.

Given a frame $\boldsymbol{X} = \langle X, R \rangle$, we define the **complex algebra** of $\boldsymbol{X}$ as $L(\boldsymbol{X}) = \langle \mathcal{P}(X), \neg^C \rangle$, where $\mathcal{P}(X)$ is the powerset Boolean algebra of X and, for $A \in \mathcal{P}(X)$,

$$\neg^C A = \{x \in X : A \subseteq R(x)\} = \{x \in X : \forall y(y \in A \Rightarrow xRy)\}.$$

Lemma 12.1. *If $\boldsymbol{W} \in \mathcal{SUA}$, then $\boldsymbol{X}(\boldsymbol{W}) \in \mathcal{R}$. If $\boldsymbol{X} \in \mathcal{R}$ then $L(\boldsymbol{X}) \in \mathcal{SUA}$.*

Theorem 12.1. *Each $\boldsymbol{W} \in \mathcal{SUA}$ is embeddable into $L(\boldsymbol{X}(\boldsymbol{W}))$.*

Proof. The embedding is defined in a standard way:

$$f(a) = \{G \in X(\boldsymbol{W}) : a \in G\}.$$

Acknowledgement. We are indebted to the referees for valuable remarks that helped to improve the final version of the paper.

References

1. Allwein, G., Dunn, J.M.: Kripke models for linear logic. J. Symb. Logic **58** (1993) 514–545.
2. Balbes, R. and Dwinger, P.: Distributive Lattices. University of Missouri Press (1974).
3. Białynicki-Birula, A., Rasiowa, H.: On constructible falsity in the constructive logic with strong negation. Colloquium Mathematicum **6** (1958) 287–310.
4. Dunn, J.M.: Star and Perp: Two Treatments of Negation. In J. Tomberlin (ed.), Philosophical Perspectives (Philosophy of Language and Logic) **7** (1993) 331–357.
5. Dunn, J.M., Hardegree, G.M.: Algebraic Methods in Philosophical Logic. Clarendon Press, Oxford (2001).
6. Düntsch, I., Orłowska, E., Radzikowska, A.M.: Lattice–based relation algebras and their representability. In: de Swart, C.C.M. et al (eds), Theory and Applications of Relational Structures as Knowledge Instruments, Lecture Notes in Computer Science **2929** Springer–Verlag (2003) 234–258.
7. Düntsch, I., Orłowska, E., Radzikowska, A.M., Vakarelov, D.: Relational representation theorems for some lattice-based structures. Journal of Relation Methods in Computer Science JoRMiCS vol.1, Special Volume, ISSN 1439-2275 (2004) 132–160.

8. Dzik, W., Orłowska, E., vanAlten, C..: Relational Representation Theorems for Lattices with Negations, to appear in : Relmics' 2006 Proceedings.
9. Goldblatt, R.: Representation for Ortholattices. Bull. London Math. Soc. **7** (1975) 45–48.
10. Johansson, I.: Der Minimalkalül, ein reduzierte intuitionistischer Formalismus. Compositio Mathematica **4** (1936) 119-136.
11. Orłowska, E., Radzikowska, A.M.: Information relations and operators based on double residuated lattices. In de Swart, H.C.M. (ed), Proceedings of the 6th Seminar on Relational Methods in Computer Science RelMiCS'2001 (2001) 185–199.
12. Orłowska, E., Radzikowska, A.M.: Double residuated lattices and their applications. In: de Swart, H.C.M. (ed), Relational Methods in Computer Science, Lecture Notes in Computer Science **2561** Springer–Verlag, Heidelberg (2002) 171–189.
13. Orłowska, E., Vakarelov, D. Lattice-based modal algebras and modal logics. In: Hajek, P., Valdes, L., Westerstahl, D. (eds), Proceedings of the 12th International Congress of Logic, Methodology and Philosophy of Science, Oviedo, August 2003, King's College London Publication (2005) 147–170.
14. Rasiowa, H.: An Algebraic Approach to Non-Classical Logics. North Holland, Studies in Logic and the Foundations of Mathematics vol. 78 (1974).
15. Rasiowa, H., Sikorski, R.: Mathematics of Metamathematics, PWN, Warszawa (1970).
16. Stone, M.H.: Topological representation of distributive lattices and Brouwerian logics. Cas. Mat. Fiz. **67** (1937) 1–25.
17. Urquhart, A.: A topological representation theorem for lattices. Algebra Universalis **8** (1978) 45–58.

Lattice-Based Relation Algebras II[*]

Ivo Düntsch[1,**], Ewa Orłowska[2], and Anna Maria Radzikowska[3]

[1] Brock University
St. Catharines, Ontario, Canada, L2S 3A1
duentsch@brocku.ca
[2] National Institute of Telecommunications
Szachowa 1, 04–894 Warsaw, Poland
orlowska@itl.waw.pl
[3] Faculty of Mathematics and Information Science
Warsaw University of Technology
Plac Politechniki 1, 00–661 Warsaw, Poland
annrad@mini.pw.edu.pl

Abstract. We present classes of algebras which may be viewed as weak relation algebras, where a Boolean part is replaced by a not necessarily distributive lattice. For each of the classes considered in the paper we prove a relational representation theorem.

1 Introduction

In the first paper on lattice-based relation algebras [8] we presented a class of lattices with the operators, referred to as LCP algebras, which was the abstract counterpart to the class of relation algebras with the specific operations of relative product and converse. In the present paper we expand the LCP class with new operators which model residua of relative product, relative sum, dual converse, and dual residua of relative sum. In the classical relation algebras based on Boolean algebras these operators are definable with the standard relational operations and the complement. In lattice-based algebras they should be specified axiomatically since there is no way to define them without a complement. We construct this extension in two steps. In Section 5 we introduce the class of LCPR algebras which extend the class LCP with the residua of product, and in Section 6 we present the class of LCPRS algebras which are obtained from LCPR algebras by adding sum, dual converse, and dual residua of sum. For each of these classes we prove a relational representation theorem in the style of Urquhart-Allwein-Dunn (see [1], [19]). Sections 2, 3, and 4 present an overview of Urquhart's representation theory for lattices and a survey of LCP algebras. The contributions of the paper fit, on the one hand, into the study of lattices

---

[*] This work was carried out in the framework of the European Commission's COST Action 274 *Theory and Applications of Relational Structures as Knowledge Instruments* (TARSKI).

[**] Ivo Düntsch gratefully acknowledges support by the Natural Sciences and Engineering Research Council of Canada.

H. de Swart et al. (Eds.): TARSKI II, LNAI 4342, pp. 267–289, 2006.
© Springer-Verlag Berlin Heidelberg 2006

with additional operators presented in a number of papers, for example in [10], [15], [17], [18], and on the other hand, into a relational approach to modeling algebraic and logical structures. A study of lattices with operators evolved from the concept of Boolean algebras with operators originated in [13]. It is continued, among others, in the context of modeling incomplete information in [3], [5], [7], and [14].

2   Doubly Ordered Sets

In this section we recall the notions introduced in [8] and some of their properties.

Definition 1. *Let X be a non–empty set and let $\leqslant_1$ and $\leqslant_2$ be two quasi orderings in X. A structure $(X, \leqslant_1, \leqslant_2)$ is called a **doubly ordered set** iff for all $x, y \in X$, if $x \leqslant_1 y$ and $x \leqslant_2 y$ then $x = y$.*   □

Definition 2. *Let $(X, \leqslant_1, \leqslant_2)$ be a doubly ordered set. We say that $A \subseteq X$ is $\leqslant_1$–**increasing** (resp. $\leqslant_2$–**increasing**) whenever for all $x, y \in X$, if $x \in A$ and $x \leqslant_1 y$ (resp. $x \leqslant_2 y$), then $y \in A$.*   □

For a doubly ordered set $(X, \leqslant_1, \leqslant_2)$, we define two mappings $l, r : 2^X \to 2^X$ by: for every $A \subseteq X$,

$$l(A) = \{x \in X : (\forall y \in X)\, x \leqslant_1 y \Rightarrow y \notin A\} \qquad (1)$$
$$r(A) = \{x \in X : (\forall y \in X)\, x \leqslant_2 y \Rightarrow y \notin A\}. \qquad (2)$$

Observe that mappings l and r can be expressed in terms of modal operators as follows: $l(A) = [\leqslant_1](-A)$ and $r(A) = [\leqslant_2](-A)$, where $-$ is the Boolean complement and $[\leqslant_i]$, $i = 1, 2$, are the necessity operators determined by relations $\leqslant_i$. Consequently, r and l are intuitionistic–like negations.

Definition 3. *Given a doubly ordered set $(X, \leqslant_1, \leqslant_2)$, a subset $A \subseteq X$ is called l–**stable** (resp. r–**stable**) iff $l(r(A)) = A$ (resp. $r(l(A)) = A$).*   □

The family of all l-stable (resp. r–stable) subsets of X will be denoted by $L(X)$ (resp. $R(X)$).

Recall the following notion from e.g. [4]:

Definition 4. *Let $(X, \leqslant_1)$ and $(Y, \leqslant_2)$ be partially ordered sets and let f and g be mappings $f : X \to Y$, $g : Y \to X$. We say that f and g are a **Galois connection** iff for all $x, y \in X$*

$$x \leqslant_1 g(y) \ \text{ iff } \ y \leqslant_2 f(x).$$ □

Lemma 1. [17] *For any doubly ordered set $(X, \leqslant_1, \leqslant_2)$ and for any $A \subseteq X$,*

(i) *$l(A)$ is $\leqslant_1$–increasing and $r(A)$ is $\leqslant_2$–increasing*
(ii) *if A is $\leqslant_1$–increasing, then $r(A) \in R(X)$*
(iii) *if A is $\leqslant_2$–increasing, then $l(A) \in L(X)$*

(iv) *if* $A \in L(X)$, *then* $r(A) \in R(X)$

(v) *if* $A \in R(X)$, *then* $l(A) \in L(X)$

(vi) *if* $A, B \in L(X)$, *then* $r(A) \cap r(B) \in R(X)$.   ∎

It is well–known that the following facts hold.

Lemma 2. *The family of* $\leqslant_i$*–increasing sets,* $i = 1, 2$, *forms a distributive lattice, where join and meet are union and intersection of sets.*   ∎

Lemma 3. *[19] For every doubly ordered set* $(X, \leqslant_1, \leqslant_2)$, *the mappings* l *and* r *form a Galois connection between the lattice of* $\leqslant_1$*–increasing subsets of* X *and the lattice of* $\leqslant_2$*–increasing subsets of* X.   ∎

In other words, Lemma 3 implies that for any $A \in L(X)$ and for any $B \in R(X)$, $A \subseteq l(B)$ iff $B \subseteq r(A)$.

Lemma 4. *[8] For every doubly ordered set* $(X, \leqslant_1, \leqslant_2)$ *and for every* $A \subseteq X$,

(i) $l(r(A)) \in L(X)$ *and* $r(l(A)) \in R(X)$

(ii) *if* A *is* $\leqslant_1$*–increasing, then* $A \subseteq l(r(A))$

(iii) *if* A *is* $\leqslant_2$*–increasing, then* $A \subseteq r(l(A))$.   ∎

Lemma 4 immediately implies:

Corollary 1. *For every doubly ordered set* $(X, \leqslant_1, \leqslant_2)$ *and for every* $A \subseteq X$,

(i) *if* $A \in L(X)$, *then* $A \subseteq l(r(A))$

(ii) *if* $A \in R(X)$, *then* $A \subseteq r(l(A))$.   ∎

Let $(X, \leqslant_1, \leqslant_2)$ be a doubly ordered set. Define two binary operations in 2^X: for all $A, B \subseteq X$,

$$A \sqcap B = A \cap B \tag{3}$$

$$A \sqcup B = l(r(A) \cap r(B)). \tag{4}$$

Observe that $\sqcup$ is defined from $\sqcap$ resembling a De Morgan law with two different negations.

Moreover, put

$$\mathbf{0} = \emptyset. \tag{5}$$

$$\mathbf{1} = X \tag{6}$$

In [19] it was shown that for a doubly ordered set $(X, \leqslant_1, \leqslant_2)$, the system $((X), \sqcap, \sqcup, \mathbf{0}, \mathbf{1})$ is a lattice. This lattice is called the ***complex algebra of*** X.

3   Urquhart's Representation of Lattices

In this paper we are interested in studying relationships between relational structures (frames) providing Kripke–style semantics of logics, and algebras based on lattices. Therefore, we do not assume any topological structure in the frames. As a result, we have a weaker form of the representation theorems than the original Urquhart result, which requires compactness.

Let $(W, \wedge, \vee, 0, 1)$ be a non–trivial bounded lattice.

Definition 5. *A **filter-ideal pair** of a bounded lattice $(W, \wedge, \vee, 0, 1)$ is a pair $x = (x_1, x_2)$ such that x_1 is a filter of W, x_2 is an ideal of W and $x_1 \cap x_2 = \emptyset$.* □

The family of all filter–ideal pairs of a lattice W will be denoted by $FIP(W)$. Let us define the following two quasi ordering relations on $FIP(W)$: for any $(x_1, x_2), (y_1, y_2) \in FIP(W)$,

$$(x_1, x_2) \preccurlyeq_1 (y_1, y_2) \iff x_1 \subseteq y_1 \tag{7}$$

$$(x_1, x_2) \preccurlyeq_2 (y_1, y_2) \iff x_2 \subseteq y_2. \tag{8}$$

Next, define

$$(x_1, x_2) \preccurlyeq (y_1, y_2) \iff (x_1, x_2) \preccurlyeq_1 (y_1, y_2) \ \& \ (x_1, x_2) \preccurlyeq_2 (y_1, y_2).$$

We say that $(x_1, x_2) \in FIP(W)$ is **maximal** iff it is maximal with respect to $\preccurlyeq$. We will write $X(W)$ to denote the family of all maximal filter–ideal pairs of the lattice W.

Observe that $X(W)$ is a binary relation on 2^W.

Proposition 1. [19] *Let $(W, \wedge, \vee, 0, 1)$ be a bounded lattice. Then for every $(x_1, x_2) \in FIP(W)$ there exists $(y_1, y_2) \in X(W)$ such that $(x_1, y_1) \preccurlyeq (y_1, y_2)$.* ■

For any $(x_1, x_2) \in FIP(W)$, the maximal filter–ideal pair (y_1, y_2) such that $(x_1, x_2) \preccurlyeq (y_1, y_2)$ will be referred to as an *extension* of (x_1, x_2).

Definition 6. *Let $(W, \wedge, \vee, 0, 1)$ be a bounded lattice. The **canonical frame of** W is the structure $(X(W), \preccurlyeq_1, \preccurlyeq_2)$.* □

Lemma 5. *For every bounded lattice W, its canonical frame $(X(W), \preccurlyeq_1, \preccurlyeq_2)$ is a doubly ordered set.* ■

Consider the complex algebra $(L(X(W)), \sqcap, \sqcup, \mathbf{0}, \mathbf{1})$ of the canonical frame of a lattice $(W, \wedge, \vee, 0, 1)$. Observe that $L(X(W))$ is an algebra of subrelations of $X(W)$.

Let us define the mapping $h : W \to 2^{X(W)}$ as follows: for every $a \in W$,

$$h(a) = \{x \in X(W) : a \in x_1\}. \tag{9}$$

Theorem 1. [19] *For every lattice* $(W, \wedge, \vee, 0, 1)$ *the following assertions hold:*

(i) *For every* $a \in W$, $r(h(a)) = \{x \in X(W) : a \in x_2\}$

(ii) $h(a)$ *is* l-*stable for every* $a \in W$

(iii) h *is a lattice embedding.* ∎

The following theorem is a weak version of the Urquhart result.

Theorem 2 (Representation theorem for lattices). *Every bounded lattice is isomorphic to a subalgebra of the complex algebra of its canonical frame.* ∎

4 LCP Algebras and Frames

In this section we recall the class LCP of lattices with the operations of product and converse introduced in [8]. We add one more axiom, **(CP0)**, to the axioms of LCP postulated in [8] and we explain its role.

Definition 7. *An* **LCP algebra** *is a system* $(W, \wedge, \vee, \smile, \otimes, 0, 1, 1')$ *such that* $(W, \wedge, \vee, 0, 1)$ *is a non–trivial bounded lattice,* $\smile$ *is a unary operation in* W *and* $\otimes$ *is a binary operation in* W *satisfying the following conditions for all* $a, b, c \in W$,

(CP.0) $0 \otimes a = a \otimes 0 = 0$

(CP.1) $a^{\smile\smile} = a$

(CP.2) $(a \vee b)^{\smile} = a^{\smile} \vee b^{\smile}$

(CP.3) $a \otimes 1' = 1' \otimes a = a$

(CP.4) $a \otimes (b \otimes c) = (a \otimes b) \otimes c$

(CP.5) $a \otimes (b \vee c) = (a \otimes b) \vee (a \otimes c)$

(CP.6) $(a \vee b) \otimes c = (a \otimes c) \vee (b \otimes c)$

(CP.7) $(a \otimes b)^{\smile} = b^{\smile} \otimes a^{\smile}$. □

It is worth noting that axiom **(CP.0)** does not follow from the remaining axioms. Consider, for example, a bounded lattice $(W, \wedge, \vee, 0, 1)$ and define the additional operations $\otimes$ and $\smile$ as follows: for all $a, b \in W$,

$$a^{\smile} = a$$
$$a \otimes b = a \vee b$$
$$1' = 0.$$

One can easily check that axioms **(CP.1)**–**(CP.7)** hold, but **(CP.0)** does not. Consequently, Lemma 24 of [8] needs repair. For its proof we refer to [1]. The crucial argument is on page 529 of [1] in the paragraph following equation (3). In line 4 of this paragraph they obtain the disjoint pair $([t), U)$, which, as they claim, can be extended to the maximal filter–ideal pair. This, however, is only possible if $t \neq 0$.

Note also that axiom **(CP.0)** follows from the relation algebra axioms and implies that $0 \neq 1'$ in every LCP algebra with at least two elements. To see that, suppose that **(CP.0)** holds and $0 = 1'$. Then $1 = 1' \otimes 1 = 0 \otimes 1 = 0$, which contradicts our hypothesis that W has at least two elements.

For any $A \subseteq W$, let us denote

$$A^{\smile} = \{a^{\smile} \in W : a \in A\}. \tag{10}$$

Lemma 6. [8] *For any LCP algebra* $(W, \wedge, \vee, ^{\smile}, \otimes, 0, 1, 1')$ *and for all subsets* $A, B \subseteq W$,

    (i) $A \subseteq B$ *iff* $A^{\smile} \subseteq B^{\smile}$

    (ii) $A^{\smile\smile} = A$.   ■

Some other properties of LCP algebras can be found in [8].

Definition 8. *An* **LCP frame** *is a relational system* $(X, \leqslant_1, \leqslant_2, C, R, S, Q, I)$ *such that* $(X, \leqslant_1, \leqslant_2)$ *is a doubly ordered set,* C *is a mapping* $C : X \to X$, R, S, *and* Q *are ternary relations on* X *and* $I \subseteq X$ *is an unary relation on* X *satisfying the following conditions for all* $x, y \in X$:

Monotonicity conditions:

    (MCP.1) $x \leqslant_1 y$ implies $C(x) \leqslant_1 C(y)$

    (MCP.2) $x \leqslant_2 y$ implies $C(x) \leqslant_2 C(y)$

    (MCP.3) $R(x, y, z) \,\&\, x' \leqslant_1 x \,\&\, y' \leqslant_1 y \,\&\, z \leqslant_1 z' \implies R(x', y', z')$

    (MCP.4) $S(x, y, z) \,\&\, x \leqslant_2 x' \,\&\, y' \leqslant_1 y \,\&\, z' \leqslant_2 z \implies S(x', y', z')$

    (MCP.5) $Q(x, y, z) \,\&\, x' \leqslant_1 x \,\&\, y \leqslant_2 y' \,\&\, z' \leqslant_2 z \implies Q(x', y', z')$

    (MCP.6) $I(x) \,\&\, x \leqslant_1 x' \implies I(x')$

Stability conditions:

    (SCP.1) $C(C(x)) = x$

    (SCP.2) $R(x, y, z) \implies \exists x'' \in X \, (x \leqslant_1 x'' \,\&\, S(x'', y, z))$

    (SCP.3) $R(x, y, z) \implies \exists y'' \in X \, (y \leqslant_1 y'' \,\&\, Q(x, y'', z))$

    (SCP.4) $S(x, y, z) \implies \exists z'' \in X \, (z \leqslant_2 z'' \,\&\, R(x, y, z''))$

    (SCP.5) $Q(x, y, z) \implies \exists z'' \in X \, (z \leqslant_2 z'' \,\&\, R(x, y, z''))$

    (SCP.6) $\exists u \in X (R(x, y, u) \,\&\, Q(x', u, z)) \implies \exists w \in X (R(x', x, w) \,\&\, S(w, y, z))$

    (SCP.7) $\exists u \in X (R(x, y, u) \,\&\, S(u, z, z')) \implies \exists w \in X (R(y, z, w) \,\&\, Q(x, w, z'))$

    (SCP.8) $I(x) \,\&\, (R(x, y, z) \text{ or } R(y, x, z)) \implies y \leqslant_1 z$

    (SCP.9) $\exists u \in X (I(u) \,\&\, S(u, x, x))$

    (SCP.10) $\exists u \in X (I(u) \,\&\, Q(x, u, x))$

    (SCP.11) $Q(x, y, z) \iff S(C(y), C(x), C(z))$.   □

In [1] there was no general concept of LCP frames. The results of [1] concern canonical frames and complex algebras of the canonical frames. In our approach canonical frames are examples of a general frame.

For an LCP frame $(X, \leqslant_1, \leqslant_2, C, R, S, Q, I)$ let us define the following mappings $^{\curlyvee} : 2^X \to 2^X$ and $\otimes_s$, $\otimes_Q$, $\boxtimes : 2^X \times 2^X \to 2^X$ by: for all $A, B \subseteq X$,

$$A^{\curlyvee} = \{C(x) : x \in A\} \tag{11}$$

$$A \otimes_Q B = \{z \in X : \forall x, y \in X (Q(x, y, z) \,\&\, x \in A \implies y \in r(B))\} \tag{12}$$

$$A \otimes_s B = \{z \in X : \forall x, y \in X (S(x, y, z) \,\&\, y \in B \implies x \in r(A))\} \tag{13}$$

$$A \boxtimes B = l(A \otimes_Q B). \tag{14}$$

Moreover, put

$$1' = l(r(I)). \tag{15}$$

The family $L(X)$ of all l–stable subsets of X is closed under the operations (11) and (14).

Lemma 7. [8] *Let* $(X, \leqslant_1, \leqslant_2, C, R, S, Q, I)$ *be an LCP frame. Then for all* $A, B \subseteq X$,

  (i)   *if A is l–stable, then so is $A^\curlyvee$*
  (ii) *if A and B are l–stable, then so is $A \boxtimes B$*
  (iii) *$1'$ is l–stable*
  (iv) *if A and B are l–stable, then $A \otimes_S B = A \otimes_Q B$.*   ∎

Definition 9. *The **complex algebra** of an LCP frame* $(X, \leqslant_1, \leqslant_2, C, R, S, Q, I)$ *is a system* $(L(X), \sqcap, \sqcup, {}^\curlyvee, \boxtimes, \mathbf{0}, \mathbf{1}, \mathbf{1}')$ *with the operations defined by (3)–(4), (11), (14) and the constants defined by (5), (6) and (15).*   □

Theorem 3. *The complex algebra of an LCP frame is an LCP algebra.*

Proof. In [8] it was shown that any complex algebra of an LCP frame satisfies the axioms **(CP.1)**–**(CP.7)**. Then it suffices to show that **(CP.0)** also holds, i.e. $\mathbf{0} \boxtimes A = A \boxtimes \mathbf{0} = \mathbf{0}$ for every $A \in L(X)$.

First, note that $l(L(X)) = \emptyset$ and $r(\mathbf{0}) = L(X)$. Next, since for every $A \subseteq X$ and for every $x, y, z \in X$ it holds $Q(x, y, z)$ & $x \in \emptyset \Longrightarrow y \in r(A)$, whence $\mathbf{0} \otimes_Q A = L(X)$. Therefore, $\mathbf{0} \boxtimes A = l(\mathbf{0} \otimes_Q A) = \mathbf{0}$. Moreover, from the definition of $\otimes_Q$ it is easily observed that $A \otimes_Q \mathbf{0} = L(X)$. Consequently, $A \boxtimes \mathbf{0} = l(A \otimes_Q \mathbf{0}) = \mathbf{0}$.   ∎

Let $(W, \wedge, \vee, \smile, \odot, 0, 1, 1')$ be an LCP algebra. We will write $FIP(X)$ (resp. $X(W)$) to denote the family of all filter–ideal pairs (resp. maximal filter–ideal pairs) of the lattice reduct of W. Note that since W is non–trivial, $X(W)$ is not empty.

Let us define a mapping $C^\star : FIP(X) \to FIP(X)$ by: for $x \in FIP(X)$,

$$C^\star(x) = (x_1{}^\smile, x_2{}^\smile). \tag{16}$$

Moreover, let us define the following three ternary relations on $X(W)$ by: for all $x, y, z \in X(W)$,

$$R^\star(x, y, z) \iff (\forall a, b \in W)\; a \in x_1 \;\&\; b \in y_1 \Longrightarrow a \otimes b \in z_1 \tag{17}$$

$$S^\star(x, y, z) \iff (\forall a, b \in W)\; a \otimes b \in z_2 \;\&\; b \in y_1 \Longrightarrow a \in x_2 \tag{18}$$

$$Q^\star(x, y, z) \iff (\forall a, b \in W)\; a \otimes b \in z_2 \;\&\; a \in x_1 \Longrightarrow b \in y_2 \tag{19}$$

Also, let

$$I^\star = \{x \in X(W) : 1' \in x_1\}. \tag{20}$$

We extend the operation $\otimes$ for subsets of X in the following way: for all $A, B \subseteq W$,

$$A \otimes B = \{a \otimes b : a \in A, b \in B\}.$$

Then it is straightforward to see that for all $x, y, z \in X(W)$,

$$R^\star(x, y, z) \quad \Longleftrightarrow \quad x_1 \otimes y_1 \subseteq z_1 \tag{21}$$

$$S^\star(x, y, z) \quad \Longleftrightarrow \quad -x_2 \otimes y_1 \subseteq -z_2 \tag{22}$$

$$Q^\star(x, y, z) \quad \Longleftrightarrow \quad x_1 \otimes -y_2 \subseteq -z_2. \tag{23}$$

In [8] we showed that for $x \in X(W)$, $C^\star(x) \in X(W)$.

Definition 10. *Let an LCP algebra* $(W, \wedge, \vee, \smile, \otimes, 0, 1, 1')$ *be given. The system* $(X(W), \preccurlyeq_1, \preccurlyeq_2, C^\star, R^\star, S^\star, Q^\star, I^\star)$ *is called the* **canonical frame of** W. $\square$

The following auxiliary lemma will be useful.

Lemma 8. [8] *Let* $(W, \wedge, \vee, \otimes, \smile, 0, 1, 1')$ *be an LCP algebra and let* Δ *and* ∇ *be a filter and an ideal of* W, *respectively. Then the set*

$$V = \{a \in W : (\{a\} \otimes \Delta) \cap \nabla \neq \emptyset\}$$

is an ideal of W.   ∎

In the following theorem we show that canonical frames satisfy the postulates assumed for the LCP frames. We only give a few exemplary proofs which were not given in [8].

Theorem 4. *The canonical frame of an LCP algebra is an LCP frame.*

Proof. Let an LCP algebra $(W, \wedge, \vee, \smile, \otimes, 0, 1, 1')$ be given and let $(X(W), \preccurlyeq_1, \preccurlyeq_2, C^\star, R^\star, S^\star, Q^\star, I^\star)$ be its canonical frame. Proceeding as in [1] one can prove that **(MCP.3)**–**(MCP.5)** and **(SCP.2)**–**(SCP.7)** hold in the canonical frame.

We show now that **(MCP.1)** is satisfied. Let $x, y \in X(W)$ be such that $x \preccurlyeq_1 y$. This means that (i) $x_1 \subseteq y_1$. Also, $C^\star(x) = (x_1^\smile, x_2^\smile)$. By Lemma 6, (i) is equivalent with $x_1^\smile \subseteq y_1^\smile$, so $C^\star(x) \preccurlyeq_1 C^\star(y)$. In the analogous way we can show that **(MCP.2)** holds.

Next we prove that **(MCP.6)** is satisfied. Let $x, x' \in X(W)$ and assume that $I^\star(x)$ and $x \preccurlyeq_1 x'$ hold. From (20) we immediately get $1' \in x_1 \subseteq x_1'$, so $I^\star(x')$ holds.

Furthermore, we show that **(SCP.1)** holds. For every $x = (x_1, x_2) \in X(W)$, we have: $C^\star(C^\star(x)) = C^\star(x_1^\smile, x_2^\smile) = (x_1^{\smile\smile}, x_2^{\smile\smile}) = (x_1, x_2) = x$ by Lemma 6(ii).

Consider now the condition **(SCP.8)**. Assume that for any $x, y, z \in X(W)$, $I^\star(x)$ holds, i.e. (ii) $1' \in x_1$, and $R^\star(x, y, z)$ or $R^\star(y, x, z)$. Let $R^\star(x, y, z)$ holds. Hence, by (ii), we get $(\forall b \in W) \, b \in y_1 \Rightarrow 1' \otimes b \in z_1$. Since $1' \otimes b = b$, we get $y_1 \subseteq z_1$, that is $y \preccurlyeq_1 z$. If $R^\star(y, x, z)$ holds, then again by (ii) we get $(\forall a \in W) \, a \in y_1 \Rightarrow a \otimes 1' \in z_1$, so since $a \otimes 1' = a$, we obtain again $y_1 \subseteq z_1$, i.e. $y \preccurlyeq_1 z$.

We show now that **(SCP.9)** holds. Let $y \in X(W)$ and consider the set $V = \{a \in W : (\{a\} \otimes y_1) \cap y_2 \neq \emptyset\}$. By Lemma 8, V is an ideal of W. Let $[1']$ be the filter generated by $1'$. We show that $[1'] \cap V = \emptyset$. Suppose that there exists

$a \in W$ such that **(iii)** $a \in [1')$ and **(iv)** $a \in V$. From **(iii)** it follows that **(v)** $1' \leqslant a$. Also, **(iv)** implies that there exists $b \in W$ such that **(vi)** $b \in y_1$ and **(vii)** $a \otimes b \in y_2$. Since $\otimes$ is isotone in both arguments, **(v)** implies $1' \otimes b \leqslant a \otimes b$. But $1' \otimes b = b$, so we have $b \leqslant a \otimes b$, which in view of **(vii)** and the fact that y_2 is an ideal gives $b \in y_2$ – a contradiction with **(vi)**.

Then $([1'), V)$ is a filter–ideal pair. Let $u = (u_1, u_2)$ be its extension to the maximal pair. Therefore, $[1') \subseteq u_1$ and $V \subseteq u_2$. Since $1' \in [1')$, we get $1' \in u_1$, so $I^\star(u)$ holds. We show now that $S^\star(u, y, y)$ holds. Let $a, b \in W$ be such that $a \otimes b \in y_2$ and $b \in y_1$. Then $a \in V$, so $a \in u_2$. Whence $S^\star(u, y, y)$ holds.

In the similar way one can check that **(SCP.10)** holds.

Finally we show that **(SCP.11)** holds. Using the axiom **(CP.7)** and the definition (10), we have for all $x, y, z \in X(W)$,

$$
\begin{aligned}
S^\star(C^\star(y), C^\star(x), C^\star(z)) \quad &\text{iff} \quad (\forall a, b \in W)\, a \otimes b \in z_2^\smile \;\&\; b \in x_1^\smile \implies a \in y_2^\smile \\
&\text{iff} \quad (\forall a, b \in W)\, (a \otimes b)^\smile \in z_2 \;\&\; b^\smile \in x_1 \implies a^\smile \in y_2 \\
&\text{iff} \quad (\forall a, b \in W)\, b^\smile \otimes a^\smile \in z_2 \;\&\; b^\smile \in x_1 \implies a^\smile \in y_2 \\
&\text{iff} \quad (\forall c, d \in W)\, c \otimes d \in z_2 \;\&\; c \in x_1 \implies d \in y_2 \\
&\text{iff} \quad Q^\star(x, y, z).
\end{aligned}
$$

This completes the proof.   ∎

We conclude this section by stating the representability of LCP algebras.

Theorem 5. *Every LCP algebra is isomorphic to a subalgebra of the complex algebra of its canonical frame.*

Proof. See [8].   ∎

In the axiomatization of relation algebras, apart from the axioms for Boolean algebras, the only axiom which contains complementation is

$$
a \otimes -(a^\smile \otimes -b) \leqslant b.
$$

This axiom is equivalent to the De Morgan equivalences

$$
(a \otimes b) \wedge c = 0 \iff (a^\smile \otimes c) \wedge b = 0 \iff (c \otimes b^\smile) \wedge a = 0 \tag{24}
$$

and could be added to the LCP axioms. However, we showed in [8] that adding (24) does not add anything new. An alternative is the *modular inequality*

$$
(a \otimes b) \wedge c \leqslant a \otimes (b \wedge (a^\smile \otimes c)). \tag{25}
$$

(25) is true for relation algebras and is also an axiom for rough relation algebras ([5]), i.e., relation algebras based on regular double Stone algebras. One consequence of (25) is that for every $a < 1'$ we have $a \otimes 1 < 1$ (here $a < b$ means $a \leqslant b$

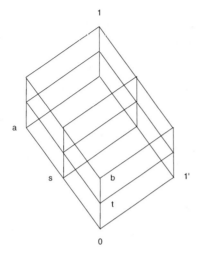

Table 1. Composition Table

$\otimes$	a	b	s	t
a	a	1	a	$t \vee 1' \vee a$
b	1	b	1	b
s	a	$s \vee 1' \vee b$	s	$s \vee t \vee 1'$
t	1	b	$s \vee t \vee 1'$	t

Fig. 1. An LCP–algebra where (25) fails

and $a \neq b$). The following example from [9] shows that not every LCP–algebra satisfies (25).

Example 1. Consider the algebra L of Fig.1. By **(CP.2)** and **(CP.5)** it is enough to define how composition and converse act on the join irreducible elements. These are $1', a, b, s, t$, and we set $a^\smile = b$, $s^\smile = t$. Composition for the non–identity irreducible elements is given in Table 1. Now consider

$$(t \otimes a) \wedge b = b \qquad \text{since } t \otimes a = 1$$
$$\not\leqslant t \otimes s \qquad \text{from the composition table}$$
$$= t \otimes [(s \vee t \vee 1') \wedge a] \qquad \text{from the lattice ordering}$$
$$= t \otimes [(s \otimes b) \wedge a]$$
$$= t \otimes [(t^\smile \otimes b) \wedge a].$$

So we may want the following inequality as an additional axiom of LCP algebras:

(CP.8) $(a \otimes b) \wedge c \leqslant a \otimes (b \wedge (a^\smile \otimes c))$.

To obtain a representation theorem for LCP algebras with **(CP.8)** is still an open problem.

The next example illustrates the constructions employed in the proof of the representation theorem.

Example 2. Consider an algebra $(W, \wedge, \vee, \smile, \otimes, 0, 1, 1')$ with $W = \{a, b, c, , 0, 1\}$, $\wedge$ and $\vee$ as on Fig.2, $a^\smile = a$ for every $a \in W$, $\otimes$ in given in Table 2, and $1' = c$. The maximal filter–ideal pairs of W are

$$x = ([a), (b]), \qquad y = ([b), (c]), \qquad z = ([c), (a]).$$

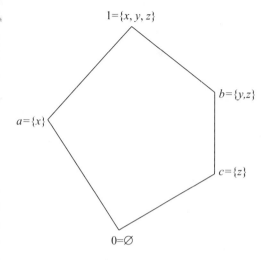

1=\{x, y, z\}

b=\{y,z\}

a=\{x\}

c=\{z\}

0=\varnothing

Fig. 2. The pentagon

Table 2. The product $\otimes$

$\otimes$	a	b	c	1
a	1	a	a	1
b	a	b	b	1
c	a	b	c	1
1	1	1	1	1

Let us find R, Q, and S. We can simplify the calculations by observing that $A \otimes B = B \otimes A$ for any $A, B \subseteq W$, since $\otimes$ is symmetric on W.
$R(x, y, z)$ iff $x_1 \otimes y_1 \subseteq z_1$:

$\quad R(x, x, v) \quad x_1 \otimes x_1 = \{1\}$, and $\{1\} \subseteq v_1$ for all $v \in FIP(W)$.
$\quad R(x, y, v) \quad x_1 \otimes y_1 = \{a, 1\}$, and $\{a, 1\} \subseteq v_1$ only for $v = x$.
$\quad R(x, z, v) \quad x_1 \otimes z_1 = \{a, 1\}$, and $\{a, 1\} \subseteq v_1$ only for $v = x$.
$\quad R(y, y, v) \quad y_1 \otimes y_1 = \{b, 1\}$, and $\{b, 1\} \subseteq v_1$ for $v \in \{y, z\}$.
$\quad R(y, z, v) \quad y_1 \otimes z_1 = \{b, 1\}$, and $\{b, 1\} \subseteq v_1$ for $v \in \{y, z\}$.
$\quad R(z, z, v) \quad z_1 \otimes z_1 = \{c, 1\}$, and $\{c, 1\} \subseteq v_1$ for $v = z$.

$S(x, y, z)$ iff $(-x_2 \otimes y_1) \cap z_2 = \emptyset$:

$\quad S(x, x, v) \quad -x_2 \otimes x_1 = \{1\}$, and $\{1\} \cap v_2 = \emptyset$ for all $v \in FIP(W)$.
$\quad S(x, y, v) \quad -x_2 \otimes y_1 = \{a, 1\}$, and $\{a, 1\} \cap v_2 = \emptyset$ for $v \in \{x, y\}$.
$\quad S(x, z, v) \quad -x_2 \otimes z_1 = \{a, 1\}$, and $\{a, 1\} \cap v_2 = \emptyset$ for $v \in \{x, y\}$.
$\quad S(y, x, v) \quad -y_2 \otimes x_1 = \{a, 1\}$, and $\{a, 1\} \cap v_2 = \emptyset$ for $v \in \{x, y\}$.

$\quad S(y, y, v) \quad -y_2 \otimes y_1 = \{a, b, 1\}$, and $\{a, 1\} \cap v_2 = \emptyset$ for $v \in \{y, z\}$.
$\quad S(y, z, v) \quad -y_2 \otimes z_1 = \{a, b, 1\}$, and $\{a, 1\} \cap v_2 = \emptyset$ for $v \in \{y, z\}$.
$\quad S(z, x, v) \quad -z_2 \otimes x_1 = \{a, 1\}$, and $\{a, 1\} \cap v_2 = \emptyset$ for $v \in \{x, y\}$.
$\quad S(z, y, v) \quad -z_2 \otimes y_1 = \{b, 1\}$, and $\{b, 1\} \cap v_2 = \emptyset$ for $v \in \{y, z\}$.
$\quad S(z, z, v) \quad -z_2 \otimes z_1 = \{b, c, 1\}$, and $\{b, c, 1\} \cap v_2 = \emptyset$ only for $v = z$.

$Q(x,y,z)$ iff $(x_1 \otimes -y_2) \cap z_2 = \emptyset$:

$\quad Q(x,x,v) \quad x_1 \otimes -x_2 = \{a,1\}$ and $\{a,1\} \cap v_2 = \emptyset$ for $v \in \{x,y\}$.

$\quad Q(x,y,v) \quad x_1 \otimes -y_2 = \{a,1\}$ and $\{a,1\} \cap v_2 = \emptyset$ for $v \in \{x,y\}$.

$\quad Q(x,z,v) \quad x_1 \otimes -z_2 = \{a,1\}$ and $\{a,1\} \cap m_2 = \emptyset$ for $v \in \{x,y\}$.

$\quad Q(y,x,v) \quad y_1 \otimes -x_2 = \{a,1\}$ and $\{a,1\} \cap v_2 = \emptyset$ for $v \in \{x,y\}$.

$\quad Q(y,y,v) \quad y_1 \otimes -y_2 = \{a,b,1\}$ and $\{a,b,1\} \cap v_2 = \emptyset$ for $v \in \{y,z\}$.

$\quad Q(y,z,v) \quad y_1 \otimes -z_2 = \{b,1\}$ and $\{a,b,1\} \cap v_2 = \emptyset$ for $v \in \{y,z\}$.

$\quad Q(z,x,v) \quad z_1 \otimes -x_2 = \{a,1\}$ and $\{a,1\} \cap v_2 = \emptyset$ for $m \in \{x,y\}$.

$\quad Q(z,y,v) \quad z_1 \otimes -y_2 = \{a,b,1\}$ and $\{a,b,1\} \cap v_2 = \emptyset$ for $v \in \{y,z\}$.

$\quad Q(z,z,v) \quad z_1 \otimes -z_2 = \{b,c,1\}$ and $\{b,c,1\} \cap v_2 = \emptyset$ only for $v = z$.

The embedding h is given by

$$h(0) = \emptyset \qquad\qquad h(a) = \{x\} \qquad\qquad h(c) = \{z\}.$$
$$h(1) = \{x,y,z\} \qquad h(b) = \{y,z\}$$

We conclude this section with the observation that the diamond lattice of Figure 3 cannot be made into an LCP algebra. We omit the proof which is straightforward, if somewhat tedious.

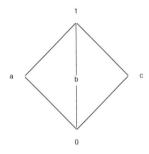

Fig. 3. The diamond lattice

5 LCPR Algebras and Frames

In this section we extend LCP algebras by adding the residuation operations. In classical relation algebras residuations are definable with composition (;), converse ($\smile$) and complement($-$) as $x/y = -(y^\smile; -x)$ and $y \backslash x = -(-x; y^\smile)$.

Definition 11. *By an **LCPR algebra** we mean a system $(W, \wedge, \vee, \smile, \otimes, \rightarrow, \leftarrow, 0, 1, 1')$ such that $(W, \wedge, \vee, \smile, \otimes, 0, 1, 1')$ is an LCP algebra and $\rightarrow$ and $\leftarrow$ are binary operations in W satisfying the following conditions for all $a, b, c \in W$,*

$\quad$ **(CPR.1)** $a \otimes b \leqslant c$ *iff* $b \leqslant a \rightarrow c$
$\quad$ **(CPR.2)** $a \otimes b \leqslant c$ *iff* $a \leqslant c \leftarrow b$.

The operations $\leftarrow$ and $\rightarrow$ are called the left and the right residuum of $\otimes$, respectively. $\qquad\square$

Note that an LCPR algebra is an extension of a residuated lattice by the converse $\smile$ operation.

The following lemma provides some basic properties of LCPR algebras.

Lemma 9. *Let* $(W, \wedge, \vee, \smile, \otimes, \rightarrow, \leftarrow, 0, 1, 1')$ *be an LCPR algebra. Then for any* $a, b, c \in W$ *and for every indexed family* $(b_i)_{i \in I}$ *of elements of* W,

(i) *if* $a \leqslant b$, *then*

$$c \otimes a \leqslant c \otimes b \text{ and } a \otimes c \leqslant b \otimes c$$
$$b \rightarrow c \leqslant a \rightarrow c \text{ and } c \rightarrow a \leqslant c \rightarrow b$$
$$a \leftarrow c \leqslant b \leftarrow c \text{ and } c \leftarrow b \leqslant c \leftarrow a$$

(ii) $a \leqslant b \text{ iff } a^{\smile} \leqslant b^{\smile}$

(iii) $(a \wedge b)^{\smile} = a^{\smile} \wedge b^{\smile}$

(iv) $a \otimes (a \rightarrow b) \leqslant b$ (iv') $(b \leftarrow a) \otimes a \leqslant b$

(v) $(a \rightarrow b) \otimes (b \rightarrow c) \leqslant a \rightarrow c$ (v') $(a \leftarrow b) \otimes (b \leftarrow c) \leqslant \leftarrow c$

(vi) $b \leqslant a \rightarrow (a \otimes b)$ (vi') $a \leqslant (a \otimes b) \leftarrow b$

(vii) $(a \rightarrow b)^{\smile} = b^{\smile} \leftarrow a^{\smile}$ (vii') $(a \leftarrow b)^{\smile} = b^{\smile} \rightarrow a^{\smile}$

(viii) *if* $\sup_i b_i$ *exists, then*

$$a \otimes \sup_i b_i = \sup_i (a \otimes b_i)$$
$$\sup_i b_i \otimes a = \sup_i (b_i \otimes a)$$

(ix) *if* $\inf_i b_i$ *exists, then*   (ix') *if* $\inf_i b_i$ *exists, then*

$$a \rightarrow \inf_i b_i = \inf_i (a \rightarrow b_i) \quad\quad \inf_i b_i \leftarrow a = \inf_i (b_i \leftarrow a)$$

(x) *if* $\sup_i b_i$ *exists, then*   (x') *if* $\sup_i b_i$ *exists, then*

$$\sup_i b_i \rightarrow a = \inf_i (b_i \rightarrow a) \quad\quad a \leftarrow \sup_i b_i = \inf_i (a \leftarrow b_i).$$

Proof. By way of example we prove **(vii)**

Let $c \in W$ such that $c \leqslant (a \rightarrow b)^{\smile}$. Then we have:

$$
\begin{array}{llll}
c \leqslant (a \rightarrow b)^{\smile} & \text{iff} & c^{\smile} \leqslant (a \rightarrow b) & \text{by } \textbf{(ii)}, \textbf{(CP.1)} \\
& \text{iff} & a \otimes c^{\smile} \leqslant b & \text{by } \textbf{(CPR.1)} \\
& \text{iff} & (a \otimes c^{\smile})^{\smile} \leqslant b^{\smile} & \text{by } \textbf{(ii)} \\
& \text{iff} & c \otimes a^{\smile} \leqslant b^{\smile} & \text{by } \textbf{(CP.1)}, \textbf{(CP.7)} \\
& \text{iff} & c \leqslant b^{\smile} \leftarrow a^{\smile} & \text{by } \textbf{(CPR.2)}.
\end{array}
$$
∎

For the recent development of residuated lattices we refer, for example, to [2], [11], [12], and [16].

LCPR frames are the same as LCP frames defined in Section 3 (Definition 8).

Let an LCPR frame $(X, \leqslant_1, \leqslant_2, C, R, S, Q, I)$ be given. We define the following two mappings $\rightarrow\!\!\!\triangleright, \triangleleft\!\!\!\leftarrow : 2^X \times 2^X \rightarrow 2^X$ as follows: for all $A, B \subseteq X$,

$$A \rightarrow\!\!\!\triangleright B = \{x \in X : (\forall y, z \in X)(R(y, x, z) \ \& \ y \in A \Longrightarrow z \in B)\} \quad (26)$$
$$B \triangleleft\!\!\!\leftarrow A = \{x \in X : (\forall y, z \in X)(R(x, y, z) \ \& \ y \in A \Longrightarrow z \in B)\}. \quad (27)$$

Lemma 10. *For any* $A, B \subseteq X$,

   (i) $A \dashrightarrow B$ *und* $A \lhd B$ *are* $\leqslant_1$*-increasing*

   (ii) *if* A *and* B *are* l*-stable, then so are* $A \dashrightarrow B$ *and* $A \lhd\!\!- B$.

Proof.

(i) Assume that for some $A, B \subseteq X$, $A \dashrightarrow B$ is not $\leqslant_1$–increasing. Then there are $x, y \in X$ such that **(i.1)** $x \in A \dashrightarrow B$ **(i.2)** $x \leqslant_1 y$, and **(i.3)** $y \notin A \dashrightarrow B$. From **(i.3)**, by the definition (26), there exist $u, w \in X$ such that **(i.4)** $R(u, y, w)$ **(i.5)** $u \in A$ and **(i.6)** $w \notin B$. Next, by **(i.2)**, **(i.4)** and the monotonicity condition **(MCP.3)**, we get $R(u, x, w)$, which together with **(i.5)** and **(i.6)** gives $x \notin A \dashrightarrow B$ – a contradiction with **(i.1)**.

   Proceeding in the similar way one can show that $A \lhd\!\!- B$ is $\leqslant_1$–increasing.

(ii) Let $A, B \subseteq X$. We show first that $A \lhd\!\!- B$ is l–stable.

By (i), $A \lhd\!\!- B$ is $\leqslant_1$–increasing, so from Lemma 4(ii), $A \lhd\!\!- B \subseteq l(r(A \lhd\!\!- B))$. Then it suffices to show that $l(r(A \lhd\!\!- B)) \subseteq A \lhd\!\!- B$.

Let $x \in X$ be such that **(ii.1)** $x \notin A \lhd\!\!- B$. We will show **(ii.2)** $x \notin l(r(A \lhd\!\!- B))$. From **(ii.1)**, by the definition (27) it follows that there exist $y, z \in X$ such that **(ii.3)** $R(x, y, z)$ **(ii.4)** $y \in B$ **(ii.5)** $z \notin A$. Since B is l–stable, **(ii.5)** means that $z \notin l(r(A))$, so there exists $z' \in X$ such that **(ii.6)** $z \leqslant_1 z'$ and **(ii.7)** $z' \in r(A)$. From **(ii.3)**, **(ii.6)** and the monotonicity condition **(MCP.3)**, $R(x, y, z')$, which by the stability condition **(SCP.1)** implies that there is $x' \in X$ such that **(ii.8)** $x \leqslant_1 x'$ and **(ii.9)** $S(x', y, z')$. We show now that **(ii.10)** $x' \in r(A \lhd\!\!- B)$. This, by **(ii.8)**, gives **(ii.2)**.

   Consider an arbitrary $x'' \in X$ satisfying **(ii.11)** $x' \leqslant_2 x''$. By **(MCP.4)**, **(ii.9)** and **(ii.11)** lead to $S(x'', y, z')$, which by **(SCP.3)** gives that there exists $z'' \in X$ such that **(ii.12)** $z' \leqslant_2 z''$ and **(ii.13)** $R(x'', y, z'')$. From Lemma 1(ii), $r(B)$ is $\leqslant_2$–increasing, so by **(ii.7)** and **(ii.12)** we get $z'' \in r(A)$, whence **(ii.14)** $z'' \notin A$. In view of the definition (27), **(ii.4)**, **(ii.13)** and **(ii.14)** imply **(ii.15)** $x'' \notin A \lhd\!\!- B$. Therefore, we have shown that for any $x'' \in X$ satisfying **(ii.11)**, the condition **(ii.15)** holds, hence **(ii.10)** was proved.

   Using the relation Q in place of S, in the analogous way we can show that $A \dashrightarrow B$ is l–stable.   □

Definition 12. *The **complex algebra** of an LCPR frame* $(X, \leqslant_1, \leqslant_2, C, R, S, Q, I)$ *is a structure* $(L(X), \sqcap, \sqcup, {}^\curlyvee, \boxtimes, \dashrightarrow, \lhd\!\!-, \mathbf{0}, \mathbf{1}, \mathbf{1'})$ *with the operations defined by (3), (4), (11), (14), (26), (27) and the constants (5), (6), and (15).*   □

We show that the complex algebra of an LCPR frame is an LCPR algebra. It is sufficient to show the following lemma.

Lemma 11. *For any LCPR frame* $(X, \leqslant_1, \leqslant_2, C, R, S, Q, I)$ *and for all* l*–stable subsets* $A, B, C \subseteq X$,

   (i) $A \boxtimes B \subseteq C$ *iff* $B \subseteq A \dashrightarrow C$

   (ii) $A \boxtimes B \subseteq C$ *iff* $A \subseteq C \lhd\!\!- B$.

Proof.

(i) ($\Leftarrow$) Assume that **(i.1)** $A \boxtimes B \subseteq C$ and **(i.2)** $B \not\subseteq A \dashrightarrow C$. From **(i.2)**, there exists $x \in X$ such that **(i.3)** $x \in B$ and **(i.4)** $x \notin A \dashrightarrow C$. By the definition (26), **(i.4)** means that for some $y, z \in X$ it holds **(i.5)** $R(y, x, z)$, **(i.6)** $y \in A$ and **(i.7)** $z \notin C$. Next, from **(i.1)** and **(i.7)** we get **(i.8)** $x \notin A \boxtimes B$. By the definition (14), $A \boxtimes B = l(A \otimes_Q B)$, but from Lemma 7**(iv)**, $A \otimes_Q B = A \otimes_S B$. Then we get **(i.8)** implies that there exist $z' \in X$ such that **(i.9)** $z \leqslant_1 z'$ and **(i.10)** $z' \in A \otimes_S B$. Furthermore, from **(i.5)** and **(i.9)**, by **(M.3)** we get $R(y, x, z')$, which by **(S.1)** implies that there is $y' \in X$ such that **(i.11)** $y \leqslant_1 y'$ and **(i.12)** $S(y', x, z')$. Also, **(i.3)**, **(i.10)** and **(i.12)** imply $y' \in r(A)$, which together with **(i.11)** gives $y \notin l(r(A))$. Since A is l–stable, this means $y \notin A$ – a contradiction with **(i.6)**.

($\Rightarrow$) Assume that **(i.13)** $A \boxtimes B \not\subseteq C$. We will show that **(i.14)** $B \not\subseteq A \dashrightarrow C$.

From **(i.13)**, there is $x \in X$ such that **(i.15)** $x \in A \boxtimes B$ and **(i.16)** $x \notin C$. Since C is l–stable, **(i.16)** gives $x \notin l(r(C))$, so there exists $x' \in X$ such that **(i.17)** $x \leqslant_1 x'$ and **(i.18)** $x' \in r(C)$. Next, from **(i.15)**, **(i.17)**, and Lemma 7**(iv)**, $x' \notin A \otimes_S B$, which means that there are $y, z \in X$ such that **(i.19)** $S(y, z, x')$, **(i.20)** $z \in B$ and **(i.21)** $y \notin r(A)$. From **(i.21)**, there is $y' \in X$ such that **(i.22)** $y \leqslant_2 y'$ and **(i.23)** $y' \in A$. By **(M.4)**, **(i.19)** and **(i.22)** imply $S(y', z, x')$. Hence, applying **(S.3)** we get that for some $x'' \in X$ such that **(i.24)** $x' \leqslant_2 x''$ it holds **(i.25)** $R(y', z, x'')$. Furthermore, by **(i.18)** and **(i.24)** it follows that $x'' \notin C$, which together with **(i.23)** and **(i.25)** gives $z' \notin A \dashrightarrow C$. Whence, in view of **(i.20)** we finally obtain **(i.14)**.

In the analogous way **(ii)** can be proved. $\square$

Therefore, we have

Theorem 6. *The complex algebra of an LCPR frame is an LCPR algebra.* $\square$

Since LCPR frames are just LCP frames, the above theorem implies the following

Corollary 2. *Any LCP algebra can be isomorphically embedded into an LCPR algebra.* $\square$

Let $(W, \wedge, \vee, \otimes, \rightarrow, \leftarrow, 0, 1)$ be an LCPR algebra. For any two subsets $A, B \subseteq W$, let us define:

$$A \leftarrow B = \{a \leftarrow b : a \in A \ \& \ b \in B\}$$
$$A \rightarrow B = \{a \rightarrow b : a \in A \ \& \ b \in B\}.$$

Lemma 12. *Let $(W, \wedge, \vee, \otimes, \rightarrow, \leftarrow, 0, 1)$ be an LCPR algebra and let Δ and Δ' be filters of W and let ∇ be an ideal of W. Define the following subsets of W:*

$$U = \{a \in W : \Delta \cap (\nabla \leftarrow \{a\} \neq \emptyset\}$$
$$U' = \{a \in W : \Delta \cap (\{a\} \rightarrow \nabla) \neq \emptyset\}$$
$$V = \{a \in W : \Delta \cap (\{a\} \leftarrow \Delta') \neq \emptyset\}$$
$$V' = \{a \in W : \Delta \cap (\Delta' \rightarrow \{a\}) \neq \emptyset\}.$$

Then U and U' are ideals of W and V and V' are filters of W.

Proof. By way of example we show that U is an ideal of W. Let $a, b \in W$ be such that (i) $a \in U$ and (ii) $b \leqslant a$. By the definition of U, (i) implies that there exists $c \in \nabla$ such that (iii) $c \leftarrow u \in \Delta$. By Lemma 9(i) we get from (ii) that $c \leftarrow a \leqslant c \leftarrow b$. Hence, by (iii), we get (iv) $c \leftarrow b \in \Delta$, since Δ is a filter. Therefore, for some $c \in \nabla$ (iv) holds, which implies $b \in U$.

Assume that (v) $a, b \in U$. It suffices to show that $a \vee b \in U$. From (v), there are $c, d \in \nabla$ such that (vi) $c \leftarrow a \in \Delta$ and (vii) $d \leftarrow b \in \Delta$. Since $c \leqslant c \vee d$ and $d \leqslant c \vee d$, by Lemma 9(i) we get $c \leftarrow a \leqslant (c \vee d) \leftarrow a$ and $d \leftarrow b \leqslant (c \vee d) \leftarrow b$. Hence, by (vi) and (vii) it follows that $(c \vee d) \leftarrow a \in \Delta$ and $(c \vee d) \leftarrow b \in \Delta$, so $((c \vee d) \leftarrow a) \wedge ((c \vee d) \leftarrow b) \in \Delta$. By Lemma 9(x'), $((c \vee d) \leftarrow a) \wedge ((c \vee d) \leftarrow b) = (c \vee d) \leftarrow (a \vee b)$. Then $(c \vee d) \leftarrow (a \vee b) \in \Delta$. Since $c, d \in \nabla$, $c \vee d \in \nabla$. So we get that for some $e = c \vee d \in \nabla$, $e \leftarrow (a \vee b) \in \Delta$, which gives $a \vee b \in U$. ∎

The canonical frame of an LCPR algebra is the same as the canonical frame of an LCP algebra (Definition 10), i.e., it is a system $(X(W), \preccurlyeq_1, \preccurlyeq_2, C^\star, R^\star, S^\star, Q^\star, I^\star)$. Given the canonical frame of an LCPR algebra, define the following auxiliary ternary relations on $X(W)$: for all $x, y, z \in X(W)$,

$$R^\star_{\leftarrow}(x, y, z) \quad \textit{iff} \quad (\forall a, b \in W) \, b \leftarrow a \in x_1 \, \& \, a \in y_1 \Longrightarrow b \in z_1 \tag{28}$$

$$R^\star_{\rightarrow}(x, y, z) \quad \textit{iff} \quad (\forall a, b \in W) \, a \in x_1 \, \& \, a \rightarrow b \in y_1 \Longrightarrow b \in z_1. \tag{29}$$

Note that

Lemma 13. $R^\star = R^\star_{\leftarrow} = R^\star_{\rightarrow}$

Proof. We show that $R^\star = R^\star_{\leftarrow}$. The proof of $R^\star = R^\star_{\rightarrow}$ is analogous.

($\subseteq$) Assume on the contrary that for some $x, y, z \in X(W)$, (i) $R^\star(x, y, z)$ and there exist $a, b \in W$ such that (ii) $b \leftarrow a \in x_1$ (iii) $a \in y_1$ (iv) $b \notin z_1$. From (i), (ii) and (iii) it follows that $(b \leftarrow a) \otimes a \in z_1$. By Lemma 9(iv'), $(b \leftarrow a) \otimes a \leqslant b$. Since z_1 is a filter, this implies $b \in z_1$ – a contradiction with (iv).

($\supseteq$) Similarly, assume that for some $x, y, z \in X(W)$, (v) $R^\star_{\leftarrow}(x, y, z)$ and there exist $a, b \in W$ such that (vi) $a \in x_1$, (vii) $b \in y_1$ and (viii) $a \otimes b \notin z_1$. By Lemma 9(vi'), $a \leqslant (a \otimes b) \leftarrow b$, so from (vi), $(a \otimes b) \leftarrow b \in x_1$, since x_1 is a filter. By (v) this gives $a \otimes b \in z_1$ – a contradiction with (viii). ∎

Theorem 7 (Representation theorem for LCPR algebras). *Any LCPR algebra is isomorphic to a subalgebra of the complex algebra of its canonical frame.*

Proof. In view of Theorem 5 it suffices to show that

(i) $h(a \leftarrow b) = h(a) \mathbin{\triangleleft\!-} h(b)$

(ii) $h(a \rightarrow b) = h(a) \mathbin{-\!\triangleright} h(b)$.

(i) ($\subseteq$) Let $x \in h(a \leftarrow b)$. By the definition (9) of the mapping h, this means that (i.1) $a \leftarrow b \in x_1$. Assume that $x \notin h(a) \mathbin{\triangleleft\!-} h(b)$. Then there are $y, z \in X(W)$ such that (i.2) $R^\star(x, y, z)$, (i.3) $y \in h(b)$ and (i.4) $z \notin h(a)$. From (i.3) we get (i.5) $b \in y_1$. By Lemma 13, $R^\star = R^\star_{\leftarrow}$, so from (i.1), (i.2), (i.5) and the definition of $R^\star_{\leftarrow}$, it follows $a \in z_1$, i.e. $z \in h(a)$, which contradicts (i.4).

($\supseteq$) Assume that **(i.6)** $x \notin h(b \leftarrow a)$. We will show that $x \notin h(b) \vartriangleleft- h(a)$. From **(i.6)** we have **(i.7)** $b \leftarrow a \notin x_1$. Define

$$U = \{c \in W : x_1 \cap ((b] \leftarrow \{c\}) \neq \emptyset\},$$

where $(b]$ stands for the ideal generated by b. By Lemma 12, U is an ideal. Suppose that $a \in U$. Then there exists $b' \in W$ such that **(i.8)** $b' \leqslant b$ and **(i.9)** $b' \leftarrow a \in x_1$. By Lemma 9(**iii'**) and **(i.8)** we get **(i.10)** $b' \leftarrow a \leqslant b \leftarrow a$. Since x_1 is a filter, **(i.9)** and **(i.10)** imply $b \leftarrow a \in x_1$, which contradicts **(i.7)**. Hence $a \notin U$. Let $[a)$ be the filter generated by a. Then $[a) \cap U = \emptyset$, so $([b), U)$ is a filter–ideal pair. Let (y_1, y_2) be its extension to the maximal filter–ideal pair. Then $[a) \subseteq y_1$ and $U \subseteq y_2$. Since $a \in y_1$, we have **(i.11)** $y \in h(a)$.
Now, consider a set:

$$V = \{c \in W : x_1 \cap (\{c\} \leftarrow y_1) \neq \emptyset\}.$$

By Lemma 12, V is a filter of W. Suppose that $b \in V$. Then there is $c' \in W$ such that **(i.12)** $c' \in y_1$ and **(i.13)** $b \leftarrow c' \in x_1$. By the definition of U, **(i.13)** implies $c' \in U \subseteq y_2$ – a contradiction with **(i.12)**. Hence $b \notin V$. Then $(V, (b])$ is a filter–ideal pair. Let (z_1, z_2) be its extension to the maximal filter–ideal pair. Then **(i.14)** $V \subseteq z_1$ and $(b] \subseteq z_2$. Since $b \in z_2$, we get $b \notin z_1$, so **(i.15)** $z \notin h(b)$.

Finally, consider $c, d \in W$ such that $c \leftarrow d \in x_1$ and $d \in y_1$. Then $c \in V$, so $c \in z_1$ by **(i.14)**. By the definition (28), $R^\star_{-}(x, y, z)$ holds, and so **(i.16)** $R^\star(x, y, z)$ by Lemma 13. Therefore, we have shown that for some $y, z \in X(W)$, **(i.11)**, **(i.15)** and **(i.16)** hold, which means by (27) that $x \notin h(b) \vartriangleleft- h(a)$.

The proof of **(ii)** is similar   ∎

6 LCPRS Algebras and Frames

In the classical relation algebras relative sum is definable with composition and complement, namely we have $x \oplus y = -(-x; -y)$. In the lattice-based relation algebras sum must be added as a new independent operator. This is the purpose of the present section.

Definition 13. *An **LCPRS algebra** is a system $(W, \wedge, \vee, \smile, \frown, \otimes, \oplus, \rightarrow, \leftarrow, \Rightarrow, \Leftarrow, 0, 1, 0', 1')$ such that $(W, \wedge, \vee, \smile, \otimes, \rightarrow, \leftarrow, 0, 1, 1')$ is an LCPR algebra, $\frown$ is an unary operations in W (dual converse), $\oplus$ is a binary operations in W (sum), and $\Rightarrow, \Leftarrow$ are binary operations in W (dual right and dual left residua of $\oplus$) satisfying for all $a, b, c \in W$,*

(CPRS.0) $1 \oplus a = a \oplus 1 = 1$
(CPRS.1) $a^{\frown\frown} = a$
(CPRS.2) $(a \wedge b)^\frown = a^\frown \wedge b^\frown$
(CPRS.3) $a \oplus 0' = 0' \oplus a = a$
(CPRS.4) $a \oplus (b \oplus c) = (a \oplus b) \oplus c$
(CPRS.5) $a \oplus (b \wedge c) = (a \oplus b) \wedge (a \oplus c)$

(CPRS.6) $(a \wedge b) \oplus c = (a \oplus c) \wedge (b \oplus c)$
(CPRS.7) $(a \oplus b)^\frown = b^\frown \oplus a^\frown$
(CPRS.8) $a \oplus b \geqslant c$ *iff* $b \geqslant a \Rightarrow c$
(CPRS.9) $a \oplus b \geqslant c$ *iff* $a \geqslant c \Leftarrow b$
(CPRS.10) $0' \wedge 1' = 0$
(CPRS.11) $0' \vee 1' = 1.$ □

Let $L = (W, \wedge, \vee, 0, 1)$ be a bounded lattice. By the *opposite lattice* we mean a lattice $L^{op} = (W, \vee, \wedge, 1, 0)$, where the meet (resp. the join) of L^{op} is the join (resp. the meet) of L and the greatest (resp. the least) element of L^{op} is the least (resp. the greatest) element of L. Observe that the algebra obtained from LCPRS algebra by deleting axioms **(CPRS.10)** and **(CPRS.11)** can be viewed as a join of an LCPR algebra based on the lattice L and an LCPR algebra based on L^{op}. In other words, we have:

Proposition 2. *Let* $(W, \wedge, \vee, \smile, \frown, \otimes, \oplus, \rightarrow, \leftarrow, \Rightarrow, \Leftarrow, 0, 1, 0', 1')$ *be an LCPRS algebra. Then* $(W, \vee, \wedge, \frown, \oplus, \Rightarrow, \Leftarrow, 1, 0, 0')$ *is an LCPR algebra.*

Proof. Straightforward from Definitions 11 and 13.   ■

Remark 1. *If follows that properties of operations* $\frown$, $\oplus$, $\Rightarrow$, *and* $\Leftarrow$ *can be easily obtained from the analogous properties of the operations* $\smile$, $\otimes$, $\rightarrow$, $\leftarrow$, *respectively.*   □

Remark 2. *Note that axioms* **(CPRS.10)** *and* **(CPRS.11)** *provide a connection between the LCPR part of an LCPRS algebra* L *and the LCPR part of* L *based on its opposite part.*   □

Definition 14. *An* **LCPRS frame** *is a system* $(X, \leqslant_1, \leqslant_2, C, \Gamma, R, S, Q, \Theta, \Upsilon, \Omega, I, J)$ *such that* $(X, \leqslant_1, \leqslant_2, C, R, S, Q, I)$ *is an LCPR frame,* Γ *is a mapping* $\Gamma : X \rightarrow X$, Θ, Υ, Ω *are ternary relations on* X *and* $J \subseteq X$ *is a unary relation on* X *such that the following conditions are satisfied for all* $x, x', y, y', z, z' \in X$,

Monotonicity conditions:

(MCPRS.1) $x \leqslant_1 x' \Longrightarrow \Gamma(x) \leqslant_1 \Gamma(x')$
(MCPRS.2) $x \leqslant_2 x' \Longrightarrow \Gamma(x) \leqslant_2 \Gamma(x')$
(MCPRS.3) $\Theta(x, y, z) \,\&\, x' \leqslant_2 x \,\&\, y' \leqslant_2 y \,\&\, z \leqslant_2 z' \Longrightarrow \Theta(x', y', z')$
(MCPRS.4) $\Upsilon(x, y, z) \,\&\, x \leqslant_1 x' \,\&\, y' \leqslant_2 y \,\&\, z' \leqslant_1 z \Longrightarrow \Upsilon(x', y', z')$
(MCPRS.5) $\Omega(x, y, z) \,\&\, x' \leqslant_2 x \,\&\, y \leqslant_1 y' \,\&\, z' \leqslant_1 z \Longrightarrow \Omega(x', y', z')$
(MCPRS.6) $J(x) \,\&\, x \leqslant_2 x' \Longrightarrow J(x')$

Stability conditions:

(SCPRS.1) $\Gamma(\Gamma(x)) = x$
(SCPRS.2) $\Theta(x, y, z) \Longrightarrow \exists x'' \in X \; (x \leqslant_2 x'' \,\&\, \Upsilon(x'', y, z))$
(SCPRS.3) $\Theta(x, y, z) \Longrightarrow \exists y'' \in X \; (y \leqslant_2 y'' \,\&\, \Omega(x, y'', z))$
(SCPRS.4) $\Upsilon(x, y, z) \Longrightarrow \exists z'' \in X \; (z \leqslant_1 z'' \,\&\, \Theta(x, y, z''))$

(SCPRS.5) $\Omega(x, y, z) \Longrightarrow \exists z'' \in X \; (z \leqslant_1 z'' \; \& \; \Theta(x, y, z''))$

(SCPRS.6) $\exists u \in X(\Theta(x, y, u) \& \Upsilon(u, z, y)) \Longrightarrow \exists w \in X(\Theta(y, z', w) \& \Omega(x, w, y))$

(SCPRS.7) $\exists u \in X(\Theta(x, y, u) \; \& \; \Omega(z, u, z')) \Longrightarrow$
$$\exists w \in X(\Theta(z, x, w) \; \& \; \Upsilon(w, y, z'))$$

(SCPRS.8) $J(x) \; \& \; (\Theta(x, y, z) \; or \; \Theta(y, x, z)) \Longrightarrow y \leqslant_2 z$

(SCPRS.9) $\exists u \in X(J(u) \; \& \; \Upsilon(u, x, x))$

(SCPRS.10) $\exists u \in X(J(u) \; \& \; \Omega(x, u, x))$

(SCPRS.11) $\Omega(x, y, z) = \Upsilon(\Gamma(y), \Gamma(x), \Gamma(z))$

(SCPRS.12) $lr(I) \cap l(J) = \emptyset$

(SCPRS.13) $r(I) \cap rl(J) = \emptyset.$ □

Let $(X, \leqslant_1, \leqslant_2)$ be a doubly ordered set. By the *opposite doubly ordered set* we mean a structure $(X, \leqslant_1^{op}, \leqslant_2^{op})$, where $\leqslant_1^{op} = \leqslant_2$ and $\leqslant_2^{op} = \leqslant_1$. Observe that the frame obtained from the LCPRS frame by deleting axioms **(SCPRS.12)** and **(SCPRS.13)** can be viewed as a join of the LCPR frame based on a doubly ordered set $(X, \leqslant_1, \leqslant_2)$ with the LCPR frame based on the opposite doubly ordered set $(X, \leqslant_1^{op}, \leqslant_2^{op})$. Therefore, we have:

Proposition 3. *Let* $(X, \leqslant_1, \leqslant_2, C, \Gamma, R, S, Q, \Theta, \Upsilon, \Omega, I, J)$ *be an LCPRS frame. Then* $(X, \leqslant_2, \leqslant_1, \Gamma, \Theta, \Upsilon, \Omega, J)$ *is an LCPR frame.*

Proof. Straightforward from the definition of LCPR frame and Definition 14. ∎

Remark 3. *From the above proposition it follows easily that the properties of the relations* Γ, Θ, Υ, Ω, *and* J *can be obtained from the properties of the relations* C, R, S, Q, *and* I, *respectively, by interchanging the roles of the orderings* $\leqslant_1$ *and* $\leqslant_2$.   □

Remark 4. *Note that axioms* **(SCPRS.12)** *and* **(SCPRS.13)** *provide a connection between the LCPR part of an LCPRS frame and its opposite part.*   □

Given an LCPRS frame $(X, \leqslant_1, \leqslant_2, C, \Gamma, R, S, Q, \Theta, \Upsilon, \Omega, I, J)$, let us define the following mappings $^{\curlywedge} : 2^X \to 2^X$ and $\oplus_\Omega$, $\oplus_\Upsilon$, $\boxplus$, $\Rightarrow$, $\Leftarrow : 2^X \times 2^X \to 2^X$ by: for all $A, B \subseteq X$,

$$A^{\curlywedge} = \{\Gamma(x) \in X : x \in A\} \tag{30}$$

$$A \oplus_\Omega B = \{z \in X : \forall x, y \in X \; (\Omega(x, y, z) \; \& \; x \in r(A) \Longrightarrow y \in B\} \tag{31}$$

$$A \oplus_\Upsilon B = \{z \in Z : \forall x, y \in X \; (\Upsilon(x, y, z) \; \& \; y \in r(B) \Longrightarrow x \in A\} \tag{32}$$

$$A \boxplus B = A \oplus_\Omega B. \tag{33}$$

$$A \Rightarrow B = \{x \in X : (\forall y, z \in X)(\Theta(y, x, z) \; \& \; y \in A \Longrightarrow z \in B)\} \tag{34}$$

$$B \Leftarrow A = \{x \in X : (\forall y, z \in X)(\Theta(x, y, z) \; \& \; y \in A \Longrightarrow z \in B)\}. \tag{35}$$

Moreover, put

$$\mathbf{0'} = l(J). \tag{36}$$

Definition 15. *Let* $(X, \leqslant_1, \leqslant_2, C, \Gamma, R, S, Q, \Theta, \Upsilon, \Omega, I, J)$ *be an LCPRS frame. The* ***complex algebra of*** X *is a structure* $(L(X), \sqcap, \sqcup, ^\curlyvee, ^\curlywedge, \boxtimes, \boxplus, \rightarrow, \leftarrow, \Rightarrow, \Leftarrow, 0, 1, 0', 1')$ *such that* $L(X)$ *is the family of all l–stable subsets of* X, *the operations* $\sqcap, \sqcup, ^\curlyvee, ^\curlywedge, \boxtimes, \boxplus, \rightarrow, \leftarrow, \Rightarrow, \Leftarrow$ *are respectively defined by (4), (3), (11), (30), (14), (33), (26), (27), (34), (35), and the constants* $0, 1, 0',$ *and* $1'$ *are given by (5), (6), (15) and (36), respectively.* □

We will show now that complex algebras of LCPRS frames are LCPRS algebras.

Theorem 8. *The complex algebra of an LCPRS frame is an LCPRS algebra.*

Proof. Since J is $\leqslant_2$–increasing by **(MCPRS.6)**, $L(J)$ is l–stable. From Theorem 6, Proposition 3, and Remark 3 it follows that we only need to show that the connecting axioms **(CPRS.10)** and **(CPRS.11)** hold, i.e.,

   (i) $0' \sqcap 1' = 0$
   (ii) $0' \sqcup 1' = 1$.

(i) $0' \sqcap 1' = lr(I) \cap l(J) = \emptyset$ by **(SCPRS.12)**.

(ii) By the definitions (15), (36), and (4), $0' \sqcup 1' = l(rlr(I) \cap rl(J))$. Also, by Lemma 4(ii), $I \subseteq lr(I)$, so $rlr(I) \subseteq r(I)$. Next, $rlr(I) \cap rl(J) \subseteq r(I) \cap rl(J) = \emptyset$ by **(SCPRS.13)**. Hence we have: $rlr(I) \cap rl(J) = \emptyset$, so $l(rlr(I) \cap rl(J)) = l(\emptyset) = X(W)$ ∎

Let $(W, \wedge, \vee, \smile, \frown, \otimes, \oplus, \rightarrow, \leftarrow, \Rightarrow, \Leftarrow, 0, 1, 0', 1')$ be an LCPRS algebra. As before, by $FIP(X)$ and (resp. $X(W)$) we denote the family of all filter–ideals pairs (resp. maximal filter–ideal pairs) of W.

Lemma 14. *Let* $(W, \wedge, \vee, \smile, \frown, \otimes, \oplus, \rightarrow, \leftarrow, \Rightarrow, \Leftarrow, 0, 1, 0', 1')$ *be an LCPRS algebra. Then for every* $a \in W$, $l(\{x \in X(W) : a \in x_2\}) = \{x \in X(W) : a \in x_1\}$.

Proof. ($\subseteq$) Let $a \notin x_1$. It follows that $x_1 \cap (a] = \emptyset$, so $(x_1, (a])$ is a filter–ideal pair. Let y be its extension to the maximal filter–ideal pair. Hence $x_1 \subseteq y_1$ and $a \in y_2$. It follows that $x \notin l(\{x \in X(W) : a \in x_2\})$.
($\supseteq$) Let $a \in x_1$. Take $y \in X(W)$ such that $x_1 \subseteq y_1$. Then $a \in y_1$, whence $a \notin y_2$. ∎

Define a mapping $\Gamma^\star : FIP(W) \rightarrow FIP(W)$ by: for every $x \in FIP(W)$,

$$\Gamma^\star(x) = (x_1^\frown, x_2^\frown). \tag{37}$$

Furthermore, let us define the following ternary relations on $X(W)$: for all $x, y, z \in X(W)$,

$$\Theta^\star(x, y, z) \iff (\forall a, b \in W)\, a \in x_2 \;\&\; b \in y_2 \Longrightarrow a \oplus b \in z_2 \tag{38}$$

$$\Omega^\star(x, y, z) \iff (\forall a, b \in W)\, a \in x_2 \;\&\; a \oplus b \in z_1 \Longrightarrow b \in y_1 \tag{39}$$

$$\Upsilon^\star(x, y, z) \iff (\forall a, b \in W)\, b \in y_2 \;\&\; a \oplus b \in z_1 \Longrightarrow a \in x_1. \tag{40}$$

Also, put

$$J^\star = \{x \in X(W) : 0' \in x_2\} \tag{41}$$

Definition 16. *Let* $(W, \wedge, \vee, \smile, \frown, \otimes, \oplus, \rightarrow, \leftarrow, \Rightarrow, \Leftarrow, 0, 1, 0', 1')$ *be an LCPRS algebra. The* **canonical frame of** W *is a structure* $(X(W), \preccurlyeq_1, \preccurlyeq_2, C^\star, \Gamma^\star, R^\star,$ $Q^\star, S^\star, \Theta^\star, \Omega^\star, \Upsilon^\star, I^\star, J^\star)$ *such that* $(X(W), \preccurlyeq_1, \preccurlyeq_2, C^\star, R^\star, Q^\star, S^\star, I^\star)$ *is the canonical frame of the LCPR part* $(W, \wedge, \vee, \smile, \otimes, \rightarrow, \leftarrow, 0, 1, 1')$ *of* W *and* $\Gamma^\star$, $\Theta^\star$, $\Omega^\star$, $\Upsilon^\star$, *and* $J^\star$ *are defined by (37)–(41).* □

Theorem 9. *The canonical frame of an LCPRS algebra is an LCPRS frame.*

Proof. We have to show that the conditions **(SCPRS.12)** and **(SCPRS.13)** hold in the canonical frame of an LCPRS algebra. The remaining conditions follow from Theorem 6, Proposition 3, and Remark 1.

We show that $lr(I^\star) \cap l(J^\star) = \emptyset$. Note that

$$
\begin{aligned}
&lr(I^\star) \cap l(J^\star) \\
&\quad = lr(\{x \in X(W) : 1' \in x_1\}) \cap l(\{x \in X(W) : 0' \in x_2\}) \\
&\quad = l(\{x \in X(W) : 1' \in x_2\}) \cap l(\{x \in X(W) : 0' \in x_2\}) \quad \text{by Theorem 1(i)} \\
&\quad = \{x \in X(W) : 1' \in x_1\} \cap \{x \in X(W) : 0' \in x_1\} \quad \text{by Lemma 14} \\
&\quad = \{x \in X(W) : 1' \in x_1 \ \& \ 0' \in x_1\} \\
&\quad \subseteq \{x \in X(W) : 1' \wedge 0' \in x_1\}
\end{aligned}
$$

However, by **(CPRS.10)**, $1' \wedge 0' = 0$. Since x_1 is a proper filter, $0 \notin x_1$, so we have $\{x \in X(W) : 1' \wedge 0' \in x_1\} = \emptyset$, and consequently $lr(I^\star) \cap l(J^\star) = \emptyset$.

Now we prove that $r(I) \cap rl(J) = \emptyset$. Observe:

$$
\begin{aligned}
&r(I) \cap rl(J) \\
&\quad = r(\{x \in X(W) : 1' \in x_1\}) \cap rl(\{x \in X(W) : 0' \in x_2\}) \\
&\quad = r(\{x \in X(W) : 1' \in x_1\}) \cap r(\{x \in X(W) : 0' \in x_1\}) \quad \text{by Lemma 14} \\
&\quad = \{x \in X(W) : 1' \in x_2\} \cap \{x \in X(W) : 0' \in x_2\} \quad \text{Theorem 1(i)} \\
&\quad = \{x \in X(W) : 1' \in x_2 \ \& \ 0' \in x_2\} \\
&\quad \subseteq \{x \in X(W) : 1' \vee 0' \in x_2\}.
\end{aligned}
$$

Since $1' \vee 0' = 1 \notin x_2$, it follows that $\{x \in X(W) : 1' \vee 0' \in x_2\} = \emptyset$. In conclusion, $r(I) \cap rl(J) = \emptyset$. ∎

We conclude this section by the following representation theorem.

Theorem 10 (Representation theorem for LCPRS algebras)

Any LCPRS algebra is isomorphic to a subalgebra of the complex algebra of its canonical frame.

Proof Taking into account Propositions 2, 3, and Remarks 1, 3 the proof is analogous to the proof of Theorem 7. ∎

7   Conclusion

In this paper we have studied not necessarily distributive lattices with operators that are the abstract counterparts to the converse and composition of binary relations. On the algebraic side, we have presented relational representation theorems for these classes of algebras. These theorems are obtained by a suitable extensions of Urquhart's representation theorem for lattices [19]; here, we have stressed the relational aspect of representability and have omitted the topological aspect.

On the logical side, with every class of algebras studied in the paper we have associated an appropriate class of frames. These frames constitute a basis of a Kripke-style semantics for the logics whose algebraic semantics is determined by the classes of algebras presented in the paper. The representation theorems would enable us to prove completeness of the logics. For a detailed elaboration of the respective relational logics one can follow the developments in [1] and [17].

References

1. Allwein, G. and Dunn, J. M. (1993). Kripke models for linear logic. *J. Symb. Logic*, 58, 514–545.
2. Blount, K. and Tsinaksis, C. (2003). The structure of residuated lattices. Int. J. of Algebra Comput. 13(4), 437–461.
3. Demri, S. and Orłowska, E. (2002). *Incomplete Information: Structure, Inference, Complexity*. EATCS Monographs in Theoretical Computer Science, Springer.
4. Dunn, J. M. (2001). Gaggle theory: an abstraction of Galois connections and residuation, with application to nagation, implication and various logical operators. Lecture Notes in Artificial Intelligence 478, Springer–Verlag, 31–51.
5. Düntsch, I. (1994). Rough relation algebras. Fundamenta Informaticae 21, 321–331.
6. Düntsch, I. and Orłowska, E. (2001). Beyond modalities: sufficiency and mixed algebras. In: Orłowska, E. and Szałas, A. (eds) Relational Methods for Computer Science Applications, Physica–Verlag, Heidelberg, 263–285.
7. Düntsch, I. and Orłowska, E. (2004). Boolean algebras arising from information systems. Annals of Pure and Applied Logic 127, No 1–3, Special issue, Provinces of Logic Determined, Essays in the memory of Alfred Tarski, edited by Adamowicz, z., Artemov, S., Niwiński, D., Orłowska, E., Romanowska, A., and Woleński, J., 77–98.
8. Düntsch, I., Orłowska, E., and Radzikowska, A. M. (2003). Lattice–based relation algebras and their representability. In *Theory and Applications of Relational Structures as Knowledge Instruments*, de Swart, H. C. M. et al (eds), Lecture Notes in Computer Science 2929, Springer–Verlag, 234–258.
9. Düntsch, I. and Winter, M. (2006). Rough relation algebras revisited. Fundamenta Informaticae. To appear.
10. Gehrke, M. and Jónsson, B. (1994). Bounded distributive lattices with operators. Mathematica Japonica 40, No 2, 207–215.
11. Hart, J. B., Rafter, L., and Tsinaksis, C. (2002). The structure of commutative residuated lattices. Internat. J. Algebra Comput. 12(4), 509–524.
12. Jipsen, P. and Tsinaksis, C. (2003). A Survey of Residuated Lattices. In Martinez, J. (ed.), *Ordered algebraic structures*, Kluwer Academic Publishers, Dordrecht, 19–56.
13. Jónsson, B. and Tarski, A. (1951). Boolean algebras with operators. Part I. American Journal of Mathematics 73, 891–936.
14. Orłowska, E. (1995). Information algebras. Lecture Notes in Computer Science 639, Proceedings of *AMAST'95*, Montreal, Canada, 50–65.

15. Orłowska, E. and Radzikowska, A. M. (2001). Information relations and opeators based on double residuated lattices. Procedings of the 6th International Workshop on Relational Methods in Computer Science RelMiCS'01, Oisterwijk, Netherlands, 185–199.

16. Orłowska, E. and Radzikowska, A. M. (2002). Double residuated lattices and their applications. *Relational Methods in Computer Science*, de Swart, H. C. M. (ed), Lecture Notes in Computer Science 2561, Springer–Verlag, Heidelberg, 171–189.

17. Orłowska, E. and Vakarelov, D. (2005). Lattice-based modal algebras and modal logics. In: Hajek, P., Valdés–Villanueva, L. M., and Westerstahl, D.(eds), *Logic, Methodology and Philosophy of Science*. Proceedings of the 12th International Congress. King's College London Publications, 147–170.

18. Sofronie-Stokkermans, V. (2000). Duality and canonical extensions of bounded distributive lattices with operators, and applications to the semantics of non-classical logics. Studia Logica 64, Part I, 93–122, Part II, 151–172.

19. Urquhart, A. (1978). A topological representation theorem for lattices. Algebra Universalis 8, 45–58.

Some Aspects of Lattice and Generalized Prelattice Effect Algebras

Zdenka Riečanová, Ivica Marinová, and Michal Zajac

Department of Mathematics, Faculty of Electrical Engineering and Information
Technology STU, Ilkovičova 3, SK-812 19 Bratislava, Slovakia
{Zdenka.Riecanova,Ivica.Marinova,Michal.Zajac}@stuba.sk

Abstract. Common generalizations of orthomodular lattices and MV-algebras are lattice effect algebras which may include noncompatible pairs of elements as well as unsharp elements. Thus elements of these structures may be carriers of states, or probability measures, when they represent properties, questions or events with fuzziness, uncertainty or unsharpness. Unbounded versions of these structures (more precisely without top elements) are generalized effect algebras which can be extended onto effect algebras. We touch only a few aspects of these structures. Namely, necessary and sufficient conditions for generalized effect algebras to obtain their effect algebraic extensions lattice ordered or MV-effect algebras. We also give one possible construction of pastings of MV-effect algebras together along an MV-effect algebra to obtain lattice effect algebras. In conclusions we give some applications of presented results about sets of sharp elements, direct and subdirect decompositions of lattice effect algebras and about smearings (resp. the existence) of states an probabilities on them.

1 Introduction

Lattice effect algebras generalize orthomodular lattices (including Boolean algebras) and MV-algebras [1]. Effect algebras, introduced in 1994 by Foulis and Bennet [4], or equivalent in some sense, D-poset introduced in 1994 by Kôpka and Chovanec [9] may be carriers of probability measures, where elements of these structures represent properties, questions or events with fuzziness, uncertainty or unsharpness. In spite of these facts there are (even finite lattice) effect algebras admitting no states and no probabilities (see [8] and [17]).

In the classical (Kolmogorovian) probability theory the set of events is a Boolean algebra (σ-algebra), assuming that every two events are simultaneously measurable (compatible) and thus this theory cannot explain events occurring, e.g., in quantum physics, as well as in many other areas. Orthomodular lattices are generalizations of Boolean algebras. They may include noncompatible pairs of elements.

Another generalization of Boolean algebras are MV-algebras, which were originally constructed to give an algebraic structure of the infinite-valued Lukasiewicz propositional logics. Hence MV-algebras may include unsharp elements (i.e., elements x and non x need not be disjoint).

H. de Swart et al. (Eds.): TARSKI II, LNAI 4342, pp. 290–317, 2006.
© Springer-Verlag Berlin Heidelberg 2006

Common generalizations of the above mentioned algebraic structures are lattice effect algebras, which may include noncompatible pairs of elements as well as unsharp elements.

On the other hand, a lattice effect algebra E is an orthomodular lattice iff every element of E is sharp and E is an MV-algebra iff every pair of elements of E is compatible. Thus a lattice effect algebra E is a Boolean algebra iff every pair of its elements is compatible and every element of E is sharp. Moreover, every maximal subset of pairwise compatible elements in a lattice effect algebra is an MV-effect algebra being a sub-lattice and a sub-effect algebra of E ([14]), called a block of E, and E is a set-theoretical union of its blocks, hence maximal sub-MV-algebras. Further, the set $\mathcal{S}(E)$ of all unsharp elements of E is an orthomodular lattice which is a sub-lattice and a sub-effect algebra of E ([6]). Finally, the set of all sharp elements of E which are compatible with every other element of E forms a Boolean algebra $\mathcal{C}(E)$ called a center of E. Hence $\mathcal{C}(E) = \mathcal{S}(E) \cap \mathcal{B}(E)$ where $\mathcal{B}(E) = \bigcap\{M \subseteq E \mid M \text{ block of } E\}$. Evidently, $\mathcal{C}(E)$ is also a sub-lattice and a sub-effect algebra of E. Thus the known facts on Boolean algebras, orthomodular lattices and MV-algebras may help us for study of lattice effect algebras.

Finally, note that lattice effect algebras are in fact bounded lattices. Their unbounded versions are generalized effect algebras (i.e., without top elements), which can be embedded into effect algebras as proper ideals with a special property (namely: from every pair of elements x and non x they contain exactly one). In this case, for a prelattice generalized effect algebra P and the effect algebra E extending P the set $\mathcal{S}(E) \cap P$ is a generalized orthomodular lattice, $\mathcal{C}(E) \cap P$ is a generalized Boolean algebra and for every block M the $M \cap P$ is a generalized MV-effect algebra, under which they all are proper ideals (with the above mentioned special property) in $\mathcal{S}(E)$, $\mathcal{C}(E)$, and M respectively (see [27]).

2   Basic Definitions and Important Examples

In 1994, Foulis and Bennett [4] have introduced a new algebraic structure, called an effect algebra. Effects represent unsharp measurements or observations on a quantum mechanical system. For modelling unsharp measurements in a Hilbert space, the set of all effects is the set of all selfadjoint operators T on a Hilbert space H with $0 \leq T \leq 1$. In a general algebraic form an effect algebra is defined as follows:

Definition 1. *A partial algebra $(E; \oplus, 0, 1)$ is called an* effect-algebra *if $0, 1$ are two distinguished elements and $\oplus$ is a partially defined binary operation on E which satisfies the following conditions for any $a, b, c \in E$:*

(Ei) $b \oplus a = a \oplus b$ *if $a \oplus b$ is defined,*

(Eii) $(a \oplus b) \oplus c = a \oplus (b \oplus c)$ *if one side is defined,*

(Eiii) *for every $a \in E$ there exists a unique $b \in E$ such that $a \oplus b = 1$ (set $a' = b$),*

(Eiv) *if $1 \oplus a$ is defined then $a = 0$.*

Unbounded versions (mutually equivalent) of effect algebras were studied by
Foulis and Bennett (cones), Kalmbach and Riečanová (abelian RI-semigroups
and RI-posets), Hedlíková and Pulmannová (cancellative positive partial abelian
semigroups). Their common definition is the following:

Definition 2. *A partial algebra* $(E; \oplus, 0)$ *is called a* generalized effect algebra
if $0 \in E$ *is a distinguished element and* $\oplus$ *is a partially defined binary operation
on* E *which satisfies the following conditions for any* $a, b, c \in E$:

(GEi) $a \oplus b = b \oplus a$, *if one side is defined*,
(GEii) $(a \oplus b) \oplus c = a \oplus (b \oplus c)$, *if one side is defined*,
(GEiii) $a \oplus 0 = a$ *for all* $a \in E$,
(GEiv) $a \oplus b = a \oplus c$ *implies* $b = c$ *(cancellation law)*,
(GEv) $a \oplus b = 0$ *implies* $a = b = 0$.

The following proposition with a trivial verification indicates the relation between effect algebras and generalized effect algebras.

Proposition 1. *If* $(E; \oplus, 0)$ *is a generalized effect algebra and there is* $1 \in E$
such that for all $a \in E$ *there is* $b \in E$ *with* $a \oplus b = 1$ *then* $(E; \oplus, 0, 1)$ *is an effect
algebra. Conversely, if* $(E; \oplus, 0, 1)$ *is an effect algebra then the partial operation*
$\oplus$ *satisfies axioms* (GEi)–(GEv) *of a generalized effect algebra.*

In every effect algebra E (generalized effect algebra E) the partial binary operation $\ominus$ and relation $\leq$ can be defined by

(ED) $a \leq c$ and $c \ominus a = b$ iff $a \oplus b$ is defined and $a \oplus b = c$ (set $b = c \ominus a$).

Then $\leq$ is a partial order on E under which 0 is the least element of E.

If E is an effect algebra and $(E; \leq)$ is a lattice (complete lattice) then E is
called a *lattice effect algebra* (*complete effect algebra*).

Definition 3. [11] *Let* $(E; \oplus, 0)$ *be an effect algebra (generalized effect algebra).
If* $Q \subseteq E$ *is such that* $0 \in Q$ *and for all* $a, b, c \in Q$ *with* $a \oplus b = c$ *when at
least two of* a, b, c *are in* Q *then* $a, b, c \in Q$, *then* Q *is called a* sub-effect algebra
(sub-generalized effect algebra*) of* E.

Note that every sub-effect algebra (sub-generalized effect algebra) Q of effect
algebra (generalized effect algebra) E is an effect algebra (generalized effect
algebra) in its own right.

Recall that a nonvoid subset I of a partially ordered set L is an *order ideal* if
$a \in L$, $b \in I$, and $a \leq b$ implies $a \in I$. If $I \neq L$ then I is called *proper*.

Definition 4. *Let* $(P; \leq, \oplus, 0)$ *be a generalized effect algebra. Let* P^* *be a set
disjoint from* P *with the same cardinality. Consider a bijection* $a \mapsto a^*$ *from* P
onto P^*. *Let* $E = P \dot\cup P^*$ *be the disjoint union of* P *and* P^*. *Define a partial
binary operation* $\oplus^*$ *on* E *by the following rules. For* $a, b \in P$

(i) $a \oplus^* b$ *is defined if and only if* $a \oplus b$ *is defined, and* $a \oplus^* b = a \oplus b$
(ii) $b^* \oplus^* a$ *and* $a \oplus^* b^*$ *are defined iff* $b \ominus a$ *is defined and then* $b^* \oplus^* a = (b \ominus a)^* =$
$a \oplus^* b^*$.

Then $E = P \dot\cup P^*$ *will be called an* effect algebraic extension of P.

Theorem 1. [3] *For every generalized effect algebra P and $E = P \dot\cup P^*$ the structure $(E; \oplus^*, 0, 0^*)$ is an effect algebra. Moreover P is an order ideal in E closed under $\oplus^*$ and the partial order induced by $\oplus^*$ when restricted to P, coincides with the partial order induced by $\oplus$.*

Since the definition of $\oplus^*$ on $E = P \dot\cup P^*$ coincides with $\oplus$- operation on P, it will cause no confusion if from now on we will use the same notation $\oplus$ also for its extension on E.

Definition 5. *Let $(P_1; \oplus_1, 0_1)$, $(P_2, \oplus_2, 0_2)$ be generalized effect algebras. A mapping $\psi : P_1 \to P_2$ is called a* generalized effect algebra morphism *iff whenever $a, b \in P_1$ with defined $a \oplus_1 b$ then we have $\psi(a \oplus_1 b) = \psi(a) \oplus_2 \psi(b)$. A bijective generalized effect algebra morphism ψ such that $\psi^{-1} : P_2 \to P_1$ is also a generalized effect algebra morphism is called a* generalized effect algebra isomorphism *of P_1 and P_2 (we write $P_1 \cong P_2$ and P_1 and P_2 are said to be* isomorphic*). If there is a sub-generalized effect algebra $Q \subseteq P_2$ and a generalized effect algebra isomorphism $\varphi : P_1 \to Q$ then φ is called a* generalized effect algebra embedding *of P_1 into P_2. Then we usually identify $Q = \varphi(P_1)$ with P_1 and we say that P_1 is up to isomorphism a* sub-generalized effect algebra *of P_2.*

Assume that $(E_1; \oplus_1, 0_1, 1_1)$ and $(E_2; \oplus_2, 0_2, 1_2)$ are effect algebras. An injection $\varphi : E_1 \to E_2$ is called an *embedding* if $\varphi(1_1) = 1_2$ and for $a, b \in E_1$ we have $a \leq b'$ iff $\varphi(a) \leq (\varphi(b))'$ in which case $\varphi(a \oplus_1 b) = \varphi(a) \oplus_2 \varphi(b)$. We can easily see that then $\varphi(E_1)$ is a sub-effect algebra of E_2 and we say that E_1 and $\varphi(E_1)$ are *isomorphic*, or that E_1 *is up to isomorphism a sub-effect algebra of* E_2. Clearly, if E_1 and E_2 are lattice effect algebras then $\varphi(E_1)$ is a sublattice of E_2. We usually identify E_1 with $\varphi(E_1)$.

Recall that a *direct product* $\prod \{ E_\kappa \mid \kappa \in H \}$ of effect algebras E_κ is a cartesian product with $\oplus$, 0, 1 defined "coordinatewise", i.e., $(a_\kappa)_{\kappa \in H} \oplus (b_\kappa)_{\kappa \in H}$ exists iff $a_\kappa \oplus_\kappa b_\kappa$ is defined for each $\kappa \in H$ and then $(a_\kappa)_{\kappa \in H} \oplus (b_\kappa)_{\kappa \in H} = (a_\kappa \oplus_\kappa b_\kappa)_{\kappa \in H}$. Moreover, $0 = (0_\kappa)_{\kappa \in H}$, $1 = (1_\kappa)_{\kappa \in H}$. An element $z \in E$ is called *central* if the intervals $[0, z]$ and $[0, z']$ with the inherited $\oplus$-operation are effect algebras in their own right and $E \cong [0, z] \times [0, z']$, see [11]. The set $C(E) = \{ z \in E \mid z$ is central$\}$ is called a *center* of E. If $C(E) = \{0, 1\}$ then E is called *irreducible*.

Definition 6. *A* subdirect product *of a family $\{ E_\kappa \mid \kappa \in H \}$ of lattice effect algebras is a sublattice-effect algebra Q (i.e., Q is simultaneously a sub-effect algebra and a sublattice) of the direct product $\prod \{ E_\kappa \mid \kappa \in H \}$ such that each restriction of the natural projection pr_{κ_i} to Q is onto E_{κ_i}. Moreover, Q is a* sub-direct product decomposition *of a lattice effect algebra E if there exists an isomorphism $\varphi : E \to Q$ (of E onto Q).*

A *horizontal sum* of finite chains $C_1 = \{0, a, 2a, \ldots, 1 = n_a a\}$ and $C_2 = \{0, b, 2b, \ldots, 1 = n_b b\}$ is a lattice effect algebra $E = C_1 \cup C_2$ with identified elements 0 and 1 and such that $C_1 \cap C_2 = \{0, 1\}$ and ka and ℓb for $k \neq n_a$ and $\ell \neq n_b$ are noncomparable, i.e., $ka \vee \ell b = 1$ and $ka \wedge \ell b = 0$. In the same manner we define a horizontal sum of any family of finite chains.

Example 1. (Standard effect algebra of real numbers) Assume that E is the interval $[0,1]$ of real numbers and define the partial binary operation $\oplus$ on $[0,1]$ as follows:

for $a, b \in [0,1]$, $a \oplus b$ is defined iff $a + b \leq 1$ and then $a \oplus b = a + b$.
Then $([0,1]; \oplus, 0, 1)$ is called a *standard effect algebra of real numbers*.

Example 2. (Generalized effect algebra of nonnegative integers)
Let $P = \{0, 1, 2, \dots\}$ be a set of all nonnegative integers with usual $+$ operation, hence $k \oplus l = k + l$ for all $k, l \in P$. Then $(P; \oplus, 0)$ is a generalized effect algebra. Moreover, the effect algebraic extension $E = P \cup P^*$ is an infinite chain with the greatest element 0^*. It is a noncomplete lattice effect algebra since $\bigvee P$ does not exist in E.

Let $G = E \cup \{s\} = P \dot\cup P^* \dot\cup \{s\}$ where $\{s\} \cap P = \{s\} \cap P^* = \emptyset$ and $a \leq s \leq b^*$ for all $a \in P$, $b^* \in P^*$. Then G is a chain being a complete lattice admitting no $\oplus$ operation satisfying the axioms of effect algebra.

Proposition 2. *On every finite chain $(P, \leq)$ there exists a unique operation $\oplus$ such that $a \leq b$ iff there is $c \in P$ with $a \oplus c = b$ ($\oplus$ is compatible with $\leq$), [26].*

Proof. Let $P = \{0, x_1, x_2, \dots, x_n\}$ and $0 < x_1 < x_2 < \dots < x_n = 1$. Here $x < y$ means that $x \leq y$ and $x \neq y$. Suppose that $\oplus$ is compatible with $\leq$ on P. Then for the derived $\ominus$ and a fixed $k \in \{1, 2, \dots n\}$ we have $x_k \ominus 0 = x_k$, $x_k \ominus x_k = 0$ and for $i > k$, $x_k \ominus x_i$ is not defined. Moreover, $0 = x_k \ominus x_k < x_k \ominus x_{k-1} < x_k \ominus x_{k-2} < \dots < x_k \ominus x_1 < x_k \ominus 0 = x_k$. It follows that $x_k \ominus x_{k-1} = x_1$, $x_k \ominus x_{k-2} = x_2, \dots$, $x_k \ominus x_1 = x_{k-1}$, $x_k \ominus 0 = x_k$ since the chain $0 < x_1 < x_2 < \dots x_{k-1} < x_k$ contains all elements between 0 and x_k. We conclude that $x_k \ominus x_i = x_{k-i}$ for every $i < k$, which gives that $x_k = x_i \oplus x_{k-i}$. So the unique $\oplus$ is $x_i \oplus x_k = x_{i+k}$ if $i + k \leq n$, otherwise $x_i \oplus x_k$ is not defined.

Note that the above proof shows that every finite chain is an effect algebra of the form

$$0 < a < a \oplus a < \dots < \underbrace{a \oplus a \oplus \dots \oplus a}_{n-\text{times}} = na = 1 \text{ (Fig. 1).}$$

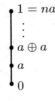

Fig. 1.

Example 3. (Distributive diamond, Fig. 2) Let $P = \{0, a, b, 1\}$. Put

$$1 = a \oplus_1 a = b \oplus_1 b, \quad x \oplus_1 0 = 0 \oplus_1 x = x \text{ for all } x \in P,$$
$$1 = a \oplus_2 b = b \oplus_2 a, \quad x \oplus_2 0 = 0 \oplus_2 x = x \text{ for all } x \in P.$$

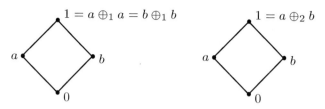

Fig. 2.

$E_1 = (P; \oplus_1, 0, 1)$ is called a distributive diamond. It is a horizontal sum of finite chains $\{0, a, 1\}$ and $\{0, b, 1\}$. $E_2 = (P; \oplus_2, 0, 1)$ is a Boolean algebra.

Evidently, partial orders derived by (ED) on P from $\oplus_1$ and $\oplus_2$ coincide.

Now, for every positive integer k, set $G_k = \prod_{n=1}^{\infty} F_n$ where $F_n = E_1$ for $n \leq k$ and $F_n = E_2$ for all $n > k$. Then G_k are mutually nonisomorphic, complete (distributive) effect algebras. However, all G_k as posets with partial orders derived by (ED) coincide (see [28]).

Example 4. (Orthomodular lattices including Boolean algebras)
Let $(L; \vee, \wedge, {}^{\perp}, 0, 1)$ be an orthomodular lattice (Boolean algebra iff L is distributive). Define $a \oplus b = a \vee b$ iff $a \leq b^{\perp}$. Then $(L; \oplus, 0, 1)$ is a lattice effect algebra.

Definition 7. *An MV-algebra is an algebra* $(M, \oplus, {}^*, 0, 1)$, *where M is a nonempty set, 0 and 1 are constant elements of M, $\oplus$ is a binary operation, and * is a unary operation satisfying the following axioms:*

(MVA1) $(a \oplus b) = (b \oplus a)$,
(MVA2) $(a \oplus b) \oplus c = a \oplus (b \oplus c)$,
(MVA3) $a \oplus 0 = a$,
(MVA4) $a \oplus 1 = 1$,
(MVA5) $(a^*)^* = a$,
(MVA6) $0^* = 1$,
(MVA7) $a \oplus a^* = 1$,
(MVA8) $(a^* \oplus b)^* \oplus b = (a \oplus b^*)^* \oplus a$.

The lattice operations $\vee$ and $\wedge$ on M can be defined by the formulas

$$a \vee b = (a^* \oplus b)^* \oplus b \qquad \text{and} \qquad a \wedge b = \left((a \oplus b^*)^* \oplus b^*\right)^*.$$

We write $a \leq b$ iff $a \vee b = b$. The relation $\leq$ is a partial ordering on M and $0 \leq a \leq 1$ for any $a \in M$. An MV-algebra is a distributive lattice with respect to the operations $\vee$ and $\wedge$.

Example 5. (MV-algebras and MV-effect algebras) Another important example of a lattice effect algebra can be derived from an MV-algebra $(M; \oplus, {}^*, 0, 1)$ if we define a partial binary operation $\hat{\oplus}$ on M by: $a \hat{\oplus} b = a \oplus b$ iff $a \leq b^*$. Then $(M; \hat{\oplus}, 0, 1)$ is a lattice effect algebra in which $a \hat{\oplus}(a^* \wedge b) = b \hat{\oplus}(b^* \wedge a)$ for all

$a, b \in E$ (then E is called an *MV-effect algebra*). In this case for every $a \in E$ we have $a \oplus a^* = 1$, i.e. $a^* = a'$.

Conversely, every lattice effect algebra $(E; \hat{\oplus}, 0, 1)$ in which $a\hat{\oplus}(a^* \wedge b) = b\hat{\oplus}(b^* \wedge a)$ for all $a, b \in E$ can be organized into an MV-algebra by putting $a \oplus b = a\hat{\oplus}(a^* \wedge b)$ for all $a, b \in E$ [2].

Finally, note that examples of lattice effect algebras which are neither orthomodular lattices nor MV-algebras are, for instance, a 0–1-pasting (horizontal sum) of two MV-algebras or a direct product of an orthomodular lattice and an MV-effect algebra. Here, instead of all these structures, we consider derived effect algebras.

Example 6. (D-lattices) If for a lattice effect algebra we consider the partial binary operation $\ominus$ derived from $\oplus$ by (ED), as a fundamental operation, then $(E; \ominus, 0, 1)$ becomes a D-lattice [9]. Conversely, from every D-lattice $(E; \ominus, 0, 1)$ we can derive a lattice effect algebra $(E; \oplus, 0, 1)$ by putting $a \oplus b = c$ iff $a \leq c$ and $b = c \ominus a$.

Definition 8. *Assume that $(E; \oplus, 0, 1)$ is an effect algebra.*

1. *If E is a lattice effect algebra then $Q \subseteq E$ is called a* sub-lattice effect algebra *of E iff Q is a sub-effect algebra of E and for every pair $a, b \in Q$, $a \vee b \in Q$.*
2. *If E is an MV-effect algebra then $Q \subseteq E$ is called a* sub-MV-effect algebra *of E iff Q is a sub-lattice effect algebra of E.*

Example 7. Let us consider the following sets of real functions:

$$E_1 = \{f : [0, 1] \to [0, 1] \mid f \text{ is a function}\}$$
$$E_2 = \{f \in E_1 \mid f \text{ is continuous}\}$$
$$E_3 = \{f \in E_1 \mid f(x) = ax^2 + bx + c; a, b, c \in (-\infty, \infty), x \in [0, 1]\}$$
$$E_4 = \{f \in E_1 \mid f(x) = kx + q; k, q \in (-\infty, \infty), x \in [0, 1]\}$$
$$E_5 = \{f \in E_1 \mid f(x) = d; d \in [0, 1], x \in [0, 1]\}.$$

We define the following partial binary operation $\oplus$ on E_1:

For $f, g \in E_1$, $f \oplus g$ is defined iff $f + g \leq 1$ and then $f \oplus g = f + g$. Moreover, we denote the function $f(x) = 0$ for all $x \in [0, 1]$ by 0 and the function $g(x) = 1$ for all $x \in [0, 1]$ by 1. Then $(E_1; \oplus, 0, 1)$ is a complete MV-effect algebra.

- For $k > n$, $(k, n \in \{1, 2, \ldots, 5\})$ E_k is a sub-effect algebra of E_n.
- E_2 is a sub-lattice effect algebra of a complete MV-effect algebra E_1.
- E_3 is not a lattice ordered effect algebra ($f \vee g$ does not exist for functions $f(x) = (x - \frac{1}{3})^2$, $g(x) = (x - \frac{2}{3})^2$, $x \in [0, 1]$, but $f \wedge g = 0$).
- E_4 is a lattice effect algebra but E_4 is not a sublattice of E_3 (E_2, E_1, respectively).
- E_5 is a complete MV-effect algebra.

We see that a sub-effect algebra of MV-effect algebra need not be an MV-effect algebra, even it need not be lattice ordered.

3   Prelattice Generalized Effect Algebras

In this section, it is of our interest to answer a question for which generalized effect algebra P the effect algebraic extension $E = P \dot\cup P^*$ (from Theorem 1) is lattice ordered, and all joins and meets existing in P (denoted by $a \vee_P b$, $a \wedge_P b$) are preserved for E. We will call such generalized effect algebra a *prelattice generalized effect algebra* (see [27]).

It is rather surprising that a prelattice effect algebra P need not be lattice ordered. *In general, a prelattice generalized effect algebra P need not be a sublattice of the lattice effect algebra $E = P \dot\cup P^*$.*

Example 8. Let $P = \{0, a, b, a \oplus a, b \oplus b\}$ be a generalized effect algebra. It is easy to check that $E = P \dot\cup P^*$ is a lattice effect algebra in spite of the fact that P is not a lattice, as, e.g., $a \vee_P b$ does not exist (see Fig. 3).

Assume that P is a generalized effect algebra. For the effect algebraic extension $E = P \dot\cup P^*$ the partial order on E, when restricted to P, coincides with the

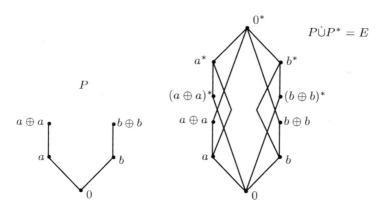

Fig. 3.

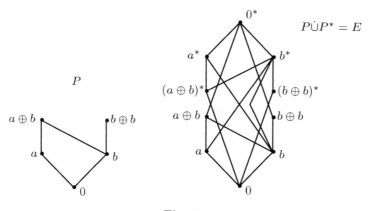

Fig. 4.

original partial order on P. In spite of this fact, the lattice operations join and meet of elements $a, b \in P$ ($a \vee_P b$, $a \wedge_P b$) need not be preserved for E, if they exist in P.

Example 9. Let $P = \{0, a, b, a \oplus b, b \oplus b\}$ be a generalized effect algebra and let $E = P \dot\cup P^*$ be its effect algebraic extension. Obviously, $a \oplus b = a \vee_P b$, while $a \vee_E b$ does not exist (see Fig. 4).

The following theorems establishe a necessary and sufficient condition for the inheritance of $a \vee_P b$, for $a, b \in P$, by the effect algebra $E = P \dot\cup P^*$ (see [27]).

Theorem 2. *Let P be a generalized effect algebra and $E = P \dot\cup P^*$. For every $a, b \in P$ with $a \vee_P b$ existing in P the following conditions are equivalent*

(i) $a \vee_E b$ *exists and* $a \vee_E b = a \vee_P b$.
(ii) *For every $c \in P$ the existence of $a \oplus c$ and $b \oplus c$ implies the existence of $(a \vee_P b) \oplus c$.*

Proof. (i)$\Rightarrow$(ii): If $a \oplus c$ and $b \oplus c$ exists in P then $a \leq c^*$ and $b \leq c^*$ which gives $a \vee_E b \leq c^*$ and hence $(a \vee_E b) \oplus c = (a \vee_P b) \oplus c$ exists in P.

(ii)$\Rightarrow$(i): By the assumptions for all $c, d \in P$ we have: $a, b \leq d$ implies $a \vee_P b \leq d$, as well as $a, b \leq c^*$ implies $a \vee_P b \leq c^*$, since $(a \vee_P b) \oplus c$ exists in P. Thus $a \vee_E b = a \vee_P b$.

Theorem 3. *Let P be a generalized effect algebra. Then $E = P \dot\cup P^*$ is a lattice effect algebra preserving joins and meets existing in P if and only if the following conditions are satisfied for all $a, b \in P$:*

(i) $a \wedge_P b$ *exists.*
(ii) *If there is $d \in P$ such that $a, b \leq d$ then $a \vee_P b$ exists.*
(iii) *For all $c \in P$ the existence of $a \vee_P b$, $a \oplus c$ and $b \oplus c$ implies the existence of $(a \vee_P b) \oplus c$.*
(iv) *Either $a \vee_p b$ exists or $\bigvee \{c \in P \mid a \oplus c$ and $b \oplus c$ are defined$\}$ exists in P.*
(v) $\bigvee \{c \in P \mid c \leq b$ and $a \oplus c$ is defined$\}$ *exists in P.*

Proof. Let $a, b \in P$. If $c \in E$ and $c \leq a, b$ then $c \in P$ and hence $a \wedge_E b$ exists iff $a \wedge_P b$ exists, in which case $a \wedge_E b = a \wedge_P b$.

Let $d \in P$ such that $a, b \leq d$. Then the existence of $a \vee_E b$ implies $a \vee_E b \leq d$, which gives $a \vee_E b \in P$ and hence there is $a \vee_P b$ and $a \vee_P b = a \vee_E b$. Conversely, by (iii), the existence of $a \vee_P b$ implies that for all $c \in P$ with $a, b \leq c^*$ we have $a \vee_P b \leq c^*$ and hence there is $a \vee_E b$ and $a \vee_E b = a \vee_P b$.

If there is no $d \in P$ with $a, b \leq d$ then $a \vee_P b$ does not exist and then $a \vee_E b$ exists iff there is $x \in P$ such that $x = \bigvee \{c \in P \mid a \oplus c$ and $b \oplus c$ are defined$\}$, in which case $a \vee_E b = x^*$. Hence $a \vee_E b$ exists by (iv).

Finally, $a^* \wedge_E b$ exists iff there is $y \in P$ such that $y = \bigvee \{c \in P \mid c \leq b$ and $a \oplus c$ is defined$\}$. In this case $a^* \wedge_E b = y$. Thus, using de Morgan laws, we obtain that E is a lattice effect algebra iff (i)–(v) are satisfied for every pair $a, b \in P$.

Note that if for $a, b \in P$ in Theorem 3 the element $a \vee_P b$ exists then the existence of $\bigvee \{c \in P \mid a \oplus c$ and $b \oplus c$ are defined$\}$ in P is not necessary to obtain $E = P \dot\cup P^*$ lattice ordered. For instance, this occurs when $P = [0, \infty)$ with the usual addition.

Definition 9. *A generalized effect algebra P satisfying conditions* (i)–(v) *of Theorem 3 is called a* prelattice generalized effect algebra.

Theorem 4. *Let P be an effect agebra and let $E = P \dot\cup P^*$. Then*

(i) 1^* *is an atom of E.*

(ii) $a \oplus 1^* = (a')^*$, *for every $a \in P$.*

(iii) $a \oplus b^* = (b \ominus a)^*$, *for all $a, b \in P$ with $a \leq b$.*

(iv) $E \cong P \times \{0, 1^*\}$.

(v) E *is a lattice effect algebra iff P is a lattice effect algebra, in which case P is a sublattice of E.*

(vi) E *is a distributive or modular lattice effect algebra or MV-effect algebra iff P has these properties.*

Proof. (i) Since for $a \in P$ the existence of $a \oplus 1$ implies $a = 0$, we obtain that the condition $a \leq 1^*$ implies $a = 0$. Moreover, for all $a \in P$ we have $1^* \leq a^*$ and hence 1^* is an atom of E.

(ii) Let $a \in P$. Then $a \leq 1 = (1^*)^*$, hence $a \oplus 1^*$ exists in E and $a \oplus 1^* = 0^* \ominus (a \oplus 1^*) = (0^* \ominus 1^*) \ominus a = (1 \ominus a) = a'$. We obtain that $(a \oplus 1^*)^* = a'$ which gives $a \oplus 1^* = (a')^*$.

(iii) If $a, b \in P$ with $a \leq b$ then $a \leq (b^*)^*$ which gives the existence of $a \oplus b^*$. Further, $(a \oplus b^*)^* = 0^* \ominus (b^* \oplus a) = (0^* \ominus b^*) \ominus a = b \ominus a$, hence $a \oplus b^* = (b \ominus a)^*$.

(iv) Let us define a map $\varphi : E \rightarrow P \times \{0, 1^*\}$ as follows: for $a \in P$ let $\varphi(a) = (a, 0)$ and $\varphi(a^*) = (a', 1^*)$. Evidently φ is a bijection of E onto $P \times \{0, 1^*\}$. Further, if $a, b \in P$ and $a \oplus b$ is defined in P then $\varphi(a \oplus b) = (a \oplus b, 0) = (a, 0) \oplus (b, 0) = \varphi(a) \oplus \varphi(b)$. If $a \in P$, $b^* \in P^*$ and $a \oplus b^*$ is defined in E then by (iii) we have $\varphi(a \oplus b^*) = \varphi((b \ominus a)^*) = ((b \ominus a)', 1^*) = (b' \oplus a, 1^*) = (b', 1^*) \oplus (a, 0) = \varphi(a) \oplus \varphi(b^*)$. If $a^*, b^* \in P^*$ then $a^* \oplus b^*$ does not exist. This proves that φ is an isomorphism.

(v) Evidently, a lattice effect algebra P satisfies conditions (i)–(v) of Theorem 3, which implies that $E = P \dot\cup P^*$ is a lattice effect algebra in which P is a sublattice. Conversely, if E is a lattice then P is a sublattice of E, since P is a bounded poset.

(vi) This follows by the fact that E is a direct product of P and the Boolean algebra $\{0, 1^*\}$.

4   Blocks of Lattice Effect Algebras

Compatibility of elements of effect algebras (D-posets) were introduced by Kôpka in [7] and Chovanec and Kôpka in [2]. Moreover, they have shown that the family of MV-effect algebras coincides with the family of lattice effect algebras with pairwise compatible elements. In this section we are going to show that every lattice effect algebra E is a set-theoretical union of MV-effect algebras (MV-algebras, see Example 5) called blocks of E. Here a block of E is any maximal subset of pairwise compatible elements of E (see [14]).

Lemma 1. *For elements of a lattice effect algebra* $(E; \oplus, 0, 1)$ *the following conditions are satisfied:*

(i) *If* $u \leq a$, $v \leq b$ *and* $a \oplus b$ *is defined then* $u \oplus v$ *is defined.*
(ii) *If* $b \oplus c$ *is defined, then* $a \leq b$ *iff* $a \oplus c \leq b \oplus c$.
(iii) *If* $a \oplus c$ *and* $b \oplus c$ *are defined, then* $(a \oplus c) \vee (b \oplus c) = (a \vee b) \oplus c$.
(iv) $a \leq b$ *iff* $b' = 1 \ominus b \leq 1 \ominus a = a'$.
(v) *If* $a \leq b'$, *then* $a \oplus b = (a \vee b) \oplus (a \wedge b)$.

The proof is left to the reader.

Definition 10. *Elements* a, b *of a lattice effect algebra* $(E; \oplus, 0, 1)$ *are compatible (denoted by* $a \leftrightarrow b$*) if* $a \vee b = a \oplus (b \ominus (a \wedge b))$.

Lemma 2. *([27]) Let* E *be a lattice effect algebra and let* $a, b \in E$. *The following conditions are equivalent:*

(i) $a \leftrightarrow b$.
(ii) $(a \ominus (a \wedge b)) \oplus (b \ominus (a \wedge b))$ *is defined.*

Proof. Since E is a lattice effect algebra, for all $a, b \in E$ we have $0 = (a \wedge b) \ominus (a \wedge b) = (a \ominus (a \wedge b)) \wedge (b \ominus (a \wedge b))$, see [3, p. 70]. Assume that $a \leftrightarrow b$. Then $a \vee b = b \oplus (a \ominus (a \wedge b)) = (a \wedge b) \oplus (b \ominus (a \wedge b)) \oplus (a \ominus (a \wedge b))$ which implies (ii). Conversely (ii) implies $(a \ominus (a \wedge b)) \oplus (b \ominus (a \wedge b)) = (a \ominus (a \wedge b)) \vee (b \ominus (a \wedge b)) \leq (a \wedge b)'$ which gives that $(a \wedge b) \oplus (a \ominus (a \wedge b)) \oplus (b \ominus (a \wedge b)) = [(a \ominus (a \wedge b)) \oplus (a \wedge b)] \vee [(b \ominus (a \wedge b)) \oplus (a \wedge b)] = a \vee b$ which implies that $a \vee b = b \oplus (a \ominus (a \wedge b))$.

Theorem 5. *[14] Let* $(E; \oplus, 0, 1)$ *be a lattice effect algebra and let* $x, y, z \in P$ *be such that* $x \leftrightarrow z$ *and* $y \leftrightarrow z$. *Then*

(i) $x \vee y \leftrightarrow z$,
(ii) *if* $x \leq y$ *then* $y \ominus x \leftrightarrow z$,
(iii) $x' = 1 \ominus x \leftrightarrow z$,
(iv) $x \wedge y \leftrightarrow z$,
(v) $x \oplus y \leftrightarrow z$.

Proof. By assumptions there exist $x \oplus (z \ominus (x \wedge z))$ and $y \oplus (z \ominus (y \wedge z))$.

(i) Since $x \wedge z, y \wedge z \leq (x \vee y) \wedge z \leq z$ we obtain $z \ominus ((x \vee y) \wedge z) \leq z \ominus (x \wedge z), z \ominus (y \wedge z)$ and hence $(x \oplus (z \ominus (x \wedge z))) \vee (y \oplus (z \ominus (y \wedge z))) \geq x \oplus (z \ominus ((x \vee y) \wedge z)) \vee (y \oplus (z \ominus ((x \vee y) \wedge z))) = (x \vee y) \oplus (z \ominus ((x \vee y) \wedge z))$ which implies that $x \vee y \leftrightarrow z$.

(ii) If $x \leq y$ then $x \wedge z \leq y \wedge z$ and $x \vee z \leq y \vee z$. It follows that there exists $w \in P$ such that $(x \wedge z) \oplus w = y \wedge z$ and $x \vee z = x \oplus (z \ominus (x \wedge z)) \leq y \vee z = y \oplus (z \ominus (y \wedge z)) = (y \wedge z) \oplus (y \ominus (y \wedge z)) \oplus (z \ominus (y \wedge z))$, thus $(x \wedge z) \oplus (x \ominus (x \wedge z)) \oplus (z \ominus (x \wedge z)) \leq (y \wedge z) \oplus (y \ominus (y \wedge z)) \oplus (z \ominus (y \wedge z))$ and since $z = (x \wedge z) \oplus (z \ominus (x \wedge z)) = (y \wedge z) \oplus (z \ominus (y \wedge z))$ we obtain $x \ominus (x \wedge z) \leq y \ominus (y \wedge z)$. The last implies that there is $e \in P$ such that $(x \ominus (x \wedge z)) \oplus e = y \ominus (y \wedge z)$. We obtain $y = (x \wedge z) \oplus w \oplus e \oplus (x \ominus (x \wedge z))$ and $y \oplus (z \ominus (y \wedge z)) = (x \wedge z) \oplus w \oplus e \oplus (x \ominus (x \wedge z)) \oplus (z \ominus (y \wedge z))$. These equalities imply that $y \ominus x = w \oplus e$ and $z = w \oplus [(x \wedge z) \oplus (z \ominus (y \wedge z))]$, since

$(x \wedge z) \oplus w = y \wedge z$. We conclude that $y \ominus x \leftrightarrow z$ since $w \oplus e \oplus [(x \wedge z) \oplus (z \ominus (y \wedge z))]$ exists.

(iii) Evidently $1 \leftrightarrow z$ and $x \leq 1$, which implies by (ii) that $x' = 1 \ominus x \leftrightarrow z$.

(iv) By (iii) $x' \leftrightarrow z$ and $y' \leftrightarrow z$ which by (i) implies that $x' \vee y' \leftrightarrow z$ and by (iii) $x \wedge y = (x' \vee y')' \leftrightarrow z$.

(v) $x \oplus y = 1 \ominus (x' \ominus y) \leftrightarrow z$ by conditions (ii) and (iii).

Lemma 3. *If $(E; \oplus, 0, 1)$ is a lattice effect algebra of pairwise compatible elements then it can be organized into an MV-algebra if we put $a^* = 1 \ominus a$, $a - b = a \ominus (a \wedge b)$ and $a \oplus b = (a^* - b)^*$ for all $a, b \in E$. Hence E is an MV-effect algebra (see Example 5).*

The common consequence of Theorem 5 and Lemma 3 is the following:

Corollary 1. *Every maximal subset M of pairwise compatible elements of a lattice effect algebra $(E; \oplus, 0, 1)$ is a sublattice and a sub-effect algebra of E. Moreover, M is an MV-effect algebra in its own right.*

Theorem 6. *[14] Every lattice effect algebra P is a set-theoretical union of MV-effect algebras, its blocks. Every subset $A \subseteq P$ of mutually compatible elements is contained in a block.*

Proof. Let $\emptyset \neq A \subseteq P$ be a set of mutually compatible elements of P and $\mathcal{A} = \{B \subseteq P \mid A \subseteq B, B \text{ is a set of mutually compatible elements}\}$. Then every chain $\mathcal{B} \subseteq \mathcal{A}$ (i.e., for $X, Y \in \mathcal{B}$ we have $X \subseteq Y$ or $Y \subseteq X$) the set $\bigcup \mathcal{B} \in \mathcal{A}$. By the maximal principle there exists a maximal element $M \in \mathcal{A}$. Moreover for $a \in P$ the set $A = \{0, a, a', 1\}$ is mutually compatible. □

In view of Lemma 3 and Corollary 1, *a lattice effect algebra E is an MV-effect algebra iff every pair of elements of E is compatible.*

It is worth noting that if E in Theorem 6 is an orthomodular lattice then every block of E is a Boolean algebra.

5 Generalized MV-Effect Algebras and Blocks of Prelattice Generalized Effect Algbras

In [27] it has been shown how generalized effect algebras can be introduced in order to obtain MV-effect algebras (MV-algebras) by their effect algebraic extensions. Moreover, every prelattice generalized effect algebra P is a set-theoretical union of generalized MV-algebras (blocks) which are maximal subsets of pairwise compatible elements in P. Further, connections between blocks of effect algebraic extension $E = P \dot\cup P^*$ of P and blocks of P have been shown.

If P is a prelattice generalized effect algebra then elements $a, b \in P$ are compatible in the lattice effect algebra $E = P \dot\cup P^*$ iff $(a \ominus (a \wedge b)) \oplus (b \ominus (a \wedge b))$ exists. Since in this case we have $(a \ominus (a \wedge b)) \oplus (b \ominus (a \wedge b)) \in P$, it makes sense to call elements a, b compatible in P. In this case $a \vee b \in P$ since $a \vee b = b \oplus (a \ominus (a \wedge b))$.

Definition 11. *Elements a, b of a prelattice generalized effect algebra P are called* compatible *if $(a \ominus (a \wedge b)) \oplus (b \ominus (a \wedge b))$ is defined.*

In what follows for $Q \subseteq P$ we will denote by Q^* the set $\{y^* \in P^* \mid y \in Q\}$.

Theorem 7. *Let P be a prelattice generalized effect algebra and $E = P \dot{\cup} P^*$. Let $M \subseteq E$ be a block of E. Then*

(i) $M \cap P^* = (M \cap P)^*$.

(ii) $M \cap P$ *is a maximal pairwise compatible subset of P and a sublattice of E. Conversely, if Q is a maximal subset of pairwise compatible elements of P then $Q \dot{\cup} Q^*$ is a block of E.*

(iii) $M \cap P$ *is a prelattice generalized effect algebra and $(M \cap P) \dot{\cup} (M \cap P)^* = M$.*

Proof. (i) For $x, y \in E$ we have $x \leftrightarrow y$ iff $x \leftrightarrow y^*$ (see Theorem 5) which gives $y \in M$ iff $y^* \in M$ and therefore $y \in M \cap P$ iff $y^* \in M \cap P^*$. It follows that $M \cap P^* = (M \cap P)^*$.

(ii) Let $x \in P$ and $x \leftrightarrow y$ for all $y \in M \cap P$. Then $x \leftrightarrow y^*$ for all $y \in M \cap P^*$ and thus $x \leftrightarrow y$ for all $y \in M$ which gives $x \in M$ by maximality of M. Further, by Theorem 5, if $x \leftrightarrow y$ and $x \leftrightarrow z$ then $x \leftrightarrow y \vee z$ and $x \leftrightarrow y \wedge z$, therefore $M \cap P$ is a sublattice of E, because $y, z \in P$ and $y \leftrightarrow z$ implies $y \vee_E z \in P$. If Q is a maximal subset of pairwise compatible elements of P then $Q \dot{\cup} Q^*$ is a maximal subset of pairwise compatible elements of E, hence $Q \dot{\cup} Q^*$ is a block of E.

(iii) $M \cap P$ is a generalized effect algebra since both M and P are generalized effect algebras. Further, P satisfies conditions (i)–(v) of Theorem 3 by the assumption that P is prelattice and M satisfies these conditions, since M is a lattice. Therefore $M \cap P$ is a prelattice generalized effect algebra. Obviously $M = (M \cap P) \dot{\cup} (M \cap P)^*$ as $M \cap P^* = (M \cap P)^*$ by (i).

Corollary 2. *Every prelattice generalized effect algebra P is a union of maximal subsets of pairwise compatible elements of P.*

Proof. If P is a prelattice effect algebra then $E = P \dot{\cup} P^*$ is a lattice effect algebra and by Theorem 6 we have $E = \bigcup \{M \subseteq E \mid M$ is a block of $E\}$. Therefore $P = \bigcup \{M \cap P \mid M$ is a block of $E\}$. The rest follows by Theorem 7, (ii).

Definition 12. *A maximal subset of pairwise compatible elements of a prelattice generalized effect algebra P is called a* block *of P. A prelattice generalized effect algebra with a unique block is called a* generalized MV-effect algebra.

Theorem 8. *For a generalized effect algebra P the following conditions are equivalent:*

(i) P *is a generalized MV-effect algebra.*

(ii) $E = P \dot{\cup} P^*$ *is an MV-effect algebra.*

(iii) P *is a prelattice generalized effect algebra and for all $a, b \in P$ the sum $(a \ominus (a \wedge b)) \oplus (b \ominus (a \wedge b))$ exists in P.*

The proof is straightforward.

Theorem 9. *A generalized effect algebra P is a generalized MV-effect algebra iff the following conditions are satisfied*

(i) *P is a lattice.*
(ii) *For all $a, b, c \in P$ the existence of $a \oplus c$ and $b \oplus c$ implies the existence of $(a \vee_P b) \oplus c$.*
(iii) *$\bigvee \{c \in P \mid a \oplus c \text{ exists and } c \leq b\}$ exists in P, for all $a, b \in P$.*
(iv) *$(a \ominus (a \wedge b)) \oplus (b \ominus (a \wedge b))$ exists for all $a, b \in P$.*

Proof. Obviously conditions (i)–(iii) imply conditions (i)–(v) of Theorem 3 hence P is a prelattice generalized effect algebra and condition (iv) implies that it has a unique block.

Conversely, if P is a generalized MV-effect algebra then obviously (ii)–(iv) are satisfied. Let $a, b \in P$ then $a \wedge_P b$ exists by Theorem 3, (i) and $a \wedge_P b = a \wedge_E b$. Further, there is $(a \ominus (a \wedge b)) \leq (a \wedge b)^*$ and since $E = P \dot{\cup} P^*$ is a lattice effect algebra we obtain

$$[(a \ominus (a \wedge b)) \oplus (b \ominus (a \wedge b))] \oplus (a \wedge b) = [(a \ominus (a \wedge b) \vee (b \ominus (a \wedge b))] \oplus (a \wedge b) = a \vee b \in P.$$

Theorem 10

(i) *Every prelattice generalized effect algebra is a union of generalized MV-effect algebras (blocks).*
(ii) *A generalized MV-effect algebra P is an MV-effect algebra iff there exists an element $1 \in P$ such that for every $a \in P$ there exists a unique $b \in P$ for which $a \oplus b = 1$.*

The proof follows by Theorems 7 and 8.

Example 10. The set $P_1 = \{0, 1, 2, 3, \ldots\}$ of nonnegative integers with the usual addition and the set $P_2 = [0, \infty)$ of nonnegative real numbers with usual addition are examples of generalized MV-effect algebras. It is easy to see that $E_1 = P_1 \dot{\cup} P_1^*$ and $E_2 = P_2 \dot{\cup} P_2^*$ are linearly (totally) ordered MV-effect algebras.

More generally, the positive cone G^+ of any lattice ordered abelian group $(G; +, 0, \leq)$ is a generalized effect algebra.

Further examples are all extendable commutative BCK-algebras directed upwards (see Section 7).

6   Sharp and Central Elements of Lattice and Generalized Prelattice Effect Algebras

The notion of sharp elements of effect algebras has been introduced by S. Gudder in [5]. In [27] it has been shown, how this notion can be introduced for generalized prelattice effect algebras in order of its inheritance from their effect algebraic extensions.

Definition 13. *Let $(E; \oplus; 0, 1)$ be an effect algebra. An element $z \in E$ is sharp if $z \wedge z' = 0$. Put $\mathcal{S}(E) = \{z \in E \mid z \wedge z' = 0\}$.*

It has been shown in [6] that in every lattice effect algebra the set $S(E)$ is a sub-lattice effect algebra. Moreover, $S(E)$ is an orthomodular lattice.

Definition 14. *An element z of a generalized effect algebra P is called a* sharp element *if for all $e \in P$ the conditions $e \leq z$ and $z \oplus e$ is defined imply that $e = 0$. Let $S(P) = \{z \in P \mid z$ a sharp element of $P\}$.*

Theorem 11. ([24]) *Let P be a prelattice generalized effect algebra and let $S(P) = \{z \in P \mid z$ is a sharp element of $P\}$. Let $E = P \dot{\cup} P^*$. Then*

(i) $S(P) = S(E) \cap P$.

(ii) *If $z_1, z_2 \in S(P)$ and $z_1 \oplus z_2$ is defined in E then $z_1 \oplus z_2 \in S(P)$.*

(iii) *$S(P)$ is a prelattice generalized effect algebra and $S(E) = S(P) \dot{\cup} (S(P))^*$, when $S(E)$ is considered as lattice effect algebra and $(S(P))^* = \{z^* \mid z \in S(P)\}$.*

(iv) *$S(P)$ is a generalized orthomodular poset being a proper ideal in the orthomodular lattice $S(E)$, closed under orthogonal joins and for every $z \in S(E)$ either $z \in S(P)$ or $0^* \ominus z \in S(P)$.*

Proof. (i) Since E is a lattice effect algebra, for $e, z \in P$ we have $e \leq z \wedge z^*$ iff $e \leq z$ and $z \oplus e$ is defined. It follows that $z \wedge z^* = 0$ iff for all $e \in P$ the conditions $e \leq z$ and $z \oplus e$ is defined imply $e = 0$. It follows that $z \in S(P)$ iff $z \in S(E) \cap P$.

(ii) If $z_1, z_2 \in S(P)$ and $z_1 \oplus z_2$ is defined in E then $z_1 \leq z_2^*$ which gives that $z_1 \wedge z_2 \leq z_2^* \wedge z_2 = 0$ and hence $z_1 \oplus z_2 = z_1 \vee z_2 \in S(E) \cap P$ because $S(P) \subseteq P$, P is closed under $\oplus$ and $S(E)$ is a sublattice of E.

(iii) Since $S(E)$ is a sublattice and a sub-effect algebra of a lattice effect algebra E we may consider $S(E)$ as a lattice effect algebra in its own right. Further, by (i) and (ii), $S(P)$ is a proper order ideal in $S(E)$ closed under $\oplus$. If we set $(S(P))^* = \{z^* \mid z \in S(P)\}$ then $S(E) = S(P) \dot{\cup} (S(P))^*$ and by [3, Proposition 1.2.7] we obtain that the effect algebra $S(P) \dot{\cup} (S(P))^*$ coincides with $S(E)$.

(iv) This follows by (ii) and (iii) and the facts that for all $z_1, z_2 \in S(E)$ with $z_1 \leq z_2^*$ we have $z_1 \wedge z_2 = 0$ and $z_1 \oplus z_2 = z_1 \vee z_2$, under which $S(E) = S(P) \dot{\cup} (S(P))^*$.

Definition 15. *Let P be a generalized MV-effect algebra. Then $Q \subseteq P$ is called a* sub-generalized MV-effect algebra *of P if Q is simultaneously a sub-generalized effect algebra and a sub-lattice of P.*

Theorem 12. *Let P be a generalized MV-effect algebra. The following conditions are equivalent:*

(i) $z \in S(P)$,

(ii) *$[0, z] = \{x \in P \mid x \leq z\}$ with $\oplus|_{[0,z]}$ ($\oplus$ restricted to $[0, z]$) is an MV-effect algebra being a sub-generalized MV-effect algebra of P.*

Proof. (i)$\Rightarrow$(ii): Let $x, y \leq z$ with $x \oplus y$ defined in P. Let $E = P \dot{\cup} P^*$. Then $z^* \in S(E)$ which gives $z \wedge z^* = 0$. By the assumptions we have $z^* \leq y^*$ which gives that $y \oplus z^*$ is defined in E and $y \oplus z^* = y \vee z^* \leq x^*$, since $y \leq z$ implies $y \wedge z^* \leq z \wedge z^* = 0$. It follows that $x \oplus y \oplus z^*$ is defined in E and hence $x \oplus y \leq z$. Thus $[0, z]$ is closed under $\oplus$ and it is a sublattice of P. It follows that $([0, z]; \oplus|_{[0,z]}, 0, z)$ is an MV-effect algebra in its own right, since for all $x, y \leq z$ we have $(x \ominus (x \wedge y)) \oplus (y \ominus (x \wedge y)) \leq z$.

(ii)$\Rightarrow$(i): Let $e \leq z$ and $e \oplus z$ be defined in P. Then $e, z \leq z \Rightarrow e \oplus z \in [0, z]$ which gives $e \oplus z \leq z$ and hence $e = 0$. Thus $z \in S(P)$.

Finally, note that the notion of a *central element* of a generalized effect algebra P has been introduced in [11]. Recall that $z \in P$ is central iff P is isomorphic to a direct product of $[0, z]$ and $Q_z = \{x \in P \mid x \wedge z = 0\}$. Moreover $z \in P$ is central element of P iff it is a central element of $E = P \dot\cup P^*$ iff E is isomorphic to the direct product $[0, z] \times [0, z^*]$ and then for Q_z defined above we have $Q_z = P \cap [0, z^*]$ (see [11, Section 5]). Thus if $C(E) = \{z \in E \mid z$ is central element of $E\}$ and $C(P) = \{z \in E \mid z$ is central element of $P\}$ then $C(P) = C(E) \cap P$.

For a lattice effect algebra E the subset $B(E) = \bigcap\{M \subseteq E \mid M$ a block of $E\}$ is called a compatibility center of E. By Theorem 7, for a prelattice generalized effect algebra P and lattice effect algebra $E = P \dot\cup P^*$ we obtain that $B(E) \cap P = \bigcap\{M \cap P \mid M$ is a block of $E\}$. We will call $B(E) \cap P$ a *compatibility center of P* and denote it by $B(P)$. Since for every lattice effect algebra E the equality $C(E) = S(E) \cap B(E)$ holds (see [12, Theorem 2.5, (iv)]), we obtain:

Theorem 13. *For every prelattice generalized effect algebra P the condition $C(P) = S(P) \cap B(P)$ is satisfied. If P is a generalized MV-effect algebra then $C(P) = S(P)$.*

As a consequence of Theorem 13 we obtain that *an element z of a prelattice generalized effect algebra P is central iff z is a sharp element of P compatible with every element of P.* Because $D \subseteq P$ is a block of P iff there is a block M of $E = P \dot\cup P^*$ such that $D = M \cap P$ and $M \cap P^* = (M \cap P)^* = D^*$ we obtain that $B(P)$ is a generalized MV-effect algebra such that the MV-algebra $B(E) = B(P) \dot\cup (B(P))^*$. Moreover, the center $C(P)$ is a proper order ideal in the Boolean algebra $C(E)$ closed under $\oplus$ and such that for every $z \in C(E)$ either $z \in C(P)$ or $0^* \ominus z \in C(P)$ and thus $C(E) = C(P) \dot\cup (C(P))^* = (C(E) \cap P) \dot\cup (C(E) \cap P^*)$, hence $C(P)$ is a generalized Boolean algebra.

7 BCK Algebras Equivalent to Generalized MV-Effect Algebras

We are going to show that Dedekind complete (or, more generally, extendable) commutative BCK-algebras directed upwards are equivalent to generalized MV-effect algebras (see [24]).

A result due to Yutani [30] says that the class of commutative BCK-algebras forms a variety, hence it is equationally definable. We present this equational base to set up these algebraic structures (see also [3, Theorem 5.1.18]).

Definition 16. *An algebra $(X; *, 0)$ of type $(2, 0)$ is a commutative BCK-algebra if the following conditions are satisfied for all $x, y, z \in X$:*

(i) $x * (x * y) = y * (y * x)$,
(ii) $(x * y) * z = x * (z * y)$,
(iii) $x * x = 0$,
(iv) $x * 0 = x$.

We can define a partial order $\leq$ (called the BCK-order) in $X = (X; *, 0)$ by $x \leq y$ iff $x * y = 0$. Then 0 is the least element of $(X; \leq)$. Condition (i) in Definition 16 is called *commutativity* and it makes X a lower semilattice with respect to the BCK-order. It means that for all $x, y \in X$ the greatest lower bound $x \wedge y$ exists in X and $x \wedge y = x * (x * y)$. We say that a commutative BCK-algebra has the *relative cancellation property* if for $a, x, y \in X$ with $a \leq x, y$ we have that $x * a = y * a$ implies $x = y$.

Theorem 14. *Let $(X; *, 0)$ be a commutative BCK-algebra having the relative cancellation property. Let a partial binary operator $\oplus$ for elements $a, b, c \in X$ be defined by*

(GEA) *$a \oplus b$ is defined and $a \oplus b = c$ iff $c \geq b$ and $c * b = a$.*

Then $(X; \oplus, 0)$ is a generalized effect algebra and the BCK-order coincides with the partial order derived from $\oplus$.

Proof. The fulfilling of the conditions (GEi)–(GEiv) has been proved in [3, p. 332]. Let us show the fulfillment of (GEv). Let $a \oplus b = 0$ then by property (VI) in Theorem 5.2.6 of [3] we obtain that $0 \leq a$ implies $0 \oplus 0 \leq a \oplus 0 \leq a \oplus b = 0 = 0 \oplus 0$ and hence $a \oplus 0 \leq 0 \oplus 0$ which gives $a \leq 0$. Hence $a = 0$ and by the same manner $b = 0$. This proves that $(X; \oplus, 0)$ is a generalized effect algebra. It follows that for $a, b, c \in X$ we have $a \oplus b$ is defined and $a \oplus b = c$ iff $b \leq c$ and $c \ominus b = a$ which implies that partial orders in $(X; *, 0)$ and $(X; \oplus, 0)$ coincide. □

Recall that a poset $(X; \leq)$ is called *directed upwards* if for every $a, b \in X$ there is $c \in X$ such that $a \leq c$ and $b \leq c$. A commutative BCK-algebra X is called directed upwards if it is directed upwards with respect to the BCK-order. In [3, Lemma 5.2.2] it has been shown that every upwards directed commutative BCK-algebra has the relative cancellative property.

Definition 17. *Let $(X; *, 0)$ be a directed upwards commutative BCK-algebra. Then X is called* extendable *if for all elements $a, b \in X$ the set $\{c \in X \mid c \leq b$ and there is $d \geq a$ with $d * a = c\}$ has the supremum in X.*

Recall that a directed upwards commutative BCK-algebra X is called *Dedekind complete* if every $P \subseteq X$ with an upper bound (lower bound) in X has the supremum (infimum) in X. Clearly, in this case X is extendable.

We are going to show that extendable upwards directed commutative BCK-algebras are unbounded versions of bounded commutative BCK-algebras and if they are Archimedean then they can be embedded densely into complete commutative BCK-algebras .

We say that a commutative BCK-algebra $(X; *, 0)$ with the relative cancellation property is *Archimedean* iff the generalized effect algebra $(X; \oplus, 0)$ derived by (GEA) is Archimedean iff for every nonzero element x there is a greatest positive integer n such that $nx = x \oplus x \oplus \ldots \oplus x$ (n times) is defined in X.

Theorem 15. (i) *Let $(X; *, 0)$ be an extendable directed upwards commutative BCK-algebra. Let a partial operation $\oplus$ on X be defined by*
(GEA) $a \oplus b$ *is defined and* $a \oplus b = c$ *iff* $c \geq b$ *and* $c * b = a$
Then $(X; \oplus, 0)$ is a generalized MV-effect algebra.
(ii) *Let $(X; \oplus, 0)$ be a generalized MV-effect algebra. Let a binary operation $*$ on X be defined by*
(BCK) $a * b = a \ominus (a \wedge b)$, *for all $a, b \in X$.*
Then $(X; *, 0)$ is an extendable directed upwards commutative BCK-algebra. In both cases (i) and (ii), the partial orders in $(X; *, 0)$ and $(X; \oplus, 0)$ coincide.

Proof. (i) By Theorem 14 $(X; \oplus, 0)$ is a generalized effect algebra and partial orders on X derived from $\oplus$ and $*$ coincide. By [3, Prop. 5.1.19], X is a distributive lattice. Hence the existence of $a \oplus c$ and $b \oplus c$ implies that $(a \oplus c) \vee (b \oplus c)$ exists in X and by [3, Proposition 5.1.19, (iv)] we have

$$((a \oplus c) \vee (b \oplus c)) \ominus c = ((a \oplus c) \ominus c) \vee ((b \oplus c) \ominus c) = a \vee b$$

which implies that $(a \oplus c) \vee (b \oplus c) = (a \vee b) \oplus c$. By Theorem 3, the effect algebraic extension $E = X \dot{\cup} X^*$ is a lattice effect algebra, since for all $a, b \in X$ we have $\{c \in X \mid c \leq b, \ a \oplus c \text{ is defined } \} = \{c \in X \mid c \leq b \text{ and there is } d \geq a \text{ with } d * a = c\}$. Hence $(X; *, 0)$ is extendable iff $(X; \oplus, 0)$ is prelattice. Moreover, by [3, Theorem 5.2.6] we have, for all $x, y \in X$

$$(x \vee y) * y = x * (x \wedge y)$$

and because $y \leq x \vee y$ and $x \wedge y \leq x$ we obtain $(x \vee y) \ominus y = x \ominus (x \wedge y)$ which gives that $x \leftrightarrow y$ and hence $(x \ominus (x \wedge y)) \oplus (y \ominus (x \wedge y))$ exists in X by Lemma 2. Thus, by Theorem 9 we obtain that $(X; \oplus, 0)$ is a generalized MV-effect algebra.

(ii) By Theorem 8 the effect algebraic extension $E = X \dot{\cup} X^*$ is an MV-effect algebra. It follows that $(E; *, 0, 1)$ with binary operation $*$ defined by (BCK) is a bounded commutative BCK-algebra (see [3, Theorem 6.33]). It follows that $(X; *, 0)$ with inherited $*$ operation is a commutative BCK-algebra. This is because X is an order ideal in E and hence for $a, b \in X$ we have $a * b = a \ominus (a \wedge b) \leq a$ which gives $a * b \in X$. Moreover, X is a lattice and hence it is directed upwards. Further $(X; \oplus, 0)$ is a prelattice which implies that $(X; *, 0)$ is extendable as we have shown in part (i). Clearly partial orders derived from $\oplus$ and $*$ coincide.

8   Embeddings of Archimedean Generalized MV-Effect Algebras into Complete MV-Effect Algebras

We can show that when the MacNeille completion $MC(P)$ of a generalized effect algebra P cannot be organized into a complete effect algebra by extending the operation $\oplus$ onto $MC(P)$ then still P may be densely embedded into a complete effect algebra (see [24]). Namely, we show these facts for Archimedean generalized MV-effect algebras (GMV-effect algebras, for short).

Definition 18. *Let P be a GMV-effect algebra. Then $Q \subseteq P$ is called a sub-GMV-effect algebra of P if Q is simultaneously a sub-generalized effect algebra and a sub-lattice of P.*

It is well known that every poset P has a MacNeille completion $MC(P)$ (i.e., completion by cuts). By Schmidt [29] a MacNeille completion $MC(P)$ of a poset P is up to isomorphism (unique over P) any complete lattice into which P can be supremum-densely and infimum-densely embedded, which means that for every $x \in MC(P)$ there exist $M, Q \subseteq P$ such that $x = \bigvee \varphi(M) = \bigwedge \varphi(Q)$, where $\varphi : P \to MC(P)$ is the embedding. In this case $MC(P)$ preserves all infima and suprema existing in P.

In what follows (for simplicity of notations) for subsets Q_1, Q_2, Q of a generalized effect algebra we will write $Q_1 \oplus Q_2$ instead of $\{p \oplus q \mid p \in Q_1, q \in Q_2\}$ as well as $Q_1 \leq Q_2$ iff $p \leq q$ for all $p \in Q_1$ and $q \in Q_2$ and then we write $Q_2 \ominus Q_1$ instead of $\{q \ominus p \mid q \in Q_2, p \in Q_1\}$. We will write $Q \leq p$ instead of $Q \leq \{p\}$.

If for an element x of a generalized effect algebra P there is a maximal natural number n such that $nx = x \oplus x \cdots \oplus x$ (n times) exists in P then n is called an *isotropic index* of x (written $n = \text{ord}(x)$, or $n = n_x$ for short). Otherwise we put $\text{ord}(x) = \infty$. Hence P is *Archimedean* if every $x \in P$ has a finite isotropic index.

Remark 1. For an Archimedean GMV-effect algebra P without maximal elements, the MacNeille completion $MC(P)$ need not be a complete effect algebra including P as a sub-GMV-effect algebra. Moreover, P need not be meet-dense in $MC(E)$, where $E = P \dot\cup P^*$.

Example 11. Let Q be a direct product of effect algebras $Q_n = \{0, 1, 2, \ldots, n\}$, $n \in N$ (see Proposition 2). Let
$$P = \{(a_n)_{n=1}^\infty \in Q \mid a_n \neq 0 \text{ for only finite number of indices}\}.$$
with $\oplus$-operation inherited from Q. Then P is a GMV-effect algebra without maximal elements. Clearly,
$$P^* = \{(a_n)_{n=1}^\infty \in Q \mid a_n \neq n \text{ for only finite number of indices}\}.$$
Further, P is Archimedean and $E = P \dot\cup P^*$ is an Archimedean MV-effect algebra with $MC(E) = \prod_{n=1}^\infty \{\{0, 1, 2, \ldots, n\} \mid n \in N\} = Q$. Moreover, E is a sub-MV-effect algebra (of finite and cofinite elements) of $MC(E)$. On the other hand $MC(P) = P \cup \{b\}$, where $b = (n)_{n=1}^\infty$. $MC(P)$ is not a complete MV-effect algebra including P as a sub-GMV effect algebra, since $\oplus$ from P cannot be extended onto $MC(P)$ to make $MC(P)$ a complete effect algebra. Further, we see that P is join-dense but not meet dense in $MC(E)$, see also [13].

Let $(E; \leq)$ be a poset and $Q \subseteq P \subseteq E$. We will denote by $\bigvee_P Q$ ($\bigvee_E Q$) the least upper bound of Q in $(P; \leq)$ (in $(E; \leq)$). Here $(P; \leq)$ is a subposet of $(E; \leq)$.

Theorem 16. *Let P be an Archimedean GMV-effect algebra and let $E = P \dot\cup P^*$ be its effect-algebraic extension. Then*

(i) $MC(E)$ *is a complete MV-effect algebra including* P, *up to isomorphism, as a sub-GMV-effect algebra.*

(ii) *If* P *has a maximal element then* $MC(E) = MC(P) \oplus \{0, 1_P^*\} \cong MC(P) \times \{0, 1_P^*\}$, *where* $1_P = \bigvee_P P$.

(iii) *If* P *does not have maximal elements then* $\bigvee_E P = 0^*$ *is the unity of* $MC(E)$, P *is, up to isomorphism, join-dense in* $MC(E)$ *and* $MC(P)$ *need not be a complete MV-effect algebra extending* P.

Proof. (i) For every $a^* \in P^*$ the $\mathrm{ord}(a^*) = 1$ because $a^* \oplus a^*$ is not defined. It follows that E is Archimedean iff P is Archimedean. Thus, by [15, Theorem 3.4] we obtain that $MC(E)$ is a complete MV-effect algebra including E, up to isomorphism, as a sub-MV-effect algebra which gives that P is, up to isomorphism a sub-GMV-effect algebra of $MC(E)$. This is because P is a prelattice and sub-generalized effect algebra of E.

(ii) Let $u \in P$ be a maximal element in P. Then for every $y \in P$ we have $y \le u \vee y \le u$, since P is a lattice. It follows that $u = \bigvee_P P$ and P is an MV-effect algebra. By [27, Theorem 2.3, (iv)] we obtain that $E = P \oplus \{0, 1_P^*\} \cong P \times \{0, 1_P^*\}$, where $1_P = u = \bigvee_P P$. Since P is an Archimedean effect algebra we obtain that $MC(P)$ is a complete effect algebra containing P, up to isomorphism, as a join-dense sub-MV-effect algebra. Moreover, $MC(E) = MC(P) \oplus \{0, 1_P^*\} \cong MC(P) \times \{0, 1_P^*\}$.

(iii) Let P have no maximal elements. Let $x \in P$ be such that $P \le x^*$. Then $P \oplus x \subseteq P$, because P is closed under $\oplus$. It follows that $P \oplus x \le x^*$ and $P \oplus x \oplus x \subseteq P, \ldots P \oplus nx \le x^*, \ldots$, where $nx = x \oplus x \oplus \ldots \oplus x$ (n times) exists for every $n \in N$. Since P is Archimedean, we obtain that $x = 0$ and $\bigvee_E P = 0^*$. Clearly, 0^* is the unity of E. Further, for every $x^* \in P^*$ we obtain by [6, Theorem 2.1] that

$$x^* = x^* \wedge 0^* = x^* \wedge \bigvee_E P = \bigvee_E (P \wedge x^*),$$

because $x^* \leftrightarrow P$. Since P is an order ideal in E we have $P \wedge x^* \subseteq P$.

Let us put $\widehat{P} = \{x \in MC(E) \mid S \subseteq P,\ x = \bigvee_{MC(E)} S\}$. As we have just proved $E \subseteq \widehat{P}$. Let us show that $\widehat{P} = MC(E)$. Assume $A \subseteq \widehat{P}$, $A \ne \emptyset$. By definition of $\widehat{P}$, for every $x \in A$ there is $S_x \subseteq P$ with $\bigvee_{MC(E)} S_x = x$. It follows that $\bigvee_{MC(E)} A = \bigvee_{MC(E)} (\bigcup_{x \in A} S_x) \in \widehat{P}$, because $\bigcup_{x \in A} S_x \subseteq P$. Thus for all $Q \subseteq E$ we have $\bigvee_{MC(E)} Q \in \widehat{P}$, because $E \subseteq \widehat{P}$. Moreover, if $x \in MC(E)$ then there is $Q_x \subseteq E$ with $\bigvee_{MC(E)} Q_x = x$ and hence $x \in \widehat{P}$. We obtain $MC(E) \subseteq \widehat{P} \subseteq MC(E)$. Example 11 shows that $MC(P)$ need not be a complete MV-effect algebra extending P.

Corollary 3. *Every Archimedean GMV-effect algebra* P *is, up to isomorphism, a join-dense sub-GMV-effect algebra of a complete MV-effect algebra* $\widehat{E}$ *preserving all suprema and infima existing in* P.

Proof. If P has a maximal element then let $\widehat{E} = MC(P)$. If P does not have maximal elements then let $\widehat{E} = MC(P \dot\cup P^*)$. By Theorem 16, P is join-dense in

$\widehat{E}$, up to isomorphism (i.e., for every $x \in \widehat{E}$ there is $Q \subseteq P$ with $x = \bigvee_{\widehat{E}} \varphi(Q)$, where $\varphi : P \to \widehat{E}$ is the embedding). We may identify P with its isomorphic image in $\widehat{E}$. Then for every nonzero element $x \in \widehat{E}$ there is a nonzero element $a \in P$ such that $a \leq x$. Assume that $M \subseteq P$ and $\bigvee_P M = b \in P$. Let $z \in \widehat{E}$ and $M \leq z$. Then $M \leq b \wedge z$. If $b \wedge z \neq b$ then $b \ominus (b \wedge z) \neq 0$ and hence there is $c \neq 0$, $c \in P$ and $c \leq b \ominus (b \wedge z) \leq b$. It follows that

$$b \wedge z = b \ominus (b \ominus (b \wedge z)) \leq b \ominus c \in P$$

which gives that $M \leq b \wedge z \leq b \ominus c$. Thus $b \leq b \ominus c$ which implies that $c = 0$, a contradiction. We obtain that $b \wedge z = b$ and hence $b \leq z$. This proves that $\bigvee_{MC(E)} M = \bigvee_P M = b$. Similarly, if $\bigwedge_P M \in P$ then $\bigwedge_P M = \bigwedge_{MC(E)} M$.

Corollary 4. *The effect algebraic extension $E = P \dot{\cup} P^*$ of an Archimedean GMV-effect algebra P preserves all suprema and infima existing in P.*

Proof. Let $Q \subseteq P$ and let $\bigvee_P Q \in P$ exist. Then $\bigvee_{MC(E)} Q = \bigvee_P Q \in P$ by Corollary 3. It follows that $\bigvee_E Q = \bigvee_P Q$. By the same manner we obtain that $\bigwedge_E Q = \bigwedge_P Q$ if $\bigwedge_P Q$ exists.

9   Pastings of MV-Effect Algebras

In Section 4 we have shown that every lattice effect algebra is a set-theoretical union of MV-effect algebras (MV-algebras, see Example 5). The converse assertion is not true, i.e., a set-theoretical union of MV-effect algebras need not be a lattice effect algebra. If a union of MV-effect algebras $(M_\kappa; \oplus_\kappa, 0_\kappa, 1_\kappa)$ is a lattice effect algebra $(E; \oplus, 0, 1)$ such that for all κ: $0_\kappa = 0$, $1_\kappa = 1$ and the restriction of $\oplus$ onto M_κ coincides with $\oplus_\kappa$, then E is called a *pasting* of M_κ (see [25]).

Every MV-effect algebra M has the Riesz decomposition property (RDP, for short): $a, b, c \in M$ with $c \leq a \oplus b$ implies that there is $a_1 \leq a$ and $b_1 \leq b$ such that $c = a_1 \oplus b_1$.

Recall that the *length* of a finite chain is the number of its elements minus 1. The *length (height)* of a lattice L is finite if the supremum over the number of elements of chains in L equals to some natural number n and then $n - 1$ is called *length of the lattice L*.

Every *finite chain* $0 < a < 2a < \ldots < 1 = n_a a$ is a distributive effect algebra in which every pair of elements is compatible, hence it is an MV-effect algebra.

An element a of an effect algebra E is called an *atom* if $0 \leq b < a$ implies $b = 0$ and E is called *atomic* if for every $x \in E$, $x \neq 0$ there is an atom $a \in E$ with $a \leq x$. Clearly every finite effect algebra is atomic.

Definition 19. *Let E be an effect algebra and let $(E_\kappa)_{\kappa \in H}$ be a family of sub-effect algebras of E such that:*

(i) $E = \bigcup_{\kappa \in H} E_\kappa$.

(ii) If $x \in E_{\kappa_1} \setminus \{0,1\}$, $y \in E_{\kappa_2} \setminus \{0,1\}$ and $\kappa_1 \neq \kappa_2$, $\kappa_1, \kappa_2 \in H$, then $x \wedge y = 0$ and $x \vee y = 1$.

Then E is called a horizontal sum of effect algebras $(E_\kappa)_{\kappa \in H}$.

Example 12. If MV-effect algebras M_1 and M_2 are finite chains of different lengths then the horizontal sum of M_1 and M_2 is the unique lattice effect algebra E such that $\{M_1, M_2\}$ is the family of all blocks of E.

Really, assume that $M_1 = \{0, a, \ldots, n_a a\}$, $M_2 = \{0, b, \ldots, n_b b\}$ and $n_a < n_b$. Contrary to our claim, assume that $ka = \ell b$ for some $\ell \neq n_b$. Then $a \leq ka = \ell b < b'$, which gives $a \leftrightarrow b$ and hence, by [14], $M_1 \cup M_2$ is the set of pairwise compatible elements, a contradiction. Thus $ka = \ell b$ implies $\ell = n_b$ which gives $k = n_a$, because $n_b b \in S(E)$ while for $k < n_a$ we have $ka \notin S(E)$. This proves that $E = M_1 \cup M_2$ is the horizontal sum of its blocks M_1 and M_2.

If M_1 and M_2 have the same length, i.e., $n_a = n_b$ then M_1 and M_2 are isomorphic MV-effect algebras (by Proposition 2) and we can identify them. In this case we will call a and b isotropically equivalent.

Definition 20. *Let M_1 and M_2 be complete atomic MV-effect algebras, let A_1 and A_2 be the sets of all atoms of M_1 and M_2, respectively, and let $D_1 \subseteq A_1$ and $D_2 \subseteq A_2$. The sets D_1 and D_2 are called* isotropically equivalent *(written $D_1 \overset{istr}{\sim} D_2$) if there is a bijection $\varphi : D_1 \to D_2$ such that $n_p = ord(p) = n_{\varphi(p)} = ord(\varphi(p))$ for all $p \in D_1$. If $D_1 = \{p\}$ and $D_2 = \{q\}$ then p and q are called* isotropically equivalent atoms.

Example 13. Assume that M_1 and M_2 are complete atomic MV-effect algebras and $E = M_1 \cup M_2$ is an effect algebra such that $\{M_1, M_2\}$ is the family of all blocks in E. Then every atom p of E is an atom of M_1 or M_2 and conversely, since $p \leftrightarrow x$ iff $p \leq x$ or $p \leq x'$. Further, if $p \in M_1 \cap M_2$ then $\{0, p, 2p, \ldots, n_p p\} \subseteq M_1 \cap M_2$, because $x \leftrightarrow p$ gives $x \leftrightarrow kp$ for all kp existing in E, and hence the isotropic indices of p in M_1 and M_2 must coincide. Moreover, if for atoms $p \neq q$ we have $n_p p = n_q q$ then $p \nleftrightarrow q$ and the interval $[0, n_p p]_E$ in E is the horizontal sum of chains $\{0, p, \ldots, n_p p\}$ and $\{0, q, \ldots, n_q q\}$. Otherwise we have $p \leftrightarrow q$, which implies $q \leq p \oplus q = p \vee q = n_p p$ and by RDP we obtain $q = p$, a contradiction. Moreover, $kp = \ell q$ for $k < n_p$ implies that $q \leq kp \leq p'$, which again gives $p \leftrightarrow q$ hence $p = q$, a contradiction.

Recall that a map $\omega : E \to [0,1] \subseteq R$ is called a (finitely additive) *state* on an effect algebra $(E; \oplus, 0, 1)$ if $\omega(1) = 1$ and $x \leq y' \Rightarrow \omega(x \oplus y) = \omega(x) + \omega(y)$, and ω is called *(o)-continuous* if $x_\alpha \overset{(o)}{\to} x \Rightarrow \omega(x_\alpha) \to \omega(x)$ in R. Here, for a net $(x_\alpha)_{\alpha \in \mathcal{E}}$ of elements of E we right $x_\alpha \overset{(o)}{\to} x$ if there exist nets $(u_\alpha)_{\alpha \in \mathcal{E}}$, $(v_\alpha)_{\alpha \in \mathcal{E}}$ such that $u_\alpha \leq x_\alpha \leq v_\alpha$ for all $\alpha \in \mathcal{E}$ and $u_\alpha \uparrow x$, $v_\alpha \downarrow x$ (i.e., $\alpha \leq \beta \Rightarrow u_\alpha \leq u_\beta$, $v_\alpha \geq v_\beta$ and $x = \bigvee \{u_\alpha \mid \alpha \in \mathcal{E}\} = \bigwedge \{v_\alpha \mid \alpha \in \mathcal{E}\}$.

Theorem 17. *Let* $(M_\kappa)_{\kappa \in H}$ *be a family of complete atomic MV-effect algebras such that there are nonempty sets* D_κ *of atoms of* M_κ, $\kappa \in H$ *satisfying* $D_{\kappa_1} \overset{istr}{\sim} D_{\kappa_2}$, *for every pair* $\kappa_1, \kappa_2 \in H$. *Let* $u_\kappa = \bigoplus\{n_p p \mid p \in D_\kappa\} \neq 1_\kappa$ *and* $[0_\kappa, u'_\kappa]_{M_\kappa} \neq \{0_\kappa, u'_\kappa\}$, $\kappa \in H$. *Then for chosen* $\kappa_0 \in H$ *and every* $\kappa \in H$:

(i) $u_\kappa \in C(M_\kappa)$.
(ii) $[0_\kappa, u_\kappa]_{M_\kappa} \cong [0_{\kappa_0}, u_{\kappa_0}]_{M_{\kappa_0}}$.
(iii) $F_\kappa = [0_\kappa, u_\kappa]_{M_\kappa} \cup [u'_\kappa, 1_\kappa]_{M_\kappa} \cong F_{\kappa_0}$ *and* F_{κ_0} *is an MV-effect algebra.*
(iv) *There is a complete atomic effect algebra* $E = \bigcup_{\kappa \in H} M_\kappa$, *whose family of all blocks coincides with* $(M_\kappa)_{\kappa \in H}$, $\bigcap_{\kappa \in H} M_\kappa = F_{\kappa_0}$ *and* $E \cong [0_{\kappa_0}, u_{\kappa_0}]_{M_{\kappa_0}} \times G$, *where* G *is the horizontal sum of all* $[0_\kappa, u'_\kappa]_{M_\kappa}$, $\kappa \in H$.
(v) $M_{\kappa_1} \cap M_{\kappa_2} = F_{\kappa_0}$, *for any pair of blocks of* E.
(vi) *There is an (o)-continuous state on* E.

Proof. (i) Let A_κ be the set of all atoms of M_κ, $\kappa \in H$. By [18, Theorem 3.3], for every $x \in M_\kappa$ there is a set $\{a_\alpha \in A_\kappa \mid \alpha \in \mathcal{E}\}$ and positive integers k_α, $\alpha \in \mathcal{E}$ such that $x = \bigoplus\{k_\alpha a_\alpha \mid \alpha \in \mathcal{E}\} = \bigvee\{k_\alpha a_\alpha \mid \alpha \in \mathcal{E}\}$. Moreover, $x \in S(M_\kappa)$ iff $k_\alpha = n_{a_\alpha} = \mathrm{ord}(a_\alpha)$ for all $\alpha \in \mathcal{E}$. Since M_κ is an MV-effect algebra, we have $S(M_\kappa) = C(M_\kappa)$, which proves that $u_\kappa \in C(M_\kappa)$.

(ii) Since M_κ is an MV-effect algebra, it satisfies Riesz decomposition property. By part (i) of the proof and RDP we have that for every $p \in D_\kappa$ the element $n_p p$ is an atom of $S(M_\kappa) = C(M_\kappa)$ and hence the interval $[0_\kappa, n_p p]_{M_\kappa}$ in M_κ is a finite chain $0_\kappa < p < 2p < \ldots < n_p p$ (see [20]). Let $\varphi : D_\kappa \to D_{\kappa_0}$ be a bijection satisfying $n_p = \mathrm{ord}(p) = \mathrm{ord}(\varphi(p)) = n_{\varphi(p)}$ for all $p \in D_\kappa$. We can extend the mapping φ onto $[0_\kappa, n_p p]_{M_\kappa}$ by putting $\varphi(kp) = k\varphi(p)$ for all $k \leq n_p$. Obviously, $\varphi : [0_\kappa, n_p p]_{M_\kappa} \to [0_{\kappa_0}, \varphi(n_p p)]_{M_{\kappa_0}}$ is an isomorphism. Further $\{n_p p \mid p \in D_\kappa\}$ is the set of all atoms of the center $C([0_\kappa, u_\kappa]_{M_\kappa})$ which, by [19, Lemma 4.3], gives $[0_\kappa, u_\kappa]_{M_\kappa} \cong \prod\{[0_\kappa, n_p p]_{M_\kappa} \mid p \in D_\kappa\} \cong \prod\{[0_{\kappa_0}, \varphi(n_p p)]_{M_{\kappa_0}} \mid p \in D_{\kappa_0}\} \cong [0_{\kappa_0}, u_{\kappa_0}]_{M_{\kappa_0}}$.

(iii) Using (ii) we obtain $F_\kappa \cong [0_\kappa, u_\kappa]_{M_\kappa} \times \{0_\kappa, u'_\kappa\} \cong [0_{\kappa_0}, u_{\kappa_0}]_{M_{\kappa_0}} \times \{0_{\kappa_0}, u'_{\kappa_0}\} \cong F_{\kappa_0}$. Since F_{κ_0} is a sub-effect algebra and a complete sub-lattice of M_{κ_0}, we obtain that F_{κ_0} is a complete atomic MV-effect algebra as well.

(iv) Since the intervals $[0_\kappa, u'_\kappa]_{M_\kappa}$ are complete sub-lattices of M_κ, these intervals, with $\oplus$ inherited from M_κ, as well as their horizontal sum G are complete atomic effect algebras. Moreover, $([0_\kappa, u'_\kappa]_{M_\kappa})_{\kappa \in H}$ is a family of blocks of G. Let us construct an effect algebra $E = \bigcup_{\kappa \in H} M_\kappa \cong [0_{\kappa_0}, u_{\kappa_0}]_{M_{\kappa_0}} \times G$ by such a way that we identify all $F_\kappa = [0_\kappa, u_\kappa]_{M_\kappa} \cup [u'_\kappa, 1_\kappa]_{M_\kappa}$, $\kappa \in H$, with the MV-effect algebra F_{κ_0} and, moreover, we make a horizontal sum of all $[0_\kappa, u'_\kappa]_{M_\kappa}$ identifying all 0_κ with 0_{κ_0} and u'_κ with u'_{κ_0}. By [21] every block of E is isomorphic to a direct product of $[0_{\kappa_0}, u_{\kappa_0}]_{M_{\kappa_0}}$ and a block of G, and conversely, since $[0_{\kappa_0}, u_{\kappa_0}]_{M_{\kappa_0}}$ is an MV-effect algebra. This proves that M is a block of E iff there is $\kappa \in H$ such that $M \cong [0_\kappa, u_\kappa]_{M_\kappa} \times [0_\kappa, u'_\kappa]_{M_\kappa}$ and hence $M = M_\kappa$.

(v) Let $\kappa_1, \kappa_2 \in H$. Then $M_{\kappa_1} \cong [0_{\kappa_0}, u_{\kappa_0}]_{M_{\kappa_0}} \times [0_{\kappa_1}, u'_{\kappa_1}]_{M_{\kappa_1}}$ and $M_{\kappa_2} \cong [0_{\kappa_0}, u_{\kappa_0}]_{M_{\kappa_0}} \times [0_{\kappa_2}, u'_{\kappa_2}]_{M_{\kappa_2}}$. Further, we have identified elements $0_{\kappa_0}, 0_{\kappa_1}$ and 0_{κ_2}, and elements $u'_{\kappa_0}, u'_{\kappa_1}$ and u'_{κ_2}. It follows that in E we have $[0_{\kappa_1}, u'_{\kappa_1}]_{M_{\kappa_1}} \cap [0_{\kappa_2}, u'_{\kappa_2}]_{M_{\kappa_2}} = \{0_{\kappa_0}, u'_{\kappa_0}\}$. We obtain that $M_{\kappa_1} \cap M_{\kappa_2} \cong [0_{\kappa_0}, u_{\kappa_0}] \times \{0_{\kappa_0}, u'_{\kappa_0}\} \cong F_{\kappa_0}$.

(vi) By (iii), $E \cong [0, u] \times G$, where $u \in C(E)$, $[0, u]$ is a complete atomic MV-effect algebra and G is the horizontal sum of a family $(E_\kappa)\kappa \in H$ of complete atomic MV-effect algebras. By [19, Theorem 5.2], on every complete atomic MV-effect algebra E_κ there is an (o)-continuous state ω_κ, $\kappa \in H$. Let us define a mapping $\omega_G : G \to [0, 1]$ by the following way: For every $x \in G$ let $\omega_G(x) = \omega_\kappa(x)$, where $\kappa \in H$ be such that $x \in E_\kappa$. Obviously ω_G is an (o)-continuous state on G. Further, let ω_0 be an (o)-continuous state on $[0, u]$. For every pair of nonnegative real numbers k_1 and k_2 with $k_1 + k_2 = 1$, the mapping $\omega = k_1\omega_0 + k_2\omega_G$ is an (o)-continuous state on E. The last follows by the facts that for $x, y \in E$ with $x \leq y'$ we have $(x \oplus y) \wedge u = (x \wedge u) \oplus (y \wedge u)$, as well as $(x \oplus y) \wedge u' = (x \wedge u') \oplus (y \wedge u')$ (see [19, Lemma 4.1]) and $x \oplus y = ((x \oplus y) \wedge u) \oplus ((x \oplus y) \wedge u')$. Further for $x_\alpha, x \in E$ such that $x_\alpha \uparrow x$, $\alpha \in \mathcal{E}$ we have $x_\alpha \wedge u \uparrow x \wedge u$, $x_\alpha \wedge u' \uparrow x \wedge u'$ and $x_\alpha = (x_\alpha \wedge u) \oplus (x_\alpha \wedge u') \uparrow (x \wedge u) \oplus (x \wedge u') = x$, $\alpha \in \mathcal{E}$, since there is $\kappa \in H$ such that for all $\alpha \in \mathcal{E}$ we have $x_\alpha \wedge u' \in E_\kappa$.

Definition 21. *The complete atomic effect algebra E, constructed in Theorem 17, is called a* pasting of MV-effect algebras $(M_\kappa)_{\kappa \in \mathcal{E}}$ *together along an MV-effect algebra*

$$[0_{\kappa_0}, u_{\kappa_0}]_{M_{\kappa_0}} \cup [u'_{\kappa_0}, 1_{\kappa_0}]_{M_{\kappa_0}} \subseteq M_{\kappa_0}, \text{ for chosen } \kappa_0 \in H.$$

Remark 2. Form the proof of Theorem 17 it is clear that if the chosen sets D_κ of atoms of M_κ are finite then the completeness of MV-effect algebras M_κ, $\kappa \in H$ can be weakened to the assumption that all M_κ are Archimedean, since then elements $u_\kappa = \bigoplus \{n_p p \mid p \in D_\kappa\}$ exist. Obviously, then $E = \bigcup_{\kappa \in H} M_\kappa$ will be an Archimedean atomic lattice effect algebra admitting an (o)-continuous state.

If M_κ, $\kappa \in H$ are complete atomic Boolean algebras and all sets D_κ of atoms of M_κ have the same cardinality then $E = \bigcup_{\kappa \in H} M_\kappa$ constructed in Theorem 17 will be a complete atomic orthomodular lattice with blocks M_κ, $\kappa \in H$. If D_κ are finite then the assumption of completeness of M_κ can be omitted. For pasting of orthomodular posets we refer the reader to Navara and Rogalewicz [10].

10   Applications, Conclusions and Open Problems

Recently studied new algebraic structures (for modeling of noncompatibility, uncertainity or unsharpness), lattice effect algebras have sub-lattice effect algebras which are well-known structures: orthomodular lattices (the sets $\mathcal{S}(E)$ of all sharp elements), Boolean algebras (centers $\mathcal{C}(E)$) and MV-algebras (blocks $M \subseteq E$ and the centers of compatibility $\mathcal{B}(E)$).

(1) The result about sets $\mathcal{S}(E)$ shows that lattice effect algebras are in fact "smeared" orthomodular lattices. Moreover, this "smearing" is done in such a good way that also existing states or probabilities (additive maps $\omega : \mathcal{S}(E) \to [0, 1] \subseteq (-\infty, \infty)$ such that $\omega(1) = 1$ and $x \leq y' \Rightarrow \omega(x \oplus y) = \omega(x) + \omega(y)$) on $\mathcal{S}(E)$ are smeared onto whole lattice effect algebras.

Smearing Theorem for States, [18]. *For every complete (o)-continuous atomic effect algebra $(E; \oplus, 0, 1)$ the following conditions are equivalent:*

(i)   *There is a state on the orthomodular lattice* $\mathcal{S}(E) = \{x \in E \mid x \wedge x' = 0\}$.
(ii) *There is a state on* E.
(iii) *There is an* (o)-*continuous state on* E.

Smearing Theorem for (o)-**Continuous States** ([23]). *Let* E *be a complete atomic effect algebra and let* $\omega : \mathcal{S}(E) \to [0, 1] \in (-\infty, \infty)$ *be an* (o)-*continuous state on* $\mathcal{S}(E)$. *Then there is a state* $\hat{\omega}$ *on* E *extending* ω

From the point of view of constructing states or probabilities on lattice effect algebras the following theorem is a crucial fact:

Theorem on Basic Decomposition of Elements ([23]). *Let* E *be a complete atomic effect algebra. Then for every* $x \in E$, $x \neq 0$, *there exists a unique* $w_x \in \mathcal{S}(E)$, *a unique set* $\{a_\alpha \mid \alpha \in \mathcal{A}\}$ *of atoms of* E *and unique positive integers* k_α, $\alpha \in \mathcal{A}$ *such that*

$$x = w_x \oplus \left(\bigoplus \{k_\alpha a_\alpha \mid \alpha \in \mathcal{A}\} \right)$$

and if $w \in \mathcal{S}(E)$ *with* $w \leq x \ominus w_\alpha$ *then* $w = 0$.

(2) For more detailed description of some important sub-families of lattice effect algebras, it is a big help to know its *direct and subdirect product decompositions* (see Definition 6).

Theorem on Subdirect Product Decompositions ([21]). *Let* $(E; \oplus, 0, 1)$ *be a lattice effect algebra and* $D \subseteq C(E)$ *with (1)* $\bigvee D = 1$ *and (2)* $d_1 \wedge d_2 = 0$ *for all* $d_1 \neq d_2$; $d_1, d_2 \in D$. *Then:*

(i)   *E is isomorphic to a subdirect product of the family of effect algebras* $\{[0, d] \mid d \in D\}$.
(ii) *Up to isomorphism,* E *is a sub-lattice of* $\hat{E} = \prod\{[0, d] \mid d \in D\}$.
(iii) *If, moreover,* E *is complete or* D *is finite then* $E \cong \hat{E}$.

Corollary A ([21]). *Under the assumptions of Subdirect Product Decompositions theorem, if, moreover* E *is complete or* D *is finite then:*

(i)   *M is a block of* E *iff* $M \cong \prod\{M_d \mid d \in D, \ M_d \text{ is a block of } E\} = \prod\{M \cap [0, d] \mid d \in D\}$.
(ii) $\mathcal{S}(E) \cong \prod\{\mathcal{S}([0, d]) \mid d \in D\}$.
(iii) $\mathcal{B}(E) \cong \prod\{\mathcal{B}([0, d]) \mid d \in D\}$.
(iv) $\mathcal{C}(E) \cong \prod\{\mathcal{C}([0, d]) \mid d \in D\}$.

Corollary B ([21]). *Under the assumptions of Subdirect Product Decompositions theorem, if* φ_D *is the embedding of* E *into* $\prod\{[0, d] \mid d \in D\}$ *defined for all* $x \in E$ *by* $\varphi_D(x) = (x \wedge d)_{d \in D}$, *then*

(i)   *M is a block of* E *iff there are blocks* M_d *of* $[0, d]$, $d \in D$ *such that* $M = \varphi_D^{-1}(\prod\{M_d \mid d \in D\})$.
(ii) $\mathcal{S}(E) = \varphi_D^{-1}(\prod\{\mathcal{S}([0, d]) \mid d \in D\})$,
(iii) $\mathcal{B}(E) = \varphi_D^{-1}(\prod\{\mathcal{B}([0, d]) \mid d \in D\})$,
(iv) $\mathcal{C}(E) = \varphi_D^{-1}(\prod\{\mathcal{C}([0, d]) \mid d \in D\})$.

(3) Using results from (2), we can obtain a detailed description of some families of Archimedean atomic lattice effect algebras. Note that a lattice effect algebra is called distributive (modular) if it is a distributive (modular) lattice.

Proposition A ([20]). *Every atomic Archimedean distributive effect algebra can be sub-directly decomposed into finite chains and distributive diamonds.*

Proposition B ([22]). *Every complete atomic modular effect algebra E can be decomposed into direct product $\prod\{[0,p] \mid p \text{ atom of } \mathcal{C}(E)\}$ where every interval $[0,p]$ has at least one of the following properties:*

(i) $[0,p]$ *is an irreducible complete atomic modular ortholattice.*
(ii) $[0,p]$ *is a finite chain.*
(iii) $[0,p]$ *is a horizontal sum of a family of Boolean algebras and chains, all of length 2.*

(4) Evidently, for families of MV-effect algebras, distributive and modular effect algebras we have

MV-effect algebras $\subseteq$ distributive effect algebras $\subseteq$ modular effect algebras.

Using their descriptions from (3) we can prove the existence of (o)-continuous states on them.

Existence Theorem for States ([22]). *Let E be a complete atomic modular effect algebra. Then:*

(i) *There is an (o)-continuous state on E.*
(ii) *There is a faithful (o)-continuous subadditive state on E iff $\mathcal{C}(E)$ is separable.*

Here subadditivity of a state w on E means that $w(a \vee b) \leq w(a) + w(b)$ for all $a, b \in E$. A state is faithful if $w(a) = 0$ implies $a = 0$.

(5) Finally, let us note that some subfamilies of lattice effect algebras are characterized by relations between subalgebras $\mathcal{S}(E)$, $\mathcal{C}(E)$ and $\mathcal{B}(E)$. Thus

(a) E is an orthomodular lattice iff $E = \mathcal{S}(E)$ and then $\mathcal{C}(E) = \mathcal{B}(E)$. Conversely, $\mathcal{C}(E) = \mathcal{B}(E)$ does not imply $E = \mathcal{S}(E)$, see Fig. 5.
(b) E is an MV-effect algebra iff $E = \mathcal{B}(E)$ and then $\mathcal{C}(E) = \mathcal{S}(E)$. Conversely, $\mathcal{C}(E) = \mathcal{S}(E)$ does not imply $E = \mathcal{B}(E)$, see Fig. 6.
(c) E is a Boolean algebra iff $E = \mathcal{C}(E)$ and then $\mathcal{C}(E) = \mathcal{S}(E) = \mathcal{B}(E)$. Conversely, $\mathcal{C}(E) = \mathcal{S}(E) = \mathcal{B}(E)$ does not imply $E = \mathcal{C}(E)$, see Fig. 6.

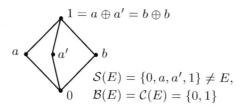

Fig. 5.

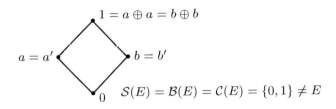

Fig. 6.

Open Problems

The following questions are still unanswered:

(1) For which lattice effect algebras we have $\mathcal{C}(E) = \mathcal{S}(E)$ and for which $\mathcal{C}(E) = \mathcal{S}(E) = \mathcal{B}(E)$?

(2) Does there exist an atomic lattice effect algebra with non-atomic $\mathcal{C}(E)$ or with non-atomic $\mathcal{S}(E)$? It is known that an atomic orthomodular lattice with a nonatomic block exists.

(3) Since for a generalized effect algebra P and its effect algebraic extension $E = P \dot\cup P^*$ we have $\mathcal{S}(E) \cap P = \mathcal{S}(P)$, $\mathcal{C}(E) \cap P = \mathcal{C}(P)$ and $\mathcal{B}(E) \cap P = \mathcal{B}(P)$ (see previous sections), we may ask which algebraic and probabilistic properties of P are preserved for (inherited from) $E = P \dot\cup P^*$.

(4) In Section 9 we have shown one possibility of pastings of MV-effect algebras. It would be interesting to study other pastings of MV-effect algebras for constructing lattice effect algebras.

References

1. Chang, C.C.: Algebraic analysis of many-valued logics. Trans. Amer. Math. Soc. **88** (1958) 467–490
2. Chovanec, F., Kôpka, F.: Boolean D-posets. Tatra Mt. Math. Publ. **10** (1997) 183–197
3. Dvurečenskij, A., Pulmannová, S.: *New Trends in Quantum Structures* (Kluwer Academic Publishers, Dordrecht, Boston, London and Ister Science, Bratislava 2000)
4. Foulis, D., Bennett, M.K.: Effect algebras and unsharp quantum logics. Foundations of Physics **24** (1994) 1331–1352
5. Gudder, S.: S-dominating effect algebras. Internat. J. Theor. Phys. **37** (1998) 915–923
6. Jenča, G., Riečanová, Z. On sharp elements in lattice effect algebras. BUSEFAL **80** (1999) 24–29
7. Kôpka, F.: Compatibility in D-posets. Internat. J. Theor. Phys. **34** (1995) 1525–1531
8. Greechie, R.J.: Orthomodular lattices admitting no states. J. Combin. Theory Ser. A **10** (1971) 119–132
9. Kôpka, F., Chovanec, F.: D-posets. Mathematica Slovaca **44** (1994) 21–34
10. Navara, M, P. and Rogalewicz, V.: The pasting construction for orthomodular posets. Mathematische Nachrichten **154** (1991) 157–168

11. Riečanová, Z.: Subalgebras, intervals and central elements of generalized effect algebras. Internat. J. Theor. Phys. **38** (1999) 3209–3220
12. Riečanová, Z.: Compatibilty and central elements in effect algebras. Tatra Mt. Math. Publ. **16** (1999) 151–158
13. Riečanová, Z.: MacNeille completions of D-posets and effect algebras. Internat. J. Theor. Phys. **39** (2000) 859 869
14. Riečanová, Z.: Generalization of blocks for D-lattices and lattice ordered effect algebras. Internat. J. Theor. Phys. **39** (2000) 231–237
15. Riečanová, Z.: Archimedean and block-finite lattice effect algebras. Demonstratio Mathematica **33** (2000) 443–452
16. Riečanová, Z.: Orthogonal sets in effect algebras. Demonstratio Mathematica **34** (2001) 525–532
17. Riečanová, Z.: Proper effect algebras admitting no states. Internat. J. Theor. Phys. **40** (2001) 1683–1691
18. Riečanová, Z.: Smearings of states defined on sharp elements onto effect algebras. Internat. J. Theor. Phys. **41** (2002) 1511–1524
19. Riečanová, Z.: Continuous effect algebra admitting order-continuous states. Fuzzy Sets and Systems **136** (2003) 41–54
20. Riečanová, Z.: Distributive atomic effect algebras. Demonstratio Mathematica **36** (2003) 247–259
21. Riečanová, Z.: Subdirect decompositions of lattice effect algebras. Internat. J. Theor. Phys. **42** (2003) 1425–1433
22. Riečanová, Z.: Modular atomic effect algebras and the existence of subadditive states. Kybernetika **40** (2004) 459–468
23. Riečanová, Z.: Basic decomposition of elements and Jauch-Piron effect algebras. Fuzzy Sets and Systems **155** (2005) 138–149
24. Riečanová, Z.: Embeddings of generalized effect algebras into complete effect algebras. Soft Computing **10** (2006) 476–482
25. Riečanová, Z.: Pastings of MV-effect algebras. Internat. J. Theor. Phys. **43** (2004) 1875–1883
26. Riečanová, Z., Bršel, D.: Contraexamples in difference posets and orthoalgebras. Internat. J. Theor. Phys. **33** (1994) 133–141
27. Riečanová, Z., Marinová. I.: Generalized homogeneous, prelattice and MV-effect algebras. Kybernetika **41** (2005) 129–141
28. Riečanová, Z., Marinová. I., Zajac, M.: From Real Numbers to Effect Algebras and Families of Fuzzy Sets. Chapter 7, Begabtenförderung im MINT-Bereich (Mathematik, Informatik, Naturwissenschaften, Technik) band 9 (2004) pp. 103–115
29. Schmidt, J.: Zur Kennzeichnung der Dedekind-Mac Neilleschen Hulle einer Geordneten Menge. Archiv d. Math. **7** (1956) 241–249
30. Yutani, H.: The class of commutative BCK-algebras is equationally definable. Math. Seminar Notes **5** (1977) 207–210

A Decision Procedure for Monotone Functions over Bounded and Complete Lattices

Domenico Cantone[1] and Calogero G. Zarba[2]

[1] Università degli Studi di Catania, Italy
cantone@dmi.unict.it
[2] Universität des Saarlandes, Germany
zarba@alan.cs.uni-sb.de

Abstract. We present a decision procedure for the quantifier-free satisfiability problem of the language **BLmf** of bounded lattices with monotone unary functions. The language contains the predicates $=$ and $\leq$, as well as the operators $\sqcap$ and $\sqcup$ over terms which may involve uninterpreted unary function symbols. The language also contains predicates for expressing increasing and decreasing monotonicity of functions, as well as a predicate for pointwise function comparison.

Our decision procedure runs in polynomial time $\mathcal{O}(m^4)$ for normalized conjunctions of m literals, thus entailing that the quantifier-free satisfiability problem for **BLmf** is $\mathcal{NP}$-complete. Furthermore, our decision procedure can be used to decide the quantifier-free satisfiability problem for the language **CLmf** of complete lattices with monotone functions. This allows us to conclude that the languages **BLmf** and **CLmf** are equivalent for quantifier-free formulae.

1 Introduction

Lattices are partial orders in which every pair of elements has a least upper bound and a greatest lower bound. They have several applications in mathematics and computer science, including model checking [7], knowledge representation [11], partial order programming [12], denotational semantics [10], rewrite systems [4], relational methods in computer science [5], computer security [9], and so on.

In this paper we introduce the language **BLmf** (*Bounded Lattices with monotone functions*) for expressing constraints over lattices and monotone functions. The language **BLmf** contains the equality predicate $=$, the ordering predicate $\leq$, and the operators $\sqcap$ (meet) and $\sqcup$ (join). The language also allows for uninterpreted unary function symbols, and has the following predicates for expressing monotonicity properties of functions:

- the predicate symbol $inc(f)$, stating that the function f is increasing;
- the predicate symbol $dec(f)$, stating that the function f is decreasing;
- the predicate symbol $const(f)$, stating that the function f is constant;
- the predicate symbol $leq(f, g)$, stating that $f(a) \leq g(a)$, for each a.

H. de Swart et al. (Eds.): TARSKI II, LNAI 4342, pp. 318–333, 2006.
© Springer-Verlag Berlin Heidelberg 2006

We prove that the quantifier-free satisfiability problem of **BLmf** is decidable. In particular, we present a decision procedure that allows one to decide the **BLmf**-satisfiability of normalized conjunctions[1] of m literals in polynomial time $\mathcal{O}(m^4)$. Such result entails at once that the quantifier-free satisfiability problem of **BLmf** is $\mathcal{NP}$-complete.

We also study the language **CLmf** of complete lattices with monotone functions. The syntax of **CLmf** is the same of that of **BLmf**. Semantically, **CLmf** differs from **BLmf** in that a model of the language **CLmf** involves a complete lattice rather than just a bounded lattice.

We show that our decision procedure for the quantifier-free satisfiability problem of **BLmf** is also a decision procedure for the quantifier-free satisfiability problem of **CLmf**. Therefore it follows that a quantifier-free formula is **BLmf**-satisfiable if and only if it is **CLmf**-satisfiable, so that the language **BLmf** and **CLmf** are equivalent for quantifier-free formulae.

1.1 Related Work

Cantone, Ferro, Omodeo, and Schwartz [1] provide a decision procedure for the quantifier-free language **POSMF** of lattices extended with unary function symbols and the predicates $inc(f)$ and $dec(f)$. The language **POSMF** does not contain the operators $\sqcap$ and $\sqcup$, and it does not contain the predicates $const(f)$ and $leq(f, g)$. The decision procedure for **POSMF** is based on a nondeterministic quadratic reduction to the quantifier-free fragment of set theory **MLS** (cf. [6]).

Sofronie-Stokkermans [13] proved that the quantifier-free languages of (a) partially ordered sets, (b) totally ordered sets, (c) dense totally ordered sets, (d) semilattices, (e) lattices, (f) distributive lattices, (g) boolean algebras, and (h) real numbers can be extended, while still preserving decidability, with one or more monotone increasing unary functions.

In a preliminary version of this paper [2], which dealt only with the quantifier-free satisfiability problem of the language **CLmf**, the authors give a flawed proof of the decidability of **CLmf**. The bug is as follows: In [2, page 9], the partial order $\langle \mathcal{A}, \leq^{\mathcal{A}} \rangle$ is not necessarily a lattice, and therefore the functions $\sqcap^{\mathcal{A}}$ and $\sqcup^{\mathcal{A}}$ are not well-defined in general. Here we fix such problem by taking the Dedekind-MacNeille completion of the partial order $\langle \mathcal{A}, \leq^{\mathcal{A}} \rangle$ (see Section 4, Proposition 28, for details).

Tarski [14] proved that the fully quantified language of lattices is undecidable.

1.2 Organization of the Paper

In Section 2 we introduce some basic notions of lattice theory, and we define the syntax and semantics of the language **BLmf**. In Section 3 we present our decision procedure for the quantifier-free satisfiability problem of **BLmf**, and we give an example of our decision procedure in action. In Section 4 we prove that

---

[1] The notion of normalized set of literals will be defined in Definition 20.

our decision procedure is correct, and we analyze its complexity. In Section 5 we discuss the language **CLmf**. In Section 6 we draw some final conclusions.

2   Preliminaries

2.1   Partial Orders

Definition 1. A PARTIAL ORDER is a pair $(A, \leq)$ where A is a nonempty set and $\leq$ is a reflexive, antisymmetric, and transitive binary relation of A. $\square$

Definition 2. Let $(A, \leq)$ be a partial order and let $\emptyset \neq X \subseteq A$. We say that y is a MAXIMUM of X with respect to $(A, \leq)$ if the following conditions hold:

- $y \in X$;
- $x \leq y$, for each $x \in X$. $\square$

When it exists, the maximum of X with respect to $(A, \leq)$ is unique. Consequently, we use the notation $max(X, A, \leq)$ to denote the unique maximum of X with respect to $(A, \leq)$ when it exists; otherwise, we let $max(X, A, \leq) = undef$.

Definition 3. Let $(A, \leq)$ be a partial order and let $\emptyset \neq X \subseteq A$. We say that y is a MINIMUM of X with respect to $(A, \leq)$ if the following conditions hold:

- $y \in X$;
- $y \leq x$, for each $x \in X$. $\square$

When it exists, the minimum of X with respect to $(A, \leq)$ is unique. Consequently, we use the notation $min(X, A, \leq)$ to denote the unique minimum of X with respect to $(A, \leq)$ when it exists; otherwise, we let $min(X, A, \leq) = undef$.

Definition 4. Let $(A, \leq)$ be a partial order and let $\emptyset \neq X \subseteq A$. We say that y is a LEAST UPPER BOUND of X with respect to $(A, \leq)$ if the following conditions hold:

- $x \leq y$, for each $x \in X$;
- if $x \leq z$, for each $x \in X$, then $y \leq z$. $\square$

When it exists, the least upper bound of X with respect to $(A, \leq)$ is unique. Consequently, we use the notation $lub(X, A, \leq)$ to denote the unique least upper bound of X with respect to $(A, \leq)$ when it exists; otherwise, we let $lub(X, A, \leq) = undef$.

Proposition 5. Let $(A, \leq)$ be a partial order and let $\emptyset \neq X \subseteq A$. Then, $max(X, A, \leq) \neq undef$ implies $max(X, A, \leq) = lub(X, A, \leq)$. $\square$

Definition 6. Let $(A, \leq)$ be a partial order and let $\emptyset \neq X \subseteq A$. We say that y is a GREATEST LOWER BOUND of X with respect to $(A, \leq)$ if the following conditions hold:

- $y \leq x$, for each $x \in X$;
- if $z \leq x$, for each $x \in X$, then $z \leq y$.   ⌐   □

When it exists, the greatest lower bound of X with respect to $(A, \leq)$ is unique. Consequently, we use the notation $glb(X, A, \leq)$ to denote the unique greatest lower bound of X with respect to $(A, \leq)$ when it exists; otherwise, we let $glb(X, A, \leq) = undef$.

Proposition 7. *Let $(A, \leq)$ be a partial order and let $\emptyset \neq X \subseteq A$. Then, $min(X, A, \leq) \neq undef$ implies $min(X, A, \leq) = glb(X, A, \leq)$.*   □

2.2   Lattices

Definition 8. A LATTICE is a tuple $(A, \leq, \sqcup, \sqcap)$ where:

- $(A, \leq)$ is a partial order;
- $glb(\{a, b\}, A, \leq) \neq undef$ and $lub(\{a, b\}, A, \leq) \neq undef$, for all $a, b \in A$;
- $a \sqcup b = lub(\{a, b\}, A, \leq)$;
- $a \sqcap b = glb(\{a, b\}, A, \leq)$.   □

Definition 9. A BOUNDED LATTICE is a tuple $(A, \leq, \sqcup, \sqcap, \mathbf{1}, \mathbf{0})$ where:

- $(A, \leq, \sqcup, \sqcap)$ is a lattice;
- $\mathbf{1} = max(A, A, \leq)$;
- $\mathbf{0} = min(A, A, \leq)$.   □

Definition 10. A COMPLETE LATTICE is a tuple $(A, \leq, \sqcup, \sqcap, \mathbf{1}, \mathbf{0})$ where:

- $(A, \leq, \sqcup, \sqcap, \mathbf{1}, \mathbf{0})$ is a bounded lattice;
- $glb(X, A, \leq) \neq undef$ and $lub(X, A, \leq) \neq undef$, for each $\emptyset \neq X \subseteq A$.   □

Remark 11. If $(A, \leq, \sqcup, \sqcap, \mathbf{1}, \mathbf{0})$ is a complete lattice, we let $lub(\emptyset, A, \leq) = \mathbf{0}$ and $glb(\emptyset, A, \leq) = \mathbf{1}$.   □

Proposition 12. *Let $(A, \leq, \sqcup, \sqcap, \mathbf{1}, \mathbf{0})$ be a complete lattice. Then, the following properties hold:*

$$a \sqcup b = b \sqcup a, \qquad\qquad a \sqcap b = b \sqcap a,$$
$$(a \sqcup b) \sqcup c = a \sqcup (b \sqcup c), \qquad (a \sqcap b) \sqcap c = a \sqcap (b \sqcap c),$$
$$a \sqcup a = a, \qquad\qquad a \sqcap a = a,$$
$$a \sqcup (a \sqcap b) = a, \qquad\qquad a \sqcap (a \sqcup b) = a.$$

Moreover, we have:

$$a \leq b \quad\leftrightarrow\quad a \sqcup b = b \quad\leftrightarrow\quad a \sqcap b = a.\qquad □$$

Proposition 13. *Let $(A, \leq, \sqcup, \sqcap, \mathbf{1}, \mathbf{0})$ be a complete lattice, and let $X, Y \subseteq A$. Then, $X \subseteq Y$ implies $lub(X) \leq lub(Y)$.*   □

Proposition 14. *Let $(A, \leq, \sqcup, \sqcap, \mathbf{1}, \mathbf{0})$ be a complete lattice, and let $X, Y \subseteq A$. Then, $X \subseteq Y$ implies $glb(Y) \leq glb(X)$.*   □

2.3   Dedekind-MacNeille Completion

The Dedekind-MacNeille completion allows one to extend a partial order $(A, \leq)$ into a complete lattice. It was introduced by MacNeille [8], who generalized the Dedekind completion [3] for constructing the set $\mathbb{R}$ of real numbers from the set $\mathbb{Q}$ of rational numbers.

Proposition 15 ([8]). *Let $(A, \leq)$ be a partial order. Then there exists a unique[2] minimal complete lattice $(B, \sqsubseteq, \sqcup, \sqcap, \mathbf{1}, \mathbf{0})$ such that:*

(a) $A \subseteq B$;
(b) $a \leq b$ iff $a \sqsubseteq b$, for each $a, b \in A$;
(c) If $lub(X, A, \leq) \neq undef$, then $lub(X, B, \sqsubseteq) = lub(X, A, \leq)$, for each $\emptyset \neq X \subseteq A$;
(d) If $glb(X, A, \leq) \neq undef$, then $glb(X, B, \sqsubseteq) = glb(X, A, \leq)$, for each $\emptyset \neq X \subseteq A$. $\qquad\square$

Definition 16. Let $(A, \leq)$ be a partial order. The DEDEKIND-MACNEILLE COMPLETION of $(A, \leq)$ is the unique complete lattice $(B, \sqsubseteq, \sqcup, \sqcap, \mathbf{1}, \mathbf{0})$ satisfying properties (a)–(d) of Proposition 15. $\qquad\square$

2.4   Syntax of BLmf

The language **BLmf** (*Bounded Lattices with monotone functions*) is a quantifier-free language containing the following symbols:

-   arbitrarily many variables $x, y, z, \ldots$;
-   the constant symbols $\mathbf{1}$ and $\mathbf{0}$;
-   the function symbols $\sqcup$ and $\sqcap$;
-   the binary predicate symbols $\leq$ and $=$;
-   arbitrarily many unary function symbols $f, g, \ldots$
-   the predicate symbol $inc(f)$;
-   the predicate symbol $dec(f)$;
-   the predicate symbol $const(f)$;
-   the predicate symbol $leq(f, g)$.

Definition 17. The set of **BLmf**-TERMS is the smallest set satisfying the following conditions:

-   Every variable is a **BLmf**-term;
-   $\mathbf{1}$ and $\mathbf{0}$ are **BLmf**-terms;
-   If s and t are **BLmf**-terms, so are $s \sqcup t$ and $s \sqcap t$;
-   If s is a **BLmf**-term and f is a function symbol, then $f(s)$ is a **BLmf**-term.

---

[2] Up to an isomorphism.

BLmf-ATOMS are of the form:

$$s = t, \qquad\qquad s \leq t, \qquad\qquad inc(f),$$
$$dec(f), \qquad\qquad const(f), \qquad\qquad leq(f, g),$$

where s, t are **BLmf**-terms and f, g are unary function symbol.

BLmf-FORMULAE are constructed from **BLmf**-atoms using the propositional connectives $\neg$, $\vee$, $\wedge$, $\rightarrow$, and $\leftrightarrow$. **BLmf**-LITERALS are **BLmf** atoms or their negations. ☐

If φ is a **BLmf**-formula, we denote with $vars(\varphi)$ the set of variables occurring in φ. If Φ is a set of **BLmf**-formulae, we let $vars(\Phi) = \bigcup_{\varphi \in \Phi} vars(\varphi)$.

2.5 Semantics of BLmf

Definition 18. A **BLmf**-INTERPRETATION $\mathcal{A}$ is a pair $(A, (\cdot)^{\mathcal{A}})$ where $A \neq \emptyset$ and $(\cdot)^{\mathcal{A}}$ interprets the symbols of the language **BLmf** as follows:

- $(A, \leq^{\mathcal{A}}, \sqcup^{\mathcal{A}}, \sqcap^{\mathcal{A}}, \mathbf{1}^{\mathcal{A}}, \mathbf{0}^{\mathcal{A}})$ is a bounded lattice;
- $=^{\mathcal{A}}$ is interpreted as the identity in A;
- each variable x is mapped to an element $x^{\mathcal{A}} \in A$;
- each unary function symbol f is mapped to a function $f^{\mathcal{A}} : A \rightarrow A$;
- $[inc(f)]^{\mathcal{A}} = true$ iff $a \leq^{\mathcal{A}} b$ implies $f^{\mathcal{A}}(a) \leq^{\mathcal{A}} f^{\mathcal{A}}(b)$, for each $a, b \in A$;
- $[dec(f)]^{\mathcal{A}} = true$ iff $a \leq^{\mathcal{A}} b$ implies $f^{\mathcal{A}}(b) \leq^{\mathcal{A}} f^{\mathcal{A}}(a)$, for each $a, b \in A$.
- $[const(f)]^{\mathcal{A}} = true$ iff $f^{\mathcal{A}}(a) = f^{\mathcal{A}}(b)$, for each $a, b \in A$.
- $[leq(f, g)]^{\mathcal{A}} = true$ iff $f^{\mathcal{A}}(a) \leq^{\mathcal{A}} g^{\mathcal{A}}(a)$, for all $a \in A$. ☐

Let φ be either a **BLmf**-formula or a **BLmf**-term, and let $\mathcal{A}$ be a **BLmf**-interpretation. We denote with $\varphi^{\mathcal{A}}$ the evaluation of φ under $\mathcal{A}$.

Definition 19. A **BLmf**-formula $\mathcal{A}$ is **BLmf**-SATISFIABLE if there exists a **BLmf**-interpretation $\mathcal{A}$ such that $\varphi^{\mathcal{A}} = true$. A set Φ of **BLmf**-formulae is **BLmf**-SATISFIABLE if there exists a **BLmf**-interpretation $\mathcal{A}$ such that $\varphi^{\mathcal{A}} = true$, for each $\varphi \in \Phi$. ☐

3 A Decision Procedure for BLmf

In this section we present a decision procedure for the quantifier-free satisfiability problem for the language **BLmf**. Without loss of generality, we restrict ourselves to normalized sets of **BLmf**-literals.

Definition 20. A set Γ of **BLmf**-literals is NORMALIZED if it satisfies the following conditions:

1. Each **BLmf**-literal in Γ is of the form

$$x = y, \qquad\qquad x \neq y, \qquad\qquad x \leq y, \qquad\qquad x \not\leq y,$$
$$x = y \sqcup z, \qquad x = y \sqcap z, \qquad x = f(y), $$
$$inc(f), \qquad\qquad dec(f), \qquad\qquad const(f), \qquad\qquad leq(f, g),$$

where:

- x, y, z can be either variables or the constant symbols **1** and **0**;
- f, g are unary function symbols.

2. For each unary function symbol f, no more than one of the following **BLmf**-literals is in Γ:

$$inc(f), \qquad\qquad dec(f), \qquad\qquad const(f). \qquad\qquad \Box$$

Proposition 21. *Every finite set of **BLmf**-literals can be converted in polynomial time into a **BLmf**-equisatisfiable normalized set of **BLmf**-literals.* $\qquad \Box$

PROOF. Let Γ be a finite set of **BLmf**-literals. By opportunely introducing fresh variables, we can convert all literals in Γ—while still preserving **BLmf**-satisfiability—to literals conforming condition 1 of Definition 20. In particular, literals of the form $\neg inc(f)$ can be replaced by a conjunction $x \leq y \wedge u = f(x) \wedge v = f(y) \wedge u \not\leq v$, where x, y, u, and v are fresh variables. Similar replacements can be performed for the literals of the form $\neg dec(f)$, $\neg const(f)$, and $\neg leq(f, g)$. Finally, condition 2 of Definition 20 can be enforced by exploiting the following equivalences:

$$inc(f) \wedge dec(f) \equiv const(f),$$
$$inc(f) \wedge const(f) \equiv const(f),$$
$$dec(f) \wedge const(f) \equiv const(f),$$
$$inc(f) \wedge dec(f) \wedge const(f) \equiv const(f). \qquad\blacksquare$$

Given a normalized set Γ of **BLmf**-literals, we define the following four pairwise disjoint sets:

- $INC(\Gamma)$ contains all unary function symbols f such that the literal $inc(f)$ is in Γ.
- $DEC(\Gamma)$ contains all unary function symbols f such that the literal $dec(f)$ is in Γ.
- $CONST(\Gamma)$ contains all unary function symbols f such that the literal $const(f)$ is in Γ.
- $NORM(\Gamma)$ contains all unary function symbols that do not belong to $INC(\Gamma) \cup DEC(\Gamma) \cup CONST(\Gamma)$.

Our decision procedure is based on the inference rules shown in Figure 1. In order to ensure termination we require the following:

- If $\mathcal{R}$ is not a fresh-variable rule, then $\mathcal{R}$ cannot be applied to a conjunction of normalized literals Γ if the conclusion of $\mathcal{R}$ is already in Γ.
- If $\mathcal{R}$ is a fresh-variable rule whose conclusion is a literal ℓ, then $\mathcal{R}$ cannot be applied to Γ if the literal $\ell\{\mathsf{fresh}/w\}$ is already in Γ, for some variable w.

Definition 22. A normalized set Γ of **BLmf**-literals is SATURATED if no inference rule in Figure 1 can be applied to Γ. $\qquad\qquad \Box$

$=$-*rules*

$$\frac{}{x = x} \qquad\qquad \frac{\begin{array}{c} x = y \\ \ell \end{array}}{\ell\{x/y\}}$$

$\leq$-*rules*

$$\frac{}{x \leq x} \qquad \frac{\begin{array}{c} x \leq y \\ y \leq x \end{array}}{x = y} \qquad \frac{\begin{array}{c} x \leq y \\ y \leq z \end{array}}{x \leq z} \qquad \frac{}{x \leq 1} \qquad \frac{}{0 \leq x}$$

$\sqcap$-*rules*

$$\frac{x = y \sqcap z}{\begin{array}{c} x \leq y \\ x \leq z \end{array}} \qquad \frac{\begin{array}{c} x = y \sqcap z \\ w \leq y \\ w \leq z \end{array}}{w \leq x}$$

$\sqcup$-*rules*

$$\frac{x = y \sqcup z}{\begin{array}{c} y \leq x \\ z \leq x \end{array}} \qquad \frac{\begin{array}{c} x = y \sqcup z \\ y \leq w \\ z \leq w \end{array}}{x \leq w}$$

Functions rules

$$\frac{\begin{array}{c} x = x' \\ y = f(x) \\ y' = f(x') \end{array}}{y = y'} \quad \frac{\begin{array}{c} inc(f) \\ x \leq x' \\ y = f(x) \\ y' = f(x') \end{array}}{y \leq y'} \quad \frac{\begin{array}{c} dec(f) \\ x \leq x' \\ y = f(x) \\ y' = f(x') \end{array}}{y' \leq y} \quad \frac{\begin{array}{c} const(f) \\ y = f(x) \\ y' = f(x') \end{array}}{y = y'} \quad \frac{\begin{array}{c} leq(f,g) \\ y = f(x) \\ y' = g(x) \end{array}}{y \leq y'}$$

Fresh-variables rules

$$\frac{\begin{array}{c} leq(f,g) \\ y = f(x) \end{array}}{\mathsf{fresh} = g(x)} \quad \frac{\begin{array}{c} leq(f,g) \\ y = g(x) \end{array}}{\mathsf{fresh} = f(x)} \quad \frac{}{\mathsf{fresh} = f(1)} \quad \frac{}{\mathsf{fresh} = f(0)}$$

Notes

- In the second $=$-rule, the literal ℓ does not contain any function symbol. Additionally, by $\ell\{x/y\}$ we mean the literal obtained by replacing *any* occurrence of x in ℓ by y.
- In the first $=$ rule, first $\leq$-rule, and last two $\leq$-rules, the variables x already occurs in Γ.
- In the fresh-variables rules, fresh stands for a newly introduced variable.

Fig. 1. Inference rules for computing $closure(\Gamma)$

If Γ is a normalized set of **BLmf**-literals, we denote with $closure(\Gamma)$ the smallest saturated set of **BLmf**-literals containing Γ. Note that the set $closure(\Gamma)$ is normalized.

The above two constraints on the applicability of the inference rules in Figure 1 imply that $closure(\Gamma)$ has at most $\mathcal{O}(m^4)$ literals, for any normalized set Γ of **BLmf**-literals with m literals.

Proposition 23. *Let Γ be a normalized set of* **BLmf***-literals with m literals. Then $closure(\Gamma)$ has at most $\mathcal{O}(m^4)$ literals.* $\square$

PROOF. Clearly, the first two fresh-variables rules introduce at most $\mathcal{O}(m^2)$-variables. The second two fresh-variables rules introduce at most $\mathcal{O}(k)$-variables, where k is the number of unary function symbols in Γ. Since $k = \mathcal{O}(m)$, it follows that all the fresh-variables rules introduce at most $\mathcal{O}(m^2)$-variables, and therefore $closure(\Gamma)$ contains at most $\mathcal{O}(m^2)$-variables. Therefore, the remaining rules can introduce at most $\mathcal{O}((m^2)^2)$ literals, which implies that $closure(\Gamma)$ contains at most $\mathcal{O}(m^4)$-literals. ∎

Definition 24. A normalized set Γ of **BLmf**-literals is CONSISTENT if it does not contain any two complementary literals $\ell, \neg\ell$; otherwise it is INCONSISTENT. $\square$

Given a finite normalized set Γ of **BLmf**-literals, our decision procedure consists of the following two steps:

Step 1. Compute $\Delta = closure(\Gamma)$.
Step 2. Output `satisfiable` if Δ is consistent; otherwise output `unsatisfiable`.

Example 25. Let Γ be the following set of **BLmf**-literals

$$\Gamma = \left\{ \begin{array}{l} inc(f), \\ dec(g), \\ leq(f,g), \\ f(\mathbf{0}) = g(\mathbf{0}), \\ f(x) \neq g(x) \end{array} \right\} .$$

We claim that Γ is **BLmf**-unsatisfiable. In fact, the first four literals imply that $f = g$, which contradicts the last literal.

We use our decision procedure in order to automatically check that Γ is **BLmf**-unsatisfiable. First, note that Γ is **BLmf**-equisatisfiable with the following normalized set Γ' of **BLmf**-literals:

$$\Gamma' = \left\{ \begin{array}{l} inc(f), \\ dec(g), \\ leq(f,g), \\ y_1 = f(\mathbf{0}), \\ y_2 = g(\mathbf{0}), \\ y_1 = y_2, \\ z_1 = f(x), \\ z_2 = g(x), \\ z_1 \neq z_2 \end{array} \right\}$$

Then, note that $closure(\Gamma')$ must contain, among others, the following literals:

$\mathbf{0} \leq x$,	by the fifth $\leq$-rule,
$y_1 \leq z_1$,	by the second functions rule,
$z_1 \leq z_2$,	by the fifth functions rule,
$z_2 \leq y_2$,	by the third functions rule,
$z_2 \leq y_1$,	by the second =-rule,
$y_1 \leq z_2$,	by the third $\leq$-rule,
$z_2 = y_1$,	by the second $\leq$-rule,
$z_1 \leq y_1$,	by the second =-rule,
$z_1 = y_1$,	by the second $\leq$-rule,
$z_1 = z_2$,	by the second =-rule.

Since $closure(\Gamma')$ contains the complementary literals $z_1 = z_2$ and $z_1 \neq z_2$, our decision procedure outputs unsatisfiable, as desired. □

4   Correctness and Complexity

Proposition 26. *Let Γ be a **BLmf**-satisfiable normalized set of **BLmf**-literals, and let Γ' be the result of extending Γ by means of an application of one of the inference rules in Figure 1. Then Γ' is **BLmf**-satisfiable.* □

PROOF. Let $\mathcal{A}$ be a **BLmf**-interpretation satisfying Γ. If Γ' involves the same variables of Γ, then it is routine to verify that $\mathcal{A}$ satisfies Γ' too. Otherwise, if Γ' is obtained from Γ by applying a fresh-variables rule, and therefore it involves a variable fresh not present in Γ, then it can easily be argued that Γ' is satisfied by a suitable variant $\mathcal{A}'$ of the **BLmf**-interpretation $\mathcal{A}$, which assigns the same values to every symbol of the language, except possibly the variable fresh. ■

Proposition 27 (Soundness). *Let Γ be a **BLmf**-satisfiable finite normalized set of **BLmf**-literals. Then $closure(\Gamma)$ is **BLmf**-satisfiable.* □

PROOF. By Propositions 23 and 26. ■

Proposition 28. *Any saturated and consistent normalized set of **BLmf**-literals is **BLmf**-satisfiable.* □

PROOF. Let Γ be a saturated and consistent set of **BLmf**-literals.

Let $X = vars(\Gamma) \cup \{\mathbf{1}, \mathbf{0}\}$, and let $\sim$ be the binary relation of X induced by the literals of the form $x = y$ in Γ. By saturation with respect to the =-rules, $\sim$ is an equivalence relation. Consequently, we can form the quotient set $A = X/\sim$.

Let $\preceq$ be the binary relation of A defined as follows:

$$[x]_\sim \preceq [y]_\sim \quad \Longleftrightarrow \quad x \leq y \text{ is in } \Gamma.$$

Since $\sim$ is an equivalence relation, $\preceq$ is well-defined. Moreover, by saturation with respect to the $\leq$-rules, $(A, \preceq)$ is a partial order with maximum $[\mathbf{1}]_\sim$ and minimum $[\mathbf{0}]_\sim$.[3]

Let $(B, \sqsubseteq, +, \cdot, \top, \bot)$ be the Dedekind-MacNeille completion of $(A, \preceq)$. Note that we have $\top = [\mathbf{1}]_\sim$ and $\bot = [\mathbf{0}]_\sim$.

We define a **BLmf**-interpretation $\mathcal{B} = (B, (\cdot)^{\mathcal{B}})$ by letting:

- $a =^{\mathcal{B}} b$ iff $a = b$.
- $a \leq^{\mathcal{B}} b$ iff $a \sqsubseteq b$;
- $a \sqcup^{\mathcal{B}} b = a + b$;
- $a \sqcap^{\mathcal{B}} b = a \cdot b$;
- $\mathbf{1}^{\mathcal{B}} = \top = [\mathbf{1}]_\sim$;
- $\mathbf{0}^{\mathcal{B}} = \bot = [\mathbf{0}]_\sim$;
- $x^{\mathcal{B}} = [x]_\sim$;
- $f^{\mathcal{B}}(a) = lub(Z_{f,a}, B, \sqsubseteq)$ where

$$Z_{f,a} = X_{f,a} \cup \bigcup_{leq(h,f) \in \Gamma} X_{h,a} \,,$$

and

$$X_{f,a} = \begin{cases} \{[y]_\sim \mid y = f(x) \text{ is in } \Gamma \text{ and } a = [x]_\sim\}, & \text{if } f \in NORM(\Gamma), \\ \{[y]_\sim \mid y = f(x) \text{ is in } \Gamma \text{ and } [x]_\sim \sqsubseteq a\}, & \text{if } f \in INC(\Gamma), \\ \{[y]_\sim \mid y = f(x) \text{ is in } \Gamma \text{ and } a \sqsubseteq [x]_\sim\}, & \text{if } f \in DEC(\Gamma), \\ \{[y]_\sim \mid y = f(x) \text{ is in } \Gamma\}, & \text{if } f \in CONST(\Gamma). \end{cases}$$

By construction, $\mathcal{B}$ is a **BLmf**-interpretation. Next, we show that $\mathcal{B}$ satisfies all **BLmf**-literals in Γ.

Literals of the form $x = y$. We have $x \sim y$, which implies $x^{\mathcal{B}} = [x]_\sim = [y]_\sim = y^{\mathcal{B}}$.

Literals of the form $x \neq y$. If it were $x^{\mathcal{B}} = y^{\mathcal{B}}$, we would have $x \sim y$, which implies that the literal $x = y$ is in Γ, a contradiction.

Literals of the form $x \leq y$. We have $[x]_\sim \preceq [y]_\sim$, so that $[x]_\sim \sqsubseteq [y]_\sim$. Therefore $[x]_\sim \leq^{\mathcal{B}} [y]_\sim$, which in turn implies $[x \leq y]^{\mathcal{B}} = true$.

Literals of the form $\neg(x \leq y)$. We have $[x]_\sim \npreceq [y]_\sim$. Hence $[x]_\sim \not\sqsubseteq [y]_\sim$, which implies $[x \leq y]^{\mathcal{B}} = false$.

Literals of the form $x = y \sqcap z$. By saturation with respect to the $\sqcap$-rules, we have that $[x]_\sim = glb(\{[y], [z]\}, A, \preceq)$. It follows that $[x]_\sim = glb(\{[y], [z]\}, B, \sqsubseteq)$.

---

[3] In general, $(A, \preceq)$ is not a lattice. As an example, consider $\Gamma = closure(\{u \leq x, v \leq x, u \leq y, v \leq y\})$. This is the flaw in [2], which we correct in this paper by taking the Dedekind-MacNeille completion of $(A, \preceq)$.

Literals of the form $x = y \sqcup z$. By saturation with respect to the $\sqcup$-rules, we have that $[x]_\sim = lub(\{[y], [z]\}, A, \preceq)$. It follows that $[x]_\sim = lub(\{[y], [z]\}, B, \sqsubseteq)$.

Literals of the form $y = f(x)$. Let the literal $y = f(x)$ be in Γ. We need to show that $y^{\mathcal{B}} = f^{\mathcal{B}}(x^{\mathcal{B}})$. This amounts to verify that $[y]_\sim = lub(Z_{f,[x]_\sim}, B, \sqsubseteq)$. Since the literal $y = f(x)$ is in Γ, we have immediately $[y]_\sim \in X_{f,[x]_\sim} \subseteq Z_{f,[x]_\sim}$. Therefore, it is enough to show that $[y']_\sim \sqsubseteq [y]_\sim$ holds, for each $[y']_\sim \in Z_{f,[x]_\sim}$.

Thus, let $[y']_\sim \in Z_{f,[x]_\sim}$. It is convenient to distinguish the following two cases.

Case 1: $[y']_\sim \in X_{f,[x]_\sim}$. We consider the following four subcases.

(1a) Let $f \in NORM(\Gamma)$. Then a literal of the form $y' = f(x')$ is in Γ, and $[x]_\sim = [x']_\sim$. Hence the literal $x = x'$ is in Γ. By saturation with respect to the rules of Figure 1, it follows that also the literal $y = y'$ is in Γ, and therefore $[y]_\sim = [y']_\sim$.

(1b) Let $f \in INC(\Gamma)$. Then a literal of the form $y' = f(x')$ is in Γ. Moreover, $[x']_\sim \sqsubseteq [x]_\sim$, which implies $[x']_\sim \preceq [x]_\sim$, so that the literal $x' \leq x$ is in Γ. By saturation, it follows that the literal $y' \leq y$ is in Γ too, which implies $[y']_\sim \preceq [y]_\sim$, and therefore $[y']_\sim \sqsubseteq [y]_\sim$.

(1c) Let $f \in DEC(\Gamma)$. Then a literal of the form $y' = f(x')$ is in Γ. Moreover, $[x]_\sim \sqsubseteq [x']_\sim$, which implies $[x]_\sim \preceq [x']_\sim$, so that the literal $x \leq x'$ is in Γ. By saturation, it follows that also the literal $y' \leq y$ must be in Γ, which implies $[y']_\sim \preceq [y]_\sim$, and therefore $[y']_\sim \sqsubseteq [y]_\sim$.

(1d) Let $f \in CONST(\Gamma)$. Then a literal of the form $y' = f(x')$ is in Γ. By saturation, it follows that the literal $y = y'$ is in Γ, and therefore $[y]_\sim = [y']_\sim$.

Case 2: $[y']_\sim \in X_{h,[x]_\sim}$, where the literal $leq(h, f)$ is in Γ, and h is a function symbol distinct from f. We consider the following four subcases.

(2a) Let $h \in NORM(\Gamma)$. Then a literal of the form $y' = h(x')$ is in Γ, such that $[x]_\sim = [x']_\sim$. By saturation, it follows that the literal $y = f(x')$ is in Γ. But then, again by saturation, the literal $y' \leq y$ must be in Γ. Therefore, $[y']_\sim \preceq [y]_\sim$, which implies $[y']_\sim \sqsubseteq [y]_\sim$.

(2b) Let $h \in INC(\Gamma)$. Then a literal of the form $y' = h(x')$ is in Γ, such that $[x']_\sim \sqsubseteq [x]_\sim$. Therefore, $[x']_\sim \preceq [x]_\sim$, so that the literal $x' \leq x$ is in Γ. By saturation, a literal of the form $y'' = h(x)$ must be in Γ. Therefore, again by saturation, the literals $y' \leq y''$ and $y'' \leq y$ are in Γ, so that also the literal $y' \leq y$ is in Γ. Hence, $[y']_\sim \preceq [y]_\sim$, which implies $[y']_\sim \sqsubseteq [y]_\sim$.

(2c) Let $h \in DEC(\Gamma)$. Then a literal of the form $y' = h(x')$ is in Γ, such that $[x]_\sim \sqsubseteq [x']_\sim$. Therefore, $[x]_\sim \preceq [x']_\sim$, so that the literal $x \leq x'$ is in Γ. By saturation, a literal of the form $y'' = h(x)$ must be in Γ. Therefore, again by saturation, the literals $y' \leq y''$ and $y'' \leq y$ are in Γ, so that also the literal $y' \leq y$ is in Γ. Hence, $[y']_\sim \preceq [y]_\sim$, which implies $[y']_\sim \sqsubseteq [y]_\sim$.

(2d) Let $h \in CONST(\Gamma)$. Then a literal of the form $y' = h(x')$ is in Γ. By saturation, a literal of the form $y'' = h(x)$ is in Γ. Therefore, again by saturation, the literals $y' = y''$ and $y'' \leq y$ are in Γ, so that also the literal $y' \leq y$ must be in Γ. Therefore, $[y']_\sim \preceq [y]_\sim$, which implies $[y']_\sim \sqsubseteq [y]_\sim$.

Literals of the form $inc(f)$. Let the literal $inc(f)$ be in Γ. We need to show that the function $f^{\mathcal{B}}$ is increasing in the lattice $(B, \sqsubseteq, +, \cdot, \top, \bot)$. Thus, let $a, b \in B$ such that $a \sqsubseteq b$. To prove that $f^{\mathcal{B}}(a) \sqsubseteq f^{\mathcal{B}}(b)$, or equivalently that $lub(Z_{f,a}, B \sqsubseteq) \sqsubseteq lub(Z_{f,b}, B, \sqsubseteq)$, it is enough to show that for each $[y]_\sim \in Z_{f,a}$ there exists $[y']_\sim \in Z_{f,b}$ such that $[y]_\sim \sqsubseteq [y']_\sim$.

Thus, let $[y]_\sim \in Z_{f,a}$. We distinguish two cases.

Case 1: $[y]_\sim \in X_{f,a}$. Then the literal $y = f(x)$ is in Γ and $[x]_\sim \sqsubseteq a \sqsubseteq b$, which implies $[y]_\sim \in X_{f,b} \subseteq Z_{f,b}$.

Case 2: $[y] \in X_{h,a}$, where the literal $leq(h, f)$ is in Γ, and h is a function symbol distinct from f. We consider the following four subcases.

(2a) $h \in NORM(\Gamma)$. Then a literal of the form $y = h(x)$ is in Γ, and $a = [x]_\sim$. It follows that the literal $y' = f(x)$ is in Γ. But then, by saturation, the literal $y \leq y'$ is in Γ. Therefore, $[y]_\sim \preceq [y']_\sim$, which implies $[y]_\sim \sqsubseteq [y']_\sim$. Moreover, $[y']_\sim \in X_{h,b} \subseteq Z_{f,b}$.

(2b) $h \in INC(\Gamma)$. Then a literal of the form $y = h(x)$ is in Γ, and $[x]_\sim \sqsubseteq a \sqsubseteq b$. Therefore, $[y] \in X_{h,b} \subseteq Z_{f,b}$.

(2c) $h \in DEC(\Gamma)$. Then a literal of the form $y = h(x)$ is in Γ, and $a \sqsubseteq [x]_\sim$. By saturation, the following literals are in Γ: $y' = h(0)$, $y'' = f(0)$, $y \leq y'$, $y' \leq y''$, and $y \leq y''$. It follows that $[y]_\sim \preceq [y'']_\sim$, which implies $[y]_\sim \sqsubseteq [y'']_\sim$. Moreover, since $[0]_\sim = 0^{\mathcal{B}} \sqsubseteq b$, we have $[y'']_\sim \in X_{f,b} \subseteq Z_{f,b}$.

(2d) $h \in CONST(\Gamma)$. Then a literal of the form $y = h(x)$ is in Γ. By saturation, the following literals are in Γ: $y' = h(0)$, $y'' = f(0)$, $y = y'$, $y' \leq y''$, and $y \leq y''$. It follows that $[y]_\sim \preceq [y'']_\sim$, which implies $[y]_\sim \sqsubseteq [y'']_\sim$. Moreover, since $[0]_\sim = 0^{\mathcal{B}} \sqsubseteq b$, we have $[y'']_\sim \in X_{f,b} \subseteq Z_{f,b}$.

Literals of the form $dec(f)$. This case is similar to the case of literals of the form $inc(f)$.

Literals of the form $const(f)$. This case is similar to the case of literals of the form $inc(f)$.

Literals of the form $leq(f, g)$. We have $Z_{f,a} \subseteq Z_{g,a}$. Therefore, $f^{\mathcal{B}}(a) = lub(Z_{f,a}, B, \sqsubseteq) \sqsubseteq lub(Z_{g,a}, B, \sqsubseteq) = g^{\mathcal{B}}(a)$. ∎

Proposition 29 (Completeness). *Let Γ be a normalized set of **BLmf**-literals, and assume that $closure(\Gamma)$ is consistent. Then Γ is **BLmf**-satisfiable.* □

PROOF. By Proposition 28, $closure(\Gamma)$ is **BLmf**-satisfiable. It follows that Γ is **BLmf**-satisfiable, since $\Gamma \subseteq closure(\Gamma)$. ∎

Proposition 30. *The satisfiability problem for finite sets of **BLmf**-literals is decidable in polynomial time.* □

PROOF. By Propositions 21 and 23. ∎

Proposition 31. *The satisfiability problem for* **BLmf***-formulae is* $\mathcal{NP}$*-comple-te.* □

PROOF. The satisfiability problem for **BLmf**-formulae is clearly $\mathcal{NP}$-hard, In order to show membership to $\mathcal{NP}$, it suffices to note that one can check whether a **BLmf**-formula is **BLmf**-satisfiable by:

1. guessing a disjunct Γ of a DNF of φ;
2. converting Γ to a conjunction of normalized literals Γ';
3. computing $closure(\Gamma')$;
4. checking whether $closure(\Gamma')$ is consistent. ∎

5 The Language CLmf

In this section we define the language **CLmf** (*complete lattices with monotone functions*) and we prove that it is equivalent to the language **BLmf** for quantifier-free formulae.

Syntactically, the language **CLmf** coincides with **BLmf**. Semantically, we have the following definition.

Definition 32. A **CLmf**-INTERPRETATION $\mathcal{A}$ is **BLmf**-interpretation in which the lattice $(A, \leq^{\mathcal{A}}, \sqcup^{\mathcal{A}}, \sqcap^{\mathcal{A}}, 1^{\mathcal{A}}, 0^{\mathcal{A}})$ is complete. □

Proposition 33. *Let φ be a quantifier-free* **BLmf***- or* **CLmf***-formula. Then φ is* **BLmf***-satisfiable if and only if φ is* **CLmf***-satisfiable.* □

PROOF. Assume first that φ is **CLmf**-satisfiable. Since every **CLmf**-interpretation is also a **BLmf**-interpretation, it follows that φ is **BLmf**-satisfiable.

Conversely, assume that φ is **BLmf**-satisfiable. Without loss of generality, we can assume that φ is a normalized set of **BLmf**-literals. Let $\psi = closure(\varphi)$. By Proposition 26, ψ is **BLmf**-satisfiable. It follows that ψ is consistent. By Proposition 28, ψ is **CLmf**-satisfiable. Since $\varphi \subseteq \psi$, it follows that φ is **CLmf**-satisfiable. ∎

6 Conclusion

We presented a decision procedure for the quantifier-free satisfiability problem of the language **BLmf** (*Bounded Lattices with monotone functions*). The language contains the predicates = and $\leq$, the operators $\sqcap$ and $\sqcup$ over terms which may involve uninterpreted unary function symbols, predicates for expressing increasing and decreasing monotonicity of functions, and a predicate for pointwise function comparison.

We proved that our decision procedure runs in polynomial time $\mathcal{O}(m^4)$ for conjunctions of literals, thus entailing that the quantifier-free satisfiability problem for **BLmf** is $\mathcal{NP}$-complete.

Finally, we defined the language **CLmf** (*Complete Lattices with monotone functions*), and we proved that the languages **CLmf** and **BLmf** are equivalent for quantifier-free formulae.

In our proofs, we used the hypothesis that lattices are bounded (see, for instance, Proposition 28, case of literals of the form $inc(f)$, subcases 2c and 2d). Thus, a possible direction of future research would be to relax this hypothesis, and study the language **Lmf** (*Lattices with monotone functions*) in which the semantics does not require lattices to be bounded. We conjecture that **Lmf** is decidable. A promising result in this direction can be found in [13], where decidability is proved after removing from **Lmf** the predicates $dec(f)$, $const(f)$, and $leq(f,g)$.

Acknowledgments

We are grateful to Viorica Sofronie-Stokkermans for fruitful discussions on lattice theory.

References

1. D. Cantone, A. Ferro, E. G. Omodeo, and J. T. Schwartz. Decision algorithms for some fragment of analysis and related areas. *Communications on Pure and Applied Mathematics*, 40(3):281–300, 1987.
2. D. Cantone and C. G. Zarba. A decision procedure for monotone functions over lattices. In F. Buccafurri, editor, *Joint Conference on Declarative Programming APPIA-GULP-PRODE*, pages 1–12, 2003.
3. R. Dedekind. *Stetigkeit und Irrationale Zahlen*. Braunschweig: F. Vieweg, 1872.
4. N. Dershowitz and J.-P. Jouannaud. Rewrite systems. In J. V. Leeuwen, editor, *Handbook of Theoretical Computer Science (vol. B): Formal Models and Semantics*, pages 243–320. MIT Press, Cambridge, MA, USA, 1990.
5. J. Desharnais, B. Möller, and G. Struth. Modal Kleene algebra and applications — A survey —. *Journal on Relational Methods in Computer Science*, 1:93–131, 2004.
6. A. Ferro, E. G. Omodeo, and J. T. Scwhartz. Decision procedures for elementary sublanguages of set theory. I. Multi-level syllogistic and some extensions. *Communications on Pure and Applied Mathematics*, 33(5):599–608, 1980.
7. S. Hazelhurst and C.-J. H. Seger. Model checking lattices: Using and reasoning about information orders for abstraction. *Logic Journal of the IGPL*, 7(3):375–411, 1999.
8. H. M. MacNeille. Partially ordered sets. *Transactions of the American Mathematical Society*, 42:416–460, 1937.
9. D. Micciancio and S. Goldwasser. *Complexity of Lattice Problems: a cryptographic perspective*, volume 671 of *The Kluwer International Series in Engineering and Computer Science*. Kluwer Academic Publishers, Boston, Massachusetts, 2002.
10. P. D. Mosses. Denotational semantics. In J. V. Leeuwen, editor, *Handbook of Theoretical Computer Science (vol. B): Formal Models and Semantics*, pages 575–631. MIT Press, Cambridge, MA, USA, 1990.

11. F. J. Oles. An application of lattice theory to knowledge representation. *Theoretical Computer Science*, 249(1):163–196, 2000.

12. M. Osorio, B. Jayaraman, and D. A. Plaisted. Theory of partial-order programming. *Science of Computer Programming*, 34(3):207–238, 1999.

13. V. Sofronie-Stokkermans. Hierarchic reasoning in local theory extensions. In R. Nieuwenhuis, editor, *Automated Deduction – CADE-20*, volume 3632 of *Lecture Notes in Computer Science*, pages 219–234. Springer, 2005.

14. A. Tarski. Undecidability of the theory of lattices and projective geometries. *Journal of Symbolic Logic*, 14(1):77–78, 1949.

The Dominance Relation on the Class of Continuous T-Norms from an Ordinal Sum Point of View

Susanne Saminger[1], Peter Sarkoci[2], and Bernard De Baets[3]

[1] Department of Knowledge-Based Mathematical Systems
Johannes Kepler University
Altenbergerstrasse 69, A-4040 Linz, Austria
susanne.saminger@jku.at
[2] Department of Mathematics, IIEAM
Slovak University of Technology
Radlinského 9, SK-81368 Bratislava, Slovakia
peter.sarkoci@stuba.sk
[3] Department of Applied Mathematics, Biometrics, and Process Control,
Ghent University,
Coupure links 653, B-9000 Gent, Belgium
Bernard.DeBaets@Ugent.be

Abstract. This paper addresses the relation of dominance on the class of continuous t-norms with a particular focus on continuous ordinal sum t-norms. Exactly, in this framework counter-examples to the conjecture that dominance is not only a reflexive and antisymmetric, but also a transitive relation could be found. We elaborate the details which have led to these results and illustrate them by several examples. In addition, to this original and comprehensive overview, we provide geometrical insight into dominance relationships involving prototypical Archimedean t-norms, the Łukasiewicz t-norm and the product t-norm.

1 Introduction

The *dominance* property was originally introduced within the framework of *probabilistic metric spaces* [42] and was soon abstracted to operations on an arbitrary partially ordered set [38]. A probabilistic metric space allows for *imprecise distances*: the distance between two objects p and q is characterized by a cumulative distribution function $F_{pq} : \mathbb{R} \to [0, 1]$. The metric in such spaces is defined in analogy to the axioms of (pseudo-)metric spaces, the most disputable axiom being the probabilistic analogue of the triangle inequality. For an important subclass of probabilistic metric spaces known as *Menger spaces* the triangle inequality reads as follows: for any three objects p, q, r and for any $x, y \geq 0$ it holds that

$$F_{pr}(x + y) \geq T(F_{pq}(x), F_{qr}(y)), \tag{1}$$

H. de Swart et al. (Eds.): TARSKI II, LNAI 4342, pp. 334–354, 2006.
© Springer-Verlag Berlin Heidelberg 2006

where $T\colon [0,1]^2 \to [0,1]$ is a *t-norm*, *i.e.* a binary operation on the unit interval which is commutative, associative, increasing in both arguments and which has neutral element 1.

The dominance property plays an important role in the construction of *Cartesian products* of probabilistic metric spaces, as it ensures that the triangle inequality holds for the resulting product space provided it holds for all factor spaces involved [38,42]. Similarly, it is responsible for the preservation of the T-transitivity property when building *fuzzy equivalence* or *fuzzy order relations* on a product space, *i.e.* $R\colon X^2 \to [0,1]$, defined by $R(\mathbf{x}, \mathbf{y}) = \mathbf{A}(R_1(x_1, y_1), \ldots, R_n(x_n, y_n))$ with $X = \prod_{i=1}^{n} X_i$, $R_i\colon X_i^2 \to [0,1]$ fuzzy relations on X_i being all T-transitive, *i.e.*

$$T(R_i(x,y), R_i(y,z)) \le R_i(x,z)$$

and $\mathbf{A}$ some aggregation operator, or when intersecting such fuzzy relations on a single space, *i.e.* $R(x,y) = T(R_1(x,y), \ldots, R_n(x,y))$ [2,3,8,32]. The dominance property was therefore introduced in the framework of aggregation operators where it enjoyed further development, again due its role in the preservation of a variety of properties, most of them expressed by some inequality, during (dis-)aggregation processes (see also [9,29]).

Besides these application points of view, the dominance property turned out to be an interesting mathematical notion *per se*. Due to the common neutral element of t-norms and their commutativity and associativity, the dominance property constitutes a reflexive and antisymmetric relation on the class of all t-norms. Whether it is also transitive has been posed as an open question already in 1983 in [38] and remained unanswered for quite some time. Several particular families of t-norms have been investigated (see, e.g., [17,34,40]) and supported the conjecture that the dominance relation would indeed be transitive, either due to its rare occurrence within the family considered or due to its abundant occurrence, in accordance with the parameter of the family. Several research teams participating in the EU COST action TARSKI have been studying various aspects of the dominance relation over the past few years. Finally, the conjecture was recently rejected [35]: the dominance relation is not transitive on the class of continuous t-norms and therefore also not on the class of t-norms in general. The counterexample was found among continuous ordinal sum t-norms.

In this contribution we discuss the dominance relation on the class of continuous t-norms and elaborate the details which have led to the counterexamples demonstrating the non-transitivity of the dominance relation in the class of t-norms. First, we provide a thorough introduction of all the necessary properties and details about t-norms. We then continue with a brief discussion of the dominance relation on the class of continuous Archimedean t-norms and provide geometrical insight in two prototypical cases. Subsequently, we turn to continuous ordinal sum t-norms and particular families of such ordinal sum t-norms. The present contribution provides a comprehensive and original overview of the state-of-the-art knowledge of the dominance relation on the class of continuous ordinal sum t-norms and as such depends on results also published in [17,30,31,35].

2   Triangular Norms

For the reader's convenience we briefly summarize basic properties of t-norms which will be necessary for a thorough understanding of the following parts. Many of the herein included results (including proofs, further details and references) can be found in [18,19,20] or in the monographs [1,17].

2.1   Basic Properties

Triangular norms (briefly t-norms) were first introduced in the context of probabilistic metric spaces [36,38,39], based on some ideas already presented in [24]. They are an indispensable tool for the interpretation of the conjunction in fuzzy logics [14] and, as a consequence, for the intersection of fuzzy sets [46]. Further, they play an important role in various further fields like decision making [11,13], statistics [26], as well as the theories of non-additive measures [21,27,41,45] and cooperative games [4].

Definition 1. *A* triangular norm *(briefly* t-norm*) is a binary operation T on the unit interval $[0, 1]$ which is commutative, associative, increasing and has 1 as neutral element, i.e. it is a function $T : [0, 1]^2 \to [0, 1]$ such that for all $x, y, z \in [0, 1]$:*

(T1) $T(x, y) = T(y, x)$,
(T2) $T(x, T(y, z)) = T(T(x, y), z)$,
(T3) $T(x, y) \le T(x, z)$ whenever $y \le z$,
(T4) $T(x, 1) = x$.

It is an immediate consequence that due to the boundary and monotonicity conditions as well as commutativity it follows that, for all $x \in [0, 1]$, any t-norm T satisfies

$$T(0, x) = T(x, 0) = 0, \tag{2}$$
$$T(1, x) = x. \tag{3}$$

Therefore, all t-norms coincide on the boundary of the unit square $[0, 1]^2$.

Example 1. The most prominent examples of t-norms are the *minimum* $T_\mathbf{M}$, the *product* $T_\mathbf{P}$, the *Lukasiewicz t-norm* $T_\mathbf{L}$ and the *drastic product* $T_\mathbf{D}$ (see Figure 1 for 3D and contour plots). They are given by:

$$T_\mathbf{M}(x, y) = \min(x, y), \tag{4}$$
$$T_\mathbf{P}(x, y) = x \cdot y, \tag{5}$$
$$T_\mathbf{L}(x, y) = \max(x + y - 1, 0), \tag{6}$$
$$T_\mathbf{D}(x, y) = \begin{cases} 0 & \text{if } (x, y) \in [0, 1[^2, \\ \min(x, y) & \text{otherwise.} \end{cases} \tag{7}$$

Since t-norms are just functions from the unit square into the unit interval, the comparison of t-norms is done in the usual way, *i.e.* pointwisely.

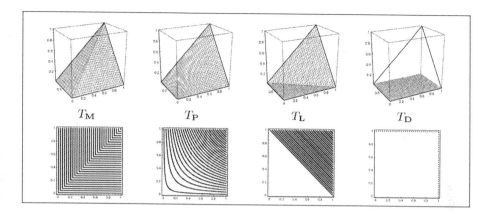

Fig. 1. 3D plots (top) and contour plots (bottom) of the four basic t-norms $T_{\mathbf{M}}$, $T_{\mathbf{P}}$, $T_{\mathbf{L}}$, and $T_{\mathbf{D}}$ (observe that there are no contour lines for $T_{\mathbf{D}}$)

Definition 2. *Let T_1 and T_2 be two t-norms. If $T_1(x, y) \leq T_2(x, y)$ for all $x, y \in [0, 1]$, then we say that T_1 is* weaker *than T_2 or, equivalently, that T_2 is* stronger *than T_1, and we write $T_1 \leq T_2$.*

Further, t-norms can be transformed by means of an order isomorphism, *i.e.* an increasing $[0, 1] \rightarrow [0, 1]$ bijection, preserving several properties (like, e.g., continuity) of the t-norm involved.

Definition 3. *Let T be a t-norm and φ an order isomorphism. Then the isomorphic transform of T under φ is the t-norm T_φ defined by*

$$T_\varphi(x, y) = \varphi^{-1}(T(\varphi(x), \varphi(y))). \tag{8}$$

Note that the drastic product $T_{\mathbf{D}}$ and the minimum $T_{\mathbf{M}}$ are the smallest and the largest t-norm, respectively. Moreover, they are the only t-norms that are invariant under arbitrary order isomorphisms.

Let us now focus on the continuity of t-norms.

Definition 4. *A t-norm T is* continuous *if for all convergent sequences $(x_n)_{n \in \mathbb{N}}$, $(y_n)_{n \in \mathbb{N}} \in [0, 1]^{\mathbb{N}}$ we have*

$$T\left(\lim_{n \to \infty} x_n, \lim_{n \to \infty} y_n\right) = \lim_{n \to \infty} T(x_n, y_n).$$

Obviously, the basic t-norms $T_{\mathbf{M}}, T_{\mathbf{P}}$ and $T_{\mathbf{L}}$ are continuous, whereas the drastic product $T_{\mathbf{D}}$ is not. Note that for a t-norm T its continuity is equivalent to the continuity in each component (see also [17,18]), *i.e.* for any $x_0, y_0 \in [0, 1]$ both the vertical section $T(x_0, \cdot) : [0, 1] \rightarrow [0, 1]$ and the horizontal section $T(\cdot, y_0) : [0, 1] \rightarrow [0, 1]$ are continuous functions in one variable.

The following classes of continuous t-norms are of particular importance.

Definition 5.   *(i) A t-norm T is called* strict *if it is continuous and strictly monotone, i.e. it fulfills for all $x, y, z \in [0, 1]$*

$$T(x, y) < T(x, z) \text{ whenever } x > 0 \text{ and } y < z.$$

(ii) A t-norm T is called nilpotent *if it is continuous and if each $x \in]0, 1[$ is a* nilpotent element *of T, i.e. there exists some $n \in \mathbb{N}$ such that*

$$T(\underbrace{x, \ldots, x}_{n \text{ times}}) = 0.$$

The product $T_\mathbf{P}$ is a strict t-norm whereas the Łukasiewicz t-norm $T_\mathbf{L}$ is a nilpotent t-norm. Both of them are *Archimedean* t-norms, *i.e.* they fulfill for all $(x, y) \in]0, 1[^2$ that there exists an $n \in \mathbb{N}$ such that

$$T(\underbrace{x, \ldots, x}_{n \text{ times}}) < y.$$

It is remarkable that continuous Archimedean t-norms can be divided into just two subclasses — the nilpotent and the strict t-norms [17,18]. Moreover, since two continuous Archimedean t-norms are isomorphic if and only if they are either both strict or both nilpotent, we can immediately formulate the following proposition (see also [17,18]).

Proposition 1. *Let T be a t-norm.*

- *T is a strict t-norm if and only if it is isomorphic to the product $T_\mathbf{P}$.*
- *T is a nilpotent t-norm if and only if it is isomorphic to the Łukasiewicz t-norm $T_\mathbf{L}$.*

Besides the above introduced properties, idempotent elements play an important role in the characterization of t-norms.

Definition 6. *Let T be a t-norm. An element $x \in [0, 1]$ is called an* idempotent element *of T if $T(x, x) = x$. We will further denote by $\mathcal{I}(T)$ the set of all idempotent elements of T. The numbers 0 and 1 (which are idempotent elements for each t-norm T) are called* trivial idempotent elements *of T, each idempotent element in $]0, 1[$ will be called a* non-trivial idempotent element *of T.*

The set of idempotent elements of the minimum $T_\mathbf{M}$ equals $[0, 1]$ (actually, $T_\mathbf{M}$ is the only t-norm with this property) whereas $T_\mathbf{P}$, $T_\mathbf{L}$, and $T_\mathbf{D}$ possess only trivial idempotent elements.

2.2   Ordinal Sum T-Norms

Ordinal sum t-norms are based on a construction principle for semigroups which goes back to A.H. Clifford [5] (see also [6,15,28]) based on ideas presented in [7,16]. It has been successfully applied to t-norms in [12,22,37].

Definition 7. *Let* $(]a_i, b_i[)_{i \in I}$ *be a family of non-empty, pairwise disjoint open subintervals of* $[0, 1]$ *and let* $(T_i)_{i \in I}$ *be a family of t-norms. The* ordinal sum $T = ((\langle a_i, b_i, T_i \rangle)_{i \in I}$ *is the t-norm defined by*

$$T(x, y) = \begin{cases} a_i + (b_i - a_i)T_i(\frac{x - a_i}{b_i - a_i}, \frac{y - a_i}{b_i - a_i}), & \text{if } (x, y) \in [a_i, b_i]^2, \\ \min(x, y), & \text{otherwise.} \end{cases}$$

We will refer to $\langle a_i, b_i, T_i \rangle$ as its *summands*, to $[a_i, b_i]$ as its *summand carriers*, and to T_i as its *summand operations* or *summand t-norms*. The index set I is necessarily finite or countably infinite. It may also be empty in which case the ordinal sum is nothing else but $T_{\mathbf{M}}$.

Note that by construction, the set of idempotent elements $\mathcal{I}(T)$ of some ordinal sum $T = ((\langle a_i, b_i, T_i \rangle)_{i \in I}$ contains the set $M = [0, 1] \setminus \bigcup_{i \in I}]a_i, b_i[$. Moreover, $\mathcal{I}(T) = M$ if and only if each T_i has only trivial idempotent elements. It is clear that an ordinal sum t-norm is continuous if and only if all of its summand t-norms are continuous.

In general, the representation of a t-norm as an ordinal sum of t-norms is not unique. For instance, for each subinterval $[a, b]$ of $[0, 1]$ we have

$$T_{\mathbf{M}} = (\emptyset) = ((\langle 0, 1, T_{\mathbf{M}} \rangle)) = ((\langle a, b, T_{\mathbf{M}} \rangle)).$$

This gives rise to the following definition.

Definition 8. *A t-norm* T *that has no ordinal sum representation different from* $(\langle 0, 1, T \rangle)$ *is called* ordinally irreducible.

Note that each continuous Archimedean t-norm, in particular also $T_{\mathbf{P}}$ and $T_{\mathbf{L}}$, has only trivial idempotent elements and is therefore ordinally irreducible. Moreover, there are no other ordinally irreducible continuous t-norms.

Based on the above information, we can now turn to the representation of continuous t-norms (see also [17,22,25,38])

Theorem 1. *A binary operation on the unit interval is a continuous t-norm if and only if it is an ordinal sum of continuous Archimedean t-norms.*

Therefore, continuous t-norms are either:

- strict, *i.e.* isomorphic to the product t-norm $T_{\mathbf{P}}$,
- nilpotent, *i.e.* isomorphic to the Łukasiewicz t-norm $T_{\mathbf{L}}$,
- the minimum $T_{\mathbf{M}}$ itself, *i.e.* $I = \emptyset$, or
- non-trivial ordinal sums with strict or nilpotent summand operations, *i.e.* $I \neq \emptyset$ and no $]a_i, b_i[$ equals $]0, 1[$.

2.3   The Dominance Property for T-Norms

Let us now focus on the dominance relation on the class of t-norms [38,42,44].

Definition 9. *We say that a t-norm T_1 dominates a t-norm T_2, or equivalently, that T_2 is dominated by T_1, and write $T_1 \gg T_2$, if for all $x, y, u, v \in [0,1]$*

$$T_1(T_2(x, y), T_2(u, v)) \geq T_2(T_1(x, u), T_1(y, v)). \tag{9}$$

Due to the fact that 1 is the common neutral element of all t-norms, dominance of one t-norm by another t-norm implies their comparability (see also [29]), *i.e.* $T_1 \gg T_2$ implies $T_1 \geq T_2$. Similarly to the ordering of t-norms, any t-norm T is dominated by itself and by $T_{\mathbf{M}}$, and dominates $T_{\mathbf{D}}$, *i.e.* for any t-norm T it holds that

$$T_{\mathbf{M}} \gg T, \qquad T \gg T, \qquad T \gg T_{\mathbf{D}}.$$

As a consequence we can immediately state that dominance is a reflexive and antisymmetric relation on the class of all t-norms. We will show later that it is not transitive, not even on the class of continuous t-norms. Hence, the dominance relation is not a partial order on the set of all t-norms.

Finally, we mention that a dominance relationship between two t-norms is preserved under isomorphic transformations [32].

Proposition 2. *A t-norm T_1 dominates a t-norm T_2 if and only if $(T_1)_\varphi$ dominates $(T_2)_\varphi$ for any order isomorphism φ.*

3   Continuous Archimedean T-Norms

3.1   Isomorphic Transformations

The problem we study here is to determine whether a first continuous Archimedean t-norm T_1 dominates a second such t-norm T_2. Since dominance is preserved under isomorphic transformations, this problem can be transformed into one of the following prototypical problems. Suppose that $T_1 \gg T_2$:

- If T_1 is nilpotent, then T_2 has to be nilpotent as well. In that case, there exist some order isomorphisms φ and ψ such that $(T_1)_\varphi = T_{\mathbf{L}}$ and $(T_2)_\psi = T_{\mathbf{L}}$ leading to
$$T_1 \gg T_2 \Leftrightarrow (T_1)_\psi \gg T_{\mathbf{L}} \Leftrightarrow T_{\mathbf{L}} \gg (T_2)_\varphi.$$

- If T_1 is strict, then T_2 can be either strict or nilpotent. In both cases, there exist order isomorphisms φ and ψ such that
$$T_1 \gg T_2 \Leftrightarrow (T_1)_\psi \gg T_{\mathbf{L}} \Leftrightarrow T_{\mathbf{P}} \gg (T_2)_\varphi$$
in case T_2 is nilpotent, and
$$T_1 \gg T_2 \Leftrightarrow (T_1)_\psi \gg T_{\mathbf{P}} \Leftrightarrow T_{\mathbf{P}} \gg (T_2)_\varphi$$
in case T_2 is strict.

Summarizing, it suffices to investigate the classes of t-norms dominating or being dominated either by $T_{\mathbf{P}}$ or by $T_{\mathbf{L}}$. In the next section, we will provide a geometrical interpretation for these particular cases. Necessary as well as sufficient conditions for aggregation operators (and therefore also t-norms) dominating one of these t-norms can be found, e.g., in [29,30,32,43].

3.2   Geometrical Interpretation

The inequality expressing dominance is difficult to grasp since it concerns four variables involved in various compositions of mappings. Providing an insightful geometrical interpretation would be more than welcome. We will present such an interpretation for the two cases discussed above: t-norms dominating or being dominated either by $T_{\mathbf{L}}$ or by $T_{\mathbf{P}}$.

Note that the inequality expressing dominance trivially holds if at least one of the arguments equals 0. Hence, we can restrict our attention to arguments $x, y, u, v \in]0, 1]$ only.

Dominance Relationships Involving $T_{\mathbf{L}}$. Let us consider some t-norm T which dominates $T_{\mathbf{L}}$, i.e. for all $x, y, u, v \in [0, 1]$ we have

$$T(T_{\mathbf{L}}(x, u), T_{\mathbf{L}}(y, v)) \geq T_{\mathbf{L}}(T(x, y), T(u, v)). \tag{10}$$

For any fixed $u, v \in]0, 1]$, we introduce new variables $a = T_{\mathbf{L}}(x, u)$ and $b = T_{\mathbf{L}}(y, v)$ ranging over $[0, u]$ and $[0, v]$, respectively. If $a = 0$ then $x + u - 1 \leq 0$; similarly, if $b = 0$ then $y + v - 1 \leq 0$. In any case, it follows that $T(x, y) + T(u, v) - 1 \leq 0$ and (10) is satisfied trivially as both sides evaluate to 0. On the other hand, if $a, b > 0$ then x and y can be recovered from the expressions $x = 1 - u + a$ and $y = 1 - v + b$. Using these new variables, the dominance inequality is transformed into

$$T(a, b) \geq T_{\mathbf{L}}(T(1 - u + a, 1 - v + b), T(u, v)) \tag{11}$$

for all $u, v \in]0, 1]$ and all $a \in [0, u]$, $b \in [0, v]$. The right-hand side can be interpreted geometrically in the following way:

- First, the graph of $T(1 - u + a, 1 - v + b)$ as a function of a and b is nothing else but a translation of the original graph such that the point $(1, 1, 1)$ is moved to the point $(u, v, 1)$.
- Using $T_{\mathbf{L}}$ to combine this function with the value $T(u, v)$ means that this translated graph is subsequently translated along the direction of the z-axis such that the original reference point $(1, 1, 1)$ is now located in the point $(u, v, T(u, v))$.
- As a consequence, parts of the resulting surface are now located outside the unit cube. Due to the definition of $T_{\mathbf{L}}$, these parts are simply truncated by 0, i.e. they are substituted by the corresponding parts of the xy-plane.

The fact that T dominates $T_{\mathbf{L}}$ means that this translated surface lies below the original one, and this for any choice of u, v. The situation in which a t-norm T is dominated by $T_{\mathbf{L}}$ has a similar interpretation, the only difference being that the translated surface should now be above the original one.

In Fig. 2, this geometrical interpretation is illustrated for the case $T_{\mathbf{M}} \gg T_{\mathbf{L}}$. For any choice of u, v (see Fig. 2 (a)) the box $[0, u] \times [0, v] \times [0, T_{\mathbf{M}}(u, v)]$ is constructed (see Fig. 2 (b)) and the original graph of $T_{\mathbf{M}}$ is translated moving

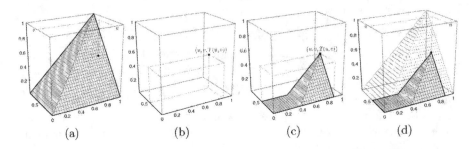

Fig. 2. Geometrical interpretation of T_M dominating T_L

the point $(1, 1, 1)$ to the point $(u, v, T_\mathrm{M}(u, v))$ (see Fig. 2 (c)). Then the translated surface is compared with the original one (see Fig. 2 (d)). One can see immediately that the new surface lies below the original one for any choice of u, v.

Dominance Relationships Involving T_P. The case of a t-norm T dominating T_P has an even simpler geometrical interpretation. First of all, $T \gg T_\mathrm{P}$ means that for all $x, y, u, v \in [0, 1]$ it holds

$$T(xu, yv) \geq T(x, y)T(u, v).$$

For any fixed $u, v \in \,]0, 1]$, we introduce new variables $a = xu$ and $b = yv$ ranging over $[0, u]$ and $[0, v]$, respectively. Using these new variables, the dominance inequality is transformed into

$$T(a, b) \geq T(\tfrac{a}{u}, \tfrac{b}{v})T(u, v) \tag{12}$$

for all $u, v \in \,]0, 1]$ and all $a \in [0, u]$, $b \in [0, v]$. The right-hand side can be interpreted geometrically in the following way.

The graph of $T(\tfrac{a}{u}, \tfrac{b}{v})T(u, v)$ as a function of a and b is exactly the graph of T linearly rescaled in order to fit into the box $[0, u] \times [0, v] \times [0, T(u, v)]$. This rescaling is obviously different for any u, v. The fact that T dominates T_P means that this rescaled graph lies below the original graph. The situation in which

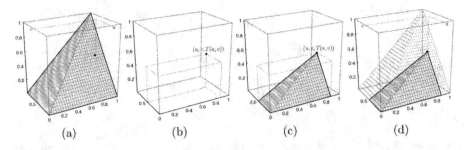

Fig. 3. Geometrical interpretation of T_M dominating T_P

a t-norm T is dominated by $T_\mathbf{P}$ has again a similar interpretation, the only difference being again that the rescaled graph should now be above the original one.

In Fig. 3, this geometrical interpretation is illustrated for the case $T_\mathbf{M} \gg T_\mathbf{P}$. For any choice of u, v (see Fig. 3 (a)) the box $[0, u] \times [0, v] \times [0, T_\mathbf{M}(u, v)]$ is constructed (see Fig. 3 (b)) and the original graph of $T_\mathbf{M}$ is rescaled in order to fit into this box (see Fig. 3 (c)). Then the rescaled surface is compared with the original one (see Fig. 3 (d)). One can see immediately that the new surface lies below the original one for any choice of u, v.

4   Continuous Non-Archimedean T-Norms

Let us now focus on dominance involving continuous non-Archimedean t-norms, *i.e.* involving non-trivial ordinal sums of continuous Archimedean t-norms.

4.1   Summand-wise Dominance

When studying the dominance relationship between two ordinal sum t-norms, we have to take into account the underlying structure of the ordinal sums. In case both ordinal sum t-norms are determined by the same family of non-empty, pairwise disjoint open subintervals, dominance between the ordinal sum t-norms is determined by the dominance between all corresponding summand t-norms [30].

Proposition 3. *Consider the two ordinal sum t-norms* $T_1 = (\langle a_i, b_i, T_{1,i} \rangle)_{i \in I}$ *and* $T_2 = (\langle a_i, b_i, T_{2,i} \rangle)_{i \in I}$. *Then* T_1 *dominates* T_2 *if and only if* $T_{1,i}$ *dominates* $T_{2,i}$ *for all* $i \in I$.

4.2   Ordinal Sum T-Norms with Different Summand Carriers

In case the structure of both ordinal sum t-norms is not the same, we are able to provide some necessary conditions which lead to a characterization of dominance between ordinal sum t-norms in general. Assume that the ordinal sum t-norms T_1 and T_2 under consideration are based on two at least partially different families of summand carriers, *i.e.* $T_1 = (\langle a_{1,i}, b_{1,i}, T_{1,i} \rangle)_{i \in I}$ and $T_2 = (\langle a_{2,j}, b_{2,j}, T_{2,j} \rangle)_{j \in J}$. W.l.o.g. we can assume that these representations are the finest possible, *i.e.* that each summand t-norm is ordinally irreducible.

Since for a continuous t-norm T the existence of a non-trivial idempotent element d is even equivalent to being representable as an ordinal sum $T = (\langle 0, d, T' \rangle, \langle d, 1, T'' \rangle)$ for some summand t-norms T' and T'' (see also [17]), it is indeed reasonable to assume that the representations of two continuous t-norms $T_1 = (\langle a_{1,i}, b_{1,i}, T_{1,i} \rangle)_{i \in I}$ and $T_2 = (\langle a_{2,j}, b_{2,j}, T_{2,j} \rangle)_{j \in J}$ are such that there exists no $T_{1,i}$, resp. $T_{2,j}$, with a non-trivial idempotent element $d \in]a_{1,i}, b_{1,i}[$, resp. $d \in]a_{2,j}, b_{2,j}[$.

Necessary Conditions Due to the Induced Order. Since any t-norm is bounded from above by T_M and dominance implies their comparability we immediately can state the following lemma [30].

Lemma 1. *If a t-norm T_1 dominates a t-norm T_2, then $T_1(x,y) = T_M(x,y)$ whenever $T_2(x,y) = T_M(x,y)$.*

Geometrically speaking, if an ordinal sum t-norm T_1 dominates an ordinal sum t-norm T_2, then it must necessarily consist of more regions where it acts as T_M than T_2. Two such cases are displayed in Fig. 4 (a) and (c). Note that no dominance relationship between T_1 and T_2 is possible in a case like illustrated in Fig. 4 (b).

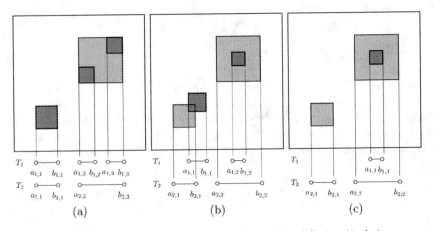

Fig. 4. Examples of two ordinal sum t-norms T_1 and T_2 differing in their summand carriers

Therefore, we can immediately state the following corollary [30].

Corollary 1. *Consider the two ordinal sum t-norms $T_1 = (\langle a_{1,i}, b_{1,i}, T_{1,i} \rangle)_{i \in I}$ and $T_2 = (\langle a_{2,j}, b_{2,j}, T_{2,j} \rangle)_{j \in J}$ with ordinally irreducible summand t-norms only. If T_1 dominates T_2 then*

$$\forall i \in I \colon \exists j \in J \colon [a_{1,i}, b_{1,i}] \subseteq [a_{2,j}, b_{2,j}] \,. \tag{13}$$

Note that each $[a_{2,j}, b_{2,j}]$ can contain several or even none of the summand carriers $[a_{1,i}, b_{1,i}]$ (see also Fig. 4 (a) and (c)). Hence, for each $j \in J$ we can consider the following subset of I:

$$I_j = \{ i \in I \mid [a_{1,i}, b_{1,i}] \subseteq [a_{2,j}, b_{2,j}] \} \,. \tag{14}$$

Based on these notions and due to Proposition 3, dominance between two ordinal sum t-norms can be reformulated in the following way [30].

Proposition 4. *Consider two ordinal sum t-norms $T_1 = (\langle a_{1,i}, b_{1,i}, T_{1,i} \rangle)_{i \in I}$ and $T_2 = (\langle a_{2,j}, b_{2,j}, T_{2,j} \rangle)_{j \in J}$ with ordinally irreducible summand operations only. Then T_1 dominates T_2 if and only if*

(i) $\cup_{j \in J} I_j = I$,

(ii) $T_1^j \gg T_{2,j}$ *for all $j \in J$ with*

$$T_1^j = (\langle \varphi_j(a_{1,i}), \varphi_j(b_{1,i}), T_{1,i} \rangle)_{i \in I_j} \tag{15}$$

and $\varphi_j \colon [a_{2,j}, b_{2,j}] \to [0,1]$, $\varphi_j(x) = \frac{x - a_{2,j}}{b_{2,j} - a_{2,j}}$.

Note that due to Proposition 4, the study of dominance between ordinal sum t-norms can be reduced to the study of dominance of a single ordinally irreducible t-norm by some ordinal sum t-norm. In particular, if all ordinal sum t-norms involved are just based on a single t-norm T^* as summand operation, it suffices to investigate the dominance of T^* by ordinal sum t-norms $T = (\langle a_i, b_i, T^* \rangle)_{i \in I}$.

Example 2. Let us now briefly elaborate the three different cases of ordinal sum t-norms displayed in Fig. 4 in more detail :

- Consider the ordinal sum t-norms T_1 and T_2 as displayed in Fig. 4 (a). Due to Proposition 3, $T_1 \gg T_2$ is equivalent to showing that $T_{1,1} \gg T_{2,1}$ and $T_1^2 \gg T_{2,2}$, where T_1^2 is the ordinal sum t-norm defined by

$$T_1^2 = (\langle \varphi_2(a_{1,2}), \varphi_2(b_{1,2}), T_{1,2} \rangle, \langle \varphi_2(a_{1,3}), \varphi_2(b_{1,3}), T_{1,3} \rangle),$$

  with $\varphi_2 \colon [a_{2,2}, b_{2,2}] \to [0,1]$, $\varphi_2(x) = \frac{x - a_{2,2}}{b_{2,2} - a_{2,2}}$.
- Having a look at the ordinal sum t-norms T_1 and T_2 as displayed in Fig. 4 (b), we immediately see that $[a_{1,1}, b_{1,1}] \not\subseteq [a_{2,1}, b_{2,1}]$ and vice versa, so that

$$T_1(x, y) = T_\mathbf{M}(x, y) \neq T_2(x, y) \text{ for some } x, y \in [a_{2,1}, a_{1,1}],$$
$$T_2(x, y) = T_\mathbf{M}(x, y) \neq T_1(x, y) \text{ for some } x, y \in [b_{2,1}, b_{1,1}].$$

  Hence, due to Lemma 1, in this case a dominance relationship is impossible.
- On the other hand, for the ordinal sum t-norms T_1 and T_2 as displayed in Fig. 4 (c), the dominance of T_2 by T_1 is still possible. Again, due to Proposition 3, $T_1 \gg T_2$ is equivalent to $T_1^2 \gg T_{2,2}$, where T_1^2 is the ordinal sum t-norm defined by

$$T_1^2 = (\langle \varphi_2(a_{1,1}), \varphi_2(b_{1,1}), T_{1,1} \rangle),$$

  with $\varphi_2 \colon [a_{2,2}, b_{2,2}] \to [0,1]$, $\varphi_2(x) = \frac{x - a_{2,2}}{b_{2,2} - a_{2,2}}$.

Necessary Conditions Due to Idempotent Elements. The idempotent elements play an important role in dominance relationships, as is expressed by the following proposition [30].

Proposition 5. *If a t-norm T_1 dominates a t-norm T_2, then the following observations hold:*

(i) $\mathcal{I}(T_1)$ *is closed under T_2;*

(ii) $\mathcal{I}(T_2) \subseteq \mathcal{I}(T_1)$.

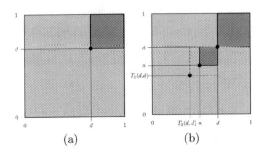

Fig. 5. Illustrations to Example 3

Note that for the representation of a continuous ordinal sum t-norm $T = (\langle a_i, b_i, T_i \rangle)_{i \in I}$ in terms of ordinally irreducible summand t-norms T_i, the set of idempotent elements is given by $\mathcal{I}(T) = [0,1] \setminus \bigcup_{i \in I}]a_i, b_i[$. Therefore, this proposition has some interesting consequences for the boundary elements of the summand carriers. Firstly, all idempotent elements of T_2 are idempotent elements of T_1, *i.e.* either endpoints of summand carriers of T_1 or elements of some domain where T_1 acts as $T_\mathbf{M}$. Secondly, for any idempotent elements d_1, d_2 of T_1 we know that also $T_2(d_1, d_2)$, is an idempotent element of T_1. Consequently, if T_1 is some ordinal sum t-norm that dominates $T_2 = T_\mathbf{P}$, resp. $T_2 = T_\mathbf{L}$, and $d \in \mathcal{I}(T_1)$ then also $d^n \in \mathcal{I}(T_1)$, resp. $\max(nd - n + 1, 0) \in \mathcal{I}(T_1)$, for all $n \in \mathbb{N}$.

Example 3. Consider a t-norm T^* with trivial idempotent elements only, *i.e.* $\mathcal{I}(T^*) = \{0, 1\}$. We are now interested in constructing ordinal sum t-norms T_1 with summand operations T^* which fulfill the necessary conditions for dominating $T_2 = T^*$ as expressed by Proposition 5. Clearly, $T_1 = (\langle d, 1, T^* \rangle)$ is a first possibility (see Fig. 5 (a)). Adding one further summand to T_1, *i.e.* building $T_1' = (\langle a, d, T^* \rangle, \langle d, 1, T^* \rangle)$, demands that $a \geq T_2(d, d)$, since otherwise $T_2(d, d) \notin \mathcal{I}(T_1')$ (see also Fig. 5 (b)).

5   Particular Continuous Ordinal Sum T-Norms

We will now focus on particular ordinal sum t-norms with either the Łukasiewicz t-norm or the product t-norm as only summand operation and study the dominance relationship between such t-norms.

5.1   Ordinal Sum T-Norms Based on $T_\mathbf{L}$

According to Proposition 5, the set of idempotent elements of a t-norm T_1 dominating a t-norm T_2 should be closed under T_2 and should contain the idempotent elements of T_2. If we restrict our attention to ordinal sum t-norms with $T_\mathbf{L}$ as only summand operation, this proposition can be turned into a characterization [33].

Proposition 6. *Consider two ordinal sum t-norms T_1 and T_2 based on $T_\mathbf{L}$, i.e. $T_1 = (\langle a_i, b_i, T_\mathbf{L}\rangle)_{i \in I}$ and $T_2 = (\langle a_j, b_j, T_\mathbf{L}\rangle)_{j \in J}$. Then T_1 dominates T_2 if and only if the following two conditions hold:*

(i) $\mathcal{I}(T_1)$ is closed under T_2;
(ii) $\mathcal{I}(T_2) \subseteq \mathcal{I}(T_1)$.

Now consider the particular case $T_2 = T_\mathbf{L}$. Clearly, the second condition is trivially fulfilled and can be omitted. In order to be able to apply the above proposition to this case, we need to understand what it means for a set to be closed under $T_\mathbf{L}$ [33].

Lemma 2. *A subset $S \subseteq [0,1]$ is closed under $T_\mathbf{L}$ if and only if the set*

$$1 - S = \{1 - x \mid x \in S\}$$

is closed under truncated addition, i.e. whenever $a, b \in 1 - S$ also $\min(a + b, 1) \in 1 - S$.

Consequently, an ordinal sum t-norm T based on $T_\mathbf{L}$ dominates $T_\mathbf{L}$ if and only if the set of its complemented idempotent elements is closed under truncated addition. Let us apply this insight to some particular families of ordinal sum t-norms based on $T_\mathbf{L}$.

The Mayor-Torrens Family. The Mayor-Torrens t-norms form a family parameterized by a single real parameter $\lambda \in [0,1]$ [23]:

$$T_\lambda^{\mathbf{MT}} = (\langle 0, \lambda, T_\mathbf{L}\rangle).$$

These t-norms are ordinal sums based on $T_\mathbf{L}$ with a single summand located in the lower left corner of the unit square (see also Fig. 6 (a)). In particular, it holds that $T_0^{\mathbf{MT}} = T_\mathbf{M}$ and $T_1^{\mathbf{MT}} = T_\mathbf{L}$. Note that $T_{\lambda_1}^{\mathbf{MT}} \geq T_{\lambda_2}^{\mathbf{MT}}$ if and only if $\lambda_1 \leq \lambda_2$. Hence, $T_{\lambda_1}^{\mathbf{MT}} \gg T_{\lambda_2}^{\mathbf{MT}}$ implies $\lambda_1 \leq \lambda_2$.

If $\lambda_1 = 0$ or $\lambda_1 = \lambda_2$, then the dominance relationship trivially holds. Suppose that $0 < \lambda_1 < \lambda_2$, then $T_{\lambda_1}^{\mathbf{MT}}$ dominates $T_{\lambda_2}^{\mathbf{MT}}$ if and only if $T_{\lambda^*}^{\mathbf{MT}} = (\langle 0, \lambda^*, T_\mathbf{L}\rangle)$ dominates $T_\mathbf{L}$ with $\lambda^* = \frac{\lambda_1}{\lambda_2}$ (see also Proposition 4). The set of idempotent elements of $T_{\lambda^*}^{\mathbf{MT}}$ is

$$\mathcal{I}(T_{\lambda^*}^{\mathbf{MT}}) = \{0\} \cup [\lambda^*, 1]$$

and therefore

$$1 - \mathcal{I}(T_{\lambda^*}^{\mathbf{MT}}) = [0, 1 - \lambda^*] \cup \{1\}.$$

For $a = 1 - \lambda^*$ and $b = \min(a, \frac{1-a}{2})$ it holds that $a, b \in 1 - \mathcal{I}(T_{\lambda^*}^{\mathbf{MT}})$, $a + b < 1$ but $a + b \notin 1 - \mathcal{I}(T_{\lambda^*}^{\mathbf{MT}})$. According to Lemma 2 and Proposition 6, there exist no dominance relationships within the Mayor-Torrens family other than $T_\mathbf{M}$ dominating all other members and self-dominance. Hence, there exists no triplet of pairwisely different t-norms $T_{\lambda_1}^{\mathbf{MT}}$, $T_{\lambda_2}^{\mathbf{MT}}$ and $T_{\lambda_3}^{\mathbf{MT}}$ fulfilling $T_{\lambda_1}^{\mathbf{MT}} \gg T_{\lambda_2}^{\mathbf{MT}}$ and $T_{\lambda_2}^{\mathbf{MT}} \gg T_{\lambda_3}^{\mathbf{MT}}$, implying that the dominance relation is (trivially) transitive, and therefore a partial order, on this family. The Hasse-diagram of $((T_\lambda^{\mathbf{MT}})_{\lambda \in [0,1]}, \ll)$ is displayed in Fig. 6 (b).

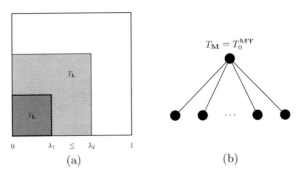

$$T_M = T_0^{MT}$$

(a)   (b)

Fig. 6. Examples of Mayor-Torrens t-norms, Hasse-diagram of $((T_\lambda^{\mathbf{MT}})_{\lambda \in [0,1]}, \ll)$

The Modified Mayor-Torrens Family. In this paragraph, we consider the family of t-norms parameterized by a single real parameter $\mu \in [0,1]$:

$$T_\mu = (\langle \mu, 1, T_{\mathbf{L}} \rangle).$$

Contrary to the Mayor-Torrens family, the summands are located in the upper right corner of the unit square. Hence, $T_0 = T_{\mathbf{L}}$ and $T_1 = T_{\mathbf{M}}$ (see also Fig. 7 (a)). Note that $T_{\mu_1} \geq T_{\mu_2}$ if and only if $\mu_1 \geq \mu_2$. Hence, $T_{\mu_1} \gg T_{\mu_2}$ implies $\mu_1 \geq \mu_2$.

If $\mu_1 = 1$ or $\mu_1 = \mu_2$, then the dominance relationship trivially holds. Assume that $\mu_2 < \mu_1 < 1$, then T_{μ_1} dominates T_{μ_2} if and only if T_{μ^*} dominates $T_{\mathbf{L}}$ with $\mu^* = \frac{\mu_1 - \mu_2}{1 - \mu_2}$. The set of idempotent elements of T_{μ^*} is

$$\mathcal{I}(T_{\mu^*}) = [0, \mu^*] \cup \{1\}$$

and therefore

$$1 - \mathcal{I}(T_{\mu^*}) = \{0\} \cup [1 - \mu^*, 1].$$

One easily verifies that the latter set is closed under truncated addition. Hence, within the modified family, it holds that $T_{\mu_1} \gg T_{\mu_2}$ whenever $\mu_1 \geq \mu_2$. In other words, this family is totally ordered by the dominance relation. The Hasse-diagram of $((T_\mu)_{\mu \in [0,1]}, \ll)$ is displayed in Fig. 7 (b).

Violation of Transitivity. We can now provide counterexamples to the conjecture that the dominance relation is transitive on the class of t-norms by considering ordinal sum t-norms based on $T_{\mathbf{L}}$ with two summands. More specifically, we consider the t-norm $T_\lambda = (\langle 0, \lambda, T_{\mathbf{L}} \rangle, \langle \lambda, 1, T_{\mathbf{L}} \rangle)$ with parameter $\lambda \in [0,1]$. We will show that for any $\lambda \in]0, \frac{1}{2}]$ it holds that

$$T_\lambda^{\mathbf{MT}} \gg T_\lambda, \qquad T_\lambda \gg T_{\mathbf{L}}, \qquad T_\lambda^{\mathbf{MT}} \not\gg T_{\mathbf{L}} \qquad (16)$$

violating the transitivity of the dominance relation.

First, both $T_\lambda^{\mathbf{MT}}$ and T_λ can be understood as ordinal sum t-norms with the same structure: $T_\lambda^{\mathbf{MT}}$ can be written as $(\langle 0, \lambda, T_{\mathbf{L}} \rangle, \langle \lambda, 1, T_{\mathbf{M}} \rangle)$, hence the common summand carriers are $[0, \lambda]$ and $[\lambda, 1]$ (see Fig. 8).

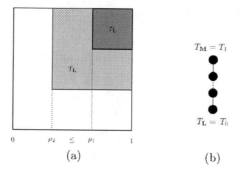

(a)   (b)

Fig. 7. Examples modified Mayor-Torrens t-norms, Hasse-diagram of $((T_\mu)_{\mu\in[0,1]},\ll)$

Since $T_L \gg T_L$ and $T_M \gg T_L$, Proposition 3 implies that $T_\lambda^{MT} \gg T_\lambda$ for any $\lambda \in [0,1]$. Second, the set of idempotent elements of T_λ is given by $\mathcal{I}(T_\lambda) = \{0,\lambda,1\}$ and thus

$$1 - \mathcal{I}(T_\lambda) = \{0, 1-\lambda, 1\}.$$

This set is closed under truncated addition if and only if $1 - \lambda \geq \frac{1}{2}$. Therefore, according to Lemma 2 and Proposition 6, it holds that T_λ dominates T_L if and only if $\lambda \in [0, \frac{1}{2}]$. Finally, in the Mayor-Torrens family it does not hold that $T_\lambda^{MT} \gg T_L = T_0^{MT}$ for any $\lambda \in \,]0,1[$. Combining all of the above shows that (16) holds if and only if $\lambda \in \,]0, \frac{1}{2}]$.

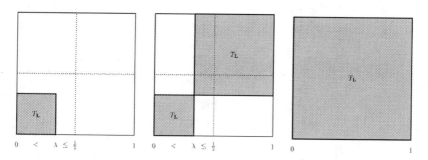

Fig. 8. Three ordinal sum t-norms based on T_L violating the transitivity of the dominance relation. From left to right: T_λ^{MT}, T_λ and T_L. Violation of transitivity occurs if and only if $\lambda \in \,]0, \frac{1}{2}]$.

5.2 Ordinal Sum T-Norms Based on T_P

We now turn to ordinal sum t-norms with T_P as only summand operation and start again with a family of t-norms with a single summand in the lower left corner of the unit square.

The Dubois-Prade Family. The Dubois-Prade t-norms form a family parameterized by a single real parameter $\lambda \in [0, 1]$ [10]:

$$T_\lambda^{\mathbf{DP}} = (\langle 0, \lambda, T_{\mathbf{P}} \rangle) .$$

The case $\lambda = 0$ corresponds to $T_{\mathbf{M}}$, the case $\lambda = 1$ to $T_{\mathbf{P}}$. Note that $T_{\lambda_1}^{\mathbf{DP}} \geq T_{\lambda_2}^{\mathbf{DP}}$ if and only if $\lambda_1 \leq \lambda_2$. Hence, $T_{\lambda_1}^{\mathbf{DP}} \gg T_{\lambda_2}^{\mathbf{DP}}$ implies $\lambda_1 \leq \lambda_2$.

If $\lambda_1 = 0$ or $\lambda_1 = \lambda_2$, then the dominance relationship trivially holds. Therefore, suppose that $0 < \lambda_1 < \lambda_2$. The set of idempotent elements of $T_{\lambda_1}^{\mathbf{DP}}$ is given by

$$\mathcal{I}(T_{\lambda_1}^{\mathbf{DP}}) = \{0\} \cup [\lambda_1, 1] .$$

It then holds that

$$0 \neq T_{\lambda_2}^{\mathbf{DP}}(\lambda_1, \lambda_1) = \lambda_2 \cdot T_{\mathbf{P}}(\tfrac{\lambda_1}{\lambda_2}, \tfrac{\lambda_1}{\lambda_2}) = \tfrac{\lambda_1}{\lambda_2} \cdot \lambda_1 < \lambda_1$$

due to the strict monotonicity of $T_{\mathbf{P}}$. Hence, $T_{\lambda_2}^{\mathbf{DP}}(\lambda_1, \lambda_1) \notin \mathcal{I}(T_{\lambda_1}^{\mathbf{DP}})$. According to Proposition 5, $T_{\lambda_1}^{\mathbf{DP}}$ does not dominate $T_{\lambda_2}^{\mathbf{DP}}$.

Consequently, the only dominance relationships in the Dubois-Prade family are $T_{\mathbf{M}}$ dominating all other members and self-dominance. The dominance relation is again (trivially) transitive, and therefore a partial order, on this family (see Fig. 9).

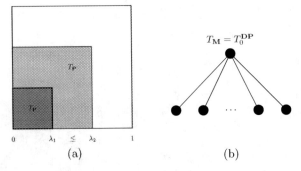

Fig. 9. Examples of Dubois-Prade t-norms, Hasse-diagram of $((T_\lambda^{\mathbf{DP}})_{\lambda \in [0,1]}, \ll)$

In contrast to dominance between ordinal sum t-norms based on $T_{\mathbf{L}}$, dominance between ordinal sum t-norms based on $T_{\mathbf{P}}$ is not fully understood. The following lemma provides one way of constructing an ordinal sum t-norm based on $T_{\mathbf{P}}$ dominating $T_{\mathbf{P}}$. It follows immediately from Proposition 5.

Lemma 3. *Let $\lambda \in]0, 1[$ and $m \in \mathbb{N}$. Then the ordinal sum t-norm $T_{\lambda, m}$ defined as*

$$T_{\lambda, m} = (\langle \lambda^n, \lambda^{n-1}, T_{\mathbf{P}} \rangle)_{n=1,2,\dots,m}$$

dominates $T_{\mathbf{P}}$.

This simple lemma allows to construct interesting examples.

The Modified Dubois-Prade Family. Similarly as for the Mayor-Torrens family, we propose a modification of the Dubois-Prade family, by locating the single summand in the upper right corner of the unit square. Explicitly, we consider the family of t-norms parameterized by a single real parameter $\lambda \in [0, 1]$:

$$T_\lambda = (\langle \lambda, 1, T_\mathbf{P} \rangle).$$

Note that these t-norms are special cases of Lemma 3 as $T_\lambda = T_{\lambda,1}$. In particular, $T_0 = T_\mathbf{P}$ and $T_1 = T_\mathbf{M}$. Note that $T_{\lambda_1} \geq T_{\lambda_2}$ if and only if $\lambda_1 \geq \lambda_2$. Hence, $T_{\lambda_1} \gg T_{\lambda_2}$ implies $\lambda_1 \geq \lambda_2$.

If $\lambda_1 = 1$ or $\lambda_1 = \lambda_2$, then the dominance relationship again trivially holds. Moreover, due to Lemma 3, the dominance relationship also holds if $\lambda_2 = 0$, *i.e.* $T_{\lambda_1} \gg T_0$. Consider the case $0 < \lambda_2 < \lambda_1 < 1$, then T_{λ_1} dominates T_{λ_2} if and only if $(\langle \frac{\lambda_1 - \lambda_2}{1 - \lambda_2}, 1, T_\mathbf{P} \rangle)$ dominates $T_\mathbf{P}$. Thanks to Lemma 3, it then follows that the modified Dubois-Prade family is totally ordered by the dominance relation (see Fig. 10 (b)).

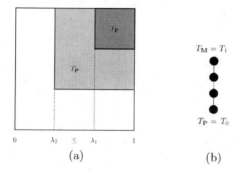

(a)   (b)

Fig. 10. Examples modified Dubois-Prade t-norms, Hasse-diagram of $((T_\lambda)_{\lambda \in [0,1]}, \ll)$

Violation of Transitivity. Also ordinal sum t-norms based on $T_\mathbf{P}$ allow us to construct a counterexample demonstrating the non-transitivity of the dominance relation. Consider the ordinal sum t-norms $T_1 = (\langle \frac{1}{4}, \frac{1}{2}, T_\mathbf{P} \rangle, \langle \frac{3}{4}, 1, T_\mathbf{P} \rangle)$ and $T_2 = T_{\frac{1}{2},2}$ (see Lemma 3). It then holds that

$$T_1 \gg T_2, \qquad T_2 \gg T_\mathbf{P}, \qquad T_1 \not\gg T_\mathbf{P}$$

violating the transitivity of the dominance relation (see Fig. 11).

Note that the t-norm T_1 can also be written as $T_1 = (\langle \frac{1}{4}, \frac{1}{2}, T_\mathbf{P} \rangle, \langle \frac{1}{2}, 1, T^* \rangle)$ with T^* the member of the modified Dubois-Prade family with parameter $\lambda = \frac{1}{2}$. Using Proposition 3 and the dominance relationships within the modified Dubois-Prade family, it follows immediately that $T_1 \gg T_2$. The dominance relationship $T_2 \gg T_\mathbf{P}$ is an immediate consequence of Lemma 3. Finally, we consider the set of idempotent elements of T_1:

$$\mathcal{I}(T_1) = \left[0, \tfrac{1}{4}\right] \cup \left[\tfrac{1}{2}, \tfrac{3}{4}\right].$$

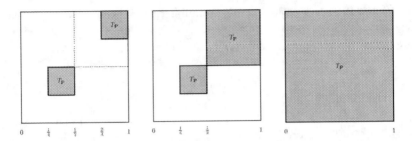

Fig. 11. Three ordinal sum t-norms based on $T_{\mathbf{P}}$ violating the transitivity of the dominance relation

It holds that $\frac{5}{8} \in \mathcal{I}(T_1)$, while

$$T_{\mathbf{P}}\left(\tfrac{5}{8}, \tfrac{5}{8}\right) = \tfrac{25}{64} \notin \mathcal{I}(T_1).$$

Proposition 5 then implies that T_1 does not dominate $T_{\mathbf{P}}$.

6   Final Remarks

The dominance relation is a reflexive and antisymmetric relation on the class of t-norms. That it is not transitive and therefore not a partial order was illustrated by several examples whereas the particular role of ordinal sums dominating either the Łukasiewicz t-norm or the product t-norm is remarkable. Note that by the isomorphism property of dominance these examples can be transformed into counterexamples involving arbitrary nilpotent resp. strict t-norms. Properties related to idempotent elements and to the induced order heavily determine the occurrence of dominance within particular families of t-norms as shown by the parameterized families in the last section.

Acknowledgement

The support of this work by the EU COST Action 274 (TARSKI: Theory and Applications of Relational Structures as Knowledge Instruments) is gratefully acknowledged. The second author was also supported by the grants VEGA 1/3012/06, VEGA 1/2005/05 and APVT-20-046402.

References

1. C. Alsina, M. Frank, and B. Schweizer. *Associative Functions: Triangular Norms and Copulas*. World Scientific Publishing Company, Singapore, 2006.
2. U. Bodenhofer. Representations and constructions of similarity-based fuzzy orderings. *Fuzzy Sets and Systems*, 137(1):113–136.

3. U. Bodenhofer. *A Similarity-Based Generalization of Fuzzy Orderings*, volume C 26 of *Schriftenreihe der Johannes-Kepler-Universität Linz*. Universitätsverlag Rudolf Trauner, 1999.

4. D. Butnariu and E. P. Klement. *Triangular Norm-Based Measures and Games with Fuzzy Coalitions*, volume 10 of *Theory and Decision Library, Series C: Game Theory, Mathematical Programming and Operations Research*. Kluwer Academic Publishers, Dordrecht, 1993.

5. A. H. Clifford. Naturally totally ordered commutative semigroups. *Amer. J. Math.*, 76:631–646, 1954.

6. A. H. Clifford and G. B. Preston. *The algebraic theory of semigroups*. American Mathematical Society, Providence, 1961.

7. A. C. Climescu. Sur l'équation fonctionelle de l'associativité. *Bull. École Polytechn. Iassy*, 1:1–16, 1946.

8. B. De Baets and R. Mesiar. *T*-partitions. *Fuzzy Sets and Systems*, 97:211–223, 1998.

9. S. Díaz, S. Montes, and B. De Baets. Transitivity bounds in additive fuzzy preference structures. *IEEE Trans. Fuzzy Systems*, 2006. in press.

10. D. Dubois and H. Prade. New results about properties and semantics of fuzzy set-theoretic operators. In P. P. Wang and S. K. Chang, editors, *Fuzzy Sets: Theory and Applications to Policy Analysis and Information Systems*, pages 59–75. Plenum Press, New York, 1980.

11. J. C. Fodor and M. Roubens. *Fuzzy Preference Modelling and Multicriteria Decision Support*. Kluwer Academic Publishers, Dordrecht, 1994.

12. M. J. Frank. On the simultaneous associativity of $F(x,y)$ and $x + y - F(x,y)$. *Aequationes Mathematicae*, 19:194–226, 1979.

13. M. Grabisch, H. T. Nguyen, and E. A. Walker. *Fundamentals of Uncertainty Calculi with Applications to Fuzzy Inference*. Kluwer Academic Publishers, Dordrecht, 1995.

14. P. Hájek. *Metamathematics of Fuzzy Logic*. Kluwer Academic Publishers, Dordrecht, 1998.

15. K. H. Hofmann and J. D. Lawson. Linearly ordered semigroups: Historic origins and A. H. Clifford's influence. volume 231 of *London Math. Soc. Lecture Notes*, pages 15–39. Cambridge University Press, Cambridge, 1996.

16. F. Klein-Barmen. Über gewisse Halbverbände und kommutative Semigruppen II. *Math. Z.*, 48:715–734, 1942–43.

17. E. P. Klement, R. Mesiar, and E. Pap. *Triangular Norms*, volume 8 of *Trends in Logic. Studia Logica Library*. Kluwer Academic Publishers, Dordrecht, 2000.

18. E. P. Klement, R. Mesiar, and E. Pap. Triangular norms. Position paper I: basic analytical and algebraic properties. *Fuzzy Sets and Systems*, 143:5–26, 2004.

19. E. P. Klement, R. Mesiar, and E. Pap. Triangular norms. Position paper II: general constructions and parameterized families. *Fuzzy Sets and Systems*, 145:411–438, 2004.

20. E. P. Klement, R. Mesiar, and E. Pap. Triangular norms. Position paper III: continuous t-norms. *Fuzzy Sets and Systems*, 145:439–454, 2004.

21. E. P. Klement and S. Weber. Generalized measures. *Fuzzy Sets and Systems*, 40:375–394, 1991.

22. C. M. Ling. Representation of associative functions. *Publ. Math. Debrecen*, 12:189–212, 1965.

23. G. Mayor and J. Torrens. On a family of t-norms. *Fuzzy Sets and Systems*, 41:161–166, 1991.

24. K. Menger. Statistical metrics. *Proc. Nat. Acad. Sci. U.S.A.*, 8:535–537, 1942.

25. P. S. Mostert and A. L. Shields. On the structure of semi-groups on a compact manifold with boundary. *Ann. of Math., II. Ser.*, 65:117–143, 1957.

26. R. B. Nelsen. *An Introduction to Copulas*, volume 139 of *Lecture Notes in Statistics*. Springer, New York, 1999.

27. E. Pap. *Null-Additive Set Functions*. Kluwer Academic Publishers, Dordrecht, 1995.

28. G. B. Preston. A. H. Clifford: an appreciation of his work on the occasion of his sixty-fifth birthday. *Semigroup Forum*, 7:32–57, 1974.

29. S. Saminger. *Aggregation in Evaluation of Computer-Assisted Assessment*, volume C 44 of *Schriftenreihe der Johannes-Kepler-Universität Linz*. Universitätsverlag Rudolf Trauner, 2005.

30. S. Saminger, B. De Baets, and H. De Meyer. On the dominance relation between ordinal sums of conjunctors. *Kybernetika*, 42(3):337–350, 2006.

31. S. Saminger, B. De Baets, and H. De Meyer. The domination relation between continuous t-norms. In *Proceedings of Joint 4th Int. Conf. in Fuzzy Logic and Technology and 11th French Days on Fuzzy Logic and Applications, Barcelona (Spain)*, pages 247–252, September 2005.

32. S. Saminger, R. Mesiar, and U. Bodenhofer. Domination of aggregation operators and preservation of transitivity. *Internat. J. Uncertain. Fuzziness Knowledge-Based Systems*, 10/s:11–35, 2002.

33. P. Sarkoci. Dominance of ordinal sums of Łukasiewicz and product t-norm. forthcoming.

34. P. Sarkoci. Domination in the families of Frank and Hamacher t-norms. *Kybernetika*, 41:345–356, 2005.

35. P. Sarkoci. Dominance is not transitive on continuous triangular norms. *Aequationes Mathematicae*, 2006. submitted.

36. B. Schweizer and A. Sklar. Statistical metric spaces. *Pacific J. Math.*, 10:313–334, 1960.

37. B. Schweizer and A. Sklar. Associative functions and abstract semigroups. *Publ. Math. Debrecen*, 10:69–81, 1963.

38. B. Schweizer and A. Sklar. *Probabilistic Metric Spaces*. North-Holland, New York, 1983.

39. A. N. Šerstnev. Random normed spaces: problems of completeness. *Kazan. Gos. Univ. Učen. Zap.*, 122:3–20, 1962.

40. H. Sherwood. Characterizing dominates on a family of triangular norms. *Aequationes Mathematicae*, 27:255–273, 1984.

41. M. Sugeno. *Theory of fuzzy integrals and its applications*. PhD thesis, Tokyo Institute of Technology, 1974.

42. R. M. Tardiff. Topologies for probabilistic metric spaces. *Pacific J. Math.*, 65:233–251, 1976.

43. R. M. Tardiff. On a functional inequality arising in the construction of the product of several metric spaces. *Aequationes Mathematicae*, 20:51–58, 1980.

44. R. M. Tardiff. On a generalized Minkowski inequality and its relation to dominates for t-norms. *Aequationes Mathematicae*, 27:308–316, 1984.

45. S. Weber. ⊥-decomposable measures and integrals for Archimedean t-conorms ⊥. *J. Math. Anal. Appl.*, 101:114–138, 1984.

46. L. A. Zadeh. Fuzzy sets. *Inform. and Control*, 8:338–353, 1965.

Aggregation on Bipolar Scales

Michel Grabisch

Université Paris I - Panthéon-Sorbonne
CERMSEM, 110-116 Bd de l'Hôpital, 75013 Paris, France
`Michel.Grabisch@lip6.fr`

Abstract. The paper addresses the problem of extending aggregation operators typically defined on $[0, 1]$ to the symmetric interval $[-1, 1]$, where the "0" value plays a particular role (neutral value). We distinguish the cases where aggregation operators are associative or not. In the former case, the "0" value may play the role of neutral or absorbant element, leading to pseudo-addition and pseudo-multiplication. We address also in this category the special case of minimum and maximum defined on some finite ordinal scale. In the latter case, we find that a general class of extended operators can be defined using an interpolation approach, supposing the value of the aggregation to be known for ternary vectors.

1   Introduction

Most of the works done on aggregation operators take the $[0, 1]$ interval as range for quantities to be aggregated, or some similar structure, i.e. a closed interval of some linearly ordered set (see, e.g., [3,17,21]). The lower and upper bounds of this interval represent the worst and best scores that can be achieved on each dimension.

We may desire to introduce a third remarkable point of the interval, say e, which will play a particular role, for example a neutral value (in some sense) or an absorbant value. This situation is already considered for uninorms [28]: e is a neutral element in the sense that, U denoting a uninorm, $U(e, x) = x$ for any $x \in [0, 1]$.

For convenience, up to a rescaling, we may always consider that we work on $[-1, 1]$, and 0 corresponds to our particular point, denoted e before. In the more general case of bounded linearly ordered sets, we will apply a symmetrization procedure.

The motivation for such a work may be only mathematical. However, there are psychological evidence that in many cases, scores or utilities manipulated by humans lie on a *bipolar scale*, that is to say, a scale with a neutral value making the frontier between good or satisfactory scores, and bad or unsatisfactory scores. With our convention, good scores are positive ones, while negative scores reflect bad scores. Most of the time, our behaviour with positive scores is not the same than with negative ones: for example, a conjunctive attitude may be turned into a disjunctive attitude when changing the sign of the scores. So, it becomes

H. de Swart et al. (Eds.): TARSKI II, LNAI 4342, pp. 355–371, 2006.
© Springer-Verlag Berlin Heidelberg 2006

important to define agregation operators being able to reflect the variety of aggregation behaviours on bipolar scales.

Let M be an aggregation operator defined on $[-1, 1]$. Clearly, the restriction of M to non negative numbers corresponds to some (usual) aggregation operator M^+ on $[0, 1]$. Similarly, its restriction to $[-1, 0]$ corresponds to a (possibly different) operator M^-, after some suitable symmetrization. However, this does not suffice to define the value of M for the mixed case, when positive and negative scores coexist. The exact way to do this is dependent on the nature of M and the meaning of 0. We shall distinguish several cases.

Let us consider first that M is associative, so that we need to consider only two arguments. For the meaning of the 0 point, we can think of two cases of interest: either 0 is a neutral value in the sense that $M(0, x) = M(x, 0) = x$ for any $x \in [-1, 1]$, or 0 is an absorbing value, i.e. $M(0, x) = M(x, 0) = 0$, for any $x \in [-1, 1]$. The first case leads naturally to pseudo-additions, while the second one leads to pseudo-multiplications. This is the topic of Section 3. The particular case of the definition of min and max on $[-1, 1]$ will be addressed in Section 4, where we deal with symmetrized linearly ordered sets.

Let us consider now (possibly) non associative aggregation operators. A first important class of operators are those under the form:

$$M(x) := \phi(M^+(x^+), M^-(x^-)) \tag{1}$$

where $x \in [-1, 1]^n$ for some n, and $x^+ := x \vee 0$, $x^- := (-x)^+$, M^+, M^- are given aggregation operators on $[0, 1]$, and ϕ is a pseudo-difference. We call such aggregation operators *separable*. A more general case is defined as follows. We say that x is a *ternary vector* if $x \in \{-1, 0, 1\}^n$ for some n. Let us suppose that the value of M for each ternary vector is given. Then we define M for every $x \in [-1, 1]^n$ by some interpolation rule between the known values. The separable case is recovered if M^+ and M^- are also obtained by some interpolation rule. As in the usual unipolar case, we will show that this type of aggregation operator is based on an integral (Section 6). We begin by a preliminary section introducing necessary definitions.

2   Basic Material

We begin by recalling definitions of t-norms, t-conorms, uninorms and nullnorms (see, e.g., [19,21] for details).

Definition 1. *A triangular norm (t-norm for short) T is a binary operation on $[0, 1]$ such that for any $x, y, z \in [0, 1]$ the following four axioms are satisfied:*

(**P1**) *commutativity:* $T(x, y) = T(y, x)$;
(**P2**) *associativity:* $T(x, T(y, z)) = T(T(x, y), z))$;
(**P3**) *monotonicity:* $T(x, y) \leq T(x, z)$ *whenever* $y \leq z$;
(**P4**) *neutral element:* $T(1, x) = x$.

Any t-norm satisfies $T(0, x) = 0$. Typical t-norms are the minimum ($\wedge$), the algebraic product ($\cdot$), and the Łukasiewicz t-norm defined by $T_{\mathbf{L}}(x, y) := (x + y - 1) \vee 0$.

Definition 2. *A* triangular conorm *(t-conorm for short)* S *is a binary operation on* $[0, 1]$ *such that, for any* $x, y, z \in [0, 1]$, *it satisfies* **P1**, **P2**, **P3** *and*

(P5) *neutral element:* $S(0, x) = x$.

Any t-conorm satisfies $S(1, x) = 1$. Typical t-conorms are the maximum $\vee$, the probabilistic sum $S_{\mathbf{P}}(x, y) := x + y - xy$, and the Łukasiewicz t-conorm defined by $S_{\mathbf{L}}(x, y) := (x + y) \wedge 1$. T-norms and t-conorms are dual operations in the sense that for any given t-norm T, the binary operation S_T defined by

$$S_T(x, y) = 1 - T(1 - x, 1 - y)$$

is a t-conorm (and similarly when starting from S). Hence, their properties are also dual. The above examples are all dual pairs of t-norms and t-conorms.

A t-norm (or a t-conorm) is said to be *strictly monotone* if $T(x, y) < T(x, z)$ whenever $x > 0$ and $y < z$. A continuous t-norm (resp. t-conorm) is *Archimedean* if $T(x, x) < x$ (resp. $S(x, x) > x$) for all $x \in]0, 1[$. A strictly monotone and continuous t-norm (resp. t-conorm) is called *strict*. Strict t-norms (resp. t-conorms) are Archimedean. Non-strict continuous Archimedean t-norms (resp. t-conorms) are called *nilpotent*.

Any continuous Archimedean t-conorm S has an additive generator s, i.e. a strictly increasing function $s : [0, 1] \to [0, +\infty]$, with $s(0) = 0$, such that, for any $x, y \in [0, 1]$:

$$S(x, y) = s^{-1}[s(1) \wedge (s(x) + s(y))]. \tag{2}$$

Similarly, any continuous Archimedean t-norm has an additive generator t that is strictly decreasing and satisfies $t(1) = 0$. Strict t-conorms are characterized by $s(1) = +\infty$, nilpotent t-conorms by a finite value of $s(1)$. Additive generators are determined up to a positive multiplicative constant. If t is an additive generator of a t-norm T, then $s(x) = t(1 - x)$ is an additive generator of its dual t-conorm S_T.

Definition 3. *[28] A* uninorm U *is a binary operation on* $[0, 1]$ *such that, for any* $x, y, z \in [0, 1]$, *it satisfies* **P1**, **P2**, **P3** *and*

(P6) *neutral element: there exists* $e \in]0, 1[$ *such that* $U(e, x) = x$.

It follows that on $[0, e]^2$ a uninorm behaves like a t-norm, while on $[e, 1]^2$ it behaves like a t-conorm. In the remaining parts, monotonicity implies that U is comprised between min and max. Associativity implies that $U(0, 1) \in \{0, 1\}$. Uninorms such that $U(0, 1) = 1$ are called disjunctive, while the others are called conjunctive.

If U is a uninorm with neutral element e, strictly monotone on $]0, 1[^2$, and continuous on $[0, 1]^2 \setminus \{(0, 1), (1, 0)\}$, there exists an additive generator u, i.e. a strictly increasing $[0,1] \to [-\infty, \infty]$ mapping u such that $u(e) = 0$ and for any $x, y \in [0,1]$:

$$U(x, y) = u^{-1}(u(x) + u(y)), \tag{3}$$

where by convention $\infty - \infty = -\infty$ if U is conjunctive, and $+\infty$ if U is disjunctive.

Definition 4. [2] A nullnorm V is a binary operation on $[0,1]$ such that for any $x, y, z \in [0,1]$, it satisfies **P1**, **P2**, **P3**, and there is an element $a \in [0,1]$ such that

$$V(x, 0) = x, \quad \forall x \leq a, \qquad V(x, 1) = x, \quad \forall x \geq a.$$

By monotonicity, $V(x, a) = a$ for all $x \in [0,1]$, hence a is an absorbant value, and V restricted to $[0, a]^2$ is a t-conorm, while its restriction to $[a, 1]^2$ is a t-norm. Remark that this is the opposite situation of uninorms. On the remaining part of $[0,1]^2$, monotonicity imposes that $V(x, y) = a$. Hence to each pair of t-norm and t-conorm corresponds a unique nullnorm, provided a is fixed.

We turn to the definition of Choquet and Sugeno integrals. We denote by $[n]$ the set $\{1, \ldots, n\}$ of the n first integers, which will be the number of arguments of our aggregation operators. Details on what follows can be found in, e.g., [16].

Definition 5. A (normalized) capacity is a function $\mu : 2^{[n]} \to [0,1]$ satisfying $\mu(\varnothing) = 0$, $\mu([n]) = 1$, and $\mu(A) \leq \mu(B)$ for every $A, B \in 2^{[n]}$ such that $A \subseteq B$.

To any capacity μ we associate its *conjugate* $\overline{\mu}$, which is a capacity defined by:

$$\overline{\mu}(A) := 1 - \mu([n] \setminus A), \quad A \subseteq [n].$$

Definition 6. Let $x \in [0,1]^n$ and μ be a capacity on $[n]$.

(i) The (discrete) Choquet integral of x w.r.t. μ is defined by:

$$\mathcal{C}_\mu(x) := \sum_{i=1}^{n} [x_{\sigma(i)} - x_{\sigma(i-1)}] \mu(\{x_{\sigma(i)}, \ldots, x_{\sigma(n)}\}),$$

with σ indicating a permutation on $[n]$ such that $x_{\sigma(1)} \leq \cdots \leq x_{\sigma(n)}$, and $x_{\sigma(0)} := 0$.

(ii) The (discrete) Sugeno integral of x w.r.t. μ is defined by:

$$\mathcal{S}_\mu(x) := \bigvee_{i=1}^{n} [x_{\sigma(i)} \wedge \mu(\{x_{\sigma(i)}, \ldots, x_{\sigma(n)}\})],$$

with same notations.

These two aggregation operators being integrals, they are called *integral-based* operators. Others can be defined, considering other integrals defined w.r.t capacities and based on pseudo-additions and pseudo-multiplications (see, e.g., Murofushi and Sugeno [23], Benvenuti et al. [1], and Sander and Siedekum [24,25,26]).

For any $A \subseteq [n]$, let us denote by $(1_A, 0_{A^c})$ the vector x of $[0,1]^n$ such that $x_i = 1$ if $i \in A$ and 0 else. These are the set of vertices of $[0,1]^n$. An important property is that for any capacity μ, $\mathcal{C}_\mu(1_A, 0_{A^c}) = \mathcal{S}_\mu(1_A, 0_{A^c}) = \mu(A)$ for all $A \subseteq [n]$. Moreover, as explained hereafter, the Choquet integral is the only linear interpolator using the fewest number of vertices of $[0,1]^n$ (see [9,15]).

Let us denote by F an aggregation operator on $[0,1]^n$ such that for any $A \subseteq [n]$, $F(1_A, 0_{A^c}) = \mu(A)$ for a given capacity μ. Let us find a linear interpolation using the fewest possible vertices of $[0,1]^n$. For a given $x \in [0,1]^n$, let us denote by $\mathcal{V}(x)$ the set of vertices used for the linear interpolation, which writes

$$F(x) = \sum_{A \subseteq [n] | (1_A, 0_{A^c}) \in \mathcal{V}(x)} \left[\alpha_0(A) + \sum_{i=1}^{n} \alpha_i(A) x_i \right] F(1_A, 0_{A^c}), \qquad (4)$$

where $\alpha_i(A) \in \mathbb{R}$, $i = 0, \ldots, n$, $\forall A \in \mathcal{V}(x)$. To keep the meaning of interpolation, we impose that the convex hull $\mathrm{conv}(\mathcal{V}(x))$ contains x, and any $x \in [0,1]^n$ should belong to a unique polyhedron $\mathrm{conv}(\mathcal{V}(x))$ (except for common facets), and continuity should be ensured. Hence, the hypercube is partitioned into q polyhedra defined by their sets of vertices $\mathcal{V}_1, \ldots, \mathcal{V}_q$, all vertices being vertices of $[0,1]^n$. Such an operation is called a *triangulation*. Note that the least possible number of vertices is $n+1$, otherwise the polyhedra would not be n-dimensional, and hence a finite number would not cover the whole hypercube.

Many different triangulations are possible, but there is one which is of particular interest, since it leads to an interpolation where all constant terms $\alpha_0(A)$ are null. This triangulation uses the $n!$ *canonical polyhedra* of $[0,1]^n$:

$$\mathrm{conv}(\mathcal{V}_\sigma) = \{x \in [0,1]^n \mid x_{\sigma(1)} \leq \cdots \leq x_{\sigma(n)}\}, \text{ for some permutation } \sigma \text{ on } [n].$$

Note that all these polyhedra have $n+1$ vertices.

Proposition 1. *The linear interpolation (4) using the canonical polyhedra is the Choquet integral w.r.t. μ. Moreover, no other triangulation using polyhedra of $n+1$ vertices can lead to an interpolation.*

As shown in [9], the Sugeno integral is the lowest possible max-min interpolation between vertices in the canonical triangulation.

3   Pseudo-additions and Multiplications

In this section, we work on interval $[-1,1]$. Our aim is to define associative operators where 0 is either a neutral or an absorbing element, which we will suppose commutative in addition. Let us denote respectively $\oplus, \otimes : [-1,1]^2 \longrightarrow [-1,1]$ these operators, and let us adopt an infix notation. In summary, they should fulfil the following requirements for any $x, y, z \in [-1,1]$:

R1 Commutativity: $x \oplus y = y \oplus x$, $x \otimes y = y \otimes x$.
R2 Associativity: $x \oplus (y \oplus z) = (x \oplus y) \oplus z$, $x \otimes (y \otimes z) = (x \otimes y) \otimes z$.
R3 $x \oplus 0 = x$, $x \otimes 0 = 0$.

Endowing $[-1, 1]$ with the usual ordering, we may require in addition that $\oplus$ is monotone in each argument. As observed by Fuchs [6], under the assumption of monotonicity, associativity implies that $\oplus$ cannot be decreasing. Indeed, suppose e.g. $\oplus$ is decreasing in first place and take $x' \leq x$. Then $x' \oplus (y \oplus z) \geq x \oplus (y \oplus z) = (x \oplus y) \oplus z \geq (x' \oplus y) \oplus z = x' \oplus (y \oplus z)$, a contradiction unless $\oplus$ is degenerate. Hence we are lead to assume **R4**. For $\otimes$, let us require for the moment that it is monotone only on $[0, 1]$, which leads to **R5**.

R4 Isotonicity for $\oplus$: $x \oplus y \leq x' \oplus y$, for any $x' \leq x$.
R5 Isotonicity on $[0, 1]^2$ for $\otimes$.

The above requirements make that we recognize $\oplus$ as a t-conorm when restricted to $[0, 1]^2$. To make $\otimes$ on $[0, 1]^2$ a t-norm, we need in addition the following:

R6 Neutral element for $\otimes$: $x \otimes 1 = x$, for all $x \in [0, 1]$.

Let us call $\oplus, \otimes$ satisfying **R1** to **R6** *pseudo-addition* and *pseudo-multiplication*.
  We address first the construction of $\oplus$. Since $[-1, 1]$ is a symmetric interval, and if 0 plays the role of a neutral element, then we should have

R7 Symmetry: $x \oplus (-x) = 0$, for all $x \in [-1, 1]$.

Under **R1, R2, R3, R4**, and **R7**, the problem of defining $\oplus$ amounts to defining an ordered group on $[-1, 1]$ (see Fuchs [6] and [20,19]). We recall here necessary notions and facts.

Definition 7. *Let $(W, \leq)$ be a linearly ordered set, having top and bottom denoted $\top, \bot$, a particular nonextremal element e, and let us consider $\oplus$ an internal binary operation on W, and $\ominus$ a unary operation such that $x \leq y$ iff $\ominus(x) \geq \ominus(y)$.*

 – *$(W, \leq, \oplus, \ominus, e)$ is an ordered Abelian group (OAG) if it satisfies for all nonextremal elements x, y, z:*
    (i) $x \oplus y = y \oplus x$
    (ii) $x \oplus (y \oplus z) = (x \oplus y) \oplus z$
    (iii) $x \oplus e = x$
    (iv) $x \oplus (\ominus(x)) = e$
    (v) $x \leq y$ implies $x \oplus z \leq y \oplus z$.
 – *$(W, \leq, \oplus, \ominus, e)$ is an extended ordered Abelian group (OAG$^+$) if in addition*
    (i) $\top \oplus x = \top, \bot \oplus x = \bot$ for all x, $\ominus(\top) = \bot$, $\ominus(\bot) = \top$.
    (ii) If x, y are non extremal, then $x \oplus y$ is non extremal.

Clearly, our concern is to find an OAG$^+$, with $W = [-1, 1]$, $\top = 1, \bot = -1$, $\ominus = -$, and $\oplus$ corresponds to our operation $\oplus$.

Definition 8. *(i) An isomorphism of an OAG (OAG$^+$) $\mathbf{W} = (W, \leq, \oplus, \ominus, e)$ onto an OAG (OAG$^+$) $\mathbf{W}' = (W', \leq', \oplus', \ominus', e')$ is a one-to-one mapping ϕ from W onto W' preserving the structure, i.e. such that*
    (i) $\phi(x \oplus y) = \phi(x) \oplus' \phi(y)$
    (ii) $\phi(\ominus x) = \ominus' \phi(x)$

   (iii) $\phi(e) = e'$
   (iv) $x \leq y$ iff $\phi(x) \leq' \phi(y)$.
(ii) **W** *is a substructure of* **W'** *if* $W \subseteq W'$ *and the structure of* **W** *is the restriction of the structure of* **W'** *to* W, *i.e.* $x \oplus y = x \oplus' y$, $\ominus x = \ominus' x$, $e = e'$, *and* $x \leq y$ *iff* $x \leq' y$, *for all* $x, y \in W$.
(iii) *An isomorphic embedding of* **W** *into* **W'** *is an isomorphism of* **W** *onto a substructure of* **W'**.

Definition 9. (i) *An OAG* **W** *is* dense *if there is no least positive element, i.e. an element* $x \in W$ *such that* $x > e$, *and there is no* $y \in W$ *such that* $e < y < x$.
(ii) *An OAG* **W** *is* completely ordered *if each non empty bounded* $X \subseteq W$ *has a least upper bound.*

Obviously, $(]-1, 1[, \leq, \oplus, -, 0)$ is dense and completely ordered, the same holds if the interval is closed.

Theorem 1. *If* **W** *is a completely ordered and dense OAG, then it is isomorphic to* $(\mathbb{R}, \leq, +, -, 0)$.

The same result holds if **W** is an OAG$^+$ and if $\mathbb{R}$ is replaced by $\overline{\mathbb{R}} := \mathbb{R} \cup \{-\infty, \infty\}$.
   This shows that necessarily, $\oplus$ has the following form:

$$x \oplus y = \phi^{-1}[\phi(x) + \phi(y)] \tag{5}$$

where $\phi : [-1, 1] \longrightarrow \overline{\mathbb{R}}$ is one-to-one, odd, increasing, and satisfies $\phi(0) = 0$. Clearly, ϕ restricted to $[0, 1]$ is the additive generator of a strict t-conorm (see Section 2), and moreover, $\oplus$ is a uninorm with $e = \frac{1}{2}$ and additive generator ϕ, up to a rescaling on $[0, 1]$ (see Section 2). These results were shown directly in [11].
   Let us turn to the case of $\otimes$. If we impose distributivity of $\otimes$ w.r.t. $\oplus$ (called **R8**), then necessarily $\otimes$ obeys the rule of sign of the usual product, i.e., for any $x, y \geq 0$, $(-x) \otimes y = -(x \otimes y)$. Indeed,

$$0 = (x \oplus (-x)) \otimes y = (x \otimes y) \oplus ((-x) \otimes y)$$

which entails $(-x) \otimes y = -(x \otimes y)$. This case corresponds to ordered rings and fields (see Fuchs [6]). Then $\otimes$ is not monotone on $[-1, 1]^2$, and is uniquely determined by its values on $[0, 1]^2$, where it is a t-norm T. In summary, under **R1**, **R2**, **R3**, **R5** and **R8**, $\otimes$ has the following form:

$$x \otimes y = \text{sign}(x \cdot y)T(|x|, |y|),$$

for some t-norm T, and

$$\text{sign}(x) := \begin{cases} 1, & \text{if } x > 0 \\ -1, & \text{if } x < 0 \\ 0, & \text{if } x = 0. \end{cases}$$

If distributivity is not needed, nothing prevents us from imposing monotonicity of $\otimes$ on the whole domain $[-1,1]^2$ (called **R5'**). Then, if we impose in addition $(-1)\otimes x = x$ for all $x \leq 0$ (called **R6'**), up to a rescaling in $[0,1]$, $\otimes$ is a nullnorm with $a = 1/2$, since $\otimes$ is associative, commutative, non decreasing, and -1 is neutral on $[-1,0]^2$, 1 is neutral on $[0,1]^2$. In sumary, under **R1**, **R2**, **R5'**, **R6** and **R6'**, $\otimes$ has the following form:

$$x \otimes y = \begin{cases} T(x,y), & \text{if } x, y \geq 0 \\ S(x+1, y+1) - 1, & \text{if } x, y \leq 0 \\ 0, & \text{else} \end{cases}$$

for some t-norm T and t-conorm S.

4   Minimum and Maximum on Symmetrized Linearly Ordered Sets

The previous section has shown that except for strict t-conorms, there is no way to build pseudo-addition fulfilling requirements **R1**, **R2**, **R3**, **R4**, and **R7**. Hence extending the maximum on $[-1,1]$ in this way is not possible. However, we will show that this is in fact almost possible. Also, since our construction works on any linearly ordered set, this section addresses the construction of aggregation operators on ordinal bipolar scales.

We consider a linearly ordered set $(L^+, \leq)$, with bottom and top denoted $\mathbb{O}$, $\mathbb{1}$ respectively, and we define $L := L^+ \cup L^-$, where L^- is a reversed copy of L^+, i.e. for any $a, b \in L^+$, we have $a \leq b$ iff $-b \leq -a$, where $-a, -b$ are the copies of a, b in L^-.

Our aim is to define extensions of minimum and maximum operators on L, denoted $\oslash, \oslash$ and called *symmetric minimum* and *symmetric maximum*, in the same spirit as above. Specifically, we should require among others:

(**C1**) $\oslash, \oslash$ coincide with $\vee, \wedge$ respectively on L^+
(**C2**) $\oslash, \oslash$ are associative and commutative on L.
(**C3**) $-a$ is the symmetric of a, i.e. $a \oslash (-a) = \mathbb{O}$.
(**C4**) $-(a \oslash b) = (-a) \oslash (-b)$, $-(a \oslash b) = (-a) \oslash b$, $\forall a, b \in L$.

Conditions **C1** and **C2** replace requirements **R1** to **R4** above, while condition **C3** is requirement **R7**. Condition **C4** tells that $\oslash, \oslash$ should behave like addition and product on real numbers. The following result shows that this task is impossible [10].

Proposition 2. *We consider conditions (**C1**), (**C3**), (**C4**), and denote by (**C4+**) condition (**C4**) when a, b are restricted to L^+. Then:*
*(1) Conditions (**C1**) and (**C3**) implies that associativity cannot hold for $\oslash$.*
*(2) Under (**C1**) and (**C4+**), $\mathbb{O}$ is neutral for $\oslash$. If we require in addition associativity, then $|a \oslash (-a)| \geq |a|$. Further, if we require isotonicity of $\oslash$, then $|a \oslash (-a)| = |a|$.*

In [7], the following definitions for $\varovee, \varowedge$ were proposed.

$$a \varovee b := \begin{cases} -(|a| \vee |b|) & \text{if } b \neq -a \text{ and } |a| \vee |b| = -a \text{ or } = -b \\ \mathbb{O} & \text{if } b = -a \\ |a| \vee |b| & \text{else.} \end{cases} \qquad (6)$$

Except for the case $b = -a$, $a \varovee b$ equals the absolutely larger one of the two elements a and b.

$$a \varowedge b := \begin{cases} -(|a| \wedge |b|) & \text{if sign } a \neq \text{sign } b \\ |a| \wedge |b| & \text{else.} \end{cases} \qquad (7)$$

The absolute value of $a \varowedge b$ equals $|a| \wedge |b|$ and $a \varowedge b < \mathbb{O}$ iff the two elements a and b have opposite signs[1]. Both operators are represented on Figure 1. These operators have the following properties [10].

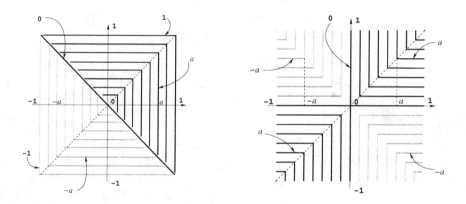

Fig. 1. Constant level curves of the symmetric maximum (left) and minimum (right)

Proposition 3. *The structure* $(L, \varovee, \varowedge)$ *has the following properties.*

(i) $\varovee$ *and* $\varowedge$ *fulfil conditions* **C1, C3** *and* **C4.**
(ii) $\mathbb{1}$ *(resp.* $\mathbb{O}$*) is the unique absorbant element of* $\varovee$ *(resp.* $\varowedge$*);*
(iii) $\varovee$ *is associative for any expression involving* $a_1, \ldots, a_n$, $a_i \in L$, *such that* $\bigvee_{i=1}^{n} a_i \neq - \bigwedge_{i=1}^{n} a_i$.
(iv) $\varowedge$ *is associative on* L.
(v) $\varowedge$ *is distributive w.r.t* $\varovee$ *in* L^+ *and* L^- *separately.*
(vi) $\varovee$ *is isotone, i.e.* $a \leq a', b \leq b'$ *implies* $a \varovee b \leq a' \varovee b'$.

The following result [10] shows that there is no "better" definition of $\varovee$ under the given conditions.

---

[1] As in Section 3, one may impose as well non-decreasingness of $\varowedge$ on $[-\mathbb{1}, \mathbb{1}]$, making $\varowedge$ a nullnorm.

Proposition 4. *Under conditions (**C1**), (**C3**) and (**C4**), no operation is asso ciative on a larger domain than* $\oslash$ *as given by (6).*

The problem of non associativity may be a severe limitation if $\oslash$ is used as a group operation to perform computation, like $\oslash_{i=1}^{n} a_i$. To overcome this difficulty, Grabisch has proposed several *computation rules* [10], which amount to eliminate situations where non associativity occurs, as given in Prop. 3. We denote them by $\langle \cdot \rangle$.

(i) The *splitting rule* $\langle \cdot \rangle_{-}^{+}$, splitting positive and negative terms:

$$\langle \mathop{\oslash}_{i=1}^{n} a_i \rangle_{-}^{+} := \left(\mathop{\oslash}_{a_i \geq 0} a_i \right) \oslash \left(\mathop{\oslash}_{a_i < 0} a_i \right).$$

(ii) The *strong rule* $\langle \cdot \rangle_0$, cancelling maximal opposite terms successively until condition (iii) in Prop. 3 is satisfied. Formally,

$$\langle \mathop{\oslash}_{a_i \in A} a_i \rangle_0 := \mathop{\oslash}_{a_i \in A \setminus \bar{A}} a_i,$$

with the convention that $\oslash_{\varnothing} a_i := \mathbb{O}$, and $A := a_1, \ldots, a_n$, while $\bar{A} := \bar{a}_1, \ldots, \bar{a}_{2k}$ is the sequence of pairs of maximal opposite terms.

(iii) The *weak rule* $\langle \cdot \rangle_=$, cancelling maximal opposite terms as before, but with duplicates, i.e. the set $\bar{A}$ contains in addition all duplicates of maximal opposite terms.

Taking for example $L = \mathbb{Z}$ and the sequence of numbers $3, 3, 3, 2, 1, 0, -2, -3, -3$, for which associativity does not hold, the result for splitting rule is 0, while we have:

$$\langle 3 \oslash 3 \oslash 3 \oslash 2 \oslash 1 \oslash 0 \oslash -2 \oslash -3 \oslash -3 \rangle_0 = 3 \oslash 2 \oslash 1 \oslash 0 \oslash -2 = 3$$
$$\langle 3 \oslash 3 \oslash 3 \oslash 2 \oslash 1 \oslash 0 \oslash -2 \oslash -3 \oslash -3 \rangle_= = 1 \oslash 0 = 1.$$

The symmetric maximum with the strong rule coincides with the limit of some family of uninorms proposed by Mesiar and Komorníková [22].

We give several simple properties of these computation rules.

Lemma 1. *All computation rules satisfy the following boundary property for any sequence* $a_1, \ldots, a_n$

$$\bigwedge_{i=1}^{n} a_i \leq \langle \mathop{\oslash}_{i=1}^{n} a_i \rangle \leq \bigvee_{i=1}^{n} a_i.$$

Lemma 2. *The rules* $\langle \cdot \rangle_{-}^{+}$ *and* $\langle \cdot \rangle_0$ *are isotone, i.e. they satisfy*

$$a_i \leq a_i', \quad i = 1, \ldots, n \text{ implies } \langle \mathop{\oslash}_{i=1}^{n} a_i \rangle \leq \langle \mathop{\oslash}_{i=1}^{n} a_i' \rangle.$$

Computation rule $\langle \cdot \rangle_=$ is not isotone, as shown by the following example: take the sequence $-3, 3, 1$ in $\mathbb{Z}$. Applying the weak rule leads to 1. Now, if 1 is raised to 3, the result becomes $\mathbb{O}$.

The sequence $a_1, \dots, a_n$ in L is said to be a *cancelling sequence for the rule* $\langle \cdot \rangle$ if $\langle \mathbb{O}_{i=1}^n a_i \rangle = \mathbb{O}$. We denote by $\mathcal{O}_{\langle \cdot \rangle}$ the set of cancelling sequences of $\langle \cdot \rangle$.

We say that computation rule $\langle \cdot \rangle_1$ is more *discriminating* than rule $\langle \cdot \rangle_2$ if $\mathcal{O}_{\langle \cdot \rangle_1} \subset \mathcal{O}_{\langle \cdot \rangle_2}$.

Lemma 3

$$\mathcal{O}_{\langle \cdot \rangle_0} \subset \mathcal{O}_{\langle \cdot \rangle_=} \subset \mathcal{O}_{\langle \cdot \rangle_-^+}.$$

5 Separable Operators

We consider here non necessarily associative operators M, in the spirit of means. We assume in this section that the underlying scale is $[-1, 1]$, otherwise specified. We denote by n the number of arguments of M.

A simple way to build bipolar aggregation operators is the following. Let M^+, M^- be given aggregation operators on $[0, 1]$. M^+ defines the aggregation for positive values, while M^- defines the aggregation of negative values:

$$M(x) = M^+(x) \text{ if } x \geq 0, \quad M(x) = -M^-(-x) \text{ if } x \leq 0.$$

For any $x \in [-1, 1]^n$, we define $x^+ := x \vee 0$ and $x^- := (-x)^+$. Note that $x = x^+ - x^-$, which suggests the following construction:

$$M(x) := \phi(M^+(x^+), M^-(x^-)), \quad \forall x \in [-1, 1]^n, \tag{8}$$

where ϕ is a *pseudo-difference*, defined as follows.

Definition 10. *Let S be a t-conorm.*

(i) The S-difference $\overset{S}{-}$ is defined by

$$a \overset{S}{-} b := \inf\{c \mid S(b, c) \geq a\}$$

for any (a, b) in $[0, 1]^2$.
(ii) The pseudo-difference associated to S is defined by

$$a \ominus_S b := \begin{cases} a \overset{S}{-} b, & \text{if } a \geq b \\ -(b \overset{S}{-} a), & \text{if } a \leq b \\ 0, & \text{if } a = b \end{cases}$$

Two simple particular cases are with $S = \vee$ and $S = S_{\mathbf{L}}$. Then for any $a, b \in [-1, 1]$

$$a \ominus_\vee b = a \mathbb{O}(-b), \qquad a \ominus_{\mathbf{L}} b = a - b,$$

as it can be easily checked. If S is a strict t-norm with additive generator s, then

$$a \ominus_S b = g^{-1}(g(a) - g(b)),$$

with $g(x) = s(x)$ for $x \geq 0$, and $g(x) = -s(-x)$ for $x \leq 0$ (see [11]).

A bipolar aggregation operator defined by Eq. (8) is called *separable*.

If $M^+ = M^-$ is a strict t-conorm S with generator s, and $\ominus_S$ is taken as pseudo-difference, we recover the construction of Section 3. Indeed, taking $n = 2$ (sufficient since associative), and g being the generator of $\ominus_S$:

$$\begin{aligned}
M(x,y) &= S(x^+, y^+) \ominus_S S(x^-, y^-) \\
&= g^{-1}(g(S(x^+, y^+)) - g(S(x^-, y^-))) \\
&= g^{-1}(g(x^+) + g(y^+) - g(x^-) - g(y^-)) \\
&= g^{-1}(g(x) + g(y))
\end{aligned}$$

which is Eq. (5), and indeed g is odd, strictly increasing, and $g(0) = 0$.

An interesting case is when M^+, M^- are integral-based operators, such as the Choquet or Sugeno integrals (see definitions in Section 2). Applying (8) with suitable pseudo-differences, we recover various definitions of integrals for real-valued functions. Specifically, let us take M^+, M^- to be Choquet integrals with respect to capacities μ^+, μ^-, and ϕ is the usual difference $\ominus_{\mathbf{L}}$. Then:

- Taking $\mu^+ = \mu^-$ we obtain the *symmetric Choquet integral* [4] or Šipoš integral [27]:
$$\check{C}_\mu(x) := C_\mu(x^+) - C_\mu(x^-).$$

- Taking $\mu^- = \overline{\mu^+}$ we obtain the *asymmetric Choquet integral* [4]:
$$C_\mu(x) := C_\mu(x^+) - C_{\overline{\mu}}(x^-).$$

- For the general case, we obtain what is called in decision making theory the *Cumulative Prospect Theory (CPT)* model.
$$\mathrm{CPT}_{\mu^+, \mu^-}(x) := C_{\mu^+}(x^+) - C_{\mu^-}(x^-).$$

We consider now that M^+, M^- are Sugeno integrals, with respect to capacities μ^+, μ^-, and ϕ is the residuated difference associated to the maximum, i.e. $\phi(x,y) := x \oslash (-y)$. Then as above,

- Taking $\mu^+ = \mu^-$ we obtain the *symmetric Sugeno integral* [8]:
$$\check{S}_\mu(x) := S_\mu(x^+) \oslash (-S_\mu(x^-)).$$

- Taking $\mu^+ = \overline{\mu^-}$ we obtain the *asymmetric Sugeno integral* [7]:
$$\hat{S}_\mu(x) := S_\mu(x^+) \oslash (-S_{\overline{\mu}}(x^-)).$$

- For the general case, we obtain what corresponds to the CPT model in an ordinal version.
$$\mathrm{OCPT}_{\mu^+, \mu^-}(x) := S_{\mu^+}(x^+) \oslash (-S_{\mu^-}(x^-)).$$

Note that the above development on Sugeno integral could have been done on any linearly ordered set L, provided L has enough structure so we can define conjugate capacities. For a general study of the Sugeno integral as well as symmetric and asymmetric versions on linearly ordered sets, see Denneberg and Grabisch [5].

6   Integral-Based Operators

Let us study the case of integral-based operators, and we will limit ourself to the Choquet and Sugeno integrals, which are the most representative.

As explained in Section 2, the Choquet integral can be defined as the "simplest" linear interpolation between vertices of $[0, 1]^n$. Extending the domain to $[-1, 1]^n$, let us try to keep a similar approach.

The basic ingredient of the interpolative view is that $\mathcal{C}_\mu(1_A, 0_{A^c}) = \mathcal{S}_\mu(1_A, 0_{A^c}) = \mu(A)$. Let us call *binary vectors* those of the form $(1_A, 0_{A^c})$. In the unipolar case, coordinates of binary vectors are the boundaries of the interval $[0, 1]$. In the bipolar case, apart boundaries, we should also consider 0, as this value plays a particular role. We thus consider *ternary vectors*, whose components are either 1, 0 or -1. We denote them $(1_A, -1_B, 0_{(A \cup B)^c})$, which means that $x_i = 1$ if $i \in A$, $x_i = -1$ if $i \in B$, and 0 elsewhere. Obviously, $A \cap B = \varnothing$, so that the set of ternary vectors is obtained when the pair (A, B) belongs to $\mathcal{Q}([n]) := \{(A, B) \mid A, B \subseteq [n], A \cap B = \varnothing\}$. The basic idea is to produce an aggregation function F which coincides with a set of fixed quantities $v(A, B)$, for $(A, B) \in \mathcal{Q}([n])$. In order to define a monotone aggregation operator, we are led to the following definition.

Definition 11. *[12,14] A (normalized) bicapacity v on $[n]$ is a function v : $\mathcal{Q}([n]) \to [-1, 1]$ satisfying $v(\varnothing, \varnothing) = 0$, $v([n], \varnothing) = 1$, $v(\varnothing, [n]) = -1$, and $v(A, B) \leq v(C, D)$ whenever $A \subseteq C$ and $B \supseteq D$.*

Applying the same interpolative approach between ternary vectors, we are led to the following (see details in [15]). Let us consider $x \in [-1, 1]^n$. Defining $N_x^+ := \{i \in [n] \mid x_i \geq 0\}$, $N_x^- := [n] \setminus N_x^+$, with similar considerations of symmetry, we obtain as linear interpolation:

$$F(x) = |x_{\sigma(1)}| F(1_{N_x^+}, -1_{N_x^-}, 0_{(N_x^+ \cup N_x^-)^c})$$

$$+ \sum_{i=2}^n (|x_{\sigma(i)}| - |x_{\sigma(i-1)}|) F(1_{\{\sigma(i),\ldots,\sigma(n)\} \cap N_x^+}, -1_{\{\sigma(i),\ldots,\sigma(n)\} \cap N_x^-}, 0_{\{\sigma(i),\ldots,\sigma(n)\}^c})$$

where σ is a permutation on $[n]$ such that $|x_{\sigma(1)}| \leq \cdots \leq |x_{\sigma(n)}|$. This expression is the Choquet integral of $|x|$ w.r.t. a set function $\nu_{N_x^+}$ defined by:

$$\nu_{N_x^+}(A) := F(1_{A \cap N_x^+}, -1_{A \cap N_x^-}, 0_{A^c}).$$

Recalling that $F(1_A, -1_B, 0_{(A \cup B)^c}) =: v(A, B)$, we finally come up with the following definition.

Definition 12. *[13] Let v be a bicapacity and $x \in [-1,1]^n$. The Choquet integral of x w.r.t v is given by*

$$\mathcal{C}_v(x) := \mathcal{C}_{\nu_{N_x^+}}(|x|)$$

where $\nu_{N_x^+}$ is a set function on $[n]$ defined by

$$\nu_{N_x^+}(C) := v(C \cap N_x^+, C \cap N_x^-),$$

and $N_x^+ := \{i \in [n] \mid x_i \geq 0\}$, $N_x^- = [n] \setminus N_x^+$.

When there is no fear of ambiguity, we drop subscript x in N_x^+, N_x^-.

It is shown in [13] that if the bicapacity v has the form $v(A, B) := \mu^+(A) - \mu^-(B)$ for all $(A, B) \in \mathcal{Q}([n])$, where μ^+, μ^- are capacities, then $\mathcal{C}_v(x) = \mathrm{CPT}_{\mu^+, \mu^-}(x)$, for all $x \in [-1, 1]^n$. Hence the Choquet integral based on a bicapacity encompasses the CPT model, and thus symmetric and asymmetric Choquet integrals.

By analogy, a definition can be proposed for the Sugeno integral w.r.t a bicapacity:

$$\mathcal{S}_v(x) := \mathcal{S}_{\nu_{N_x^+}}(|x|)$$

with the same notations as above. However, since $\nu_{N_x^+}$ may assume negative values, it is necessary to extend the definition of Sugeno integral as follows:

$$\mathcal{S}_\nu(x) := \left\langle \overset{n}{\underset{i=1}{\oslash}} [x_{\sigma(i)} \oslash \nu(\{\sigma(i), \ldots, \sigma(n)\})] \right\rangle_-^+$$

where $x \in [0, 1]^n$, ν is any real-valued set function such that $\nu(\varnothing) = 0$, and σ is a permutation on $[n]$ such that x becomes non decreasing. $\langle \cdot \rangle_-^+$ indicates the splitting rule defined in Section 4. Then, the Sugeno integral for bicapacities can be rewritten as

$$\mathcal{S}_v(x) = \left\langle \overset{n}{\underset{i=1}{\oslash}} \left[|x_{\sigma(i)}| \oslash v(\{\sigma(i), \ldots, \sigma(n)\} \cap N^+, \{\sigma(i), \ldots, \sigma(n)\} \cap N^-) \right] \right\rangle_-^+. \tag{9}$$

This formula is similar to the one proposed by Greco *et al.* [18].

The following result shows that the Sugeno integral w.r.t. a bicapacity encompasses the OCPT model.

Proposition 5. *Let v be a bicapacity of the form $v(A, B) := \mu^+(A) \oslash (-\mu^-(B))$, where $\mu^+ \mu^-$ are capacities. Then the Sugeno integral reduces to*

$$\mathcal{S}_v(x) := \mathcal{S}_{\mu^+}(x^+) \oslash (-\mathcal{S}_{\mu^-}(x^-)) = \mathrm{OCPT}_{\mu^+, \mu^-}(x), \quad \forall x \in [-1, 1]^n.$$

Note that if $\mu^+ = \mu^-$ (v could then be called a $\vee$-*symmetric bicapacity*), then $\mathcal{S}_v$ is the symmetric Sugeno integral.

Proof. Denote by σ a permutation on $[n]$ such that $|x|$ is non-decreasing, and put $A_{\sigma(i)} := \{\sigma(i), \ldots, \sigma(n)\}$. Since x^+, x^-, μ^+, μ^- are non negative, we have

$$\mathcal{S}_{\mu^+}(x^+) = \bigvee_{i=1}^{n} \left[x^+_{\sigma(i)} \wedge \mu^+(A_{\sigma(i)} \cap N^+) \right]$$

$$\mathcal{S}_{\mu^-}(x^-) = \bigvee_{i=1}^{n} \left[x^-_{\sigma(i)} \wedge \mu^-(A_{\sigma(i)} \cap N^-) \right].$$

Using the definition of v, we get

$$\mathcal{S}_v(x) = \left\langle \bigotimes_{i=1}^{n} \left[|x_{\sigma(i)}| \oslash [\mu^+(A_{\sigma(i)} \cap N^+) \oslash (-\mu^-(A_{\sigma(i)} \cap N^-))] \right] \right\rangle_-^+.$$

Due to the definition of $\langle \cdot \rangle_-^+$, we have to show that if $\mathcal{S}_{\mu^+}(x^+)$ is larger (resp. smaller) than $\mathcal{S}_{\mu^-}(x^-)$, then the maximum of positive terms is equal to $\mathcal{S}_{\mu^+}(x^+)$ and is larger in absolute value than the maximum of negative terms (resp. the maximum of absolute value of negative terms is equal to $\mathcal{S}_{\mu^-}(x^-)$ and is larger in absolute value than the maximum of positive terms).

Let us consider $\sigma(i) \in N^+$. Two cases can happen.

– if $\mu^+(A_{\sigma(i)} \cap N^+) > \mu^-(A_{\sigma(i)} \cap N^-)$, then the corresponding term reduces to $x^+_{\sigma(i)} \oslash \mu^+(A_{\sigma(i)} \cap N^+)$. This term is identical to the ith term in $\mathcal{S}_{\mu^+}(x^+)$.

– if not, the ith term in $\mathcal{S}_v(x)$ reduces to $-x^+_{\sigma(i)} \oslash \mu^-(A_{\sigma(i)} \cap N^-)$. Due to monotonicity of μ^+, this will be also the case for all subsequent indices $\sigma(i+1), \ldots \sigma(i+k)$, provided they belong to N^+. Moreover, assuming $\sigma(i+k+1) \in N^-$, we have

$$|x_{\sigma(i+k+1)}| \oslash \left[\mu^+(A_{\sigma(i+k+1)} \cap N^+) \oslash (-\mu^-(A_{\sigma(i+k+1)} \cap N^-)) \right]$$
$$= -|x_{\sigma(i+k+1)}| \oslash \mu^-(A_{\sigma(i+k+1)} \cap N^-)$$
$$\leq |x_{\sigma(j)}| \oslash \mu^-(\underbrace{A_{\sigma(j)} \cap N^-}_{A_{\sigma(i+k+1)} \cap N^-}), \quad \forall j = i, \ldots, i+k.$$

Hence, in the negative part of $\mathcal{S}_v(x)$, the term in $\sigma(i+k+1)$ remains, while all terms in $\sigma(i), \ldots, \sigma(i+k)$ are cancelled, and it coincides with the $(i+k+1)$th term in $\mathcal{S}_{\mu^-}(x^-)$. On the other hand, in $\mathcal{S}_{\mu^+}(x^+) \oslash (-\mathcal{S}_{\mu^-}(x^-))$, the term in $\sigma(i)$ in $\mathcal{S}_{\mu^+}(x^+)$ is smaller in absolute value than the term in $\sigma(i+k+1)$ of $\mathcal{S}_{\mu^-}(x^-)$, so that the term in $\sigma(i)$ cannot be the result of the computation, and thus it can be discarded from $\mathcal{S}_{\mu^+}(x^+)$.

A similar reasoning can be done with $\sigma(i) \in N^-$. This proves that $\mathcal{S}_v(x)$ and $\mathcal{S}_{\mu^+}(x^+) \oslash (-\mathcal{S}_{\nu_-}(x^-))$ are identical.

Acknowledgments

The author acknowledges support from the COST Action 274 TARSKI. Parts of this work have been done in collaboration with B. de Baets and J. Fodor, participant to the same COST Action.

References

1. P. Benvenuti, R. Mesiar, and D. Vivona. Monotone set functions-based integrals. In E. Pap, editor, *Handbook of Measure Theory*, pages 1329–1379. Elsevier Science, 2002.

2. T. Calvo, B. De Baets, and J. Fodor. The functional equations of alsina and frank for uninorms and nullnorms. *Fuzzy Sets and Systems*, 101:15–24, 2001.

3. T. Calvo, G. Mayor, and R. Mesiar, editors. *Aggregation operators: new trends and applications*. Studies in Fuzziness and Soft Computing. Physica Verlag, 2002.

4. D. Denneberg. *Non-Additive Measure and Integral*. Kluwer Academic, 1994.

5. D. Denneberg and M. Grabisch. Measure and integral with purely ordinal scales. *J. of Mathematical Psychology*, 48:15–27, 2004.

6. L. Fuchs. *Partially Ordered Algebraic Systems*. Addison-Wesley, 1963.

7. M. Grabisch. Symmetric and asymmetric fuzzy integrals: the ordinal case. In *6th Int. Conf. on Soft Computing (Iizuka'2000)*, Iizuka, Japan, October 2000.

8. M. Grabisch. The symmetric Sugeno integral. *Fuzzy Sets and Systems*, 139:473–490, 2003.

9. M. Grabisch. The Choquet integral as a linear interpolator. In *10th Int. Conf. on Information Processing and Management of Uncertainty in Knowledge-Based Systems (IPMU 2004)*, pages 373–378, Perugia, Italy, July 2004.

10. M. Grabisch. The Möbius function on symmetric ordered structures and its application to capacities on finite sets. *Discrete Mathematics*, 287(1-3):17–34, 2004.

11. M. Grabisch, B. De Baets, and J. Fodor. The quest for rings on bipolar scales. *Int. J. of Uncertainty, Fuzziness, and Knowledge-Based Systems*, 12(4):499–512, 2004.

12. M. Grabisch and Ch. Labreuche. Bi-capacities for decision making on bipolar scales. In *EUROFUSE Workshop on Informations Systems*, pages 185–190, Varenna, Italy, September 2002.

13. M. Grabisch and Ch. Labreuche. Fuzzy measures and integrals in MCDA. In J. Figueira, S. Greco, and M. Ehrgott, editors, *Multiple Criteria Decision Analysis*, pages 563–608. Kluwer Academic Publishers, 2004.

14. M. Grabisch and Ch. Labreuche. Bi-capacities. Part I: definition, Möbius transform and interaction. *Fuzzy Sets and Systems*, 151:211–236, 2005.

15. M. Grabisch and Ch. Labreuche. Bi-capacities: towards a generalization of Cumulative Prospect Theory. *J. of Mathematical Psychology*, submitted.

16. M. Grabisch, T. Murofushi, and M. Sugeno. *Fuzzy Measures and Integrals. Theory and Applications (edited volume)*. Studies in Fuzziness. Physica Verlag, 2000.

17. M. Grabisch, S. A. Orlovski, and R. R. Yager. Fuzzy aggregation of numerical preferences. In R. Słowiński, editor, *Fuzzy Sets in Decision Analysis, Operations Research and Statistics*, The Handbooks of Fuzzy Sets Series, D. Dubois and H. Prade (eds), pages 31–68. Kluwer Academic, 1998.

18. S. Greco, B. Matarazzo, and R. Słowinski. Bipolar Sugeno and Choquet integrals. In *EUROFUSE Workshop on Informations Systems*, pages 191–196, Varenna, Italy, September 2002.

19. P. Hájek. *Metamathematics of fuzzy logic*. Kluwer Academic Publishers, 1998.

20. P. Hájek, T. Havránek, and R. Jiroušek. *Uncertain Information Processing in Expert Systems*. CRC Press, 1992.

21. E. P. Klement, R. Mesiar, and E. Pap. *Triangular Norms*. Kluwer Academic Publishers, Dordrecht, 2000.

22. R. Mesiar and M. Komorniková. Triangular norm-based aggregation of evidence under fuzziness. In B. Bouchon-Meunier, editor, *Aggregation and Fusion of Imperfect Information*, Studies in Fuzziness and Soft Computing, pages 11–35. Physica Verlag, 1998.

23. T. Murofushi and M. Sugeno. Fuzzy t-conorm integrals with respect to fuzzy measures : generalization of Sugeno integral and Choquet integral. *Fuzzy Sets & Systems*, 42:57–71, 1991.

24. W. Sander and J. Siedekum. Multiplication, distributivity and fuzzy integral I. *Kybernetika*, 41:397–422, 2005.

25. W. Sander and J. Siedekum. Multiplication, distributivity and fuzzy integral II. *Kybernetika*, 41:469–496, 2005.

26. W. Sander and J. Siedekum. Multiplication, distributivity and fuzzy integral III. *Kybernetika*, 41:497–518, 2005.

27. J. Šipoš. Integral with respect to a pre-measure. *Math. Slovaca*, 29:141–155, 1979.

28. R. Yager and A. Rybalov. Uninorm aggregation operators. *Fuzzy Sets and Systems*, 80:111–120, 1996.

Author Index

Lecture Notes in Artificial Intelligence (LNAI)

Vol. 4140: J.S. Sichman, H. Coelho, S.O. Rezende (Eds.), Advances in Artificial Intelligence - IBERAMIA-SBIA 2006. XXIII, 635 pages. 2006.

Vol. 4139: T. Salakoski, F. Ginter, S. Pyysalo, T. Pahikkala (Eds.), Advances in Natural Language Processing. XVI, 771 pages. 2006.

Vol. 4133: J. Gratch, M. Young, R. Aylett, D. Ballin, P. Olivier (Eds.), Intelligent Virtual Agents. XIV, 472 pages. 2006.

Vol. 4130: U. Furbach, N. Shankar (Eds.), Automated Reasoning. XV, 680 pages. 2006.

Vol. 4120: J. Calmet, T. Ida, D. Wang (Eds.), Artificial Intelligence and Symbolic Computation. XIII, 269 pages. 2006.

Vol. 4118: Z. Despotovic, S. Joseph, C. Sartori (Eds.), Agents and Peer-to-Peer Computing. XIV, 173 pages. 2006.

Vol. 4114: D.-S. Huang, K. Li, G.W. Irwin (Eds.), Computational Intelligence, Part II. XXVII, 1337 pages. 2006.

Vol. 4108: J.M. Borwein, W.M. Farmer (Eds.), Mathematical Knowledge Management. VIII, 295 pages. 2006.

Vol. 4106: T.R. Roth-Berghofer, M.H. Göker, H.A. Güvenir (Eds.), Advances in Case-Based Reasoning. XIV, 566 pages. 2006.

Vol. 4099: Q. Yang, G. Webb (Eds.), PRICAI 2006: Trends in Artificial Intelligence. XXVIII, 1263 pages. 2006.

Vol. 4095: S. Nolfi, G. Baldassarre, R. Calabretta, J.C.T. Hallam, D. Marocco, J.-A. Meyer, O. Miglino, D. Parisi (Eds.), From Animals to Animats 9. XV, 869 pages. 2006.

Vol. 4093: X. Li, O.R. Zaïane, Z. Li (Eds.), Advanced Data Mining and Applications. XXI, 1110 pages. 2006.

Vol. 4092: J. Lang, F. Lin, J. Wang (Eds.), Knowledge Science, Engineering and Management. XV, 664 pages. 2006.

Vol. 4088: Z.-Z. Shi, R. Sadananda (Eds.), Agent Computing and Multi-Agent Systems. XVII, 827 pages. 2006.

Vol. 4087: F. Schwenker, S. Marinai (Eds.), Artificial Neural Networks in Pattern Recognition. IX, 299 pages. 2006.

Vol. 4068: H. Schärfe, P. Hitzler, P. Øhrstrøm (Eds.), Conceptual Structures: Inspiration and Application. XI, 455 pages. 2006.

Vol. 4065: P. Perner (Ed.), Advances in Data Mining. XI, 592 pages. 2006.

Vol. 4062: G.-Y. Wang, J.F. Peters, A. Skowron, Y. Yao (Eds.), Rough Sets and Knowledge Technology. XX, 810 pages. 2006.

Vol. 4049: S. Parsons, N. Maudet, P. Moraitis, I. Rahwan (Eds.), Argumentation in Multi-Agent Systems. XIV, 313 pages. 2006.

Vol. 4048: L. Goble, J.-J.C.. Meyer (Eds.), Deontic Logic and Artificial Normative Systems. X, 273 pages. 2006.

Vol. 4045: D. Barker-Plummer, R. Cox, N. Swoboda (Eds.), Diagrammatic Representation and Inference. XII, 301 pages. 2006.

Vol. 4031: M. Ali, R. Dapoigny (Eds.), Advances in Applied Artificial Intelligence. XXIII, 1353 pages. 2006.

Vol. 4029: L. Rutkowski, R. Tadeusiewicz, L.A. Zadeh, I M. Zurada (Eds.), Artificial Intelligence and Soft Computing – ICAISC 2006. XXI, 1235 pages. 2006.

Vol. 4027: H.L. Larsen, G. Pasi, D. Ortiz-Arroyo, T. Andreasen, H. Christiansen (Eds.), Flexible Query Answering Systems. XVIII, 714 pages. 2006.

Vol. 4021: E. André, L. Dybkjær, W. Minker, H. Neumann, M. Weber (Eds.), Perception and Interactive Technologies. XI, 217 pages. 2006.

Vol. 4020: A. Bredenfeld, A. Jacoff, I. Noda, Y. Takahashi (Eds.), RoboCup 2005: Robot Soccer World Cup IX. XVII, 727 pages. 2006.

Vol. 4013: L. Lamontagne, M. Marchand (Eds.), Advances in Artificial Intelligence. XIII, 564 pages. 2006.

Vol. 4012: T. Washio, A. Sakurai, K. Nakajima, H. Takeda, S. Tojo, M. Yokoo (Eds.), New Frontiers in Artificial Intelligence. XIII, 484 pages. 2006.

Vol. 4008: J.C. Augusto, C.D. Nugent (Eds.), Designing Smart Homes. XI, 183 pages. 2006.

Vol. 4005: G. Lugosi, H.U. Simon (Eds.), Learning Theory. XI, 656 pages. 2006.

Vol. 4002: A. Yli-Jyrä, L. Karttunen, J. Karhumäki (Eds.), Finite-State Methods and Natural Language Processing. XIV, 312 pages. 2006.

Vol. 3978: B. Hnich, M. Carlsson, F. Fages, F. Rossi (Eds.), Recent Advances in Constraints. VIII, 179 pages. 2006.

Vol. 3963: O. Dikenelli, M.-P. Gleizes, A. Ricci (Eds.), Engineering Societies in the Agents World VI. XII, 303 pages. 2006.

Vol. 3960: R. Vieira, P. Quaresma, M.d.G.V. Nunes, N.J. Mamede, C. Oliveira, M.C. Dias (Eds.), Computational Processing of the Portuguese Language. XII, 274 pages. 2006.

Vol. 3955: G. Antoniou, G. Potamias, C. Spyropoulos, D. Plexousakis (Eds.), Advances in Artificial Intelligence. XVII, 611 pages. 2006.

Vol. 3949: F.A. Savacı (Ed.), Artificial Intelligence and Neural Networks. IX, 227 pages. 2006.

Vol. 3946: T.R. Roth-Berghofer, S. Schulz, D.B. Leake (Eds.), Modeling and Retrieval of Context. XI, 149 pages. 2006.

Vol. 3944: J. Quiñonero-Candela, I. Dagan, B. Magnini, F. d'Alché-Buc (Eds.), Machine Learning Challenges. XIII, 462 pages. 2006.

Vol. 3937: H. La Poutré, N.M. Sadeh, S. Janson (Eds.), Agent-Mediated Electronic Commerce. X, 227 pages. 2006.

Vol. 3932: B. Mobasher, O. Nasraoui, B. Liu, B. Masand (Eds.), Advances in Web Mining and Web Usage Analysis. X, 189 pages. 2006.

Vol. 3930: D.S. Yeung, Z.-Q. Liu, X.-Z. Wang, H. Yan (Eds.), Advances in Machine Learning and Cybernetics. XXI, 1110 pages. 2006.

Vol. 3918: W.-K. Ng, M. Kitsuregawa, J. Li, K. Chang (Eds.), Advances in Knowledge Discovery and Data Mining. XXIV, 879 pages. 2006.